F	future value, future worth	MARR_R	real dollar MARR
f	inflation rate per year	**MAUT**	multi-attribute utility theory
FW	future worth	**MCDM**	multi-criterion decision making
g	growth rate for geometric gradient	N	number of periods, useful life of an asset
i	actual interest rate	P	present value, present worth, purchase price, principal amount
I	interest amount		
i'	real interest rate		
I_c	compound interest amount	**PCM**	pairwise comparison matrix
i_e	effective interest rate	**PW**	present worth
i_s	interest rate per subperiod	$p(x)$	probability distribution
I_s	simple interest amount	$\text{Pr}\{X = x_i\}$	alternative expression of probability distribution
i°	growth adjusted interest rate		
IRR	internal rate of return	r	nominal interest rate, rating for a decision matrix
IRR_A	actual dollar IRR		
IRR_R	real dollar IRR	$R_{0,N}$	real dollar equivalent to A_N relative to year 0, the base year
i^*	internal rate of return		
i_e^*	external rate of return		
i_{ea}^*	approximate external rate of return	RI	random index
		S	salvage value
$I_{0,N}$	the value of a global price index at year N, relative to year 0	**TBF**	tax benefit factor
		t	tax rate
		UCC	undepreciated capital cost
m	number of subperiods in a period	X	random variable
		$\mathbf{w}$	an eigenvector
MARR	minimum acceptable rate of return	λ_{max}	the maximun eigenvalue
		λ	an eigenvalue
MARR_A	actual dollar MARR	π_{01}	Laspeyres price index

FOURTH EDITION

GLOBAL ENGINEERING ECONOMICS

Financial Decision Making for Engineers

FOURTH EDITION

GLOBAL ENGINEERING ECONOMICS

Financial Decision Making for Engineers

Niall M. Fraser | **Elizabeth M. Jewkes** | **Irwin Bernhardt** | **May Tajima**

Open Options Corporation   University of Waterloo   University of Waterloo – retired   University of Western Ontario

PEARSON

Prentice
Hall

Toronto

Library and Archives Canada Cataloguing in Publication

Global engineering economics: financial decision making for engineers / Niall M. Fraser ... [et al.]. — 4th ed.

Includes index.
Previous editions published under title: Engineering economics in Canada.
ISBN 978-0-13-207161-1

1. Engineering economy.   I. Fraser, Niall M. (Niall Morris), 1952–

TA177.4.F725 2008   658.15   C2008-903776-6

ISBN-13: 978-0-13-207161-1
ISBN-10: 0-13-207161-4

Vice-President, Editorial Director: Gary Bennett
Acquisitions Editor: Cathleen Sullivan
Marketing Manager: Michelle Bish
Developmental Editor: Maurice Esses
Production Editor: Imee Salumbides
Copy Editor: Laurel Sparrow
Proofreader: Susan Bindernagel
Production Coordinator: Sarah Lukaweski
Compositor: Integra
Permissions Researcher: Sandy Cooke
Art Director: Julia Hall
Cover and Interior Designer: Anthony Leung
Cover Images: Veer Inc. (architects looking at drafts in office) and Getty Images (electronic currency exchange sign in building)

1 2 3 4 5   12 11 10 09 08

Printed and bound in the United States of America.

Brief Contents

Contents

CHAPTER 6   Depreciation and Financial Accounting   171

CHAPTER 7   Replacement Decisions   220

Preface

Courses on engineering economics are found in engineering curricula throughout the world. The courses generally deal with deciding among alternative engineering projects with respect to expected costs and benefits. For example, in Canada, the Canadian Engineering Accreditation Board requires that all accredited professional engineering programs provide studies in engineering economics. Many engineers have found that a course in engineering economics can be as useful in their practice as any of their more technical courses.

There are several stages to making a good decision. One stage is being able to determine whether a solution to a problem is technically feasible. This is one of the roles of the engineer, who has specialized training to make such technical judgments. Another stage is deciding which of several technically feasible alternatives is best. Deciding among alternatives often does not require the technical competence needed to determine which alternatives are feasible, but it is equally important in making the final choice. Some engineers have found that choosing among alternatives can be more difficult than deciding what alternatives exist.

The role of engineers in society is changing. In the past, engineers tended to have a fairly narrow focus, concentrating on the technical aspects of a problem and on strictly computational aspects of engineering economics. As a result, many engineering economics texts focused on the mathematics of the subject. Today, engineers are more likely to be the decision makers, and they need to be able to take into account strategic and policy issues.

Society has changed in other ways in recent years. In particular, the world has become more interlinked. An engineer may be trained in one part of the world and end up practising somewhere completely different. The mathematics of engineering economics, like all of an engineer's technical skills, is the same everywhere.

This book is designed for teaching a course on engineering economics to match engineering practice today. It recognizes the role of the engineer as a decision maker who has to make and defend sensible decisions. Such decisions must not only take into account a correct assessment of costs and benefits; they must also reflect an understanding of the environment in which the decisions are made.

This book is a direct descendant of a book entitled *Engineering Economics in Canada*, and in some senses is the fourth edition of that book. But given the increasing globalization of many engineering activities, the title and the contents have been updated. This is appropriate because the contents are applicable to engineers everywhere. For Canadian users of the previous editions, this text retains all of the valued features that made it your text of choice. For new users, it is a proven text that can support a course taught anywhere in the world.

This book also relates to students' everyday lives. In addition to examples and problems with an engineering focus, there are a number that involve decisions that many students might face, such as renting an apartment, getting a job, or buying a car.

Content and Organization

Because the mathematics of finance has not changed dramatically over the past number of years, there is a natural order to the course material. Nevertheless, a modern view of the role of the engineer flavours this entire book and provides a new, balanced exposure to the subject.

Chapter 1 frames the problem of engineering decision making as one involving many issues. Manipulating the cash flows associated with an engineering project is an important process for which useful mathematical tools exist. These tools form the bulk of the remaining chapters. However, throughout the text, students are kept aware of the fact that the eventual decision depends not only on the cash flows, but also on less easily quantifiable considerations of business policy, social responsibility, and ethics.

Chapters 2 and 3 present tools for manipulating monetary values over time. Chapters 4 and 5 show how the students can use their knowledge of manipulating cash flows to make comparisons among alternative engineering projects. Chapter 6 provides an understanding of the environment in which the decisions are made by examining depreciation and the role it plays in the financial functioning of a company and in financial accounting.

Chapter 7 deals with the analysis of replacement decisions. Chapters 8 and 9 are concerned with taxes and inflation, which affect decisions based on cash flows. Chapter 10 provides an introduction to public-sector decision making.

Most engineering projects involve estimating future cash flows as well as other project characteristics. Since estimates can be in error and the future unknown, it is important for engineers to take uncertainty and risk into account as completely as possible. Chapter 11 deals with uncertainty, with a focus on sensitivity analysis. Chapter 12 deals with risk, using some of the tools of probability analysis.

Chapter 13 picks up an important thread running throughout the book: a good engineering decision cannot be based only on selecting the least-cost alternative. The increasing influence on decision making of health and safety issues, environmental responsibility, and human relations, among others, makes it necessary for the engineer to understand some of the basic principles of multi-criterion decision making.

New to This Edition

In addition to clarifying explanations, improving readability, updating material, and correcting errors, we have made the following important changes for this new global edition:

■ Throughout the text, the context of the examples and problems has been changed from a Canadian orientation to a global environment. Similarly, the currencies used vary—about 60% of the examples use dollars (Australian, Canadian, or American) and other currencies such as euros or pounds make up the remainder of the examples.

■ Chapter 8 has been completely rewritten to demonstrate the impact of taxes on engineering decisions independent of the tax regime involved. Detailed examples are given for Australia, Canada, the United Kingdom, and the United States.

■ About half of the **Mini-Cases**, which supplement the chapter material with a real-world example, have been replaced to address issues from around the world.

■ The **Net Value** boxes, which provide chapter-specific examples of how the internet can be used as a source of information and guidance for decision making, have been updated to highlight the global perspective of the book. In many cases, web addresses specific to countries around the world are provided.

- A new **More Challenging Problem** has been added to each chapter. These are thought-provoking questions that encourage students to stretch their understanding of the subject matter.
- Additional Problems for Chapters 2–13, with selected solutions, are presented in the Student CD-ROM that accompanies this book. Students can use those problems for more practice. And instructors can use those problems whose solutions are provided only in the Instructor's Solutions Manual for assignments.

Special Features

We have created special features for this book in order to facilitate the learning of the material and an understanding of its applications.

- **Engineering Economics in Action boxes** near the beginning and end of each chapter recount the fictional experiences of a young engineer. These vignettes reflect and support the chapter material. The first box in each chapter usually portrays one of the characters trying to deal with a practical problem. The second box demonstrates how the character has solved the problem by applying material discussed in the chapter. All of these vignettes are linked to form a narrative that runs throughout the book. The main character is Naomi, a recent engineering graduate. In the first chapter, she starts her job in the engineering department at Global Widget Industries and is given a decision problem by her supervisor. Over the course of the book, Naomi learns about engineering economics on the job as the students learn from the book. There are several characters, who relate to one another in various ways, exposing the students to practical, ethical, and social issues as well as mathematical problems.

Engineering Economics in Action, Part 6A:
The Pit Bull

Naomi liked to think of Terry as a pit bull. Terry had this endearing habit of finding some detail that irked him, and not letting go of it until he was satisfied that things were done properly. Naomi had seen this several times in the months they had worked together. Terry would sink his teeth into some quirk of Global Widgets' operating procedures and, just like a fighting dog, not let go until the fight was over.

This time, it was about the disposal of some computing equipment. Papers in hand, he quietly approached Naomi and earnestly started to explain his concern. "Naomi, I don't know what Bill Fisher is doing, but something's definitely not right here. Look at this."

Terry displayed two documents to Naomi. One was an accounting statement showing the book value of various equipment, including some CAD/CAM computers that had been sold for scrap the previous week. The other was a copy of a sales receipt from a local salvage firm for that same equipment.

"I don't like criticizing my fellow workers, but I really am afraid that Bill might be doing something wrong." Bill Fisher was the buyer responsible for capital equipment at Global Widgets, and he also disposed of surplus assets. "You know the CAD/CAM workstations they had in engineering design? Well, they were replaced recently and sold. Here is the problem. They were only three years old, and our own accounting department estimated their value as about $5000 each." Terry's finger pointed to the evidence on the accounting statement. "But here," his finger moving to the guilty figure on the sales receipt, "they were actually sold for $300 each!" Terry sat back in his chair. "How about that!"

Naomi smiled. Unfortunately, she would have to pry his teeth out of this one. "Interesting observation, Terry. But you know, I think it's probably OK. Let me explain."

■ **Close-Up boxes** in the chapters present additional material about concepts that are important but not essential to the chapter.

CLOSE-UP 6.1	Depreciation Methods
Method	**Description**
Straight-line	The book value of an asset diminishes by an equal *amount* each year.
Declining-balance	The book value of an asset diminishes by an equal *proportion* each year.
Sum-of-the-years'-digits	An accelerated method, like declining-balance, in which the depreciation rate is calculated as the ratio of the remaining years of life to the sum of the digits corresponding to the years of life.
Double-declining-balance	A declining-balance method in which the depreciation rate is calculated as $2/N$ for an asset with a service life of N years.
150%-declining-balance	A declining-balance method in which the depreciation rate is calculated as $1.5/N$ for an asset with a service life of N years.
Units-of-production	Depreciation rate is calculated per unit of production as the ratio of the units produced in a particular year to the total estimated units produced over the asset's lifetime.

■ In each chapter, a **Net Value box** provides a chapter-specific example of how the internet can be used as a source of information and guidance for decision making.

N E T   V A L U E   6 . 1

Securities Regulators

Countries that trade in corporate stocks and bonds (collectively called *securities*) generally have a regulatory body to protect investors, ensure that trading is fair and orderly, and facilitate the acquisition of capital by businesses.

In Canada, the Canadian Securities Administrators (CSA) coordinates regulators from each of Canada's provinces and territories, and also educates Canadians about the securities industry, the stock markets, and how to protect investors from scams by providing a variety of educational materials on securities and investing.

In the United States, the Securities Exchange Commission (SEC) regulates securities for the country. The SEC has been particularly active in enforcement in recent years. In the United Kingdom, the Financial Services Authority (FSA) regulates the securities industry as well as other financial services such as banks. In Australia, the regulator is the Australian Securities and Investments Commission (ASIC).

Australia **www.asic.gov.au**

Canada **www.csa-acvm.ca**

United Kingdom **www.fsa.gov.uk**

United States **www.sec.gov**

■ At the end of each chapter, a **Mini-Case**, complete with discussion questions, relates interesting stories about how familiar companies have used engineering economic principles in practice, or how engineering economics principles can help us understand broader real-world issues.

MINI-CASE 6.1

Business Expense or Capital Expenditure?

From the *Stabroek News* (Guyana), September 15, 2007:

Commissioner-General of the Guyana Revenue Authority (GRA), Khurshid Sattaur, says that the GRA has developed an extra-statutory ruling to allow for the depreciation of software over a two-year period.

A release from the Office of the Commissioner-General yesterday quoted Sattaur as saying that "the Revenue Authority for the purposes of Section 17 of the *Income Tax Act* Chapter 81:01 and the *Income Tax (Depreciate Rates) Regulation 1992* as amended by the *Income (Depreciate Rates) (Amendments) 1999* allow wear and tear at a rate of 50% per annum."

He noted that while Guyana's income tax laws adequately provide for the depreciation of computer equipment (hardware) for 50% write-off over two years, the law does not make provision for software.

According to Sattaur, the GRA feels that businesses have been taking advantage of the inadequate provisions in the law and have been treating the software as an expense to be written off in the first year or in the year of acquisition.

Therefore, the release stated, the GRA is allowing wear and tear on website development costs and software, whether or not they form part of the installed software over a two-year period.

Discussion

Calculating depreciation is made difficult by many factors. First, the value of an asset can change over time in many complicated ways. Age, wear and tear, and functional changes all have their effects, and are often unpredictable. A 30-year old VW Beetle, for example, can suddenly increase in value because a new Beetle is introduced by the

■ Two **Extended Cases** are provided, one directly following Chapter 6 and the other directly following Chapter 11. They concern complex situations that incorporate much of the material in the preceding chapters. Unlike chapter examples, which are usually directed at a particular concept being taught, the Extended Cases require the students to integrate what they have learned over a number of chapters. They can be used for assignments, class discussions, or independent study.

EXTENDED CASE: PART 1

Welcome to the Real World

A.1 Introduction

Clem looked up from his computer as Naomi walked into his office. "Hi, Naomi. Sit down. Just let me save this stuff."

After a few seconds Clem turned around, showing a grin. "I'm working on our report for the last quarter's operations. Things went pretty well. We exceeded our targets on defect reductions and on reducing overtime. And we shipped everything required—over 90% on time."

Naomi caught a bit of Clem's exuberance. "Sounds like a report you don't mind writing."

"Yeah, well, it was a team job. Everyone did good work. Talking about doing good work, I should have told you this before, but I didn't think about it at the right time. Ed Burns and Anna Kulkowski were really impressed with the report you did on the forge project."

Naomi leaned forward. "But they didn't follow my recommendation to get a new manual forging press. I assumed there was something wrong with what I did."

Dave Sullivan came in with long strides and dropped into a chair. "Good morning, everybody. It is still morning, barely. Sorry to be late. What's up?"

Clem looked at Dave and started talking. "What's up is this. I want you and Naomi to look into our policy about buying or making small aluminum parts. We now use about 200 000 pieces a month. Most of these, like bolts and sleeves, are cold-formed.

"Prabha Vaidyanathan has just done a market projection for us. If she's right, our demand for these parts will continue to grow. Unfortunately, she wasn't very precise about the *rate* of growth. Her estimate was for anything between 5% and 15% a year. We now contract this work out. But even if growth is only 5%, we may be at the level where it pays for us to start doing this work ourselves.

"You remember we had a couple of engineers from Hamilton Tools looking over our processes last week? Well, they've come back to us with an offer to sell us a cold-former. They have two possibilities. One is a high-volume job that is a version

Additional Pedagogical Features

- Each chapter begins with a **list of the major sections** to provide an overview of the material that follows.
- **Key terms** are boldfaced where they are defined in the body of the text. For easy reference, all of these terms are defined in a Glossary near the back of the book.
- Additional material is presented in **chapter appendices** at the ends of Chapters 3, 4, 5, 8, 9, and 13.
- Numerous worked-out **Examples** are given throughout the chapters. Although the decisions have often been simplified for clarity, most of them are based on real situations encountered in the authors' consulting experiences.
- Worked-out **Review Problems** near the end of each chapter provide more complex examples that integrate the chapter material.
- A concise prose **Summary** is given for each chapter.
- Each chapter has 30 to 50 **Problems** of various levels of difficulty covering all of the material presented. Like the worked-out Examples, many of the Problems have been adapted from real situations. A **More Challenging Problem** is presented at the end of each problem set. As mentioned earlier, **Addditional Problems** (with selected solutions) for Chapters 2–13 are provided on the Student CD-ROM packaged with the book.

- A **spreadsheet icon** like the one shown here indicates where Examples or Problems involve spreadsheets, which are available on the Instructor's Resource CD-ROM. The use of computers by engineers is now as commonplace as the use of slide rules was 30 years ago. Students using this book will likely be very familiar with spreadsheet software. Consequently, such knowledge is assumed rather than taught in this book. The spreadsheet Examples and Problems are presented in such a manner that they can be done using any popular spreadsheet program, such as Excel, Lotus 1-2-3, or Quattro Pro.
- **Tables of interest factors** are provided in Appendix A, Appendix B, and Appendix C.
- **Answers to Selected Problems** are provided in Appendix D.
- For convenience, a **List of Symbols** used in the book is given on the inside of the front cover, and a **List of Formulas** is given on the inside of the back cover.

Course Designs

This book is ideal for a one-term course, but with supplemental material it can also be used for a two-term course. It is intended to meet the needs of students in all engineering programs, including, but not limited to, aeronautical, chemical, computer, electrical, industrial, mechanical, mining, and systems engineering. Certain programs emphasizing public projects may wish to supplement Chapter 10, "Public Sector Decision Making," with additional material.

A course based on this book can be taught in the first, second, third, or fourth year of an engineering program. The book is also suitable for college technology programs. No more than high school mathematics is required for a course based on this text. The probability theory required to understand and apply the tools of risk analysis is provided in Chapter 12. Prior knowledge of calculus or linear algebra is not needed, except for working through the appendix to Chapter 13.

This book is also suitable for self-study by a practitioner or anybody interested in the economic aspects of decision making. It is easy to read and self-contained, with many clear examples. It can serve as a permanent resource for practising engineers or anyone involved in decision making.

Companion Website (www.pearsoned.ca/fraser)

We have created a robust Companion Website to accompany the book. It contains the following items for instructors and students:

- **Practice Quizzes** for each chapter. Students can try these self-test questions, send their answers to an electronic grader, and receive instant feedback.
- **Excel spreadsheets** for selected Examples and Problems (designated by a spreadsheet icon in the book).
- **Weblinks**
- **Interest Tables**
- **Glossary Flashcards,** which afford students the opportunity to test themselves about key terms.

Instructor's Resource CD-ROM

We have also carefully prepared an Instructor's Resource CD-ROM to assist instructors in delivering the couse. It contains the following items:

- An **Instructor's Solutions Manual,** which contains full solutions to all the Problems in the book, full solutions to all the Additional Problems on the Student CD-ROM, model solutions for the Extended Cases in the book, teaching notes for the Mini-Cases, and Excel spreadsheets for selected examples and problems (designated by a spreadsheet icon in the book).
- A **Computerized Testbank** (Pearson TestGen), which allows instructors to view and edit the questions, generate tests, print the tests in a variety of formats, administer tests on a local area network, and have the tests graded electronically.
- **PowerPoint© Slides** for each chapter, which can be used to help present material in the classroom.

Acknowledgments

The authors wish to acknowledge the contributions of a number of individuals who assisted in the development of this text. First and foremost are the hundreds of engineering students at the University of Waterloo who have provided us with feedback on passages they found hard to understand, typographical errors, and examples that they thought could be improved. There are too many individuals to name in person, but we are very thankful to each of them for their patience and diligence.

Converting a text with a very Canadian focus to one that has a global perspective required myriad changes to place names, currencies, and so forth. Peggy Fraser was very helpful in making sure that every detail was taken care of, with the able assistance of Andrea Forwell.

Other individuals who have contributed strongly to previous editions of the book include Irwin Bernhardt, Peter Chapman, David Fuller, J.B. Moore, Tim Nye, Ron Pelot, Victor Waese, and Yuri Yevdokimov.

During the development process for the new edition, Pearson Education Canada arranged for the anonymous review of parts of the manuscript by a number of very able reviewers. These reviews were extremely beneficial to us, and many of the best ideas incorporated in the final text originated with these reviewers. We can now thank them by name:

Karen Bradbury, University of Warwick
Eric Croiset, University of Waterloo

Faiza Enanny, Marine Institute
Johan Fourie, British Columbia Institute of Technology
Maruf Hasan, The University of New South Wales
Dr. Leonard Lye, Memorial University of Newfoundland
Ron Mackinnon, University of British Columbia
Paul Missios, Ryerson University
Juan Pernia, Lakehead University
Amr I. Shabaka, University of Windsor
Ted Stathopoulos, Concordia University
Claude Théoret, University of Ottawa
Zhigang Tian, Concordia University
Ayman M.A. Youssef, University of Windsor

Finally, we want to express our appreciation to the various editors at Pearson Education Canada for their professionalism and support during the writing of this book. Helen Smith, our developmental editor for most of this edition, was able support for the author team. We remain grateful to Maurice Esses, who played a particularly strong role in bringing the first and second editions to completion and for guiding us through the completion of this edition.

To all of the above, thank you again for your help. To those we may have forgotten to thank, our appreciation is just as great, even if our memories fail us. Without doubt, some errors remain in this text in spite of the best efforts of everyone involved. To help us improve for the next edition, if you see an error, please let us know.

Niall M. Fraser
Elizabeth M. Jewkes
May Tajima

A Great Way to Learn and Instruct Online

The Pearson Education Canada Companion Website is easy to navigate and is organized to correspond to the chapters in this textbook. Whether you are a student in the classroom or a distance learner you will discover helpful resources for in-depth study and research that empower you in your quest for greater knowledge and maximize your potential for success in the course.

Companion Website

[www.pearsoned.ca/fraser]

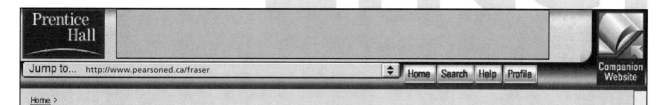

Prentice Hall

Jump to... http://www.pearsoned.ca/fraser   Home   Search   Help   Profile

Companion Website

Home >

Companion Website

Global Engineering Economics: Financial Decision Making for Engineers, Fourth Edition, by Fraser, Jewkes, Bernhardt, and Tajima

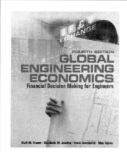

Student Resources

The modules in this section provide students with tools for learning course material. These modules include:

- Multiple Choice Quizzes
- Spreadsheets
- Glossary Flashcards
- Weblinks
- Interest Tables

In the quiz modules students can send answers to the grader and receive instant feedback on their progress through the Results Reporter. Coaching comments and references to the textbook may be available to ensure that students take advantage of all available resources to enhance their learning experience.

CHAPTER 1

Engineering Decision Making

Engineering Economics in Action, Part 1A:
Naomi Arrives

Naomi's first day on the job wasn't really her first day on the job. Ever since she had received the acceptance letter three weeks earlier, she had been reading and rereading all her notes about the company. Somehow she had arranged to walk past the plant entrance going on errands that never would have taken her that exact route in the past. So today wasn't the first time she had walked through that tidy brick entrance to the main offices of Global Widget Industries—she had done it the same way in her imagination a hundred times before.

Clement Sheng, the engineering manager who had interviewed Naomi for the job, was waiting for her at the reception desk. His warm smile and easy manner did a lot to break the ice. He suggested that they could go through the plant on the way to her desk. She agreed enthusiastically. "I hope you remember the engineering economics you learned in school," he said.

Naomi did, but rather than sound like a know-it-all, she replied, "I think so, and I still have my old textbook. I suppose you're telling me I'm going to use it."

"Yes. That's where we'll start you out, anyhow. It's a good way for you to learn how things work around here. We've got some projects lined up for you already, and they involve some pretty big decisions for Global Widgets. We'll keep you busy."

1.1 | Engineering Decision Making

Engineering is a noble profession with a long history. The first engineers supported the military, using practical know-how to build bridges, fortifications, and assault equipment. In fact, the term *civil engineer* was coined to make the distinction between engineers who worked on civilian projects and engineers who worked on military problems.

In the beginning, all engineers had to know was the technical aspects of their jobs. Military commanders, for example, would have wanted a strong bridge built quickly. The engineer would be challenged to find a solution to the technical problem, and would not have been particularly concerned about the costs, safety, or environmental impacts of the project. As years went by, however, the engineer's job became far more complicated.

All engineering projects use resources, such as raw materials, money, labour, and time. Any particular project can be undertaken in a variety of ways, with each one calling for a different mix of resources. For example, a standard light bulb requires inexpensive raw materials and little labour, but it is inefficient in its use of electricity and does not last very long. On the other hand, a high-efficiency light bulb uses more expensive raw materials and is more expensive to manufacture, but consumes less electricity and lasts longer. Both products provide light, but choosing which is better in a particular situation depends on how the costs and benefits are compared.

Historically, as the kinds of projects engineers worked on evolved and technology provided more than one way of solving technical problems, engineers were faced more often with having to choose among alternative solutions to a problem. If two solutions both dealt with a problem effectively, clearly the less expensive one was preferred. The practical science of engineering economics was originally developed specifically to deal with determining which of several alternatives was, in fact, the most economical.

Choosing the cheapest alternative, though, is not the entire story. Though a project might be technically feasible and the most reasonably priced solution to a problem, if the money isn't available to do it, it can't be done. The engineer has to become aware of the financial constraints on the problem, particularly if resources are very limited. In addition, an engineering project can meet all other criteria, but may cause detrimental environmental effects. Finally, any project can be affected by social and political constraints. For example, a large irrigation project called the Garrison Diversion Unit in North Dakota was effectively cancelled because of political action by Canadians and environmental groups, even though over $2 000 000 000 had been spent.

Engineers today must make decisions in an extremely complex environment. The heart of an engineer's skill set is still technical competence in a particular field. This permits the determination of possible solutions to a problem. However, necessary to all engineering is the ability to choose among several technically feasible solutions and to defend that choice credibly. The skills permitting the selection of a good choice are common to all engineers and, for the most part, are independent of which engineering field is involved. These skills form the discipline of engineering economics.

1.2 | What Is Engineering Economics?

Just as the role of the engineer in society has changed over the years, so has the nature of engineering economics. Originally, engineering economics was the body of knowledge that allowed the engineer to determine which of several alternatives was economically best—the least expensive, or perhaps the most profitable. In order to make this determination properly, the engineer needed to understand the mathematics governing the relationship between time and money. Most of this book deals with teaching and using this knowledge. Also, for many kinds of decisions the costs and benefits are the most important factors affecting the decision, so concentrating on determining the economically "best" alternative is appropriate.

In earlier times, an engineer would be responsible for making a recommendation on the basis of technical and analytic knowledge, including the knowledge of engineering economics, and then a manager would decide what should be done. A manager's decision would often be different from the engineer's recommendation, because the manager would take into account issues outside the engineer's range of expertise. Recently, however, the trend has been for managers to become more reliant on the technical skills of the engineers, or for the engineers themselves to be the managers. Products are often very complex; manufacturing processes are fine-tuned to optimize productivity; and even understanding the market sometimes requires the analytic skills of an engineer. As a result, it is often only the engineer who has sufficient depth of knowledge to make a competent decision.

Consequently, understanding how to compare costs, although still of vital importance, is not the only skill needed to make suitable engineering decisions. One must also be able to take into account all the other considerations that affect a decision, and to do so in a reasonable and defensible manner.

Engineering economics, then, can be defined as the science that deals with techniques of quantitative analysis useful for selecting a preferable alternative from several technically viable ones.

The evaluation of costs and benefits is very important, and it has formed the primary content of engineering economics in the past. The mathematics for doing this evaluation, which is well developed, still makes up the bulk of studies of engineering economics. However, the modern engineer must be able to recognize the limits and applicability of these economic calculations, and must be able to take into account the inherent complexity of the real world.

In recent years, the scope of the engineer has been extending geographically as well. In the past it was generally sufficient for an engineer to understand the political, social, and economic context where he or she lived in order to make sensible technical decisions. Now, however, companies are global, manufactured goods have components made in different countries, the environmental consequences of engineering decisions extend across countries and continents, and engineers may find themselves working anywhere in the world. Consequently, the modern practice of engineering economics must include the ability to work in different currencies, under varying rates of inflation and different tax regimes.

1.3 | Making Decisions

All decisions, except perhaps the most routine and automatic ones or those that are institutionalized in large organizations, are made, in the end, on the basis of belief as opposed to logic. People, even highly trained engineers, do what feels like the right thing to do. This is not to suggest that one should trust only one's intuition and not one's intellect, but rather to point out something true about human nature and the function of engineering economics studies.

Figure 1.1 is a useful illustration of how decisions are made. At the top of the pyramid are preferences, which directly control the choices made. Preferences are the beliefs about what is best, and are often hard to explain coherently. They sometimes have an emotional basis and include criteria and issues that are difficult to verbalize.

The next tier is composed of politics and people. Politics in this context means the use of power (intentional or not) in organizations. For example, if the owner of a factory has a strong opinion that automation is important, this has a great effect on engineering decisions on the plant floor. Similarly, an influential personality can affect decision making. It's difficult to make a decision without the support, either real or imagined, of other people. This support can be manipulated, for example, by a persuasive salesperson or a persistent lobbyist. Support might just be a general understanding communicated through subtle messages.

The next tier is a collection of "facts." The facts, which may or may not be valid or verifiable, contribute to the politics and the people, and indirectly to the preferences. At

Figure 1.1 Decision Pyramid

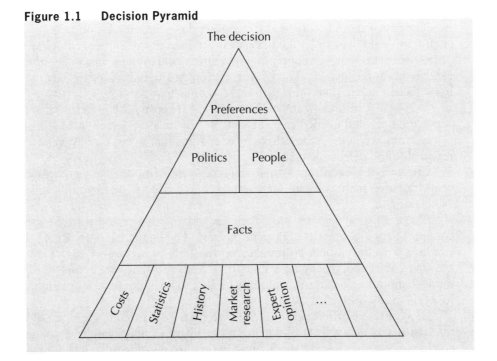

the bottom of the pyramid are the activities that contribute to the facts. These include the history of previous similar decisions, statistics of various sorts, and, among other things, a determination of costs.

In this view of decisions, engineering economics is not very important. It deals essentially with facts and, in particular, with determining costs. Many other facts affect the final decision, and even then the decision may be made on the basis of politics, personality, or unstated preferences. However, this is an extreme view.

Although preferences, politics, and people can outweigh facts, usually the relationship is the other way around. The facts tend to control the politics, the people, and the preferences. It is facts that allow an individual to develop a strong opinion, which then may be used to influence others. Facts accumulated over time create intuition and experience that control our "gut feeling" about a decision. Facts, and particularly the activities that develop the facts, form the foundation for the pyramid in Figure 1.1. Without the foundation, the pyramid would collapse.

Engineering economics is important because it facilitates the establishment of verifiable facts about a decision. The facts are important and necessary for the decision to be made. However, the decision eventually made may be contrary to that suggested by analysis. For example, a study of several methods of treating effluent might determine that method A is most efficient and moderately priced, but method B might in fact be chosen because it requires a visible change to the plant which, it is felt, will contribute to the company's image in environmental issues. Such a final decision is appropriate because it takes into account facts beyond those dealt with in the economic analysis.

Engineering Economics in Action, Part 1B:
Naomi Settles In

As Naomi and Clement were walking, they passed the loading docks. A honk from behind told them to move over so that a forklift could get through. The operator waved in passing and continued on with the task of moving coils of sheet metal into the warehouse. Naomi noticed shelves and shelves of packaging material, dies, spare parts, and other items that she didn't recognize. She would find out more soon enough. They continued to walk. As they passed a welding area, Clem pointed out the newest recycling project at Global Widgets: the water used to degrease the metal was now being cleaned and recycled rather than being used only once.

Naomi became aware of a pervasive, pulsating noise emanating from somewhere in the distance. Suddenly the corridor opened up to the main part of the plant, and the noise became a bedlam of clanging metal and thumping machinery. Her senses were assaulted. The ceiling was very high, and there were rows of humpbacked metal monsters unlike any presses she had seen before. The tang of mill oil overwhelmed her sense of smell, and she felt the throbbing from the floor knocking her bones together. Clem handed her hearing and eye protectors.

"These are our main press lines." Clem was yelling right into Naomi's ear, but she had to strain to hear. "We go right from sheet metal to finished widgets in 12 operations." A passing forklift blew propane exhaust at her, momentarily replacing the mill-oil odour with hot-engine odour. "Engineering is off to the left there."

As they went through the double doors into the engineering department, the din subsided and the ceiling came down to normal height. Removing the safety equipment, they stopped for a moment to get some juice at the vending machines. As Naomi looked around, she saw computers on desks more or less sectioned off by acoustic room dividers. As Clem led her farther, they stopped long enough for him to introduce Naomi to Carole Brown, the receptionist and secretary. Just past Carole's desk and around the corner was Naomi's desk. It was a nondescript metal desk with a long row of empty shelving above. Clem said that her computer would arrive within the week. Naomi noticed that the desk next to hers was empty, too.

→

"Am I sharing with someone?" she asked.

"Well, you will be. That's for your co-op student."

"My co-op student?"

"Yep. He's a four-month industrial placement from the university. Don't worry, we have enough to do to keep you both busy. Why don't you take a few minutes to settle in, while I take care of a couple of things. I'll be back in, say, 15 minutes. I'll take you over to human resources. You'll need a security pass, and I'm sure they have lots of paperwork for you to fill out."

Clem left. Naomi sat down and opened the briefcase she had carefully packed that morning. Alongside the brownbag lunch was an engineering economics textbook. She took it out and placed it on the empty shelf above the desk. "I thought I might need you," she said to herself. "Now, let's get this place organized!"

1.4 | Dealing With Abstractions

The world is far more complicated than can ever be described in words, or even thought about. Whenever one deals with reality, it is done through models or abstractions. For example, consider the following description:

> Naomi watched the roll of sheet metal pass through the first press. The die descended and punched six oval shapes from the sheet. These "blanks" dropped through a chute into a large metal bin. The strip of sheet metal jerked forward into the die and the press came down again. Pounding like a massive heart 30 times a minute, the machine kept the operator busy full-time just providing the giant coils of metal, removing the waste skeleton scrap, and stacking blanks in racks for transport to the next operation.

This gives a description of a manufacturing process that is reasonably complete, in that it permits one to visualize the process. But it is not absolutely complete. For example, how large and thick were the blanks? How big was the metal bin? How heavy was the press? How long did it take to change a die? These questions might be answered, but no matter how many questions are asked, it is impossible to express all of the complexity of the real world. It is also undesirable to do so.

When one describes something, one does so for a purpose. In the description, one selects those aspects of the real world that are relevant to that purpose. This is appropriate, since it would be very confusing if a great deal of unnecessary information were given every time something was talked or written about. For example, if the purpose of the above description was to explain the exact nature of the blanks, there would be considerably less emphasis on the process, and many more details about the blanks themselves.

This process of simplifying the complexities of the real world is necessary for any engineering analysis. For example, in designing a truss for a building, it is usually assumed that the members exhibit uniform characteristics. However, in the real world these members would be pieces of lumber with individual variations: some would be stronger than average and some would be weaker. Since it is impractical to measure the characteristics of each piece of wood, a simplification is made. As another example, the various components of an electric circuit, such as resistors and capacitors, have values that differ from their nominal specifications because of manufacturing tolerances, but such differences are often ignored and the nominal values are the ones used in calculations.

Figure 1.2 illustrates the basic process of modelling that applies in so much of what humans do, and applies especially to engineering. The world is too complicated to express completely, as represented by the amorphous shape at the top of the figure. People extract from the real world a simplification (in other words, a model) that captures information

Figure 1.2   The Modelling Process

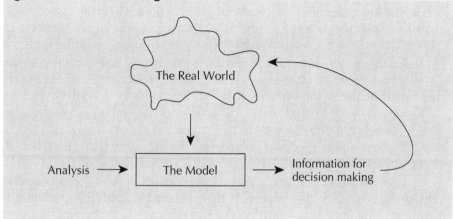

useful and appropriate for a given purpose. Once the model is developed, it is used to analyze a situation, and perhaps make some predictions about the real world. The analysis and the predictions are then related back to the real world to make sure the model is valid. As a result, the model might need some modification, so that it more accurately reflects the relevant features of the real world.

The process illustrated in Figure 1.2 is exactly what is done in engineering economics. The model is often a mathematical one that simplifies a more complicated situation, but does so in a reasonable way. The analysis of the model provides some information, such as which solution to a problem is cheapest. This information must always be related back to the real problem, however, to take into account the aspects of the real world that may have been ignored in the original modelling effort. For example, the economic model might not have included taxes or inflation, and an examination of the result might suggest that taxes and inflation should not be ignored. Or, as already pointed out, environmental, political, or other considerations might modify any conclusions drawn from the mathematical model.

EXAMPLE 1.1

Naomi's brother Ben has been given a one-year assignment in Alaska, and he wants to buy a car just for the time he is there. He has three choices, as illustrated in Table 1.1. For each alternative, there is a purchase price, an operating cost (including gas, insurance, and repairs), and an estimated resale value at the end of the year. Which should Ben buy?

Table 1.1   Buying a Car

	1968 Corvette	2004 Toyota Corolla	2001 BMW 5-Series
Purchase	$12 000	$7000	$20 000
Operation	$200/month	$100/month	$150/month
Resale	$13 000	$5000	$20 000

The next few chapters of this book will show how to take the information from Table 1.1 and determine which alternative is economically best. As it turns out, under most circumstances, the Corvette is best. However, in constructing a model of the decision, we must make a number of important assumptions.

For example, how can one be sure of the resale value of something until one actually tries to sell it? Along the same lines, who can tell what the actual maintenance costs will be? There is a lot of uncertainty about future events that is generally ignored in these kinds of calculations. Despite this uncertainty, estimates can provide insights into the appropriate decision.

Another problem for Ben is getting the money to buy a car. Ben is fairly young, and would find it very difficult to raise even $7000, perhaps impossible to raise $20 000. The Corvette might be the best value, but if the money isn't available to take advantage of the opportunity it doesn't matter. In order to do an economic analysis, we may assume that he has the money available.

If an economic model is judged appropriate, does that mean Ben should buy the Corvette? Maybe not.

A person who has to drive to work every morning would probably not want to drive an antique car. It is too important that the car be reliable (especially in Alaska in the winter). The operating costs for the Corvette are high, reflecting the need for more maintenance than with the other cars, and there are indirect effects of low reliability that are hard to capture in dollars.

If Ben were very tall, he would be extremely uncomfortable in the compact Toyota Corolla, so that, even if it were economically best, he would hesitate to resign himself to driving with his knees on either side of the steering wheel.

Ben might have strong feelings about the environmental record of one of the car manufacturers, and might want to avoid driving that car as a way of making a statement.

Clearly, there are so many intangibles involved in a decision like this that it is impossible for anyone but Ben himself to make such a personal choice. An outsider can point out to Ben the results of a quantitative analysis, given certain assumptions, but cannot authoritatively determine the best choice for Ben.■

1.5 | The Moral Question: Three True Stories

Complex decisions often have an ethical component. Recognizing this component is important for engineers, since society relies on them for so many things. The following three anecdotes concern real companies—although names and details have been altered for anonymity—and illustrate some extreme examples of the forces acting on engineering decision making.

EXAMPLE 1.2

The process of making sandpaper is similar to that of making a photocopy. A two-metre-wide roll of paper is coated with glue and given a negative electric charge. It is then passed over sand (of a particular type) that has a positive charge. The sand is attracted to the paper and sticks on the glue. The fact that all of the bits of sand have the same type of charge makes sure that the grains are evenly spaced. The paper then passes through a long, heated chamber to cure the glue. Although the process sounds fairly simple, the machine that does this, called a maker, is very complicated and expensive. One such machine, costing several million dollars, can support a factory employing hundreds of workers.

Preston Sandpapers, a subsidiary of a large firm, was located in a small town. Its maker was almost 30 years old and desperately needed replacement. However, rather than replace it, the parent company might have chosen to close down the plant and transfer production to one of the sister plants located in a different country.

The chief engineer had a problem. The costs for installing a new maker were extremely high, and it was difficult to justify a new maker economically. However, if he could not do so, the plant would close and hundreds of workers would be out of a job, including perhaps himself. What he chose to do was lie. He fabricated figures, ignored important costs, and exaggerated benefits to justify the expenditures. The investment was made, and the plant is still operating.■

EXAMPLE 1.3

Hespeler Meats is a medium-sized meat processor specializing in deli-style cold cuts and European process meats. Hoping to expand their product offerings, they decided to add a line of canned pâtés. They were eligible for a government grant to cover some of the purchase price of the necessary production equipment.

Government support for manufacturing is generally fairly sensible. Support is usually not given for projects that are clearly very profitable, since the company should be able to justify such an expense itself. On the other hand, support is also usually not given for projects that are clearly not very profitable, because taxpayers' money should not be wasted. Support is directed at projects that the company would not otherwise undertake, but that have good potential to create jobs and expand the economy.

Hespeler Meats had to provide a detailed justification for the canned pâté project in order to qualify for the government grant. Their problem was that they had to predict both the expenditures and the receipts for the following five years. This was a product line with which they had no experience, and which, in fact, had not been offered on that continent by any meat processor. They had absolutely no idea what their sales would be. Any numbers they picked would be guesses, but to get the grant they had to give numbers.

What they did was select an estimate of sales that, given the equipment expenditures expected, fell exactly within that range of profitability that made the project suitable for government support. They got the money. As it turned out, the product line was a flop, and the canning equipment was sold as scrap five years later.■

EXAMPLE 1.4

When a large metal casting is made, as for the engine block of a car, it has only a rough exterior and often has flash—ragged edges of metal formed where molten metal seeped between the two halves of the mould. The first step in finishing the casting is to grind off the flash, and to grind flat surfaces so that the casting can be held properly for subsequent machining.

Galt Casting Grinders (GCG) made the complex specialized equipment for this operation. It had once commanded the world market for this product, but lost market share to competitors. The competitors did not have a better product than GCG, but they were able to increase market share by adding fancy display panels with coloured lights, dials, and switches that looked very sophisticated.

GCG's problem was that their idea of sensible design was to omit the features the competitors included (or the customers wanted). GCG reasoned that these features added nothing to the capability of the equipment, but did add a lot to the manufacturing cost and to the maintenance costs that would be borne by the purchaser. They had no doubt that it was unwise, and poor engineering design, to make such unnecessarily complicated displays, so they made no changes.

GCG went bankrupt several years later.■

In each of these three examples, the technical issues are overwhelmed by the non-technical ones. For Preston Sandpapers, the chief engineer was pressured by his social responsibility and self-interest to lie and to recommend a decision not justified by the facts. In the Hespeler Meats case, the engineer had to choose between stating the truth—that future sales were unknown—which would deny the company a very useful grant, and selecting a convenient number that would encourage government support. For Galt Casting Grinders, the issue was marketing. They did not recognize that a product must be more than technically good; it must also be saleable.

Beyond these principles, however, there is a moral component to each of these anecdotes. As guardians of knowledge, engineers have a vital responsibility to society to behave ethically and responsibly in all ways. When so many different issues must be taken into account in engineering decision making, it is often difficult to determine what course of action is ethical.

For Preston Sandpapers, most people would probably say that what the chief engineer did was unethical. However, he did not exploit his position simply for personal gain. He was, to his mind, saving a town. Is the principle of honesty more important than several hundred jobs? Perhaps it is, but when the job holders are friends and family it is understandable that unethical choices might be made.

For Hespeler Meats, the issue is subtler. Is it ethical to choose figures that match the ideal ones to gain a government grant? It is, strictly speaking, a lie, or at least misleading, since there is no estimate of sales. On the other hand, the bureaucracy demands that some numbers be given, so why not pick ones that suit your case?

In the Galt Casting Grinders case, the engineers apparently did no wrong. The ethical question concerns the competitors' actions. Is it ethical to put features on equipment that do no good, add cost, and decrease reliability? In this case and for many other products, this is often done, ethical or not. If it is unethical, the ethical suppliers will sometimes go out of business.

There are no general answers to difficult moral questions. Practising engineers often have to make choices with an ethical component, and can sometimes rely on no stronger foundation than their own sense of right and wrong. More information about ethical issues for engineers can be obtained from professional engineering associations.

NET VALUE 1.1

Professional Engineering Associations and Ethical Decisions

Engineering associations maintain websites that can be a good source of information about engineering practice worldwide. At the time of publication, a selection of such sites includes:

Australia: **www.engineersaustralia.org.au**

Canada: **www.engineerscanada.ca**

China–Hong Kong: **www.hkie.org.hk**

Indonesia: **www.pii.or.id**

Ireland: **www.iei.ie**

Malaysia: **www.iem.org.my**

New Zealand: **www.ipenz.org.nz**

South Africa: **www.ecsa.co.za**

United Kingdom: **www.engc.org.uk**

United States: **www.nspe.org**

Check out sections such as Member Discipline and Complaints (Australia), and **nspe.org/ethics** (United States). Understanding ethics as enforced by the engineering associations is an excellent basis for making your own ethical decisions.

1.6 | Uncertainty, Sensitivity Analysis, and Currencies

Whenever people predict the future, errors occur. Sometimes predictions are correct, whether the predictions are about the weather, a ball game, or company cash flow. On the other hand, it would be unrealistic to expect anyone always to be right about things that haven't happened yet.

Although one cannot expect an engineer to predict the future precisely, approximations are very useful. A weather forecaster can dependably say that it will not snow in July in France, for example, even though it may be more difficult to forecast the exact temperature. Similarly, an engineer may not be able to precisely predict the scrap rate of a testing process, but may be able to determine a range of likely rates to help in a decision-making process.

Engineering economics analyses are quantitative in nature, and most of the time the quantities used in economic evaluations are estimates. The fact that we don't have precise values for some quantities may be very important, since decisions may have expensive consequences and significant health and environmental effects. How can the impact of this uncertainty be minimized?

One way to control this uncertainty is to make sure that the information being used is valid and as accurate as possible. The GIGO rule—"garbage in, garbage out"—applies here. Nothing is as useless or potentially dangerous as a precise calculation made from inaccurate data. However, even accurate data from the past is of only limited value when predicting the future. Even with sure knowledge of past events, the future is still uncertain.

Sensitivity analysis involves assessing the effect of uncertainty on a decision. It is very useful in engineering economics. The idea is that, although a particular value for a parameter can be known with only a limited degree of certainty, a range of values can be assessed with reasonable certainty. In sensitivity analysis, the calculations are done several times, varying each important parameter over its range of possible values. Usually only one parameter at a time is changed, so that the effect of each change on the conclusion can be assessed independently of the effect of other changes.

In Example 1.1, Naomi's brother Ben had to choose a car. He made an estimate of the resale value of each of the alternative cars, but the *actual* resale amount is unknown until the car is sold. Similarly, the operating costs are not known with certainty until the car is driven for a while. Before concluding that the Corvette is the right car to buy (on economic grounds at least), Ben should assess the sensitivity of this decision by varying the resale values and operating costs within a range from the minimum likely amount to the maximum likely amount. Since these calculations are often done on spreadsheets, this assessment is not hard to do, even with many different parameters to vary.

Sensitivity analysis is an integral part of all engineering economics decisions because data regarding future activities are always uncertain. In this text, emphasis is usually given to the structure and formulation of problems rather than to verifying whether the result is robust. In this context, *robust* means that the same decision will be made over a wide range of parameter values. It should be remembered that no decision is properly made unless the sensitivity of that decision to variation in the underlying data is assessed.

A related issue is the number of significant digits in a calculation. Modern calculators and computers can carry out calculations to a large number of decimal places of precision. For most purposes, such precision is meaningless. For example, a cost calculated as $1.0014613076 is of no more use than $1.00 in most applications. It is useful, though, to carry as many decimal places as convenient to reduce the magnitude of accumulated rounding-off errors.

In this book, all calculations have been done to as many significant digits as could conveniently be carried, even though the intermediate values are shown with three to six digits. As a rule, only three significant digits are assumed in the final value. For decision making purposes, this is plenty.

Finally, a number of different currencies are used in this book. The dollar ($) is the most common currency used because it can represent the currency of Australia, Canada, the United States, and several other countries equally well. But to illustrate the independence of the mathematics from the currency and to highlight the need to be able to work in different currencies, a number of other currency examples are used. For reference, Table 1.2 shows some of the currencies that are used in the text, their symbols, and their country of origin.

Table 1.2   Currencies of Various Countries

Country	Currency name	Symbol
Australia	Dollar	$
Canada	Dollar	$
China	Yuan	元
Europe	Euro	€
India	Rupee	Rs
Japan	Yen	¥
South Africa	Rand	R
South Korea	Won	₩
United Kingdom	Pound	£
United States	Dollar	$

1.7 | How This Book Is Organized

There are 12 chapters remaining in this book. The first block, consisting of Chapters 2 to 5, forms the core material of the book. Chapters 2 and 3 of that block provide the mathematics needed to manipulate monetary values over time. Chapters 4 and 5 deal with comparing alternative projects. Chapter 4 illustrates present worth, annual worth, and payback period comparisons, and Chapter 5 covers the internal rate of return (IRR) method of comparison.

The second block, Chapters 6 to 8, broadens the core material. It covers depreciation and analysis of a company's financial statements, when to replace equipment (replacement analysis), and taxation.

The third block, Chapters 9 to 13, provides supporting material for the previous chapters. Chapter 9 concerns the effect of inflation on engineering decisions, and Chapter 10 explores how decision making is done for projects owned by or affecting the public, rather than an individual or firm. Chapter 11 deals with handling uncertainty about important information through sensitivity analysis, while Chapter 12 deals with situations where exact parameter values are not known, but probability distributions for them are known. Finally, Chapter 13 provides some formal methods for taking into account the intangible components of an engineering decision.

Each chapter begins with a story about Naomi and her experiences at Global Widgets. There are several purposes to these stories. They provide an understanding of engineering practice that is impossible to convey with short examples. In each chapter, the story has been chosen to make clear why the ideas being discussed are important. It is also hoped that the stories make the material taught a little more interesting.

There is a two-part Extended Case in the text. Part 1, located between Chapters 6 and 7, presents a problem that is too complicated to include in any particular chapter, but that reflects a realistic situation likely to be encountered in engineering practice. Part 2, located between Chapters 11 and 12, builds on the first case to use some of the more sophisticated ideas presented in the later chapters.

Throughout the text are boxes that contain information associated with, and complementary to, the text material. One set of boxes contains Close-Ups, which focus on topics of relevance to the chapter material. These appear in each chapter in the appropriate section. There are also Net Value boxes, which tie the material presented to internet resources. The boxes are in the middle sections of each chapter. Another set of boxes presents Mini-Cases, which appear at the end of each chapter, following the problem set. These cases report how engineering economics is used in familiar companies, and include questions designed for classroom discussion or individual reflection.

End-of-chapter appendices contain relevant but more advanced material. Appendices at the back of the book provide tables of important and useful values and answers to selected chapter-end problems.

Engineering Economics in Action, Part 1C:
A Taste of What Is to Come

Naomi was just putting on her newly laminated security pass when Clem came rushing in. "Sorry to be late," he puffed. "I got caught up in a discussion with someone in marketing. Are you ready for lunch?" She certainly was. She had spent the better part of the morning going through the benefits package offered by Global Widgets and was a bit overwhelmed by the paperwork. Dental plan options, pension plan beneficiaries, and tax forms swam in front of her eyes. The thought of food sounded awfully good.

As they walked to the lunchroom, Clem continued to talk. "Maybe you will be able to help out once you get settled in, Naomi."

"What's the problem?" asked Naomi. Obviously Clem was still thinking about his discussion with this person from marketing.

"Well," said Clem, "currently we buy small aluminum parts from a subcontractor. The cost is quite reasonable, but we should consider making the parts ourselves, because our volumes are increasing and the fabrication process would not be difficult for us to bring in-house. We might be able to make the parts at a lower cost. Of course, we'd have to buy some new equipment. That's why I was up in the marketing department talking to Prabha."

"What do you mean?" asked Naomi, still a little unsure. "What does this have to do with marketing?"

Clem realized that he was making a lot of assumptions about Naomi's knowledge of Global Widgets. "Sorry," he said, "I need to explain. I was up in marketing to ask for some demand forecasts so that we would have a better handle on the volumes of these aluminum parts we might need in the next few years. That, combined with some digging on possible equipment costs, would allow us to do an analysis of whether we should make the parts in-house or continue to buy them."

Things made much more sense to Naomi now. Her engineering economics text was certainly going to come in handy.

PROBLEMS

1.1 In which of the following situations would engineering economics analysis play a strong role, and why?

    (a) Buying new equipment

    (b) Changing design specifications for a product

    (c) Deciding on the paint colour for the factory floor

    (d) Hiring a new engineer

    (e) Deciding when to replace old equipment with new equipment of the same type

    (f) Extending the cafeteria business hours

    (g) Deciding which invoice forms to use

    (h) Changing the 8-hour work shift to a 12-hour one

    (i) Deciding how much to budget for research and development programs

    (j) Deciding how much to donate for the town's new library

    (k) Building a new factory

    (l) Downsizing the company

1.2 Starting a new business requires many decisions. List five examples of decisions that might be assisted by engineering economics analysis.

1.3 For each of the following items, describe how the design might differ if the costs of manufacturing, use, and maintenance were not important. On the basis of these descriptions, is it important to consider costs in engineering design?

    (a) A car

    (b) A television set

    (c) A light bulb

    (d) A book

1.4 Leslie and Sandy, recently married students, are going to rent their first apartment. Leslie has carefully researched the market and has decided that, all things considered, there is only one reasonable choice. The two-bedroom apartment in the building at the corner of University and Erb Streets is the best value for the money, and is also close to school. Sandy, on the other hand, has just fallen in love with the top half of a duplex on Dunbar Road. Which apartment should they move into? Why? Which do you think they will move into? Why?

1.5 Describe the process of using the telephone as you might describe it to a six-year-old using it for the first time to call a friend from school. Describe using the telephone to an electrical engineer who just happens never to have seen one before. What is the correct way to describe a telephone?

1.6 **(a)** Karen has to decide which of several computers to buy for school use. Should she buy the least expensive one? Can she make the best choice on price alone?

(b) Several computers offer essentially the same features, reliability, service, etc. Among these, can she decide the best choice on price alone?

1.7 For each of the following situations, describe what you think you *should* do. In each case *would* you do this?

(a) A fellow student, who is a friend, is copying assignments and submitting them as his own work.

(b) A fellow student, who is *not* a friend, is copying assignments and submitting them as her own work.

(c) A fellow student, who is your only competitor for an important academic award, is copying assignments and submitting them as his own work.

(d) A friend wants to hire you to write an essay for school for her. You are dead broke and the pay is excellent.

(e) A friend wants to hire you to write an essay for school for him. You have lots of money, but the pay is excellent.

(f) A friend wants to hire you to write an essay for school for her. You have lots of money, and the pay is poor.

(g) Your car was in an accident. The insurance adjuster says that the car was totalled and they will give you only the "blue book" value for it as scrap. They will pick up the car in a week. A friend points out that in the meantime you could sell the almost-new tires and replace them with bald ones from the scrap yard, and perhaps sell some other parts, too.

(h) The CD player from your car has been stolen. The insurance adjuster asks you how much it was worth. It was a very cheap one, of poor quality.

(i) The engineer you work for has told you that the meter measuring effluent discharged from a production process exaggerates, and the measured value must be halved for recordkeeping.

(j) The engineer you work for has told you that part of your job is to make up realistic-looking figures reporting effluent discharged from a production process.

(k) You observe unmetered and apparently unreported effluent discharged from a production process.

(l) An engineer where you work is copying directly from a manufacturer's brochure machine-tool specifications to be included in a purchase request. These specifications limit the possible purchase to the particular one specified.

(m) An engineer where you work is copying directly from a manufacturer's brochure machine-tool specifications to be included in a purchase request. These specifications limit the possible purchase to the particular one specified. You know that the engineer's best friend is the salesman for that manufacturer.

1.8 Ciel is trying to decide whether now is a good time to expand her manufacturing plant. The viability of expansion depends on the economy (an expanding economy means more sales), the relative value of the currency (a lower-valued currency means more exports), and changes in international trade agreements (lower tariffs also mean more exports). These factors may be highly unpredictable, however. What two things can she do to help make sure she makes a good decision?

1.9 Trevor started a high-tech business two years ago, and now wants to sell out to one of his larger competitors. Two different buyers have made firm offers. They are similar in all but two respects. They differ in price: the Investco offer would result in Trevor's walking away with $2 000 000, while the Venture Corporation offer would give him $3 000 000. The other way they differ is that Investco says it will recapitalize Trevor's company to increase growth, while Trevor thinks that Venture Corporation will close down the business so that it doesn't compete with several of Venture Corporation's other divisions. What would you do if you were Trevor, and why?

1.10 Telekom Company is considering the development of a new type of cell phone based on a brand new, emerging technology. If successful, Telekom will be able to offer a cell phone that works over long distances and even in mountainous areas. Before proceeding with the project, however, what uncertainties associated with the new technology should they be aware of? Can sensitivity analysis help address these uncertainties?

More Challenging Problem

1.11 In Example 1.1 it is stated that in most circumstances, the Corvette is the right economic choice. Under what circumstances would either the Toyota or the BMW be the right choice?

MINI-CASE 1.1

Imperial Oil v. Quebec

In 1979, Imperial Oil sold a former petroleum depot in Levis, Quebec, which had been operating since the early 1920s. The purchaser demolished the facilities, and sold the land to a real estate developer. The developer conducted a cleanup that was approved by the Quebec Ministry of the Environment, which issued a certificate of authorization in 1987. Following this, the site was developed and a number of houses were built. However, years later, residents of the subdivision sued the environment ministry claiming there was remaining pollution.

The ministry, under threat of expensive lawsuits, then ordered Imperial Oil to identify the pollution, recommend corrective action, and potentially pay for the cost of cleanup. In response, Imperial Oil initiated judicial proceedings against the ministry, claiming violation of principles of natural justice and conflict of interest on its part.

In February 2003, the Supreme Court of Canada ruled that the ministry had the right to compel Imperial Oil to do the cleanup and save public money, because Imperial was the originator of the pollution and the minister did not personally benefit.

Source: *Imperial Oil v. Quebec (Minister of the Environment)*, [2003] 2 S.C.R. 624, 2003 SCC 58, Canadian Legal Information Institute (CanLII) site, www.canlii.org/ca/cas/scc/2003/2003scc58.html, accessed September 20, 2004.

Discussion

There is often strong motivation for companies to commit environmental offences. Companies primarily focus on profits, and preventing environmental damage is always a cost. It benefits society to have a clean environment, but almost never benefits the company directly. Sometimes, in spite of the efforts of upper management to be good citizens, the search for profit may result in environmental damage. In a large company it can be difficult for one person to know what is happening everywhere in the firm.

Older companies have an additional problem. An older company may have been producing goods in a certain way for years, and have established ways to dispose of waste. Changes in society make those traditional methods unacceptable. Even when a source of pollution has been identified, it may not be easy to fix. There may be decades of accumulated damage to correct. A production process may not be easily changed in a way that would allow the company to stay in business. Loss of jobs and the effect on the local economy may create strong political pressure to keep the company running.

The government has an important role to offset the profit motive for large companies, for example, by taking action through the courts. Unfortunately, that alone will not be enough to prevent some firms from continuing to cause environmental damage. Economics and politics will occasionally win out.

Questions

1. There are probably several companies in your city or country that are known to pollute. Name some of these. For each:

   (a) What sort of damage do they do?

   (b) How long have they been doing it?

   (c) Why is this company still permitted to pollute?

   (d) What would happen if this company were forced to shut down? Is it ethically correct to allow the company to continue to pollute?

2. Does it make more sense to fine a company for environmental damage or to fine management personally for environmental damage caused by a company? Why?

3. Should the fines for environmental damage be raised enough so that no company is tempted to pollute? Why or why not?

4. Governments can impose fines, give tax breaks, and take other actions that use economics to control the behaviour of companies. Is it necessary to do this whenever a company that pursues profits might do some harm to society as a whole? Why might a company do the socially correct thing even if profits are lost?

CHAPTER 2

Time Value of Money

Engineering Economics in Action, Part 2A:
A Steal For Steel

"Naomi, can you check this for me?" Terry's request broke the relative silence as Naomi and Terry worked together one Tuesday afternoon. "I was just reviewing our J-class line for Clem, and it seems to me that we could save a lot of money there."

"OK, tell me about it." Since Naomi and Terry had met two weeks earlier, just after Naomi started her job, things had being going very well. Terry, an engineering student at the local university, was on a four-month co-op work term industrial placement at Global Widgets.

"Well, mostly we use the heavy rolled stock on that line. According to the pricing memo we have for that kind of steel, there is a big price break at a volume that could supply our needs for six months. We've been buying this stuff on a week-by-week basis. It just makes sense to me to take advantage of that price break."

"Interesting idea, Terry. Have you got data about how we have ordered before?"

"Yep, right here."

"Let's take a closer look."

"Well," Terry said, as he and Naomi looked over his figures, "the way we have been paying doesn't make too much sense. We order about a week's supply. The cost of this is added to our account. Every six months we pay off our account. Meanwhile, the supplier is charging us 2% of our outstanding amount at the end of each month!"

"Well, at least it looks as if it might make more sense for us to pay off our bills more often," Naomi replied.

"Now look at this. In the six months ending last December, we ordered steel for a total cost of $1 600 000. If we had bought this steel at the beginning of July, it would have only cost $1 400 000. That's a saving of $200 000!"

"Good observation, Terry, but I don't think buying in advance is the right thing to do. If you think about it . . . "

2.1 | Introduction

Engineering decisions frequently involve evaluating tradeoffs among costs and benefits that occur at different times. A typical situation is when we invest in a project today in order to obtain benefits from the project in the future. This chapter discusses the economic methods used to compare benefits and costs that occur at different times. The key to making these comparisons is the use of an interest rate. In Sections 2.2 to 2.5, we illustrate the comparison process with examples and introduce some interest and interest rate terminology. Section 2.6 deals with cash flow diagrams, which are graphical representations of the magnitude and timing of cash flows over time. Section 2.7 explains the equivalence of benefits and costs that occur at different times.

2.2 | Interest and Interest Rates

Everyone is familiar with the idea of interest from their everyday activities:

From the furniture store ad: *Pay no interest until next year!*

From the bank: *Now 2.6% daily interest on passbook accounts!*

Why are there interest rates? If people are given the choice between having money today and the same amount of money one year from now, most would prefer the money today. If they had the money today, they could do something productive with it in hopes of benefit in the future. For example, they could buy an asset like a machine today, and could

N E T   V A L U E   2 . 1

Interest Rates

When there is news about the interest rate rising or falling, it most commonly refers to the prime interest rate charged by banks for loans, historically to their most creditworthy customers. Such news attracts attention because prime rates are the reference point for interest rates charged on many mortgage, personal, and business loans. Prime rates are generally the same between major banks in each country. Furthermore, they tend to be indexed according to the overnight rate targeted by each country's central bank, which is the approximate rate charged for short-term or overnight loans between banks, needed to fulfill

their reserve funding requirements. Overnight rates are examined and targets adjusted at regular intervals several times annually. The following central bank websites provide additional information about interest rates and policies and procedures affecting them.

Australia: **www.rba.gov.au**

Canada: **www.bankofcanada.ca**

China: **www.pbc.gov.cn**

England: **www.bankofengland.co.uk**

Europe: **www.ecb.europa.eu**

India: **www.rbi.org.in**

United States: **www.federalreserve.gov**

use it to make money from their initial investment. Or they may want to buy a consumer good like a new home theatre system and start enjoying it immediately. What this means is that one dollar today is worth more than one dollar in the future. This is because a dollar today can be invested for productive use, while that opportunity is lost or diminished if the dollar is not available until some time in the future.

The observation that a dollar today is worth more than a dollar in the future means that people must be compensated for lending money. They are giving up the opportunity to invest their money for productive purposes now on the promise of getting more money in the future. The compensation for loaning money is in the form of an interest payment, say I. More formally, **interest** is the difference between the amount of money lent and the amount of money later repaid. It is the compensation for giving up the use of the money for the duration of the loan.

An amount of money today, P (also called the *principal amount*), can be related to a *future amount F* by the interest amount I or interest rate i. This relationship is illustrated graphically in Figure 2.1 and can be expressed as $F = P + I$. The interest I can also be expressed as an interest rate i with respect to the principal amount so that $I = Pi$. Thus

$$F = P + Pi$$
$$= P(1 + i)$$

Figure 2.1   Present and Future Worth

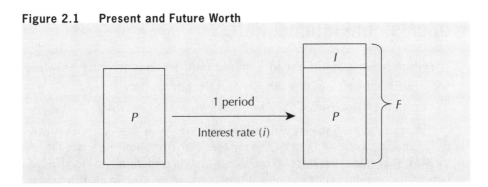

EXAMPLE 2.1

Samuel bought a one-year guaranteed investment certificate (GIC) for $5000 from a bank on May 15 last year. (The US equivalent is a certificate of deposit, or CD. A GIC is also similar to a guaranteed growth bond in the UK.) The bank was paying 10% on one-year guaranteed investment certificates at the time. One year later, Samuel cashed in his certificate for $5500.

We may think of the interest payment that Samuel got from the bank as compensation for giving up the use of money. When Samuel bought the guaranteed investment certificate for $5000, he gave up the opportunity to use the money in some other way during the following year. On the other hand, the bank got use of the money for the year. In effect, Samuel lent $5000 to the bank for a year. The $500 interest was payment by the bank to Samuel for the loan. The bank wanted the loan so that it could use the money for the year. (It may have lent the money to someone else at a higher interest rate.) ■

This leads to a formal definition of interest rates. Divide time into periods like days, months, or years. If the right to P at the beginning of a time period exchanges for the right to F at the end of the period, where $F = P(1 + i)$, i is the **interest rate** per time period. In this definition, P is called the **present worth** of F, and F is called the **future worth** of P.

EXAMPLE 2.1 RESTATED

Samuel invested $5000 with the bank on May 15 last year. The bank was paying 10% on one-year fixed term investments at the time. The agreement gave Samuel the right to claim $5500 from the bank one year later.

Notice in this example that there was a transaction between Samuel and the bank on May 15 last year. There was an exchange of $5000 on May 15 a year ago for the right to collect $5500 on May 15 this year. The bank got the $5000 last year and Samuel got the right to collect $5500 one year later. Evidently, having a dollar on May 15 last year was worth more than the right to collect a dollar a year later. Each dollar on May 15 last year was worth the right to collect 5500/5000 = 1.1 dollars a year later. This 1.1 may be written as 1 + 0.1 where 0.1 is the interest rate. The interest rate, then, gives the rate of exchange between money at the beginning of a period (one year in this example) and the right to money at the end of the period. ■

The dimension of an interest rate is currency/currency/time period. For example, a 9% interest rate means that for every dollar lent, 0.09 dollars (or other unit of money) is paid in interest for each time period. The value of the interest rate depends on the length of the time period. Usually, interest rates are expressed on a yearly basis, although they may be given for periods other than a year, such as a month or a quarter. This base unit of time over which an interest rate is calculated is called the **interest period**. Interest periods are described in more detail in Close-Up 2.1. The longer the interest period, the higher the interest rate must be to provide the same return.

Interest concerns the lending and borrowing of money. It is a parameter that allows an exchange of a larger amount of money in the future for a smaller amount of money in the present, and vice versa. As we will see in Chapter 3, it also allows us to evaluate very complicated exchanges of money over time.

Interest also has a physical basis. Money can be invested in financial instruments that pay interest, such as a bond or a savings account, and money can also be invested directly in industrial processes or services that generate wealth. In fact, the money invested in financial instruments is also, indirectly, invested in productive activities by the organization

CLOSE-UP 2.1   Interest Periods

The most commonly used interest period is one year. If we say, for example, "6% interest" without specifying an interest period, the assumption is that 6% interest is paid for a one-year period. However, interest periods can be of any duration. Here are some other common interest periods:

Interest Period	Interest Is Calculated:
Semiannually	Twice per year, or once every six months
Quarterly	Four times a year, or once every three months
Monthly	12 times per year
Weekly	52 times per year
Daily	365 times per year
Continuous	For infinitesimally small periods

providing the instrument. Consequently, the root source of interest is the productive use of money, as this is what makes the money actually increase in value. The actual return generated by a specific productive investment varies enormously, as will be seen in Chapter 4.

2.3 | Compound and Simple Interest

We have seen that if an amount, P, is lent for one interest period at the interest rate, i, the amount that must be repaid at the end of the period is $F = P(1 + i)$. But loans may be for several periods. How is the quantity of money that must be repaid computed when the loan is for N interest periods? The usual way is "one period at a time." Suppose that the amount P is borrowed for N periods at the interest rate i. The amount that must be repaid at the end of the N periods is $P(1 + i)^N$, that is

$$F = P(1 + i)^N \tag{2.1}$$

This is derived as shown in Table 2.1.

This method of computing interest is called *compounding*. Compounding assumes that there are N sequential one-period loans. At the end of the first interest period, the borrower owes $P(1 + i)$. This is the amount borrowed for the second period. Interest is required on this larger amount. At the end of the second period $[P(1 + i)](1 + i)$ is owed. This is the amount borrowed for the third period. This continues so that at the end of the

Table 2.1   Compound Interest Computations

Beginning of Period	Amount Lent		Interest Amount	Amount Owed at Period End
1	P	+	Pi	$= P + Pi = P(1 + i)$
2	$P(1 + i)$	+	$P(1 + i)i$	$= P(1 + i) + P(1 + i)i = P(1 + i)^2$
3	$P(1 + i)^2$	+	$P(1 + i)^2 i$	$= P(1 + i)^2 + P(1 + i)^2 i = P(1 + i)^3$
$\vdots$	$\vdots$			
N	$P(1 + i)^{N-1}$ +		$[P(1 + i)^{N-1}]i$	$= P(1 + i)^N$

$(N - 1)$th period, $P(1 + i)^{N-1}$ is owed. The interest on this over the Nth period is $[P(1 + i)^{N-1}]i$. The total interest on the loan over the N periods is

$$I_c = P(1 + i)^N - P \tag{2.2}$$

I_c is called **compound interest**. It is the standard method of computing interest where interest accumulated in one interest period is added to the principal amount used to calculate interest in the next period. The interest period when compounding is used to compute interest is called the **compounding period**.

EXAMPLE 2.2

If you were to lend $100 for three years at 10% per year compound interest, how much interest would you get at the end of the three years?

If you lend $100 for three years at 10% compound interest per year, you will earn $10 in interest in the first year. That $10 will be lent, along with the original $100, for the second year. Thus, in the second year, the interest earned will be $11 = $110(0.10). The $11 is lent for the third year. This makes the loan for the third year $121, and $12.10 = $121(0.10) in interest will be earned in the third year. At the end of the three years, the amount you are owed will be $133.10. The interest received is then $33.10. This can also be calculated from Equation (2.2):

$$I_c = 100(1 + 0.1)^3 - 100 = 33.10$$

Table 2.2 summarizes the compounding process. ■

Table 2.2   Compound Interest Computations for Example 2.2

Beginning of Year	Amount Lent		Interest Amount		Amount Owed at Year-End
1	100	+	100 × 0.1	=	$110
2	110	+	110 × 0.1	=	$121
3	121	+	121 × 0.1	=	$133.10

If the interest payment for an N-period loan at the interest rate i per period is computed without compounding, the interest amount, I_s, is called *simple interest*. It is computed as

$$I_s = PiN$$

Simple interest is a method of computing interest where interest earned during an interest period is not added to the principal amount used to calculate interest in the next period. Simple interest is rarely used in practice, except as a method of calculating approximate interest.

EXAMPLE 2.3

If you were to lend $100 for three years at 10% per year simple interest, how much interest would you get at the end of the three years?

The total amount of interest earned on the $100 over the three years would be $30. This can be calculated by using $I_s = PiN$:

$$I_s = PiN = 100(0.10)(3) = 30 \quad ■$$

Figure 2.2   Compound and Simple Interest at 24% Per Year for 20 Years

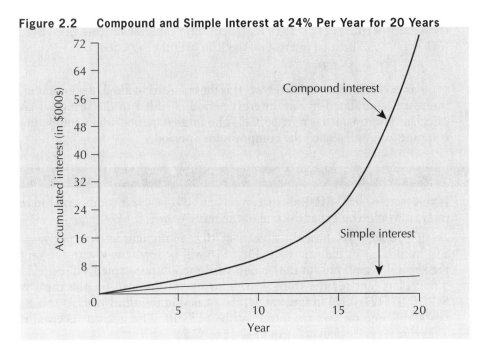

Interest amounts computed with simple interest and compound interest will yield the same results only when the number of interest periods is one. As the number of periods increases, the difference between the accumulated interest amounts for the two methods increases exponentially.

When the number of interest periods is significantly greater than one, the difference between simple interest and compound interest can be very great. In April 1993, a couple in Nevada, USA, presented the state government with a $1000 bond issued by the state in 1865. The bond carried an annual interest rate of 24%. The couple claimed the bond was now worth several trillion dollars (*Newsweek*, August 9, 1993, p. 8). If one takes the length of time from 1865 to the time the couple presented the bond to the state as 127 years, the value of the bond could have been $732 trillion = $1000(1 + 0.24)^{127}$.

If, instead of compound interest, a simple interest rate given by $iN = (24\%)(127) = 3048\%$ were used, the bond would be worth only $31\ 480 = \$1000(1 + 30.48)$. Thus, the difference between compound and simple interest can be dramatic, especially when the interest rate is high and the number of periods is large. The graph in Figure 2.2 shows the difference between compound interest and simple interest for the first 20 years of the bond example. As for the couple in Nevada, the $1000 bond was worthless after all—a state judge ruled that the bond had to have been cashed by 1872.

The conventional approach for computing interest is the compound interest method rather than simple interest. Simple interest is rarely used, except perhaps as an intuitive (yet incorrect!) way of thinking of compound interest. We mention simple interest primarily to contrast it with compound interest and to indicate that the difference between the two methods can be large.

2.4 | Effective and Nominal Interest Rates

Interest rates may be stated for some period, like a year, while the computation of interest is based on shorter compounding subperiods such as months. In this section we consider the relation between the *nominal* interest rate that is stated for the full period and the *effective* interest rate that results from the compounding based on the subperiods. This relation between nominal and effective interest rates must be understood to answer questions such as: How would you choose between two investments, one bearing 12% per year interest compounded yearly and another bearing 1% per month interest compounded monthly? Are they the same?

Nominal interest rate is the conventional method of stating the annual interest rate. It is calculated by multiplying the interest rate per compounding period by the number of compounding periods per year. Suppose that a time period is divided into m equal subperiods. Let there be stated a nominal interest rate, r, for the full period. By convention, for nominal interest, the interest rate for each subperiod is calculated as $i_s = r/m$. For example, a nominal interest rate of 18% per year, compounded monthly, is the same as

0.18/12 = 0.015 or 1.5% per month

Effective interest rate is the actual but not usually stated interest rate, found by converting a given interest rate with an arbitrary compounding period (normally less than a year) to an equivalent interest rate with a one-year compounding period. What is the effective interest rate, i_e, for the full period that will yield the same amount as compounding at the end of each subperiod, i_s? If we compound interest every subperiod, we have

$$F = P(1 + i_s)^m$$

We want to find the effective interest rate, i_e, that yields the same future amount F at the end of the full period from the present amount P. Set

$$P(1 + i_s)^m = P(1 + i_e)$$

Then

$$(1 + i_s)^m = 1 + i_e$$
$$i_e = (1 + i_s)^m - 1 \tag{2.3}$$

Note that Equation (2.3) allows the conversion between the interest rate over a compounding subperiod, i_s, and the effective interest rate over a longer period, i_e, by using the number of subperiods, m, in the longer period.

EXAMPLE 2.4

What interest rate per year, compounded yearly, is equivalent to 1% interest per month, compounded monthly?

Since the month is the shorter compounding period, we let $i_s = 0.01$ and $m = 12$. Then i_e refers to the effective interest rate per year. Substitution into Equation (2.3) then gives

$$i_e = (1 + i_s)^m - 1$$
$$= (1 + 0.01)^{12} - 1$$
$$= 0.126825$$
$$\approx 0.127 \text{ or } 12.7\%$$

An interest rate of 1% per month, compounded monthly, is equivalent to an effective rate of approximately 12.7% per year, compounded yearly. The answer to our previously posed question is that an investment bearing 12% per year interest, compounded yearly, pays less than an investment bearing 1% per month interest, compounded monthly.■

Interest rates are normally given as nominal rates. We may get the effective (yearly) rate by substituting $i_s = r/m$ into Equation (2.3). We then obtain a direct means of computing an effective interest rate, given a nominal rate and the number of compounding periods per year:

$$i_e = \left(1 + \frac{r}{m}\right)^m - 1 \tag{2.4}$$

This formula is suitable only for converting from a nominal rate r to an annual effective rate. If the effective rate desired is for a period longer than a year, then Equation (2.3) must be used.

EXAMPLE 2.5

Leona the loan shark lends money to clients at the rate of 5% interest per week! What is the nominal interest rate for these loans? What is the effective annual interest rate?

The nominal interest rate is 5% × 52 = 260%. Recall that nominal interest rates are usually expressed on a yearly basis. The effective yearly interest rate can be found by substitution into Equation (2.3):

$$i_e = (1 + 0.05)^{52} - 1 = 11.6$$

Leona charges an effective annual interest rate of about 1160% on her loans.■

EXAMPLE 2.6

The Cardex Credit Card Company charges a nominal 24% interest on overdue accounts, compounded daily. What is the effective interest rate?

Assuming that there are 365 days per year, we can calculate the interest rate per day using either Equation (2.3) with $i_s = r/m = 0.24/365 = 0.0006575$ or by the use of Equation (2.4) directly. The effective interest rate (per year) is

$$i_e = (1 + 0.0006575)^{365} - 1$$
$$= 0.271 \text{ or } 27.1\%$$

With a nominal rate of 24% compounded daily, the Cardex Credit Card Company is actually charging an effective rate of about 27.1% per year.■

Although there are laws which may require that the effective interest rate be disclosed for loans and investments, it is still very common for nominal interest rates to be quoted. Since the nominal rate will be less than the effective rate whenever the number of compounding periods per year exceeds one, there is an advantage to quoting loans using the nominal rates, since it makes the loan look more attractive. This is particularly true when interest rates are high and compounding occurs frequently.

2.5 | Continuous Compounding

As has been seen, compounding can be done yearly, quarterly, monthly, or daily. The periods can be made even smaller, as small as desired; the main disadvantage in having very small periods is having to do more calculations. If the period is made infinitesimally small, we say that interest is compounded *continuously*. There are situations in which very frequent compounding makes sense. For instance, an improvement in materials handling may reduce downtime on machinery. There will be benefits in the form of increased output that may be used immediately. If there are several additional runs a day, there will be benefits several times a day. Another example is trading on the stock market. Personal and corporate investments are often in the form of mutual funds. Mutual funds represent a changing set of stocks and bonds, in which transactions occur very frequently, often many times a day.

A formula for **continuous compounding** can be developed from Equation (2.3) by allowing the number of compounding periods per year to become infinitely large:

$$i_e = \lim_{m \to \infty} \left(1 + \frac{r}{m}\right)^m - 1$$

By noting from a definition of the natural exponential function, e, that

$$\lim_{m \to \infty} \left(1 + \frac{r}{m}\right)^m = e^r$$

we get

$$i_e = e^r - 1 \tag{2.5}$$

EXAMPLE 2.7

Cash flow at the Arctic Oil Company is continuously reinvested. An investment in a new data logging system is expected to return a nominal interest of 40%, compounded continuously. What is the effective interest rate earned by this investment?

The nominal interest rate is given as $r = 0.40$. From Equation (2.5),

$$i_e = e^{0.4} - 1$$
$$= 1.492 - 1 \cong 0.492 \text{ or } 49.2\%$$

The effective interest rate earned on this investment is about 49.2%.■

Although continuous compounding makes sense in some circumstances, it is rarely used. As with effective interest and nominal interest, in the days before calculators and computers, calculations involving continuous compounding were difficult to do. Consequently, discrete compounding is, by convention, the norm. As illustrated in Figure 2.3, the difference between continuous compounding and discrete compounding is relatively insignificant, even at a fairly high interest rate.

Figure 2.3   Growth in Value of $1 at 30% Interest for Various Compounding Periods

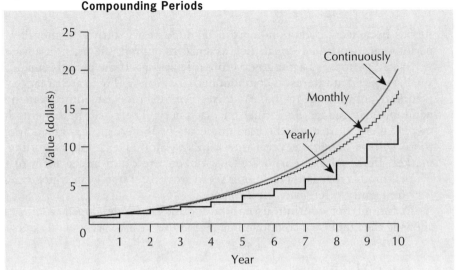

2.6 | Cash Flow Diagrams

Sometimes a set of cash flows can be sufficiently complicated that it is useful to have a graphical representation. A **cash flow diagram** is a graph that summarizes the timing and magnitude of cash flows as they occur over time.

On a cash flow diagram, the graph's vertical axis is not shown explicitly. The horizontal (X) axis represents time, measured in periods, and the vertical (Y) axis represents the size and direction of the cash flows. Individual cash flows are indicated by arrows pointing up or down from the horizontal axis, as indicated in Figure 2.4. The arrows that point up represent positive cash flows, or receipts. The downward-pointing arrows represent negative cash flows, or disbursements. See Close-Up 2.2 for some conventions pertaining to the beginning and ending of periods.

Figure 2.4   Cash Flow Diagram

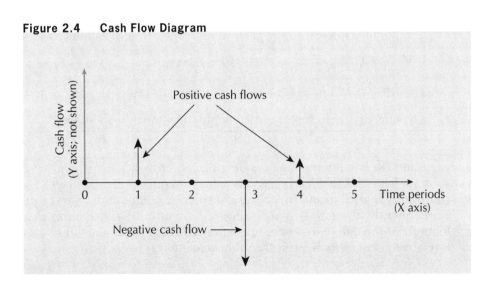

CLOSE-UP 2.2   Beginning and Ending of Periods

As illustrated in a cash flow diagram (see Figure 2.5), the end of one period is exactly the same point in time as the beginning of the next period. Now is time 0, which is the end of period –1 and also the beginning of period 1. The end of period 1 is the same as the beginning of period 2, and so on. N years from now is the end of period N and the beginning of period $(N + 1)$.

Figure 2.5   Beginning and Ending of Periods

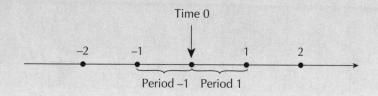

EXAMPLE 2.8

Consider Ashok, a recent university graduate who is trying to summarize typical cash flows for each month. His monthly income is $2200, received at the end of each month. Out of this he pays for rent, food, entertainment, telephone charges, and a credit card bill for all other purchases. Rent is $700 per month (including utilities), due at the end of each month. Weekly food and entertainment expenses total roughly $120, a typical telephone bill is $40 (due at the end of the first week in the month), and his credit card purchases average $300. Credit card payments are due at the end of the second week of each month.

Figure 2.6 shows the timing and amount of the disbursements and the single receipt over a typical month. It assumes that there are exactly four weeks in a month, and it is now just past the end of the month. Each arrow, which represents a cash flow, is labelled with the amount of the receipt or disbursement.

When two or more cash flows occur in the same time period, the amounts may be shown individually, as in Figure 2.6, or in summary form, as in Figure 2.7. The level of detail used depends on personal choice and the amount of information the diagram is intended to convey.

We suggest that the reader make a practice of using cash flow diagrams when working on a problem with cash flows that occur at different times. Just going through the steps in setting up a cash flow diagram can make the problem structure clearer. Seeing the pattern of cash flows in the completed diagram gives a "feel" for the problem.■

Figure 2.6   Cash Flow Diagram for Example 2.8

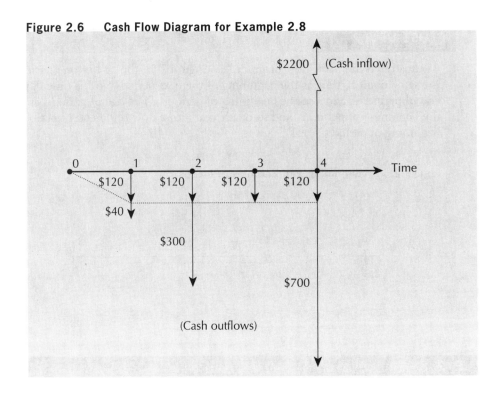

Figure 2.7   Cash Flow Diagram for Example 2.8 in Summary Form

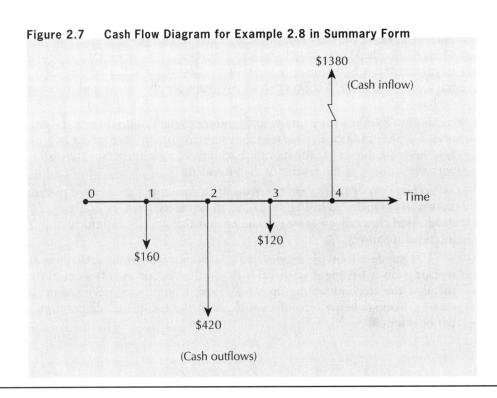

2.7 | Equivalence

We started this chapter by pointing out that many engineering decisions involve costs and benefits that occur at different times. Making these decisions requires that the costs and benefits at different times be compared. To make these comparisons, we must be able to say that certain values at different times are *equivalent*. **Equivalence** is a condition that exists when the value of a cost at one time is equivalent to the value of the related benefit received at a different time. In this section we distinguish three concepts of equivalence that may underlie comparisons of costs and benefits at different times.

With **mathematical equivalence**, equivalence is a consequence of the mathematical relationship between time and money. This is the form of equivalence used in $F = P(1 + i)^N$.

With **decisional equivalence**, equivalence is a consequence of indifference on the part of a decision maker among available choices.

With **market equivalence**, equivalence is a consequence of the ability to exchange one cash flow for another at zero cost.

Although the mathematics governing money is the same regardless of which form of equivalence is most appropriate for a given situation, it can be important to be aware of what assumptions must be made for the mathematical operations to be meaningful.

2.7.1 Mathematical Equivalence

Mathematical equivalence is simply a mathematical relationship. It says that two cash flows, P_t at time t and F_{t+N} at time $t + N$, are equivalent with respect to the interest rate, i, if $F_{t+N} = P_t(1 + i)^N$. Notice that if F_{t+N+M} (where M is a second number of periods) is equivalent to P_t, then

$$F_{t+N+M} = P_t(1 + i)^{N+M}$$
$$= F_{t+N}(1 + i)^M$$

so that F_{t+N} and F_{t+N+M} are equivalent to each other. The fact that mathematical equivalence has this property permits complex comparisons to be made among many cash flows that occur over time.

2.7.2 Decisional Equivalence

For any individual, two cash flows, P_t at time t and F_{t+N} at time $t + N$, are equivalent if the individual is indifferent between the two. Here, the implied interest rate relating P_t and F_{t+N} can be calculated from the decision that the cash flows are equivalent, as opposed to mathematical equivalence in which the interest rate determines whether the cash flows are equivalent. This can be illustrated best through an example.

EXAMPLE 2.9

Bildmet is an extruder of aluminum shapes used in construction. The company buys aluminum from Alpure, an outfit that recovers aluminum from scrap. When Bildmet's purchasing manager, Greta Kehl, called in an order for 1000 kilograms of metal on August 15, she was told that Alpure was having production difficulties and was running behind schedule. Alpure's manager, Masaaki Sawada, said that he could ship the order immediately if Bildmet required it. However, if Alpure shipped Bildmet's order, they would not be able to fill an order for another user whom Mr. Sawada was anxious to impress with Alpure's reliability. Mr. Sawada suggested that, if Ms. Kehl would wait a week until August 22, he would show his appreciation by shipping 1100 kilograms then at the same cost to Bildmet as 1000 kilograms now. In either case, payment would be due at the end of the month. Should Ms. Kehl accept Alpure's offer?

The rate of exchange, 1100 to 1000 kilograms, may be written as $(1 + 0.1)$ to 1, where the $0.1 = 10\%$ is an interest rate for the one-week period. (This is equivalent to an effective interest rate of more than 14 000% per year!) Whether or not Ms. Kehl accepts the offer from Alpure depends on her situation. There is some chance of Bildmet's running out of metal if they don't get supplied for a week. This would require Ms. Kehl to do some scrambling to find other sources of metal in order to ship to her own customers on time. Ms. Kehl would prefer the 1000 kilograms on the 15th to 1000 kilograms on the 22nd. But there is some minimum amount, larger than 1000 kilograms, that she would accept on the 22nd in exchange for 1000 kilograms on the 15th. This amount would take into account both measurable costs and immeasurable costs such as inconvenience and anxiety.

Let the minimum rate at which Ms. Kehl would be willing to make the exchange be 1 kilogram on the 15th for $(1 + x)$ kilograms on the 22nd. In this case, if $x < 10\%$, Ms. Kehl should accept Alpure's offer of 1100 kilograms on the 22nd. ∎

In Example 2.9, the aluminum is a capital good that can be used productively by Bildmet. There is value in that use, and that value can be measured by Greta's willingness to postpone receiving the aluminum. It can be seen that interest is not necessarily a function of exchanges of money at different points in time. However, money is a convenient measure of the worth of a variety of goods, and so interest is usually expressed in terms of money.

2.7.3 Market Equivalence

Market equivalence is based on the idea that there is a market for money that permits cash flows in the future to be exchanged for cash flows in the present, and vice versa. Converting a future cash flow, F, to a present cash flow, P, is called borrowing money, while converting P to F is called lending or investing money. The market equivalence of two cash flows P and F means that they can be exchanged, one for the other, at zero cost.

The interest rate associated with an individual's borrowing money is usually a lot higher than the interest rate applied when that individual lends money. For example, the interest rate a bank pays on deposits is lower than what it charges to lend money to clients. The difference between these interest rates provides the bank with income. This means that, for an individual, market equivalence does not exist. An individual can exchange a present worth for a future worth by investing money, but if he or she were to try to borrow against that future worth to obtain money now, the resulting present worth would be less than the original amount invested. Moreover, every time either borrowing or lending occurred, transaction costs (the fees charged or cost incurred) would further diminish the capital.

EXAMPLE 2.10

This morning, Averill bought a $5000 one-year guaranteed investment certificate (GIC) at his local bank. It has an effective interest rate of 7% per year. At the end of a year, the GIC will be worth $5350. On the way home from the bank, Averill unexpectedly discovered a valuable piece of art he had been seeking for some time. He wanted to buy it, but all his spare capital was tied up in the GIC. So he went back to the bank, this time to negotiate a one-year loan for $5000, the cost of the piece of art. He figured that, if the loan came due at the same time as the GIC, he would simply pay off the loan with the proceeds of the GIC.

Unfortunately, Averill found out that the bank charges 10% effective interest per year on loans. Considering the proceeds from the GIC of $5350 one year from now, the amount the bank would give him today is only $5350/1.1 = $4864 (roughly), less any fees applicable to the loan. He discovered that, for him, market equivalence does not hold. He cannot exchange $5000 today for $5350 one year from now, and vice versa, at zero cost.■

Large companies with good records have opportunities that differ from those of individuals. Large companies borrow and invest money in so many ways, both internally and externally, that the interest rates for borrowing and for lending are very close to being the same, and also the transaction costs are negligible. They can shift funds from the future to the present by raising new money or by avoiding investment in a marginal project that would earn only the rate that they pay on new money. They can shift funds from the present to the future by undertaking an additional project or investing externally.

But how large is a "large company"? Established businesses of almost any size, and even individuals with some wealth and good credit, can acquire cash and invest at about the same interest rate, provided that the amounts are small relative to their total assets. For these companies and individuals, market equivalence is a reasonable model assuming that market equivalence makes calculations easier and still generally results in good decisions.

For most of the remainder of this book, we will be making two broad assumptions with respect to equivalence: first, that market equivalence holds, and second, that decisional equivalence can be expressed entirely in monetary terms. If these two assumptions are reasonably valid, mathematical equivalence can be used as an accurate model of how costs and benefits relate to one another over time. In several sections of the book, when we cover how firms raise capital and how to incorporate non-monetary aspects of a situation into a decision, we will discuss the validity of these two assumptions. In the meantime, mathematical equivalence is used to relate cash flows that occur at different points in time.

REVIEW PROBLEMS

REVIEW PROBLEM 2.1

Atsushi has had £800 stashed under his mattress for 30 years. How much money has he lost by not putting it in a bank account at 8% annual compound interest all these years?

ANSWER

Since Atsushi has kept the £800 under his mattress, he has not earned any interest over the 30 years. Had he put the money into an interest-bearing account, he would have far more today. We can think of the £800 as a present amount and the amount in 30 years as the future amount.

Given: $P = £800$

$i = 0.08$ per year

$N = 30$ years

Formula: $F = P(1 + i)^N$

$= 800(1 + 0.08)^{30}$

$= 8050.13$

Atsushi would have £8050.13 in the bank account today had he deposited his £800 at 8% annual compound interest. Instead, he has only £800. He has suffered an opportunity cost of £8050.13 – £800 = £7250.13 by not investing the money.∎

REVIEW PROBLEM 2.2

You want to buy a new computer, but you are $1000 short of the amount you need. Your aunt has agreed to lend you the $1000 you need now, provided you pay her $1200 two years from now. She compounds interest monthly. Another place from which you can borrow $1000 is the bank. There is, however, a loan processing fee of $20, which will be included in the loan amount. The bank is expecting to receive $1220 two years from now based on monthly compounding of interest.

(a) What monthly rate is your aunt charging you for the loan? What is the bank charging?

(b) What effective annual rate is your aunt charging? What is the bank charging?

(c) Would you prefer to borrow from your aunt or from the bank?

ANSWER

(a) *Your aunt*

Given: $P = \$1000$

$F = \$1200$

$N = 24$ months (since compounding is done monthly)

Formula: $F = P(1 + i)^N$

The formula $F = P(1 + i)^N$ must be solved in terms of i to answer the question.

$i = \sqrt[N]{F/P} - 1$

$= \sqrt[24]{1200/1000} - 1$

$= 0.007626$

Your aunt is charging interest at a rate of approximately 0.76% per month.

The bank

Given: $P = \$1020$ (since the fee is included in the loan amount)

$F = \$1220$

$N = 24$ months (since compounding is done monthly)

$i = \sqrt[N]{F/P} - 1$

$= \sqrt[24]{1220/1020} - 1$

$= 0.007488$

The bank is charging interest at a rate of approximately 0.75% per month.

(b) The effective annual rate can be found with the formula $i_e = (1 + r/m)^m - 1$, where r is the nominal rate per year and m is the number of compounding periods per year. Since the number of compounding periods per year is 12, notice that r/m is simply the interest rate charged per month.

Your aunt

$\quad i = 0.007626$ per month

Then

$$\begin{aligned} i_e &= (1 + r/m)^m - 1 \\ &= (1 + 0.007626)^{12} - 1 \\ &= 0.095445 \end{aligned}$$

The effective annual rate your aunt is charging is approximately 9.54%.

The bank

$\quad i = 0.007488$ per month

Then

$$\begin{aligned} i_e &= (1 + r/m)^m - 1 \\ &= (1 + 0.007488)^{12} - 1 \\ &= 0.09365 \end{aligned}$$

The effective annual rate for the bank is approximately 9.37%.

(c) The bank appears to be charging a lower interest rate than does your aunt. This can be concluded by comparing the two monthly rates or the effective annual rates the two charge. If you were to base your decision only upon who charged the lower interest rate, you would pick the bank, despite the fact they have a fee. However, although you are borrowing $1020 from the bank, you are getting only $1000 since the bank immediately gets its $20 fee. The cost of money for you from the bank is better calculated as

Given: $\quad P = \$1000$

$\qquad\qquad F = \$1220$

$\qquad\qquad N = 24$ months (since compounding is done monthly)

$$\begin{aligned} i &= \sqrt[N]{F/P} - 1 \\ &= \sqrt[24]{1220/1000} - 1 \\ &= 0.00832 \end{aligned}$$

From this point of view, the bank is charging interest at a rate of approximately 0.83% per month, and you would be better off borrowing from your aunt.■

REVIEW PROBLEM 2.3

At the end of four years, you would like to have $5000 in a bank account to purchase a used car. What you need to know is how much to deposit in the bank account now. The account pays daily interest. Create a spreadsheet and plot the necessary deposit today as a function of interest rate. Consider nominal interest rates ranging from 5% to 15% per year, and assume that there are 365 days per year.

ANSWER

From the formula $F = P(1 + i)^N$, we have $5000 = P(1 + i)^{365 \times 4}$. This gives

$$P = 5000 \times \frac{1}{(1 + i)^{365 \times 4}}$$

Table 2.3 is an excerpt from a sample spreadsheet. It shows the necessary deposit to accumulate $5000 over four years at a variety of interest rates. The following is the calculation for cell B2 (i.e., the second row, second column):

$$5000 \times \frac{1}{\left[1 + \left(\dfrac{A2}{365}\right)\right]^{365 \times 4}}$$

Table 2.3 Necessary Deposits for a Range of Interest Rates

Interest Rate (%)	Necessary Deposit ($)
0.05	4114
0.06	3957
0.07	3805
0.08	3660
0.09	3520
0.10	3385
0.11	3256
0.12	3131
0.13	3011
0.14	2896
0.15	2785

Figure 2.8   Graph for Review Problem 2.3

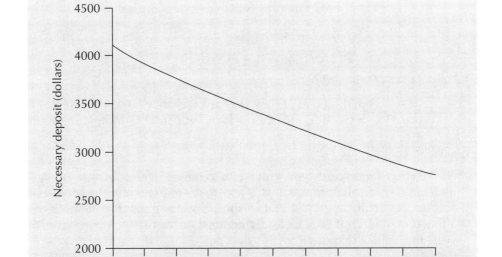

The specific implementation of this formula will vary, depending on the particular spreadsheet program used. Figure 2.8 is a diagram of the necessary deposits plotted against interest rates.■

SUMMARY

This chapter has provided an introduction to interest, interest rate terminology, and interest rate conventions. Through a series of examples, the mechanics of working with simple and compound interest, nominal and effective interest rates, and continuous compounding were illustrated. Cash flow diagrams were introduced in order to represent graphically monetary transactions at various points in time. The final part of the chapter contained a discussion of various forms of cash flow equivalence: mathematical, decisional, and market. With the assumption that mathematical equivalence can be used as an accurate model of how costs and benefits relate to one another over time, we now move on to Chapter 3, in which equivalence formulas for a variety of cash flow patterns are presented.

Engineering Economics in Action, Part 2B:
You Just Have to Know When

Naomi and Terry were looking at the steel orders for the J-class line. Terry thought money could be saved by ordering in advance. "Now look at this," Terry said. "In the six months ending last December, we ordered steel for a total cost of $1 600 000. If we had bought this steel at the beginning of July, it would have cost only $1 400 000. That's a savings of $200 000!"

"Good observation, Terry, but I don't think buying in advance is the right thing to do. If you think about it, the rate of return on our $1 400 000 would be 200 000/1 400 000 or about 14.3% over six months."

"Yes, but that's over 30% effective interest, isn't it? I'll bet we only make 8% or 10% for money we keep in the bank."

"That's true, but the money we would use to buy the steel in advance we don't have sitting in the bank collecting interest. It would have to come from somewhere else, either money we borrow from the bank, at about 14% plus administrative costs, or from our shareholders."

"But it's still a good idea, right?"

"Well, you are right and you are wrong. Mathematically, you could probably show the advantage of buying a six-month supply in advance. But we wouldn't do it for two reasons. The first one has to do with where the money comes from. If we had to pay for six months of steel in advance, we would have a capital requirement beyond what we could cover through normal cash flows. I'm not sure the bank would even lend us that much money, so we would probably have to raise it through equity, that is, selling more shares in the company. This would cost a lot, and throw all your calculations way off."

"Just because it's such a large amount of money?"

"That's right. Our regular calculations are based on the assumption that the capital requirements don't take an extraordinary effort."

"You said there were two reasons. What's the other one?"

"The other reason is that we just wouldn't do it."

"Huh?"

"We just wouldn't do it. Right now the steel company's taking the risk—if we can't pay, they are in trouble. If we buy in advance, it's the other way around—if our widget orders dropped, we would be stuck with some

→

pretty expensive raw materials. We would also have the problem of where to store the steel, and other practical difficulties. It makes sense mathematically, but I'm pretty sure we just wouldn't do it."

Terry looked a little dejected. Naomi continued, "But your figures make sense. The first thing to do is find out why we are carrying that account so long before we pay it off. The second thing to do is see if we can't get that price break, retroactively. We are good customers, and I'll bet we can convince them to give us the price break anyhow, without changing our ordering pattern. Let's talk to Clem about it."

"But, Naomi, why use the mathematical calculations at all, if they don't work?"

"But they do work, Terry. You just have to know when."

PROBLEMS

For additional practice, please see the problems (with selected solutions) provided on the Student CD-ROM that accompanies this book.

2.1 Using 12% simple interest per year, how much interest will be owed on a loan of $500 at the end of two years?

2.2 If a sum of £3000 is borrowed for six months at 9% simple interest per year, what is the total amount due (principal and interest) at the end of six months?

2.3 What principal amount will yield $150 in interest at the end of three months when the interest rate is 1% simple interest per month?

2.4 If 2400 rupees in interest is paid on a two-year simple-interest loan of 12 000 rupees, what is the interest rate per year?

2.5 Simple interest of $190.67 is owed on a loan of $550 after four years and four months. What is the annual interest rate?

2.6 How much will be in a bank account at the end of five years if €2000 is invested today at 12% interest per annum, compounded yearly?

2.7 How much is accumulated in each of these savings plans over two years?

(a) Deposit 1000 yuan today at 10% compounded annually.

(b) Deposit 900 yuan today at 12% compounded monthly.

2.8 Greg wants to have $50 000 in five years. The bank is offering five-year investment certificates that pay 8% nominal interest, compounded quarterly. How much money should he invest in the certificates to reach his goal?

2.9 Greg wants to have $50 000 in five years. He has $20 000 today to invest. The bank is offering five-year investment certificates that pay interest compounded quarterly. What is the minimum nominal interest rate he would have to receive to reach his goal?

2.10 Greg wants to have $50 000. He will invest $20 000 today in investment certificates that pay 8% nominal interest, compounded quarterly. How long will it take him to reach his goal?

2.11 Greg will invest $20 000 today in five-year investment certificates that pay 8% nominal interest, compounded quarterly. How much money will this be in five years?

2.12 You bought an antique car three years ago for 500 000 yuan. Today it is worth 650 000 yuan.

   (a) What annual interest rate did you earn if interest is compounded yearly?

   (b) What monthly interest rate did you earn if interest is compounded monthly?

2.13 You have a bank deposit now worth $5000. How long will it take for your deposit to be worth more than $8000 if

   (a) the account pays 5% actual interest every half-year, and is compounded every half-year?

   (b) the account pays 5% nominal interest, compounded semiannually?

2.14 Some time ago, you put £500 into a bank account for a "rainy day." Since then, the bank has been paying you 1% per month, compounded monthly. Today, you checked the balance, and found it to be £708.31. How long ago did you deposit the £500?

2.15 (a) If you put $1000 in a bank account today that pays 10% interest per year, how much money could be withdrawn 20 years from now?

   (b) If you put $1000 in a bank account today that pays 10% *simple* interest per year, how much money could be withdrawn 20 years from now?

2.16 How long will it take any sum to double itself,

   (a) with an 11% simple interest rate?

   (b) with an 11% interest rate, compounded annually?

   (c) with an 11% interest rate, compounded continuously?

2.17 Compute the effective annual interest rate on each of these investments.

   (a) 25% nominal interest, compounded semiannually

   (b) 25% nominal interest, compounded quarterly

   (c) 25% nominal interest, compounded continuously

2.18 For a 15% effective annual interest rate, what is the nominal interest rate if

   (a) interest is compounded monthly?

   (b) interest is compounded daily (assume 365 days per year)?

   (c) interest is compounded continuously?

2.19 A Studebaker automobile that cost $665 in 1934 was sold as an antique car at $14 800 in 1998. What was the rate of return on this "investment"?

2.20 Clifford has X euros right now. In 5 years, X will be €3500 if it is invested at 7.5%, compounded annually. Determine the present value of X. If Clifford invested X euros at 7.5%, compounded daily, how much would the value of X be in 10 years?

2.21 You have just won a lottery prize of $1 000 000 collectable in 10 yearly installments of 100 000 rupees, starting today. Why is this prize not really $1 000 000? What is it really worth today if money can be invested at 10% annual interest, compounded monthly? Use a spreadsheet to construct a table showing the present worth of each installment, and the total present worth of the prize.

2.22 Suppose in Problem 2.21 that you have a large mortgage you want to pay off now. You propose an alternative, but equivalent, payment scheme. You would like $300 000 today, and the balance of the prize in five years when you intend to purchase a large piece of waterfront property. How much will the payment be in five years? Assume that annual interest is 10%, compounded monthly.

2.23 You are looking at purchasing a new computer for your four-year undergraduate program. Brand 1 costs $4000 now, and you expect it will last throughout your program without any upgrades. Brand 2 costs $2500 now and will need an upgrade at the end of two years, which you expect to be $1700. With 8% annual interest, compounded monthly, which is the less expensive alternative, if they provide the same level of service and will both be worthless at the end of the four years?

2.24 The Kovalam Bank advertises savings account interest as 6% compounded daily. What is the effective interest rate?

2.25 The Bank of Brisbane is offering a new savings account that pays a nominal 7.99% interest, compounded continuously. Will your money earn more in this account than in a daily interest account that pays 8%?

2.26 You are comparing two investments. The first pays 1% interest per month, compounded monthly, and the second pays 6% interest per six months, compounded every six months.

   (a) What is the effective semiannual interest rate for each investment?

   (b) What is the effective annual interest rate for each investment?

   (c) On the basis of interest rate, which investment do you prefer? Does your decision depend on whether you make the comparison based on an effective six-month rate or an effective one-year rate?

2.27 The Crete Credit Union advertises savings account interest as 5.5% compounded weekly and chequing account interest at 7% compounded monthly. What are the effective interest rates for the two types of accounts?

2.28 Victory Visa, Magnificent Master Card, and Amazing Express are credit card companies that charge different interest on overdue accounts. Victory Visa charges 26% compounded daily, Magnificent Master Card charges 28% compounded weekly, and Amazing Express charges 30% compounded monthly. On the basis of interest rate, which credit card has the best deal?

2.29 April has a bank deposit now worth $796.25. A year ago, it was $750. What was the nominal monthly interest rate on her account?

2.30 You have $50 000 to invest in the stock market and have sought the advice of Adam, an experienced colleague who is willing to advise you, for a fee. Adam has told you that he has found a one-year investment for you that provides 15% interest, compounded monthly.

   (a) What is the effective annual interest rate, based on a 15% nominal annual rate and monthly compounding?

   (b) Adam says that he will make the investment for you for a modest fee of 2% of the investment's value one year from now. If you invest the $50 000 today, how much will you have at the end of one year (before Adam's fee)?

   (c) What is the effective annual interest rate of this investment including Adam's fee?

2.31 May has 2000 yuan in her bank account right now. She wanted to know how much it would be in one year, so she calculated and came up with 2140.73 yuan. Then she realized she had made a mistake. She had wanted to use the formula for monthly compounding, but instead, she had used the continuous compounding formula. Redo the calculation for May and find out how much will actually be in her account a year from now.

2.32 Hans now has $6000. In three months, he will receive a cheque for $2000. He must pay $900 at the end of each month (starting exactly one month from now). Draw a single cash flow diagram illustrating all of these payments for a total of six monthly periods. Include his cash on hand as a payment at time 0.

2.33 Margaret is considering an investment that will cost her $500 today. It will pay her $100 at the end of each of the next 12 months, and cost her another $300 one year from today. Illustrate these cash flows in two cash flow diagrams. The first should show each cash flow element separately, and the second should show only the net cash flows in each period.

2.34 Heddy is considering working on a project that will cost her $20 000 today. It will pay her $10 000 at the end of each of the next 12 months, and cost her another $15 000 at the end of each quarter. An extra $10 000 will be received at the end of the project, one year from now. Illustrate these cash flows in two cash flow diagrams. The first should show each cash flow element separately, and the second should show only the net cash flow in each period.

2.35 Illustrate the following cash flows over 12 months in a cash flow diagram. Show only the net cash flow in each period.

Cash Payments	$20 every three months, starting now
Cash Receipts	Receive $30 at the end of the first month, and from that point on, receive 10% more than the previous month at the end of each month

2.36 There are two possible investments, A and B. Their cash flows are shown in the table below. Illustrate these cash flows over 12 months in two cash flow diagrams. Show only the net cash flow in each period. Just looking at the diagrams, would you prefer one investment to the other? Comment on this.

	Investment A	Investment B
Payments	$2400 now and a closing fee of $200 at the end of month 12	$500 every two months, starting two months from now
Receipts	$250 monthly payment at the end of each month	Receive $50 at the end of the first month, and from that point on, receive $50 more than the previous month at the end of each month

2.37 You are indifferent between receiving 100 rand today and 110 rand one year from now. The bank pays you 6% interest on deposits and charges you 8% for loans. Name the three types of equivalence and comment (with one sentence for each) on whether each exists for this situation and why.

2.38 June has a small house on a small street in a small town. If she sells the house now, she will likely get €110 000 for it. If she waits for one year, she will likely get more, say, €120 000. If she sells the house now, she can invest the money in a one-year guaranteed growth bond that pays 8% interest, compounded monthly. If she keeps the house, then the interest on the mortgage payments is 8% compounded daily. June is indifferent between the two options: selling the house now and keeping the house for another year. Discuss whether each of the three types of equivalence exists in this case.

2.39 Using a spreadsheet, construct graphs for the loan described in part (a) below.

(a) Plot the amount owed (principal plus interest) on a simple interest loan of $100 for N years for $N = 1, 2, \ldots 10$. On the same graph, plot the amount owed on a compound interest loan of $100 for N years for $N = 1, 2, \ldots 10$. The interest rate is 6% per year for each loan.

(b) Repeat part (a), but use an interest rate of 18%. Observe the dramatic effect compounding has on the amount owed at the higher interest rate.

2.40 (a) At 12% interest per annum, how long will it take for a penny to become a million dollars? How long will it take at 18%?

(b) Show the growth in values on a spreadsheet using 10-year time intervals.

2.41 Use a spreadsheet to determine how long it will take for a £100 deposit to double in value for each of the following interest rates and compounding periods. For each, plot the size of the deposit over time, for as many periods as necessary for the original sum to double.

(a) 8% per year, compounded monthly

(b) 11% per year, compounded semiannually

(c) 12% per year, compounded continuously

2.42 Construct a graph showing how the effective interest rate for the following nominal rates increases as the compounding period becomes shorter and shorter. Consider a range of compounding periods of your choice from daily compounding to annual compounding.

(a) 6% per year

(b) 10% per year

(c) 20% per year

2.43 Today, an investment you made three years ago has matured and is now worth 3000 rupees. It was a three-year deposit that bore an interest rate of 10% per year, compounded monthly. You knew at the time that you were taking a risk in making such an investment because interest rates vary over time and you "locked in" at 10% for three years.

(a) How much was your initial deposit? Plot the value of your investment over the three-year period.

(b) Looking back over the past three years, interest rates for similar one-year investments did indeed vary. The interest rates were 8% the first year, 10% the second, and 14% the third. Plot the value of your initial deposit over time as if you had invested at this set of rates, rather than for a constant 10% rate. Did you lose out by having locked into the 10% investment? If so, by how much?

More Challenging Problem

2.44 Marlee has a choice between X pounds today or Y pounds one year from now. X is a fixed value, but Y varies depending on the interest rate. At interest rate i, X and Y are mathematically equivalent for Marlee. At interest rate j, X and Y have decisional equivalence for Marlee. At interest rate k, X and Y have market equivalence for Marlee. What can be said about the origins, nature, and comparative values of i, j, and k?

MINI-CASE 2.1

Student Credit Cards

Most major banks offer a credit card service for students. Common features of the student credit cards include no annual fee, a $500 credit limit, and an annual interest rate of 19.7% (in Canada as of 2007). Also, the student cards often come with many of the perks available for the general public: purchase security or travel-related insurance, extended warranty protection, access to cash advances, etc. The approval process for getting a card is relatively simple for university and college students so that they can start building a credit history and enjoy the convenience of having a credit card while still in school.

The printed information does not use the term *nominal* or *effective*, nor does it define the compounding period. However, it is common in the credit card business for the annual interest rate to be divided into daily rates for billing purposes. Hence, the quoted annual rate of 19.7% is a nominal rate and the compounding period is daily. The actual effective interest rate is then $(1 + 0.197/365)^{365} - 1 = 0.2177$ or 21.77%.

Discussion

Interest information must be disclosed by law, but lenders and borrowers have some latitude as to how and where they disclose it. Moreover, there is a natural desire to make the interest rate look lower than it really is for borrowers, and higher than it really is for lenders.

In the example of student credit cards, the effective interest rate is 21.77%, roughly 2% higher than the stated interest rate. The actual effective interest rate could even end up being higher if fees such as late fees, over-the-limit fees, and transaction fees are charged.

Questions

1. Go to your local bank branch and find out the interest rate paid for various kinds of savings accounts, chequing accounts, and loans. For each interest rate quoted, determine if it is a nominal or effective rate. If it is nominal, determine the compounding period and calculate the effective interest rate.

2. Have a contest with your classmates to see who can find the organization that will lend money to a student like you at the cheapest effective interest rate, or that will take investments which provide a guaranteed return at the highest effective interest rate. The valid rates must be generally available, not tied to particular behaviour by the client, and not secured to an asset (like a mortgage).

3. If you borrowed $1000 at the best rate you could find and invested it at the best rate you could find, how much money would you make or lose in a year? Explain why the result of your calculation could not have the opposite sign.

C H A P T E R

3

Cash Flow
Analysis

Engineering Economics in Action, Part 3A: Apples and Oranges

Review Problems
Summary

Engineering Economics in Action, Part 3B: No Free Lunch

Problems

Mini-Case 3.1

Appendix 3A: Continuous Compounding and Continuous Cash Flows

Appendix 3B: Derivation of Discrete Compound Interest Factors

Engineering Economics in Action, Part 3A:
Apples and Oranges

The flyer was slick, all right. The information was laid out so anybody could see that leasing palletizing equipment through the Provincial Finance Company (PFC) made much more sense than buying it. It was something Naomi could copy right into her report to Clem.

Naomi had been asked to check out options for automating part of the shipping department. Parts were to be stacked and bound on plastic pallets, then loaded onto trucks and sent to one of Global Widgets' sister companies. The saleswoman for the company whose equipment seemed most suitable for Global Widgets' needs included the leasing flyer with her quote.

Naomi looked at the figures again. They seemed to make sense, but there was something that didn't seem right to her. For one thing, if it was cheaper to lease, why didn't everybody lease everything? She knew that some things, like automobiles and airplanes, are often leased instead of bought, but generally companies buy assets. Second, where was the money coming from to give the finance company a profit? If the seller was getting the same amount and the buyer was paying less, how could PFC make money?

"Got a recommendation on that palletizer yet, Naomi?" Clem's voice was cheery as he suddenly appeared at her doorway. Naomi knew that the shipping department was the focus of Clem's attention right now and he wanted to get improvements in place as soon as possible.

"Yes, I do. There's really only one that will do the job, and it does it well at a good price. There is something I'm trying to figure out, though. Christine sent me some information about leasing it instead of buying it, and I'm trying to figure out where the catch is. There has got to be one, but I can't see it right now."

"OK, let me give you a hint: apples and oranges. You can't add them. Now, let's get the paperwork started for that palletizer. The shipping department is just too much of a bottleneck." Clem disappeared from her door as quickly as he had arrived, leaving Naomi musing to herself.

"Apples and *oranges*? *Apples* and oranges? Ahh . . . apples and oranges, of course!"

3.1 | Introduction

Chapter 2 showed that interest is the basis for determining whether different patterns of cash flows are equivalent. Rather than comparing patterns of cash flows from first principles, it is usually easier to use functions that define *mathematical* equivalence among certain common cash flow patterns. These functions are called *compound interest factors*. We discuss a number of these common cash flow patterns along with their associated compound interest factors in this chapter. These compound interest factors are used throughout the remainder of the book. It is, therefore, particularly important to understand their use before proceeding to subsequent chapters.

This chapter opens with an explanation of how cash flow patterns that engineers commonly use are simplified approximations of complex reality. Next, we discuss four simple, discrete cash flow patterns and the compound interest factors that relate them to each other. There is then a brief discussion of the case in which the number of time periods considered is so large that it is treated as though the relevant cash flows continue indefinitely. Appendix 3A discusses modelling cash flow patterns when the interval between disbursements or receipts is short enough that we may view the flows as being continuous. Appendix 3B presents mathematical derivations of the compound interest factors.

3.2 | Timing of Cash Flows and Modelling

The actual timing of cash flows can be very complicated and irregular. Unless some simple approximation is used, comparisons of different cash flow sequences will be very difficult and impractical. Consider, for example, the cash flows generated by a relatively simple operation like a service station that sells gasoline and supplies, and also services cars. Some cash flows, like sales of gasoline and minor supplies, will be almost continuous during the time the station is open. Other flows, like receipts for the servicing of cars, will be on a daily basis. Disbursements for wages may be on a weekly basis. Some disbursements, like those for a manager's salary and for purchases of gasoline and supplies, may be monthly. Disbursements for insurance and taxes may be quarterly or semiannual. Other receipts and disbursements, like receipts for major repairs or disbursements for used parts, may be irregular.

An analyst trying to make a comparison of two projects with different, irregular timings of cash flows might have to record each of the flows of the projects, then, on a one-by-one basis, find summary equivalent values like present worth that would be used in the comparison. This activity would be very time-consuming and tedious if it could be done, but it probably could not be done because the necessary data would not exist. If the projects were potential rather than actual, the cash flows would have to be predicted. This could not be done with great precision for either size or timing of the flows. Even if the analysis were of the past performances of ongoing operations, it is unlikely that it would be worthwhile to maintain a databank that contained the exact timing of all cash flows.

Because of the difficulties of making precise calculations of complex and irregular cash flows, engineers usually work with fairly simple models of cash flow patterns. The most common type of model assumes that all cash flows and all compounding of cash flows occur at the ends of conventionally defined periods like months or years. Models that make this assumption are called **discrete models**. In some cases, analysts use models that assume cash flows and their compounding occur continuously over time; such models are called **continuous models**. Whether the analyst uses discrete modelling or continuous modelling, the model is usually an approximation. Cash flows do not occur only at the ends of conventionally defined periods, nor are they actually continuous. We shall emphasize discrete models throughout the book because they are more common and more readily understood by persons of varied backgrounds. Discrete cash flow models are discussed in the main body of this chapter, and continuous models are presented in Appendix 3A, at the end of this chapter.

3.3 | Compound Interest Factors for Discrete Compounding

Compound interest factors are formulas that define mathematical equivalence for specific common cash flow patterns. The compound interest factors permit cash flow analysis to be done more conveniently because tables or spreadsheet functions can be used instead of complicated formulas. This section presents compound interest factors for four discrete cash flow patterns that are commonly used to model the timing of receipts and disbursements in engineering economic analysis. The four patterns are:

1. A single disbursement or receipt
2. A set of equal disbursements or receipts over a sequence of periods, referred to as an **annuity**

3. A set of disbursements or receipts that change by a constant *amount* from one period to the next in a sequence of periods, referred to as an **arithmetic gradient series**

4. A set of disbursements or receipts that change by a constant *proportion* from one period to the next in a sequence of periods, referred to as a **geometric gradient series**

The principle of discrete compounding requires several assumptions:

1. Compounding periods are of equal length.

2. Each disbursement and receipt occurs at the end of a period. A payment at time 0 can be considered to occur at the end of period -1.

3. Annuities and gradients coincide with the ends of sequential periods. (Section 3.8 suggests several methods for dealing with annuities and gradients that do not coincide with the ends of sequential periods.)

Mathematical derivations of six of the compound interest factors are given in Appendix 3B at the end of this chapter.

3.4 | Compound Interest Factors for Single Disbursements or Receipts

In many situations, a single disbursement or receipt is an appropriate model of cash flows. For example, the salvage value of production equipment with a limited service life will be a single receipt at some future date. An investment today to be redeemed at some future date is another example.

Figure 3.1 illustrates the general form of a single disbursement or receipt. Two commonly used factors relate a single cash flow in one period to another single cash flow in a different period. They are the *compound amount factor* and the *present worth factor*.

The **compound amount factor**, denoted by $(F/P,i,N)$, gives the future amount, F, that is equivalent to a present amount, P, when the interest rate is i and the number of periods is N. The value of the compound amount factor is easily seen as coming from Equation (2.1), the compound interest equation, which relates present and future values,

$$F = P(1 + i)^N$$

In the symbolic convention used for compound interest factors, this is written

$$F = P(1 + i)^N = P(F/P,i,N)$$

Figure 3.1   Single Receipt at End of Period *N*

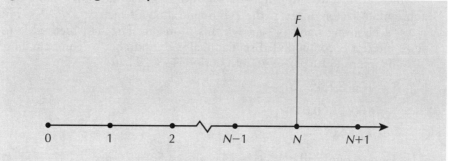

so that the compound amount factor is

$$(F/P,i,N) = (1 + i)^N$$

A handy way of thinking of the notation is (reading from left to right): "What is F, given P, i, and N?"

The compound amount factor is useful in determining the future value of an investment made today if the number of periods and the interest rate are known.

The **present worth factor**, denoted by $(P/F,i,N)$, gives the present amount, P, that is equivalent to a future amount, F, when the interest rate is i and the number of periods is N. The present worth factor is the inverse of the compound amount factor, $(F/P,i,N)$. That is, while the compound amount factor gives the future amount, F, that is equivalent to a present amount, P, the present worth factor goes in the other direction. It gives the present worth, P, of a future amount, F. Since $(F/P,i,N) = (1 + i)^N$,

$$(P/F,i,N) = \frac{1}{(1 + i)^N}$$

The compound amount factor and the present worth factor are fundamental to engineering economic analysis. Their most basic use is to convert a single cash flow that occurs at one point in time to an equivalent cash flow at another point in time. When comparing several individual cash flows which occur at different points in time, an analyst would apply the compound amount factor or the present worth factor, as necessary, to determine the equivalent cash flows at a common reference point in time. In this way, each of the cash flows is stated as an amount at one particular time. Example 3.1 illustrates this process.

Although the compound amount factor and the present worth factor are relatively easy to calculate, some of the other factors discussed in this chapter are more complicated, and it is therefore desirable to have an easier way to determine their values. The compound interest factors are sometimes available as functions in calculators and spreadsheets, but often these functions are provided in an awkward format that makes them relatively difficult to use. They can, however, be fairly easily programmed in a calculator or spreadsheet.

A traditional and still useful method for determining the value of a compound interest factor is to use tables. Appendix A at the back of this book lists values for all the compound interest factors for a selection of interest rates for discrete compounding periods. The desired compound interest factor can be determined by looking in the appropriate table.

EXAMPLE 3.1

How much money will be in a bank account at the end of 15 years if $100 is invested today and the nominal interest rate is 8% compounded semiannually?

Since a present amount is given and a future amount is to be calculated, the appropriate factor to use is the compound amount factor, $(F/P,i,N)$. There are several ways of choosing i and N to solve this problem. The first method is to observe that, since interest is compounded semiannually, the number of compounding periods, N, is 30. The interest rate per six-month period is 4%. Then

$$F = 100(F/P,4\%,30)$$
$$= 100(1 + 0.04)^{30}$$
$$= 324.34$$

The bank account will hold $324.34 at the end of 15 years.

Alternatively, we can obtain the same results by using the interest factor tables.

$$F = 100(3.2434) \qquad \text{(from Appendix A)}$$

$$= 324.34$$

A second solution to the problem is to calculate the *effective* yearly interest rate and then compound over 15 years at this rate. Recall from Equation (2.4) that the effective interest rate per year is

$$i_e = \left(1 + \frac{r}{m}\right)^m - 1$$

where i_e = the effective annual interest rate

r = the nominal rate per year

m = the number of periods in a year

$i_e = (1 + 0.08/2)^2 - 1 = 0.0816$

where $r = 0.08$

$m = 2$

When the effective yearly rate for each of 15 years is applied to the future worth computation, the future worth is

$$F = P(F/P,i,N)$$

$$= P(1 + i)^N$$

$$= 100(1 + 0.0816)^{15}$$

$$= 324.34$$

Once again, we conclude that the balance will be $324.34. ■

3.5 | Compound Interest Factors for Annuities

The next four factors involve a series of uniform receipts or disbursements that start at the end of the first period and continue over N periods, as illustrated in Figure 3.2. This pattern of cash flows is called an annuity. Mortgage or lease payments and maintenance contract fees are examples of the annuity cash flow pattern. Annuities may also be used to model series of cash flows that fluctuate over time around some average value. Here the average value would be the constant uniform cash flow. This would be done if the fluctuations were unknown or deemed to be unimportant for the problem.

The **sinking fund factor**, denoted by $(A/F,i,N)$, gives the size, A, of a repeated receipt or disbursement that is equivalent to a future amount, F, if the interest rate is i and the number of periods is N. The name of the factor comes from the term **sinking fund**. A sinking fund is an interest-bearing account into which regular deposits are made in order to accumulate some amount.

The equation for the sinking fund factor can be found by decomposing the series of disbursements or receipts made at times 1, 2, . . . , N, and summing to produce a total future value. The formula for the sinking fund factor is

$$(A/F,i,N) = \frac{i}{(1 + i)^N - 1}$$

Figure 3.2   Annuity Over *N* Periods

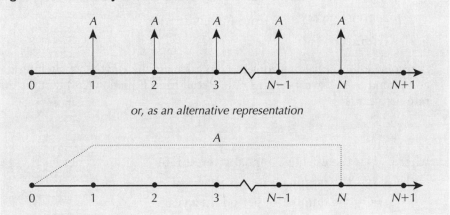

or, as an alternative representation

The sinking fund factor is commonly used to determine how much has to be set aside or saved per period to accumulate an amount F at the end of N periods at an interest rate i. The amount F might be used, for example, to purchase new or replacement equipment, to pay for renovations, or to cover capacity expansion costs. In more general terms, the sinking fund factor allows us to convert a single future amount into a series of equal-sized payments, made over N equally spaced intervals, with the use of a given interest rate i.

The **uniform series compound amount factor**, denoted by $(F/A,i,N)$, gives the future value, F, that is equivalent to a series of equal-sized receipts or disbursements, A, when the interest rate is i and the number of periods is N. Since the uniform series compound amount factor is the inverse of the sinking fund factor,

$$(F/A,i,N) = \frac{(1 + i)^N - 1}{i}$$

The **capital recovery factor**, denoted by $(A/P,i,N)$, gives the value, A, of the equal periodic payments or receipts that are equivalent to a present amount, P, when the interest rate is i and the number of periods is N. The capital recovery factor is easily derived from the sinking fund factor and the compound amount factor:

$$(A/P,i,N) = (A/F,i,N)(F/P,i,N)$$
$$= \frac{i}{(1 + i)^N - 1}(1 + i)^N$$
$$= \frac{i(1 + i)^N}{(1 + i)^N - 1}$$

The capital recovery factor can be used to find out, for example, how much money must be saved over N future periods to "recover" a capital investment of P today. The capital recovery factor for the purchase cost of something is sometimes combined with the sinking fund factor for its salvage value after N years to compose the **capital recovery formula**. See Close-Up 3.1.

The **series present worth factor**, denoted by $(P/A,i,N)$, gives the present amount, P, that is equivalent to an annuity with disbursements or receipts in the amount, A, where the interest rate is i and the number of periods is N. It is the reciprocal of the capital recovery factor:

$$(P/A,i,N) = \frac{(1 + i)^N - 1}{i(1 + i)^N}$$

| CLOSE-UP 3.1 | Capital Recovery Formula |

Industrial equipment and other assets are often purchased at a cost of P on the basis that they will incur savings of A per period for the firm. At the end of their useful life, they will be sold for some salvage value S. The expression to determine A for a given P and S combines the capital recovery factor (for P) with the sinking fund factor (for S):

$$A = P(A/P,i,N) - S(A/F,i,N)$$

Since

$$(A/F,i,N) = \frac{i}{(1+i)^N - 1} = \frac{i}{(1+i)^N - 1} + i - i$$

$$= \frac{i}{(1+i)^N - 1} + \frac{i[(1+i)^N - 1]}{(1+i)^N - 1} - i$$

$$= \frac{i + i(1+i)^N - i}{(1+i)^N - 1} - i = \frac{i(1+i)^N}{(1+i)^N - 1} - i$$

$$= (A/P,i,N) - i$$

then

$$A = P(A/P,i,N) - S[(A/P,i,N) - i]$$

$$= (P - S)(A/P,i,N) + Si$$

This is the capital recovery formula, which can be used to calculate the savings necessary to justify a capital purchase of cost P and salvage value S after N periods at interest rate i.

The capital recovery formula is also used to determine an annual amount which captures the loss in value of an asset over the time it is owned. Chapter 7 treats this use of the capital recovery formula more fully.

EXAMPLE 3.2

The Hanover Go-Kart Klub has decided to build a clubhouse and track five years from now. It must accumulate €50 000 by the end of five years by setting aside a uniform amount from its dues at the end of each year. If the interest rate is 10%, how much must be set aside each year?

Since the problem requires that we calculate an annuity amount given a future value, the solution can be obtained using the sinking fund factor where $i = 10\%$, $F = $ €50 000, $N = 5$, and A is unknown.

$$A = 50\ 000(A/F,10\%,5)$$

$$= 50\ 000(0.1638)$$

$$= 8190.00$$

The Go-Kart Klub must set aside €8190 at the end of each year to accumulate €50 000 in five years.■

EXAMPLE 3.3

A car loan requires 30 monthly payments of $199.00, starting *today*. At an annual rate of 12% compounded monthly, how much money is being lent?

This cash flow pattern is referred to as an **annuity due**. It differs from a standard annuity in that the first of the N payments occurs at time 0 (now) rather than at the end of the first time period. Annuities due are uncommon—not often will one make the first payment on a loan on the date the loan is received! Unless otherwise stated, it is reasonable to assume that any annuity starts at the end of the first period.

Two simple methods of analyzing an annuity due will be used for this example.

Method 1. Count the first payment as a present worth and the next 29 payments as an annuity:

$$P = 199 + A(P/A,i,N)$$

where $A = 199$, $i = 12\%/12 = 1\%$, and $N = 29$.

$$P = 199 + 199(P/A,1\%,29)$$
$$= 199 + 199(25.066)$$
$$= 199 + 4988.13$$
$$= 5187.13$$

The present worth of the loan is the current payment, $199, plus the present worth of the subsequent 29 payments, $4988.13, a total of about $5187.

Method 2. Determine the present worth of a standard annuity at time -1, and then find its worth at time 0 (now). The worth at time -1 is

$$P_{-1} = A(P/A,i,N)$$
$$= 199(P/A,1\%,30)$$
$$= 199(25.807)$$
$$= 5135.79$$

Then the present worth now (time 0) is

$$P_0 = P_{-1}(F/P,i,N)$$
$$= 5135.79(F/P,1\%,1)$$
$$= 5135.79(1.01)$$
$$= 5187.15$$

The second method gives the same result as the first, allowing a small margin for the effects of rounding.∎

It is worth noting here that although it is natural to think about the symbol P as meaning a cash flow at time 0, the present, and F as meaning a cash flow in the future, in fact these symbols can be more general in meaning. As illustrated in the last example, we can consider any point in time to be the "present" for calculation purposes, and similarly any point in time to be the "future," provided P is some point in time earlier than F. This observation gives us substantial flexibility in analyzing cash flows.

EXAMPLE 3.4

Clarence bought a flat for £94 000 in 2002. He made a £14 000 down payment and negotiated a mortgage from the previous owner for the balance. Clarence agreed to pay the previous owner £2000 per month at 12% nominal interest, compounded monthly. How long did it take him to pay back the mortgage?

Clarence borrowed only £80 000, since he made a £14 000 down payment. The £2000 payments form an annuity over N months where N is unknown. The interest rate per month is 1%. We must find the value of N such that

$$P = A(P/A,i,N) = A\left(\frac{(1 + i)^N - 1}{i(1 + i)^N}\right)$$

or, alternatively, the value of N such that

$$A = P(A/P,i,N) = P\left(\frac{i(1 + i)^N}{(1 + i)^N - 1}\right)$$

where $P = £80\ 000$, $A = £2000$, and $i = 0.01$.

By substituting the known quantities into either expression, some manipulation is required to find N. For illustration, the capital recovery factor has been used.

$$A = P\left(\frac{i(1 + i)^N}{(1 + i)^N - 1}\right)$$

$$2000 = 80\ 000\left(\frac{0.01(1.01)^N}{1.01^N - 1}\right)$$

$$2.5 = \frac{(1.01)^N}{(1.01)^N - 1}$$

$$2.5/1.5 = (1.01)^N$$

$$N[ln(1.01)] = ln(2.5/1.5)$$

$$N = 51.34 \text{ months}$$

It will take Clarence four years and four months to pay off the mortgage. He will make 51 full payments of £2000 and will be left with only a fraction of a full payment for his 52nd and last monthly installment. Problem 3.34 asks what his final payment will be. Note also that mortgages can be confusing because of the different terms used. See Close-Up 3.2.■

In Example 3.4, it was possible to use the formula for the compound interest factor to solve for the unknown quantity directly. It is not always possible to do this when the number of periods or the interest rate is unknown. We can proceed in several ways. One possibility is to determine the unknown value by trial and error with a spreadsheet. Another approach is to find the nearest values using tables, and then to interpolate linearly to determine an approximate value. Some calculators will perform the interpolation automatically. See Close-Up 3.3 and Figure 3.3 for a reminder of how linear interpolation works.

CLOSE-UP 3.2   Mortgages

Mortgages can be a little confusing because of the terminology used. In particular, the word *term* is used in different ways in different countries. It can mean either the duration over which the original loan is calculated to be repaid (called the **amortization period** in Canada). It can also mean the duration over which the loan agreement is valid (otherwise called the *maturity*). The interest rate is a nominal rate, usually compounded monthly.

For example, Salim has just bought a house for $135 000. He paid $25 000 down, and the rest of the cost has been obtained from a mortgage. The mortgage has a nominal interest rate of 9.5% compounded monthly with a 20-year term (amortization period). The maturity (term) of the mortgage is three years. What are Salim's monthly payments? How much does he owe after three years?

Salim's monthly payments can be calculated as

$$A = (135\,000 - 25\,000)(A/P,9.5/12\%,[20 \times 12])$$

$$= 110\,000(A/P,0.7917\%,240)$$

$$= 110\,000(0.00932)$$

$$= 1025.20$$

Salim's monthly payments would be about $1025.20. After three years he would have to renegotiate his mortgage at whatever was the current interest rate at that time. The amount owed would be

$$F = 110\,000(F/P,9.5/12\%,36) - 1025.20(F/A,9.5/12\%,36)$$

$$= 110\,000(1.3283) - 1025.20(41.47)$$

$$= 103\,598$$

After three years, Salim still owes $103 598.

CLOSE-UP 3.3   Linear Interpolation

Linear interpolation is the process of approximating a complicated function by a straight line in order to estimate a value for the independent variable based on two sample pairs of independent and dependent variables and an instance of the dependent variable. For example, the function f in Figure 3.3 relates the dependent variable y to the independent variable x. Two sample points, (x_1, y_1) and (x_2, y_2), and an instance of y, y^*, are known, but the actual shape of f is not. An estimate of the value for x^* can be made by drawing a straight line between (x_1, y_1) and (x_2, y_2).

Because the line between (x_1, y_1) and (x_2, y_2) is assumed to be straight, the following ratios must be equal:

$$\frac{x^* - x_1}{x_2 - x_1} = \frac{y^* - y_1}{y_2 - y_1}$$

Isolating the x^* gives the linear interpolation formula:

$$x^* = x_1 + (x_2 - x_1)\left[\frac{y^* - y_1}{y_2 - y_1}\right]$$

Figure 3.3   Linear Interpolation

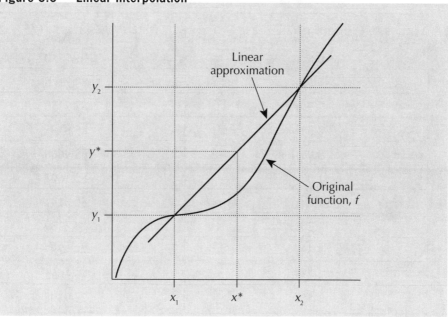

EXAMPLE 3.5

Clarence paid off an £80 000 mortgage completely in 48 months. He paid £2000 per month, and at the end of the first year made an extra payment of £7000. What interest rate was he charged on the mortgage?

Using the series present worth factor and the present worth factor, this can be formulated for an unknown interest rate:

$$80\ 000 = 2000(P/A,i,48) + 7000(P/F,i,12)$$

$$2(P/A,i,48) + 7(P/F,i,12) = 80$$

$$2\left[\frac{(1+i)^{48}-1}{i(1+i)^{48}}\right] + 7\left[\frac{1}{(1+i)^{12}}\right] = 80 \tag{3.1}$$

Solving such an equation for i directly is generally not possible. However, using a spreadsheet as illustrated in Table 3.1 can establish some close values for the left-hand side of Equation (3.1), and a similar process can be done using either tables or a calculator. Using a spreadsheet program or calculator, trials can establish a value for the unknown interest rate to the desired number of significant digits.

Once the approximate values for the interest rate are found, linear interpolation can be used to find a more precise answer. For instance, working from the values of the interest rate which give the LHS (left-hand side) value closest to the RHS (right-hand side) value of 80, which are 1.1% and 1.2%,

$$i = 1.1 + (1.2 - 1.1)\left[\frac{80 - 80.4141}{78.7209 - 80.4141}\right]$$

$$= 1.1 + 0.02 = 1.12\% \text{ per month}$$

The nominal interest rate was $1.12 \times 12 = 13.44\%$.
The effective interest rate was $(1.0112)^{12} - 1 = 14.30\%$. ∎

Table 3.1   Trials to Determine an Unknown Interest Rate

Interest Rate i	$2(P/A,i,48) + 7(P/F,i,12)$
0.5%	91.7540
0.6%	89.7128
0.7%	87.7350
0.8%	85.8185
0.9%	83.9608
1.0%	82.1601
1.1%	80.4141
1.2%	78.7209
1.3%	77.0787
1.4%	75.4855
1.5%	73.9398

Another interesting application of compound interest factors is calculating the value of a bond. See Close-Up 3.4.

CLOSE-UP 3.4   Bonds

Bonds are investments that provide an annuity and a future value in return for a cost today. They have a *par* or *face* value, which is the amount for which they can be redeemed after a certain period of time. They also have a *coupon rate*, meaning that they pay the bearer an annuity, usually semiannually, calculated as a percentage of the face value. For example, a coupon rate of 10% on a bond with an $8000 face value would pay an annuity of $400 each six months. Bonds can sell at more or less than the face value, depending on how buyers perceive them as investments.

To calculate the worth of a bond today, sum together the present worth of the face value (a future amount) and the coupons (an annuity) at an appropriate interest rate. For example, if money can earn 12% compounded semiannually, a bond maturing in 15 years with a face value of $5000 and a coupon rate of 7% is today worth

$P = 5000(P/F,6\%,30) + (5000 \times 0.07/2)(P/A,6\%,30)$

$= 5000(0.17411) + 175(13.765)$

$= 3279.43$

The bond is worth about $3279 today.

3.6 Conversion Factor for Arithmetic Gradient Series

An arithmetic gradient series is a series of receipts or disbursements that starts at zero at the end of the first period and then increases by a constant *amount* from period to period. Figure 3.4 illustrates an arithmetic gradient series of receipts. Figure 3.5 shows

Figure 3.4   Arithmetic Gradient Series of Receipts

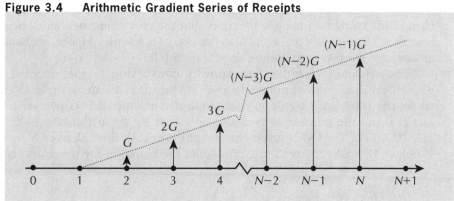

Figure 3.5   Arithmetic Gradient Series of Disbursements

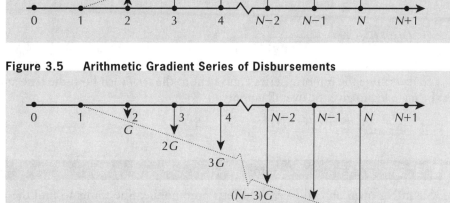

an arithmetic gradient series of disbursements. As an example, we may model a pattern of increasing operating costs for an aging machine as an arithmetic gradient series if the costs are increasing by (approximately) the same amount each period. Note carefully that the first non-zero cash flow of a gradient occurs at the end of the *second* compounding period, not the first.

The sum of an annuity plus an arithmetic gradient series is a common pattern. The annuity is a base to which the arithmetic gradient series is added. This is shown in Figure 3.6. A constant-amount increase to a base level of receipts may occur where the increase in receipts is due to adding capacity and where the ability to add capacity is

Figure 3.6   Arithmetic Gradient Series With Base Annuity

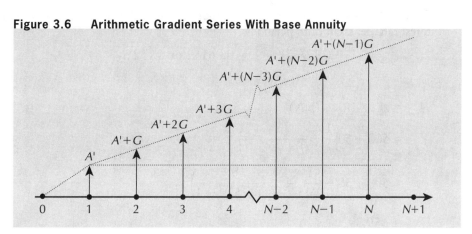

limited. For example, a company that specializes in outfitting warehouses for grocery chains can expand by adding work crews. But the crews must be trained by managers who have time to train only one crew member every six months. Hence, we would have a base amount and a constant amount of growth in cash flows each period.

The **arithmetic gradient to annuity conversion factor**, denoted by $(A/G,i,N)$, gives the value of an annuity, A, that is equivalent to an arithmetic gradient series where the constant increase in receipts or disbursements is G per period, the interest rate is i, and the number of periods is N. That is, the arithmetic gradient series, $0G$, $1G$, $2G$, . . . , $(N-1)G$ is given and the uniform cash flow, A, over N periods is found. Problem 3.29 asks the reader to show that the equation for the arithmetic gradient to annuity factor is

$$(A/G,i,N) = \frac{1}{i} - \frac{N}{(1+i)^N - 1}$$

There is often a base annuity A' associated with a gradient, as illustrated in Figure 3.6. To determine the uniform series equivalent to the *total* cash flow, the base annuity A' must be included to give the overall annuity:

$$A_{\text{tot}} = A' + G(A/G,i,N)$$

EXAMPLE 3.6

Susan Ng owns an eight-year-old Jetta automobile. She wants to find the present worth of repair bills over the four years that she expects to keep the car. Susan has the car in for repairs every six months. Repair costs are expected to increase by $50 every six months over the next four years, starting with $500 six months from now, $550 six months later, and so on. What is the present worth of the repair costs over the next four years if the interest rate is 12% compounded monthly?

First, observe that there will be $N = 8$ repair bills over four years and that the base annuity payment, A', is $500. The arithmetic gradient component of the bills, G, is $50, and hence the arithmetic gradient series is $0, $50, $100, and so on. The present worth of the repair bills can be obtained in a two-step process:

Step 1. Find the total uniform annuity, A_{tot}, equivalent to the sum of the base annuity, $A' = \$500$, and the arithmetic gradient series with $G = \$50$ over $N = 8$ periods.

Step 2. Find the present worth of A_{tot}, using the series present worth factor.

The 12% nominal interest rate, compounded monthly, is 1% per month. The effective interest rate per six-month period is

$$i_{6\text{month}} = (1 + 0.12/12)^6 - 1 = 0.06152 \text{ or } 6.152\%$$

Step 1

$$A_{\text{tot}} = A' + G(A/G,i,N)$$

$$= 500 + 50\left(\frac{1}{i} - \frac{N}{(1+i)^N - 1}\right)$$

$$= 500 + 50\left(\frac{1}{0.06152} - \frac{8}{(1.06152)^8 - 1}\right)$$

$$= 659.39$$

Step 2

$$P = A_{tot}(P/A,i,N) = A_{tot}\left(\frac{(1+i)^N - 1}{i(1+i)^N}\right)$$

$$= 659.39\left(\frac{(1.06152)^8 - 1}{0.06152(1.06152)^8}\right)$$

$$= 4070.09$$

The present worth of the repair costs is about $4070.∎

3.7 | Conversion Factor for Geometric Gradient Series

A geometric gradient series is a series of cash flows that increase or decrease by a constant *percentage* each period. The geometric gradient series may be used to model inflation or deflation, productivity improvement or degradation, and growth or shrinkage of market size, as well as many other phenomena.

In a geometric series, the base value of the series is A and the "growth" rate in the series (the rate of increase or decrease) is referred to as g. The terms in such a series are given by $A, A(1 + g), A(1 + g)^2, \ldots, A(1 + g)^{N-1}$ at the ends of periods $1, 2, 3, \ldots, N$, respectively. If the rate of growth, g, is positive, the terms are increasing in value. If the rate of growth, g, is negative, the terms are decreasing. Figure 3.7 shows a series of receipts where g is positive. Figure 3.8 shows a series of receipts where g is negative.

The **geometric gradient to present worth conversion factor**, denoted by $(P/A,g,i,N)$, gives the present worth, P, that is equivalent to a geometric gradient series where the base receipt or disbursement is A, and where the rate of growth is g, the interest rate is i, and the number of periods is N.

Figure 3.7   Geometric Gradient Series for Receipts With Positive Growth

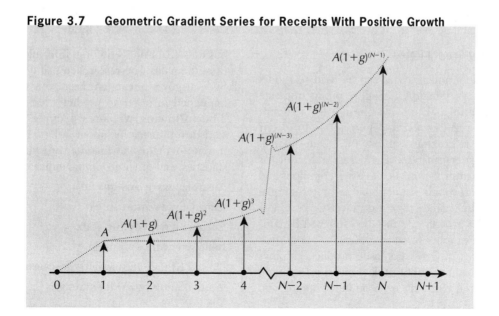

Figure 3.8   Geometric Gradient Series for Receipts With Negative Growth

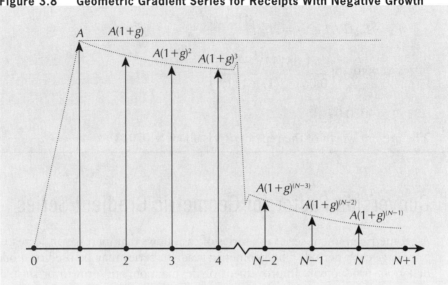

The present worth of a geometric series is

$$P = \frac{A}{1 + i} + \frac{A(1 + g)}{(1 + i)^2} + \ldots + \frac{A(1 + g)^{N - 1}}{(1 + i)^N}$$

where A = the base amount
g = the rate of growth
i = the interest rate
N = the number of periods
P = the present worth

NET VALUE 3.1

Estimating Growth Rates

Geometric gradient series can be used to model the effects of inflation, deflation, production rate change, and market size change on a future cash flow. When using a geometric gradient series, the relevant growth rate must be estimated. The internet can be a useful research tool for collecting information such as expert opinions and statistics on trends for national and international activities by product type and industry. For example, a sales growth rate may be estimated by considering a number of factors: economic condition indicators (e.g., gross domestic product, employment, consumer spending), population growth, raw material cost, and even online shopping and

business-to-business trading. Information on all of these items can be researched using the internet. Federal governments are keen to assess what has happened in order to predict what will happen. The following websites provide a wealth of statistical information about inflation rates, trends in employment and income, international trade statistics, and other economic indicators.

Australia: **www.abs.gov.au**

Canada: **www.statcan.ca**

Europe: **www.ecb.int/stats**

India: **www.mospi.nic.in**

United Kingdom: **www.statistics.gov.uk**

United States: **www.fedstats.gov**

We can define a **growth adjusted interest rate**, $i°$, as

$$i° = \frac{1 + i}{1 + g} - 1$$

so that

$$\frac{1}{1 + i°} = \frac{1 + g}{1 + i}$$

Then the geometric gradient series to present worth conversion factor is given by

$$(P/A,g,i,N) = \frac{(P/A,i°,N)}{1 + g} \text{ or}$$

$$(P/A,g,i,N) = \left(\frac{(1 + i°)^N - 1}{i°(1 + i°)^N}\right)\frac{1}{1 + g}$$

Care must be taken in using the geometric gradient to present worth conversion factor. Four cases may be distinguished:

1. $i > g > 0$. *Growth is positive, but less than the rate of interest.* The growth adjusted interest rate, $i°$, is positive. Tables or functions built into software may be used to find the conversion factor.

2. $g > i > 0$. *Growth is positive and greater than the interest rate.* The growth adjusted interest rate, $i°$, is negative. It is necessary to compute the conversion factor directly from the formula.

3. $g = i > 0$. *Growth is positive and exactly equal to the interest rate.* The growth adjusted interest rate $i° = 0$. As with any case where the interest rate is zero, the present worth of the series with constant terms, $A/(1 + g)$, is simply the sum of all the N terms

$$P = N\left(\frac{A}{1 + g}\right)$$

4. $g < 0$. *Growth is negative.* In other words, the series is decreasing. The growth adjusted interest rate, $i°$, is positive. Tables or functions built into software may be used to find the conversion factor.

EXAMPLE 3.7

Tru-Test is in the business of assembling and packaging automotive and marine testing equipment to be sold through retailers to "do-it-yourselfers" and small repair shops. One of their products is tire pressure gauges. This operation has some excess capacity. Tru-Test is considering using this excess capacity to add engine compression gauges to their line. They can sell engine pressure gauges to retailers for $8 per gauge. They expect to be able to produce about 1000 gauges in the first month of production. They also expect that, as the workers learn how to do the work more efficiently, productivity will rise by 0.25% per month for the first two years. In other words, each month's output of gauges will be 0.25% more than the previous month's. The interest rate is 1.5% per month. All gauges are sold in the month in which they are produced, and receipts from sales are at the end of each month. What is the present worth of the sales of the engine pressure gauges in the first two years?

We first compute the growth-adjusted interest rate, i°:

$$i^\circ = \frac{1 + i}{1 + g} - 1 = \frac{1.015}{1.0025} - 1 = 0.01247$$

$$i^\circ \approx 1.25\%$$

We then make use of the geometric gradient to present worth conversion factor with the uniform cash flow $A = \$8000$, the growth rate $g = 0.0025$, the growth adjusted interest rate $i^\circ = 0.0125$, and the number of periods $N = 24$.

$$P = A(P/A,g,i,N) = A\left(\frac{(P/A,i^\circ,N)}{1 + g}\right)$$

$$P = 8000\left(\frac{(P/A,1.25\%,24)}{1.0025}\right)$$

From the interest factor tables we get

$$P = 8000\left(\frac{20.624}{1.0025}\right)$$

$$P = 164\ 580$$

The present worth of sales of engine compression gauges over the two-year period would be about \$165 000. Recall that we worked with an *approximate* growth-adjusted interest rate of 1.25% when the correct rate was a bit less than 1.25%. This means that \$164 580 is a slight understatement of the present worth.■

EXAMPLE 3.8

Emery's company, Dry-All, produces control systems for drying grain. Proprietary technology has allowed Dry-All to maintain steady growth in the U.S. market in spite of numerous competitors. Company dividends, all paid to Emery, are expected to rise at a rate of 10% per year over the next 10 years. Dividends at the end of this year are expected to total \$110 000. If all dividends are invested at 10% interest, how much will Emery accumulate in 10 years?

If we calculate the growth adjusted interest rate, we get

$$i^\circ = \frac{1.1}{1.1} - 1 = 0$$

and it is natural to think that the present worth is simply the first year's dividends multiplied by 10. However, recall that in the case where $g = i$ the present worth is given by

$$P = N\left(\frac{A}{1 + g}\right) = 10\left(\frac{110\ 000}{1.1}\right) = 1\ 000\ 000$$

Intuitively, dividing by $(1 + g)$ compensates for the fact that growth is considered to start after the end of the first period, but the interest rate applies to all periods. We want the future worth of this amount after 10 years:

$$F = 1\ 000\ 000\ (F/P,10\%,10) = 1\ 000\ 000\ (2.5937) = 2\ 593\ 700$$

Emery will accumulate \$2 593 700 in dividends and interest.■

3.8 | Non-Standard Annuities and Gradients

As discussed in Section 3.3, the standard assumption for annuities and gradients is that the payment period and compounding period are the same. If they are not, the formulas given in this chapter cannot be applied directly. There are three methods for dealing with this situation:

1. Treat each cash flow in the annuity or gradient individually. This is most useful when the annuity or gradient series is not large.
2. Convert the non-standard annuity or gradient to standard form by changing the compounding period.
3. Convert the non-standard annuity to standard form by finding an equivalent standard annuity for the compounding period. This method cannot be used for gradients.

EXAMPLE 3.9

How much is accumulated over 20 years in a fund that pays 4% interest, compounded yearly, if $1000 is deposited at the end of every fourth year?

The cash flow diagram for this set of payments is shown in Figure 3.9.

Figure 3.9   Non-Standard Annuity for Example 3.9

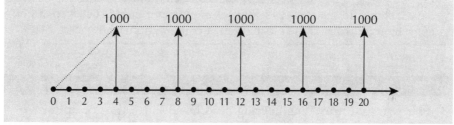

Method 1: Consider the annuities as separate future payments.

Formula: $F = P(F/P,i,N)$

Known values: $P = \$1000$, $i = 0.04$, $N = 16, 12, 8, 4,$ and 0

Year	Future Value			
4	$1000(F/P,4\%,16)$	=	1000(1.8729)	= 1873
8	$1000(F/P,4\%,12)$	=	1000(1.6010)	= 1601
12	$1000(F/P,4\%,8)$	=	1000(1.3685)	= 1369
16	$1000(F/P,4\%,4)$	=	1000(1.1698)	= 1170
20	1000			= 1000
	Total future value			= 7013

About $7013 is accumulated over the 20 years.

Method 2: Convert the compounding period from yearly to every four years. This can be done with the effective interest rate formula.

$$i_e = (1 + 0.04)^4 - 1$$
$$= 16.99\%$$

The future value is then

$$F = 1000(F/A,16.99\%,5) = 1000(7.013)$$

$$= 7013$$

Method 3: Convert the annuity to an equivalent yearly annuity. This can be done by considering the first payment as a future value over the first four-year period, and finding the equivalent annuity over that period, using the sinking fund factor:

$$A = 1000(A/F,4\%,4)$$

$$= 1000(0.23549)$$

$$= 235.49$$

In other words, a $1000 deposit at the end of the four years is equivalent to four equal deposits of $235.49 at the end of each of the four years. This yearly annuity is accumulated over the 20 years.

$$F = 235.49(F/A,4\%,20)$$

$$= 235.49(29.777)$$

$$= 7012$$

Note that each method produces the same amount, allowing for rounding. When you have a choice in methods as in this example, your choice will depend on what you find convenient, or what is the most efficient computationally.■

EXAMPLE 3.10

This year's electrical engineering class has decided to save up for a class party. Each of the 90 people in the class is to contribute $0.25 per day which will be placed in a daily interest (7 days a week, 365 days a year) savings account that pays a nominal 8% interest. Contributions will be made *five* days a week, Monday through Friday, beginning on Monday. The money is put into the account at the beginning of each day, and thus earns interest for the day. The class party is in 14 weeks (a full 14 weeks of payments will be made), and the money will be withdrawn on the Monday morning of the 15th week. How much will be saved, assuming everybody makes payments on time?

There are several good ways to solve this problem. One way is to convert each day's contribution to a weekly amount on Sunday evening/Monday morning, and then accumulate the weekly amounts over the 14 weeks:

Total contribution per day is $0.25 \times 90 = 22.50$

The interest rate per day is $\dfrac{0.08}{365} = 0.000219$

The effective interest rate for a 1-week period is

$$i = (1 + 0.08/365)^7 - 1 = 0.00154$$

Value of one week's contribution on Friday evening (*annuity due* formula):

$$22.50 \times (F/P,0.08/365,1) \times (F/A,0.08/365,5)$$

On Sunday evening this is worth

$$[22.50(F/P,0.08/365,1)(F/A,0.08/365,5)] \times (F/P,0.08/365,2)$$
$$= 22.50(F/P,0.08/365,3)(F/A,0.08/365,5)$$

Then the total amount accumulated by Monday morning of the 15th week is given by:

$$[22.50(F/P,0.08/365,3)(F/A,0.08/365,5)](F/A, (1 + 0.08/365)^7 - 1,14)$$
$$= [22.50(1.000\ 658)(5.002\ 19)](14.1406)$$
$$= 1592.56$$

The total amount saved would be $1592.56.∎

3.9 | Present Worth Computations When $N \rightarrow \infty$

We have until now assumed that the cash flows of a project occur over some fixed, finite number of periods. For long-lived projects, it may be reasonable to model the cash flows as though they continued indefinitely. The present worth of an infinitely long uniform series of cash flows is called the **capitalized value** of the series. We can get the capitalized value of a series by allowing the number of periods, N, in the series present worth factor to go to infinity:

$$P = \lim_{N \rightarrow \infty} A(P/A,i,N)$$

$$= A \lim_{N \rightarrow \infty} \left[\frac{(1 + i)^N - 1}{i(1 + i)^N} \right]$$

$$= A \lim_{N \rightarrow \infty} \left[\frac{1 - \frac{1}{(1 + i)^N}}{i} \right]$$

$$= \frac{A}{i}$$

EXAMPLE 3.11

The town of South Battleford is considering building a bypass for truck traffic around the downtown commercial area. The bypass will provide merchants and shoppers with benefits that have an estimated value of $500 000 per year. Maintenance costs will be $125 000 per year. If the bypass is properly maintained, it will provide benefits for a very long time. The actual life of the bypass will depend on factors like future economic conditions that cannot be forecast at the time the bypass is being considered. It is, therefore, reasonable to model the flow of benefits as though they continued indefinitely. If the interest rate is 10%, what is the present worth of benefits minus maintenance costs?

$$P = \frac{A}{i} = \frac{500\ 000 - 125\ 000}{0.1} = 3\ 750\ 000$$

The present worth of benefits net of maintenance costs is $3 750 000.∎

REVIEW PROBLEMS

REVIEW PROBLEM 3.1

The benefits of a revised production schedule for a seasonal manufacturer will not be realized until the peak summer months. Net savings will be $1100, $1200, $1300, $1400, and $1500 at the ends of months 5, 6, 7, 8, and 9, respectively. It is now the beginning of month 1. Assume 365 days per year, 30 days per month. What is the present worth (PW) of the savings if nominal interest is

    (a) 12% per year, compounded monthly?

    (b) 12% per year, compounded daily?

ANSWER

    (a) $A = \$1100$

        $G = \$100$

        $i = 0.12/12 = 0.01$ per month $= 1\%$

$$
\begin{aligned}
\text{PW(end of period 4)} &= (P/A,1\%,5)[1100 + 100(A/G,1\%,5)] \\
&= 4.8528[1100 + 100(1.9801)] \\
&= 6298.98
\end{aligned}
$$

$$
\begin{aligned}
\text{PW(at time 0)} &= \text{PW(end of period 4)}(P/F,1\%,4) \\
&= 6298.98/(1.01)^4 = 6053.20
\end{aligned}
$$

The present worth is about $6053.

    (b) Effective interest rate $i = (1 + 0.12/365)^{30} - 1 = 0.0099102$

$$
\begin{aligned}
\text{PW(at time 0)} &= \text{PW(end of period 4)}(P/F,i,4) \\
&= (P/A,i,5)[1100 + 100(A/G,i,5)](P/F,i,4) \\
&= 4.8547[1100 + 100(1.98023)](0.9613) \\
&= 6057.80
\end{aligned}
$$

The present worth is about $6058. ■

REVIEW PROBLEM 3.2

It is January 1 of this year. You are starting your new job tomorrow, having just finished your engineering degree at the end of last term. Your take-home pay for this year will be $36 000. It will be paid to you in equal amounts at the end of each month, starting at the end of January. There is a cost-of-living clause in your contract that says that each subsequent January you will get an increase of 3% in your yearly salary (i.e., your take-home pay for next year will be 1.03 × $36 000). In addition to your salary, a wealthy relative regularly sends you a $2000 birthday present at the end of each June.

    Recognizing that you are not likely to have any government pension, you have decided to start saving 10% of your monthly salary and 50% of your birthday present for your retirement. Interest is 1% per month, compounded monthly. How much will you have saved at the end of five years?

ANSWER

Yearly pay is a geometric gradient; convert your monthly salary into a yearly amount by the use of an effective yearly rate. The birthday present can be dealt with separately.

Salary:

The future worth (FW) of the salary at the end of the first year is

$\quad$ FW(salary, year 1) $= 3000(F/A,1\%,12) = 38\,040.00$

$\quad$ This forms the base of the geometric gradient; all subsequent years increase by 3% per year. Savings are 10% of salary, which implies that $A = \$3804.00$.

$\quad A = \$3804.00 \qquad g = 0.03$

$\quad$ Effective yearly interest rate $i_e = (1 + 0.01)^{12} - 1 = 0.1268$ per year

$$i^\circ = \frac{1 + i_e}{1 + g} - 1 = \frac{1 + 0.1268}{1 + 0.03} - 1 = 0.093981$$

$\quad$ PW(gradient) $\qquad = A\,(P/A,i^\circ,5)/(1 + g) = 3804(3.8498)/1.03$

$$= 14\,218$$

$\quad$ FW(gradient, end of five years) $=$ PW(gradient)$(F/P,i_e,5)$

$$= 14\,218(1.1268)^5 = 25\,827$$

Birthday Present:

The present arrives in the middle of each year. To get the total value of the five gifts, we can find the present worth of an annuity of five payments of $2000(0.5) as of six months prior to employment:

$\quad$ PW(-6 months) $= 2000(0.5)(P/A,i_e,5) = 3544.90$

The future worth at $5 \times 12 + 6 = 66$ months later is

$\quad$ FW(end of five years) $= 3544.9(1.01)^{66} = 6836$

$\quad$ Total amount saved $= \$6836 + \$25\,827 = \$32\,663$ ■

REVIEW PROBLEM 3.3

The Easy Loan Company advertises a "10%" loan. You need to borrow £1000, and the deal you are offered is the following: You pay £1100 (£1000 plus £100 interest) in 11 equal £100 amounts, starting one month from today. In addition, there is a £25 administration fee for the loan, payable immediately, and a processing fee of £10 per payment. Furthermore, there is a £20 non-optional closing fee to be included in the last payment. Recognizing fees as a form of interest payment, what is the actual effective interest rate?

ANSWER

Since the £25 administration fee is paid immediately, you are only getting £975. The remaining payments amount to an annuity of £110 per month, plus a £20 future payment 11 months from now.

$\quad$ Formulas: $\quad P = A(P/A,i,N), P = F(P/F,i,N)$

$\quad$ Known values: $P = £975, A = £110, F = £20, N = 11$

$$975 = 110(P/A,i,11) + 20(P/F,i,11)$$

At $i = 4\%$

$$110(P/A,4\%,11) + 20(P/F,4\%,11)$$
$$= 110(8.7603) + 20(0.64958)$$
$$= 976.62$$

At $i = 5\%$

$$110(P/A,5\%,11) + 20(P/F,5\%,11)$$
$$= 110(8.3062) + 20(0.58469)$$
$$= 925.37$$

Linearly interpolating gives

$$i = 4 + (5 - 4)(975 - 976.62)/(925.37 - 976.62)$$
$$= 4.03$$

The effective interest rate is then

$$i = (1 + 0.0403)^{12} - 1$$
$$= 60.69\% \text{ per annum } (!)$$

Although the loan is advertised as a "10%" loan, the actual effective rate is over 60%.∎

REVIEW PROBLEM 3.4

Ming wants to retire as soon as she has enough money invested in a special bank account (paying 14% interest, compounded annually) to provide her with an annual income of $25 000. She is able to save $10 000 per year, and the account now holds $5000. If she just turned 20, and expects to die in 50 years, how old will she be when she retires? There should be no money left when she turns 70.

ANSWER

Let Ming's retirement age be $20 + x$ so that

$$5000(F/P,14\%,x) + 10\ 000(F/A,14\%,x) = 25\ 000(P/A,14\%,50 - x)$$

Dividing both sides by 5000,

$$(F/P,14\%,x) + 2(F/A,14\%,x) - 5(P/A,14\%,50 - x) = 0$$

At $x = 5$

$$(F/P,14\%,5) + 2(F/A,14\%,5) - 5(P/A,14\%,45)$$
$$= 1.9254 + 2(6.6101) - 5(7.1232) = -20.4704$$

At $x = 10$

$$(F/P,14\%,10) + 2(F/A,14\%,10) - 5(P/A,14\%,40)$$
$$= 3.7072 + 2(19.337) - 5(7.1050) = 6.8562$$

Linearly interpolating,

$$x = 5 + 5 \times (20.4704)/(6.8562 + 20.4704)$$
$$= 8.7$$

Ming can retire at age $20 + 8.7 = 28.7$ years old.∎

SUMMARY

In Chapter 3 we considered ways of modelling patterns of cash flows that enable easy comparisons of the worths of projects. The emphasis was on discrete models. Four basic patterns of discrete cash flows were considered:

1. Flows at a single point
2. Flows that are constant over time
3. Flows that grow or decrease at a constant arithmetic rate
4. Flows that grow or decrease at a constant geometric rate

Compound interest factors were presented that defined mathematical equivalence among the basic patterns of cash flows. A list of these factors with their names, symbols, and formulas appears in Table 3.2. The chapter also addressed the issue of how to analyze non-standard annuities and gradients as well as the ideas of capital recovery and capitalized value.

For those who are interested in continuous compounding and continuous cash flows, Appendix 3A contains a summary of relevant notation and interest factors.

Table 3.2   Summary of Useful Formulas for Discrete Models

Name	Symbol and Formula
Compound amount factor	$(F/P,i,N) = (1 + i)^N$
Present worth factor	$(P/F,i,N) = \dfrac{1}{(1 + i)^N}$
Sinking fund factor	$(A/F,i,N) = \dfrac{i}{(1 + i)^N - 1}$
Uniform series compound amount factor	$(F/A,i,N) = \dfrac{(1 + i)^N - 1}{i}$
Capital recovery factor	$(A/P,i,N) = \dfrac{i(1 + i)^N}{(1 + i)^N - 1}$
Series present worth factor	$(P/A,i,N) = \dfrac{(1 + i)^N - 1}{i(1 + i)^N}$
Arithmetic gradient to annuity conversion factor	$(A/G,i,N) = \dfrac{1}{i} - \dfrac{N}{(1 + i)^N - 1}$
Geometric gradient to present worth conversion factor	$(P/A,g,i,N) = \dfrac{(P/A,i^\circ,N)}{1 + g}$ $(P/A,g,i,N) = \left(\dfrac{(1 + i^\circ)^N - 1}{i^\circ(1 + i^\circ)^N}\right)\dfrac{1}{1 + g}$ $i^\circ = \dfrac{1 + i}{1 + g} - 1$
Capitalized value formula	$P = \dfrac{A}{i}$
Capital recovery formula	$A = (P - S)(A/P,i,N) + Si$

Engineering Economics in Action, Part 3B:
No Free Lunch

This time it was Naomi who stuck her head in Clem's doorway. "Here's the recommendation on the shipping palletizer. Oh, and thanks for the hint on the leasing figures. It cleared up my confusion right away."

"No problem. What did you figure out?" Clem had his "mentor" expression on his face, so Naomi knew he was expecting a clear explanation of the trick used by the leasing company.

"Well, as you hinted, they were adding apples and oranges. They listed the various costs for each choice over time, including interest charges, taxes, and so on. But then, for the final comparison, they added up these costs. When they added, leasing was cheaper."

"So what's wrong with that?" Clem prompted.

"They're adding apples and oranges. We're used to thinking of money as being just money, without remembering that money always has a 'when' associated with it. If you add money at different points in time, you might as well be adding apples and oranges; you have a number but it doesn't mean anything. In order to compare leasing with buying, you first have to change the cash flows into the same money, that is, at the same point in time. That's a little harder to do, especially when there's a complicated set of cash flows."

"So were you able to do it?"

"Yes. I identified various components of the cash flows as annuities, gradients, and present and future worths. Then I converted all of these to a present worth for each alternative and summed them. This is the correct way to compare them. If you do that, buying is cheaper, even when borrowing money to do so. And of course it has to be—that leasing company has to pay for those slick brochures somehow. There's no free lunch."

Clem nodded. "I think you've covered it. Mind you, there are some circumstances where leasing is worthwhile. For example, we lease our company cars to save us the time and trouble of reselling them when we're finished with them. Leasing can be good when it's hard to raise the capital for very large purchases, too. But almost always, buying is better. And you know, it amazes me how easy it is to fall for simplistic cash flow calculations that fail to take into account the time value of money. I've even seen articles in the newspaper quoting accountants who make the same mistake, and you'd think they would know better."

"Engineers can make that mistake, too, Clem. I almost did."

PROBLEMS

For additional practice, please see the problems (with selected solutions) provided on the Student CD-ROM that accompanies this book.

3.1 St. Agatha Kennels provides dog breeding and boarding services for a nearby city. Most of the income is derived from boarding, with typical boarding stays being one or two weeks. Customers pay at the end of the dog's stay. Boarding is offered only during the months of May to September. Other income is received from breeding golden retrievers, with two litters of about eight dogs each being produced per year, spring and fall. Expenses include heating, water, and sewage, which are paid monthly, and food, bought in bulk every spring. The business has been neither growing nor shrinking over the past few years.

Joan, the owner of the kennels, wants to model the cash flows for the business over the next 10 years. What cash flow elements (e.g., single payments, annuities, gradients) would she likely consider, and how would she estimate their value? Consider the present to be the first of May. For example, one cash flow element is food. It would be modelled as an *annuity due* over 10 years, and estimated by the amount paid for food over the last few years.

3.2 It is September, the beginning of his school year, and Marco has to watch his expenses while he is going to school. Over the next eight months, he wants to estimate his cash flows. He pays rent once a month. He takes the bus to and from school. A couple of times a week he goes to the grocery store for food, and eats lunch in the cafeteria at school every school day. At the end of every four-month term, he will have printing and copying expenses because of reports that will be due. After the first term, over the Christmas holidays, he will have extra expenses for buying presents, but will also get some extra cash from his parents. What cash flow elements (e.g., single payments, annuities, gradients) would Marco likely consider in his estimates? How would he estimate them?

3.3 How much money will be in a bank account at the end of 15 years if 100 rand are deposited today and the interest rate is 8% compounded annually?

3.4 How much should you invest today at 12% interest to accumulate 1 000 000 rand in 30 years?

3.5 Martin and Marcy McCormack have just become proud parents of septuplets. They have savings of $5000. They want to invest their savings so that they can partially support the children's university education. Martin and Marcy hope to provide $20 000 for each child by the time the children turn 18. What must the annual rate of return be on the investment for Martin and Marcy to meet their goal?

3.6 You have $1725 to invest. You know that a particular investment will double your money in five years. How much will you have in 10 years if you invest in this investment, assuming that the annual rate of return is guaranteed for the time period?

3.7 Morris paid £500 a month for 20 years to pay off the mortgage on his Glasgow house. If his down payment was £5000 and the interest rate was 6% compounded monthly, what was the purchase price of the house?

3.8 An investment pays $10 000 every five years, starting in seven years, for a total of four payments. If interest is 9%, how much is this investment worth today?

3.9 An industrial juicer costs $45 000. It will be used for five years and then sold to a remarketer for $25 000. If interest is 15%, what net yearly savings are needed to justify its purchase?

3.10 Fred wants to save up for an automobile. What amount must he put in his bank account each month to save €10 000 in two years if the bank pays 6% interest compounded monthly?

3.11 It is May 1. You have just bought $2000 worth of furniture. You will pay for it in 24 equal monthly payments, starting at the end of May next year. Interest is 6% nominal per year, compounded monthly. How much will your payments be?

3.12 What is the present worth of the total of 20 payments, occurring at the end of every four months (the first payment is in four months), which are $400, $500, $600, increasing arithmetically? Interest is 12% nominal per year, compounded continuously.

3.13 What is the total value of the sum of the present worths of all the payments and receipts mentioned in Problem 2.32, at an interest rate of 0.5% per month?

3.14 How much is accumulated in each of the following savings plans over two years?

  (a) $40 at the end of each month for 24 months at 12% compounded monthly

  (b) $30 at the end of the first month, $31 at the end of the second month, and so forth, increasing by $1 per month, at 12% compounded monthly

3.15 What interest rate will result in $5000 seven years from now, starting with $2300 today?

3.16 Refer back to the Hanover Go-Kart Klub problem of Example 3.2. The members determined that it is possible to set aside only €7000 each year, and that they will have to put off building the clubhouse until they have saved the €50 000 necessary. How long will it take to save a total of €50 000, assuming that the interest rate is 10%? (*Hint:* Use logarithms to simplify the sinking fund factor.)

3.17 Gwen just bought a satellite dish, which provides her with exactly the same service as cable TV. The dish cost $2000, and the cable service she has now cancelled cost her $40 per month. How long will it take her to recoup her investment in the dish, if she can earn 12% interest, compounded monthly, on her money?

3.18 Yoko has just bought a new computer ($2000), a printer ($350), and a scanner ($210). She wants to take the monthly payment option. There is a monthly interest of 3% on her purchase.

(a) If Yoko pays $100 per month, how long does it take to complete her payments?

(b) If Yoko wants to finish paying in 24 months, how much will her monthly payment be?

3.19 Rinku has just finished her first year of university. She wants to tour Europe when she graduates in three years. By having a part-time job through the school year and a summer job during the summer, she plans to make regular weekly deposits into a savings account, which bears 18% interest, compounded monthly.

(a) If Rinku deposits $15 per week, how much will she save in three years? How about $20 per week?

(b) Find out exactly how much Rinku needs to deposit every week if she wants to save $5000 in three years.

3.20 Seema is looking at an investment in upgrading an inspection line at her plant. The initial cost would be $140 000 with a salvage value of $37 000 after five years. Use the capital recovery formula to determine how much money must be saved every year to justify the investment, at an interest rate of 14%.

3.21 Trenny has asked her assistant to prepare estimates of cost of two different sizes of power plants. The assistant reports that the cost of the 100 MW plant is $200 000 000, while the cost of the 200 MW plant is $360 000 000. If Trenny has a budget of only $300 000 000, estimate how large a power plant she could afford using linear interpolation.

3.22 Enrique has determined that investing $500 per month will enable him to accumulate $11 350 in 12 years, and that investing $800 per month will enable him to accumulate $18 950 over the same period. Estimate, using linear interpolation, how much he would have to invest each month to accumulate exactly $15 000.

3.23 A UK lottery prize pays £1000 at the end of the first year, £2000 the second, £3000 the third, and so on for 20 years. If there is only one prize in the lottery, 10 000 tickets are sold, and you could invest your money elsewhere at 15% interest, how much is each ticket worth, on average?

3.24 Joseph and three other friends bought a $110 000 house close to the university at the end of August last year. At that time they put down a deposit of $10 000 and took out a mortgage for the balance. Their mortgage payments are due at the end of each month (September 30, last year, was the date of the first payment) and are based on the assumption that Joseph and friends will take 20 years to pay off the debt. Annual nominal interest is 12%, compounded monthly. It is now February. Joseph and friends have made all their fall-term payments and have just made the January 31 payment for this year. How much do they still owe?

3.25 A new software package is expected to improve productivity at Grand Insurance. However, because of training and implementation costs, savings are not expected to occur until the third year of operation. At that time, savings of $10 000 are expected, increasing by $1000 per year for the following five years. After this time (eight years from implementation), the software will be abandoned with no scrap value. How much is the software worth today, at 15% interest?

3.26 Clem is saving for a car in a bank account that pays 12% interest, compounded monthly. The balance is now $2400. Clem will be saving $120 per month from his salary, and once every four months (starting in four months) he adds $200 in dividends from an investment. Bank fees, currently $10 per month, are expected to increase by $1 per month henceforth. How much will Clem have saved in two years?

3.27 Yogajothi is thinking of investing in a rental house. The total cost to purchase the house, including legal fees and taxes, is £115 000. All but £15 000 of this amount will be mortgaged. He will pay £800 per month in mortgage payments. At the end of two years, he will sell the house, and at that time expects to clear £20 000 after paying off the remaining mortgage principal (in other words, he will pay off all his debts for the house, and still have £20 000 left). Rents will earn him £1000 per month for the first year, and £1200 per month for the second year. The house is in fairly good condition now so that he doesn't expect to have any maintenance costs for the first six months. For the seventh month, Yogajothi has budgeted £200. This figure will be increased by £20 per month thereafter (e.g., the expected month 7 expense will be £200, month 8, £220, month 9, £240, etc.). If interest is 6% compounded monthly, what is the present worth of this investment? Given that Yogajothi's estimates of revenue and expenses are correct, should Yogajothi buy the house?

3.28 You have been paying off a mortgage in quarterly payments at a 24% nominal annual rate, compounded quarterly. Your bank is now offering an alternative payment plan, so you have a choice of two methods—continuing to pay as before or switching to the new plan. Under the new plan, you would make monthly payments, 30% of the size of your current payments. The interest rate would be 24% nominal, compounded monthly. The time until the end of the mortgage would not change, regardless of the method chosen.

(a) Which plan would you choose, given that you naturally wish to minimize the level of your payment costs? (*Hint:* Look at the costs over a three-month period.)

(b) Under which plan would you be paying a higher effective yearly interest rate?

3.29 Derive the arithmetic gradient conversion to a uniform series formula. (*Hint:* Convert each period's gradient amount to its future value, and then look for a substitution from the other compound amount factors.)

3.30 Derive the geometric gradient to present worth conversion factor. (*Hint:* Divide and multiply the present worth of a geometric series by $[1 + g]$ and then substitute in the growth-adjusted interest rate.)

3.31 Reginald is expecting steady growth of 10% per year in profits from his new company. All profits are going to be invested at 20% interest. If profits for this year (at the end of the year) total ¥1 000 000, how much will be saved at the end of 10 years?

3.32 Reginald is expecting steady growth in profits from his new company of 20% per year. All profits are going to be invested at 10% interest. If profits for this year (at the end of the year) total ¥1 000 000, how much will be saved at the end of 10 years?

3.33 Ruby's business has been growing quickly over the past few years, with sales increasing at about 50% per year. She has been approached by a buyer for the business. She has decided she will sell it for 1/2 of the value of the estimated sales for the next five years. This year she will sell products worth $1 456 988. Use the geometric gradient factor to calculate her selling price for an interest rate of 5%.

3.34 In Example 3.4, Clarence bought a £94 000 house with a £14 000 down payment and took out a mortgage for the remaining £80 000 at 12% nominal interest, compounded monthly. We determined that he would make 51 £2000 payments and then a final payment. What is his final payment?

3.35 A new wave-soldering machine is expected to save Brisbane Circuit Boards $15 000 per year through reduced labour costs and increased quality. The device will have a life of eight years, and have no salvage value after this time. If the company can generally expect to get 12% return on its capital, how much could it afford to pay for the wave-soldering machine?

3.36 Gail has won a lottery that pays her $100 000 at the end of this year, $110 000 at the end of next year, $120 000 the following year, and so on, for 30 years. Leon has offered Gail $2 500 000 today in exchange for all the money she will receive. If Gail can get 8% interest on her savings, is this a good deal?

3.37 Gail has won a lottery that pays her $100 000 at the end of this year, and increases by 10% per year thereafter for 30 years. Leon has offered Gail $2 500 000 today in exchange for all the money she will receive. If Gail can get 8% interest on her savings, is this a good deal?

3.38 Tina has saved €20 000 from her summer jobs. Rather than work for a living, she plans to buy an annuity from a trust company and become a beachcomber in Fiji. An annuity will pay her a certain amount each month for the rest of her life, and is calculated at 7% interest, compounded monthly, over Tina's 55 remaining years. Tina calculates that she needs at least €5 per day to live in Fiji, and she needs €1200 for air fare. Can she retire now? How much would she have available to spend each day?

3.39 A regional municipality is studying a water supply plan for its tri-city and surrounding area to the end of year 2055. To satisfy the water demand, one suggestion is to construct a pipeline from a major lake some distance away. Construction would start in 2015 and take five years at a cost of $20 million per year. The cost of maintenance and repairs starts after completion of construction and for the first year is $2 million, increasing by 1% per year thereafter. At an interest rate of 6%, what is the present worth of this project?

Assume that all cash flows take place at year-end. Consider the present to be the end of 2010/beginning of 2011. Assume that there is no salvage value at the end of year 2055.

 3.40 Clem has a $50 000 loan. The interest rate offered is 8% compounded annually, and the repayment period is 15 years. Payments are to be received in equal installments at the end of each year. Construct a spreadsheet (you must use a spreadsheet program) similar to the following table that shows the amount received each year, the portion that is interest, the portion that is unrecovered capital, and the amount that is outstanding (i.e., unrecovered). Also, compute the total recovered capital which must equal the original capital amount; this can serve as a check on your solution. Design the spreadsheet so that the capital amount and the interest rate can be changed by updating only one cell for each. Construct:

(a) the completed spreadsheet for the amount, interest rate, and repayment period indicated

(b) the same spreadsheet, but for $75 000 at 10% interest (same repayment period)

(c) a listing showing the formulas used

Sample Capital Recovery Calculations				
Capital amount			$50 000.00	
Annual interest rate			8.00%	
Number of years to repay			15	
Payment Periods	**Annual Payment**	**Interest Received**	**Recovered Capital**	**Unrecovered Capital**
0				$50 000.00
1	$5841.48	$4000.00	$1841.48	48 158.52
2				
.				
.				
.				
15				0.00
Total			$50 000.00	

3.41 A French software genius has been offered €10 000 per year for the next five years and then €20 000 per year for the following 10 years for the rights to his new video game. At 9% interest, how much is this worth today?

3.42 A bank offers a personal loan called "The Eight Per Cent Plan." The bank adds 8% to the amount borrowed; the borrower pays back 1/12 of this total at the end of each month for a year. On a loan of $500, the monthly payment is 540/12 = $45. There is also an administrative fee of $45, payable now. What is the actual effective interest rate on a $500 loan?

3.43 Coastal Shipping is setting aside capital to fund an expansion project. Funds earmarked for the project will accumulate at the rate of $50 000 per month until the project is completed. The project will take two years to complete. Once the project starts, costs will be incurred monthly at the rate of $150 000 per month over 24 months. Coastal currently has $250 000 saved. What is the minimum number of months it will have to wait before it can start if money is worth 18% nominal, compounded monthly? Assume that

1. Cash flows are all at the ends of months

2. The first $50 000 savings occurs one month from today

3. The first $150 000 payment occurs one month after the start of the project

4. The project must start at the beginning of a month

3.44 A company is about to invest in a joint venture research and development project with another company. The project is expected to last eight years, but yearly payments the company makes will begin immediately (i.e., a payment is made today, and the last payment is eight years from today). Salaries will account for $40 000 of each payment. The remainder of each payment will cover equipment costs and facility overhead. The initial (immediate) equipment and facility cost is $26 000. Each subsequent year the figure will drop by $3000 until a cost of $14 000 is reached, after which the costs will remain constant until the end of the project.

(a) Draw a cash flow diagram to illustrate the cash flows for this situation.

(b) At an interest rate of 7%, what is the total future worth of all project payments at the end of the eight years?

3.45 Shamsir's small business has been growing slowly. He has noticed that his monthly profit increases by 1% every two months. Suppose that the profit at the end of this month is 10 000 rupees. What is the present value of all his profit over the next two years? Annual nominal interest is 18%, compounded monthly.

3.46 Xiaohang is conducting a biochemical experiment for the next 12 months. In the first month, the expenses are estimated to be 15 000 yuan. As the experiment progresses, the expenses are expected to increase by 5% each month. Xiaohang plans to pay for the experiment by a government grant, which is received in six monthly installments, starting a month after the experiment completion date. Determine the amount of the monthly installment so that the total of the six installments pays for all expenses incurred during the experiment. Annual nominal interest is 12%, compounded monthly.

3.47 City engineers are considering several plans for building municipal aqueduct tunnels. They use an interest rate of 8%. One plan calls for a full-capacity tunnel that will meet the needs of the city forever. The cost is $3 000 000 now, and $100 000 every 10 years thereafter for repairs. What is the total present worth of the costs of building and maintaining the aqueduct?

3.48 The city of Brussels is installing a new swimming pool in the municipal recreation centre. One design being considered is a reinforced concrete pool which will cost €1 500 000 to install. Thereafter, the inner surface of the pool will need to be refinished and painted every 10 years at a cost of €200 000 per refinishing. Assuming that the pool will have essentially an infinite life, what is the present worth of the costs associated with this pool design? The city uses a 5% interest rate.

3.49 Goderich Automotive (GA) wants to donate a vacant lot next door to its plant to the city for use as a public park and ball field. The city will accept only if GA will also donate enough cash to maintain the park indefinitely. The estimated maintenance costs are $18 000 per year and interest is 7%. How much cash must GA donate?

3.50 A 7%, 20-year municipal bond has a $10 000 face value. I want to receive at least 10% compounded semiannually on this investment. How much should I pay for the bond?

3.51 A Paradorian bond pays $500 (Paradorian dollars) twice each year and $5000 five years from now. I want to earn at least 300% *annual* (effective) interest on this investment (to compensate for the very high inflation in Parador). How much should I pay for this bond now?

3.52 If money is worth 8% compounded semiannually, how much is a bond maturing in nine years, with a face value of £10 000 and a coupon rate of 9%, worth today?

3.53 A bond with a face value of $5000 pays quarterly interest of 1.5% each period. Twenty-six interest payments remain before the bond matures. How much would you be willing to pay for this bond today if the next interest payment is due now and you want to earn 8% compounded quarterly on your money?

More Challenging Problem

3.54 What happens to the present worth of an arithmetic gradient series as the number of periods approaches infinity? Consider all four cases:

(a) $i > g > 0$

(b) $g > i > 0$

(c) $g = i = 0$

(d) $g < 0$

MINI-CASE 3.1

The Gorgon LNG Project in Western Australia

Historically, natural gas was considered a relatively uninteresting by-product of oil production. It was often burned at the site of an oil well, just to get rid of it. But oil supplies are now dwindling while demand has increased from both developed and developing countries. Also, interest in clean-burning fuels has grown. Consequently, in recent years natural gas has become a key global resource.

One of the largest accessible supplies of natural gas is located off the northwest coast of Australia. Within this area, a huge field containing about 40 trillion cubic feet of gas—Australia's largest known undeveloped gas resource—was discovered in the mid-1990s. It was named "Gorgon" after a vicious monster from Greek mythology.

Western Australia is very isolated and unpopulated. There is no local industrial or residential market for the gas. However, it is relatively conveniently located to the burgeoning markets of Asia. China in particular is rapidly industrializing and has a seemingly insatiable demand for energy.

There are significant technical challenges to developing the Gorgon gas field. The gas has to be drawn from under as much as 600 m of water and shipped 65 km to shore. The processing site is on Barrow Island, a protected nature reserve. The gas itself contains large quantities of carbon dioxide, which has to be removed and reinjected into the ground so that it doesn't contribute to global warming. Finally, the gas has to be cooled to −161°C. This reduces it to 1/600 of its volume, and it becomes liquefied natural gas, or LNG. It is then transported in liquid form by specially designed ships to customers, who turn it back into gas.

However, given the vast reserves and the eager demand, it was felt by the owners—a joint venture of three large oil companies (Chevron, ExxonMobil, and Shell)—that in spite of the challenges the project was feasible. Serious project planning was done from 2000 through early 2003, allowing for the economic viability of a project design to be confirmed. On September 8, 2003, a well defined project plan was approved by the Western Australian government.

Long-term commercial agreements were concluded by the joint venture partners with customers that included China's national oil company. Everything looked like it was on track for the $8 billion project (all costs are in Australian dollars) to successfully provide profits for its joint venture partners. Thousands of jobs were to be created and millions of dollars of tax revenue would be provided to the Australian government. Start-up was planned for 2006 and gas would flow for 60 years to fuel Asian industrialization.

However, it didn't work out that way. As of 2008, over $1 billion has been spent in drilling, appraisal, and engineering design costs to date. The total project costs have ballooned from $8 billion in 2003 to as much as $30 billion in 2008. Consequently, the economics of the project are in disarray. At least one of the joint venture partners is looking to sell out its share, and there are suggestions that the project will be put off indefinitely.

Discussion

One might assume that the Gorgon project is an example of professional incompetence. It is not. Many very capable people were careful to think through all of the costs in a deep and detailed manner. The joint venture partners are very experienced at large projects like this.

However, no matter how careful the planning process might be, one can never fully understand how the future will turn out. There were two key issues that strongly affected the Gorgon project.

The first was the environmental effects of the project. The site of the gas processing plant is to be on Barrow Island, a nature reserve. In spite of the fact that processing facilities had been located on the island for decades, under modern sensitivities for environmental impacts, this caused a great deal of unexpected administrative overhead. Another environmental concern stemmed from the large amount of carbon dioxide that had to be properly dealt with. These environmental issues were dealt with, and although they caused delay and extra cost, the Australian federal government finally gave its environmental approval to the project on October 10, 2007.

However, the most important problem was a confluence of global trends aggravated by Western Australia's remote location. The global trends include the industrialization of China, with a corresponding increased demand for oil and gas. This has consequently caused oil companies to attempt to increase supply. This in turn has caused the proliferation of oil and gas developments all over the globe. Consequently, there is tremendous demand for the manufacturing capacity, technical expertise, and specialized equipment needed for developing a gas field like Gorgon, raising the costs of acquiring these in a timely manner.

At the same time, construction projects of other kinds were proliferating, both within and outside of Australia. With a very small labour pool and an isolated location, costs of labour in Western Australia skyrocketed, far beyond even the most pessimistic expectations. A similar situation has happened in Canada's tar sands in Alberta.

In order to compare alternatives economically, it is necessary to determine future cash flows. Of course nobody knows for sure what the future holds, so future cash flows are always estimates. In some cases the estimates will turn out to be significantly different from the actual cash flows, and therefore result in an incorrect decision.

There are sophisticated ways to deal with uncertainty about future cash flows, some of which are discussed in Chapters 11 and 12. In many cases it makes sense to assume that the future cash flows are treated as if they were certain because there is no particular reason to think they are not. On the other hand, in some cases even estimating future cash flows can be very difficult or impossible.

Questions

1. For each of the following, comment on how sensible it is to estimate the precise value of future cash flows:
   (a) Your rent for the next six months
   (b) Your food bill for the next six months

   (c) Your medical bills for the next six months
   (d) A company's payroll for the next six months
   (e) A company's raw material costs for the next six months
   (f) A company's legal costs for liability lawsuits for the next six months
   (g) Your country's costs for funding university research for the next six months
   (h) Your country's costs for employment insurance for the next six months
   (i) Your country's costs for emergency management for the next six months

2. Your company is looking at the possibility of buying a new widget grinder for the widget line. The future cash flows associated with the purchase of the grinder are fairly predictable, except for one factor. A significant benefit is achieved with the higher production volume of widgets, which depends on a contract to be signed with a particular important customer. This won't happen for several months, but you must make the decision about the widget grinder now. Discuss some sensible ways of dealing with this issue.

Appendix 3A Continuous Compounding and Continuous Cash Flows

We now consider compound interest factors for continuous models. Two forms of continuous modelling are of interest. The first form has discrete cash flows with continuous compounding. In some cases, projects generate discrete, end-of-period cash flows within an organization in which other cash flows and compounding occur many times per period. In this case, a reasonable model is to assume that the project's cash flows are discrete, but compounding is continuous. The second form has continuous cash flows with continuous compounding. Where a project generates many cash flows per period, we could model this as continuous cash flows with continuous compounding.

We first consider models with discrete cash flows with continuous compounding. We can obtain formulas for compound interest factors for these projects from formulas for discrete compounding simply by substituting the effective continuous interest rate with continuous compounding for the effective rate with discrete compounding.

Recall that for a given nominal interest rate, r, when the number of compounding periods per year becomes infinitely large, the effective interest rate per period is $i_e = e^r - 1$. This implies $1 + i_e = e^r$ and $(1 + i_e)^N = e^{rN}$. The various compound interest factors for continuous compounding can be obtained by substituting $e^r - 1$ for i in the formulas for the factors.

For example, the series present worth factor with discrete compounding is

$$(P/A,i,N) = \frac{(1 + i)^N - 1}{i(1 + i)^N}$$

If we substitute $e^r - 1$ for i and e^{rN} for $(1 + i)^N$, we get the series present worth factor for continuous compounding

$$(P/A,r,N) = \frac{e^{rN} - 1}{(e^r - 1)e^{rN}}$$

Similar substitutions can be made in each of the other compound interest factor formulas to get the compound interest factor for continuous rather than discrete compounding. The formulas are shown in Table 3A.1. Tables of values for these formulas are available in Appendix B at the end of the book.

Table 3A.1   Compound Interest Formulas for Discrete Cash Flow With Continuous Compounding

Name	Symbol and Formula
Compound amount factor	$(F/P,r,N) = e^{rN}$
Present worth factor	$(P/F,r,N) = \dfrac{1}{e^{rN}}$
Sinking fund factor	$(A/F,r,N) = \dfrac{e^r - 1}{e^{rN} - 1}$
Uniform series compound amount factor	$(F/A,r,N) = \dfrac{e^{rN} - 1}{e^r - 1}$
Capital recovery factor	$(A/P,r,N) = \dfrac{(e^r - 1)e^{rN}}{e^{rN} - 1}$
Series present worth factor	$(P/A,r,N) = \dfrac{e^{rN} - 1}{(e^r - 1)e^{rN}}$
Arithmetic gradient to annuity conversion factor	$(A/G,r,N) = \dfrac{1}{e^r - 1} - \dfrac{N}{e^{rN} - 1}$

EXAMPLE 3A.1

Yoram Gershon is saving to buy a new sound system. He plans to deposit $100 each month for the next 24 months in the Bank of Montrose. The nominal interest rate at the Bank of Montrose is 0.5% per month, compounded continuously. How much will Yoram have at the end of the 24 months?

We start by computing the uniform series compound amount factor for continuous compounding. Recall that the factor for discrete compounding is

$$(F/A,i,N) = \frac{(1 + i)^N - 1}{i}$$

Substituting $e^r - 1$ for i and e^{rN} for $(1 + i)^N$ gives the series compound amount, when compounding is continuous, as

$$(F/A,r,N) = \frac{e^{rN} - 1}{e^r - 1}$$

The amount Yoram will have at the end of 24 months, F, is given by

$$F = 100\left(\frac{e^{(0.005)24} - 1}{e^{0.005} - 1}\right)$$

$$= 100\left(\frac{1.127497 - 1}{1.00501 - 1}\right)$$

$$= 2544.85$$

Yoram will have about $2545 saved at the end of the 24 months.■

The formulas for *continuous cash flows with continuous compounding* are derived using integral calculus. The continuous *series present worth* factor, denoted by $(P/\bar{A},r,T)$, for a continuous flow, $\bar{A}$, over a period length, T, where the nominal interest rate is r, is given by

$$P = \bar{A}\left(\frac{e^{rT} - 1}{re^{rT}}\right)$$

so that

$$(P/\bar{A},r,T) = \frac{e^{rT} - 1}{re^{rT}}$$

It is then easy to derive the formula for the continuous *uniform series compound amount factor*, denoted by $(F/\bar{A},r,T)$, by multiplying the series present worth factor by e^{rT} to get the future worth of a present value, P.

$$(F/\bar{A},r,T) = \frac{e^{rT} - 1}{r}$$

We can get the continuous *capital recovery factor*, denoted by $(\bar{A}/P,r,T)$, as the inverse of the continuous series present worth factor. The *continuous sinking fund factor* $(F/\bar{A},r,T)$ is the inverse of the continuous uniform series compound amount factor. A summary of the formulas for continuous cash flow and continuous compounding is shown in Table 3A.2. Tables of values for these formulas are available in Appendix C at the back of the book.

Table 3A.2   Compound Interest Formulas for Continuous Cash Flow With Continuous Compounding

Name	Symbol and Formula
Sinking fund factor	$(\bar{A},F,r,T) = \dfrac{r}{e^{rT} - 1}$
Uniform series compound amount factor	$(F/\bar{A},r,T) = \dfrac{e^{rT} - 1}{r}$
Capital recovery factor	$(\bar{A},/P,r,T) = \dfrac{re^{rT}}{e^{rT} - 1}$
Series present worth factor	$(P/\bar{A},r,T) = \dfrac{e^{rT} - 1}{re^{rT}}$

EXAMPLE 3A.2

Savings from a new widget grinder are estimated to be $10 000 per year. The grinder will last 20 years and will have no scrap value at the end of that time. Assume that the savings are generated as a continuous flow. The *effective* interest rate is 15% compounded continuously. What is the present worth of the savings?

From the problem statement, we know that $\bar{A} = \$10\ 000$, $i = 0.15$, and $T = 20$. From the relation $i = e^r - 1$, for $i = 0.15$ the interest rate to apply for continuously compounding is $r = 0.13976$. The present worth computations are

$$P = \bar{A}\,(P/\bar{A},r,T)$$

$$= 10\ 000\left(\frac{e^{(0.13976)20} - 1}{(0.13976)e^{(0.13976)20}}\right)$$

$$= 67\ 180$$

The present worth of the savings is $67 180. Note that if we had used discrete compounding factors for the present worth computations we would have obtained a lower value.

$$P = A(P/A,i,N)$$

$$= 10\ 000(6.2593)$$

$$= 62\ 593\blacksquare$$

REVIEW PROBLEM 3A.1 FOR APPENDIX 3A

Mr. Big is thinking of buying the MQM Grand Hotel in Las Vegas. The hotel has continuous net receipts totalling $120 000 000 per year (Vegas hotels run 24 hours per day). This money could be immediately reinvested in Mr. Big's many other ventures, all of which earn a nominal 10% interest. The hotel will likely be out of style in about eight years, and could then be sold for about $200 000 000. What is the maximum Mr. Big should pay for the hotel today?

ANSWER

$$P = 120\ 000\ 000(P/A,10\%,8) + 200\ 000\ 000e^{-(0.1)(8)}$$

$$= 120\ 000\ 000\ \frac{e^{(0.1)(8)} - 1}{(0.1)e^{(0.1)(8)}} + 200\ 000\ 000e^{-(0.1)(8)}$$

$$= 701\ 184\ 547$$

Mr. Big should not pay more than about $700 000 000.$\blacksquare$

PROBLEMS FOR APPENDIX 3A

3A.1 An investment in new data logging technology is expected to generate extra revenue continuously for Western Petroleum Services. The initial cost is $300 000, but extra revenues total $75 000 per year. If the effective interest rate is 10% compounded continuously, does the present worth of the savings over five years exceed the original purchase cost? By how much does one exceed the other?

3A.2 Desmond earns €25 000 continuously over a year from an investment that pays 8% nominal interest, compounded continuously. How much money does he have at the end of the year?

3A.3 Gina intently plays the stock market, so that any capital she has can be considered to be compounding continuously. At the end of 2009, Gina had $10 000. How much did she have at the beginning of 2009, if she earned a nominal interest rate of 18%?

3A.4 Gina (from Problem 3A.3) had earned a nominal interest rate of 18% on the stock market every year since she started with an initial investment of $100. What year did she start investing?

Appendix 3B   Derivation of Discrete Compound Interest Factors

This appendix derives six discrete compound interest factors presented in this chapter. All of them can be derived from the compound interest equation

$$F = P(1 + i)^N$$

3B.1 | Compound Amount Factor

In the symbolic convention used for compound interest factors, the compound interest equation can be written

$$F = P(1 + i)^N = P(F/P,i,N)$$

so that the compound amount factor is

$$(F/P,i,N) = (1 + i)^N \tag{3B.1}$$

3B.2 | Present Worth Factor

The present worth factor, $(P/F,i,N)$, converts a future amount F to a present amount P:

$$P = F(P/F,i,N)$$
$$\Rightarrow F = P\left(\frac{1}{(P/F,i,N)}\right)$$

Thus the present worth factor is the reciprocal of the compound amount factor. From Equation (3B.1),

$$(P/F,i,N) = \frac{1}{(1 + i)^N}$$

3B.3 | Sinking Fund Factor

If a series of payments A follows the pattern of a standard annuity of N payments in length, then the future value of the payment in the jth period, from Equation (3B.1), is:

$$F = A(1 + i)^{N-j}$$

The future value of all of the annuity payments is then

$$F = A(1 + i)^{N-1} + A(1 + i)^{N-2} + \ldots + A(1 + i)^1 + A$$

Factoring out the annuity amount gives

$$F = A[(1 + i)^{N-1} + (1 + i)^{N-2} + \ldots + (1 + i)^1 + 1] \tag{3B.2}$$

Multiplying Equation (3B.2) by $(1 + i)$ gives

$$F(1 + i) = A[(1 + i)^{N-1} + (1 + i)^{N-2} + \ldots + (1 + i)^1 + 1](1 + i)$$

$$F(1 + i) = A[(1 + i)^N + (1 + i)^{N-1} + \ldots + (1 + i)^2 + (1 + i)] \tag{3B.3}$$

Subtracting Equation (3B.2) from Equation (3B.3) gives

$$F(1 + i) - F = A[(1 + i)^N - 1]$$

$$Fi = A[(1 + i)^N - 1]$$

$$A = F\left[\frac{i}{(1 + i)^N - 1}\right]$$

Thus the sinking fund factor is given by

$$(A/F,i,N) = \frac{i}{(1 + i)^N - 1} \tag{3B.4}$$

3B.4 | Uniform Series Compound Amount Factor

The uniform series compound amount factor, $(F/A,i,N)$, converts an annuity A into a future amount F:

$$F = A(F/A,i,N)$$

$$\Rightarrow A = F\left(\frac{1}{(F/A,i,N)}\right)$$

Thus the uniform series compound amount factor is the reciprocal of the sinking fund factor. From Equation (3B.4),

$$(F/A,i,N) = \frac{(1 + i)^N - 1}{i}$$

3B.5 | Capital Recovery Factor

If a series of payments A follows the pattern of a standard annuity of N payments in length, then the present value of the payment in the jth period is

$$P = A\frac{1}{(1 + i)^j}$$

The present value of the total of all the annuity payments is

$$P = A\left(\frac{1}{(1 + i)}\right) + A\left(\frac{1}{(1 + i)^2}\right) + \ldots + A\left(\frac{1}{(1 + i)^{N-1}}\right) + A\left(\frac{1}{(1 + i)^N}\right)$$

Factoring out the annuity amount gives

$$P = A\left[\left(\frac{1}{(1+i)}\right) + \left(\frac{1}{(1+i)^2}\right) + \ldots + \left(\frac{1}{(1+i)^{N-1}}\right) + \left(\frac{1}{(1+i)^N}\right)\right] \qquad (3B.5)$$

Multiplying both sides of Equation (3B.5) by $(1 + i)$ gives

$$P(1+i) = A\left[1 + \left(\frac{1}{(1+i)}\right) + \ldots + \left(\frac{1}{(1+i)^{N-2}}\right) + \left(\frac{1}{(1+i)^{N-1}}\right)\right] \qquad (3B.6)$$

Subtracting Equation (3B.5) from Equation (3B.6) gives

$$Pi = A\left[1 - \left(\frac{1}{(1+i)^N}\right)\right]$$

$$P = A\left[\frac{(1+i)^N - 1}{i(1+i)^N}\right]$$

$$A = P\left[\frac{i(1+i)^N}{(1+i)^N - 1}\right]$$

Thus the capital recovery factor is given by

$$(A/P,i,N) = \frac{i(1+i)^N}{(1+i)^N - 1} \qquad (3B.7)$$

3B.6 | Series Present Worth Factor

The series present worth factor, $(P/A,i,N)$, converts an annuity A into a present amount P:

$$P = A(P/A,i,N)$$

$$\Rightarrow A = P\left(\frac{1}{(P/A,i,N)}\right)$$

Thus the uniform series compound amount factor is the reciprocal of the sinking fund factor. From Equation (3B.7),

$$(P/A,i,N) = \frac{(1+i)^N - 1}{i(1+i)^N}$$

3B.7 | Arithmetic and Geometric Gradients

The derivation of the arithmetic gradient to annuity conversion factor and the geometric gradient to present worth conversion factor are left as problems for the student. See Problems 3.29 and 3.30.

CHAPTER 4

Comparison Methods Part 1

Engineering Economics in Action, Part 4A:
What's Best?

Naomi waved hello as she breezed by Carole Brown, the receptionist, on her way in from the parking lot one Monday morning. She stopped as Carole caught her eye. "Clem wants to see you right away. Good morning."

After a moment of socializing, Clem got right to the point. "I have a job for you. Put aside the vehicle-life project for a couple of days."

"OK, but you wanted a report by Friday."

"This is more important. You know that drop forging hammer in the South Shop? The beast is about 50 years old. I don't remember the exact age. We got it used four years ago. We were having quality control problems with the parts we were buying on contract and decided to bring production in-house. Stinson Brothers sold it to us cheap when they upgraded their forging operation. Fundamentally the machine is still sound, but the guides are worn out. The production people are spending too much time fiddling with it instead of turning out parts. Something has to be done. I have to make a recommendation to Ed Burns and Anna Kulkowski, who are going to be making decisions on investments for the next quarter. I'd like you to handle it." Ed Burns was the manager of manufacturing, and Anna Kulkowski was, among other things, the president of Global Widgets.

"What's the time frame?" Naomi asked. She was shifting job priorities in her mind and deciding what she would need to postpone.

"I want a report by tomorrow morning. I'd like to have a chance to review what you've done and submit a recommendation to Burns and Kulkowski for their Wednesday meeting." Clem sat back and gave Naomi his best big smile.

Naomi's return smile was a bit weak, as she was preoccupied with trying to sort out where to begin.

Clem laughed and continued with, "It's really not so bad. Dave Sullivan has done most of the work. But he's away and can't finish. His father-in-law had a heart attack on Friday, and he and Helena have gone to Florida to see him."

"What's involved?" asked Naomi.

"Not much, really. Dave has estimated all the cash flows. He's put everything on a spreadsheet. Essentially, there are three major possibilities. We can refurbish and upgrade the existing machine. We can get a manually operated mechanical press that will use less energy and be a lot quieter. Or we can go for an automated mechanical press.

"Since there is going to be down time while we are changing the unit, we might also want to replace the materials-handling equipment at the same time. If we get the automated press, there is the possibility of going the whole hog and integrating materials handling with the press. But even if we automate, we could stay with a separate materials-handling setup.

"Basically, you're looking at a fairly small first cost to upgrade the current beast, versus a large first cost for the automated equipment. But, if you take the high-first-cost route, you will get big savings down the road. All you have to do is decide what's best."

4.1 | Introduction

The essential idea of investing is to give up something valuable now for the expectation of receiving something of greater value later. An investment may be thought of as an exchange of resources now for an expected flow of benefits in the future. Business firms, other organizations, and individuals all have opportunities to make such exchanges. A company may be able to use funds to install equipment that will reduce labour costs in the future. These funds might otherwise have been used on another project or returned to the shareholders or owners. An individual may be able to study to become an engineer. Studying requires that time be given up that could have been used to earn money or to travel. The benefit of study, though, is the expectation of a good income from an interesting job in the future.

Not all investment opportunities *should* be taken. The company considering a labour-saving investment may find that the value of the savings is less than the cost of installing the equipment. Not all investment opportunities *can* be taken. The person spending the next four years studying engineering cannot also spend that time getting degrees in law and science.

Engineers play a major role in making decisions about investment opportunities. In many cases, they are the ones who estimate the expected costs of and returns from an investment. They then must decide whether the expected returns outweigh the costs to see if the opportunity is potentially acceptable. They may also have to examine competing investment opportunities to see which is best. Engineers frequently refer to investment opportunities as **projects**. Throughout the rest of this text, the term *project* will be used to mean *investment opportunity*.

In this chapter and in Chapter 5, we deal with methods of evaluating and comparing projects, sometimes called **comparison methods**. We start in this chapter with a scheme for classifying groups of projects. This classification system permits the appropriate use of any of the comparison methods. We then turn to a consideration of several widely used methods for evaluating opportunities. The **present worth method** compares projects by looking at the present worth of all cash flows associated with the projects. The **annual worth method** is similar, but converts all cash flows to a uniform series, that is, an annuity. The **payback period method** estimates how long it takes to "pay back" investments. The study of comparison methods is continued in Chapter 5, which deals with the internal rate of return.

We have made six assumptions about all the situations presented in this chapter and in Chapter 5:

1. We have assumed that costs and benefits are always measurable in terms of money. In reality, costs and benefits need not be measurable in terms of money. For example, providing safe working conditions has many benefits, including improvement of worker morale. However, it would be difficult to express the value of improved worker morale objectively in dollars and cents. Such other benefits as the pleasure gained from appreciating beautiful design may not be measurable quantitatively. We shall consider qualitative criteria and multiple objectives in Chapter 13.

2. We have assumed that future cash flows are known with certainty. In reality, future cash flows can only be estimated. Usually the farther into the future we try to forecast, the less certain our estimates become. We look at methods of assessing the impact of uncertainty and risks in Chapters 11 and 12.

3. We have assumed that cash flows are unaffected by inflation or deflation. In reality, the purchasing power of money typically declines over time. We shall consider how inflation affects decision making in Chapter 9.

4. Unless otherwise stated, we have assumed that sufficient funds are available to implement all projects. In reality, cash constraints on investments may be very important, especially for new enterprises with limited ability to raise capital. We look at methods of raising capital in Appendix 4A.

5. We have assumed that taxes are not applicable. In reality, taxes are pervasive. We shall show how to include taxes in the decision-making process in Chapter 8.

6. Unless otherwise stated, we shall assume that all investments have a cash outflow at the start. These outflows are called *first costs*. We also assume that projects with first costs have cash inflows after the first costs that are at least as great in total as the first costs. In reality, some projects have cash inflows at the start, but involve a commitment of cash outflows at a later period. For example, a consulting engineer may receive an advance payment from a client, a cash inflow, to cover

some of the costs of a project, but to complete the project the engineer will have to make disbursements over the project's life. We shall consider evaluation of such projects in Chapter 5.

4.2 | Relations Among Projects

Companies and individuals are often faced with a large number of investment opportunities at the same time. Relations among these opportunities can range from the simple to the complex. We can distinguish three types of connections among projects that cover all the possibilities. Projects may be

1. independent,
2. mutually exclusive, or
3. related but not mutually exclusive.

The simplest relation between projects occurs when they are **independent**. Two projects are independent if the expected costs and the expected benefits of each project do not depend on whether the other one is chosen. A student considering the purchase of a vacuum cleaner and the purchase of a personal computer would probably find that the expected costs and benefits of the computer did not depend on whether he or she bought the vacuum cleaner. Similarly, the benefits and costs of the vacuum cleaner would be the same, whether or not the computer was purchased. If there are more than two projects under consideration, they are said to be independent if all possible pairs of projects in the set are independent. When two or more projects are independent, evaluation is simple. Consider each opportunity one at a time, and accept or reject it on its own merits.

Projects are **mutually exclusive** if, in the process of choosing one, all other alternatives are excluded. In other words, two projects are mutually exclusive if it is impossible to do both or it clearly would not make sense to do both. For example, suppose Bismuth Realty Company wants to develop downtown office space on a specific piece of land. They are considering two potential projects. The first is a low-rise poured-concrete building. The second is a high-rise steel-frame structure with the same capacity as the low-rise building, but it has a small park at the entrance. It is impossible for Bismuth to have both buildings on the same site.

As another example, consider a student about to invest in a computer printer. She can get an inkjet printer or a laser printer, but it would not make sense to get both. She would consider the options to be mutually exclusive.

The third class of projects consists of those that are **related but not mutually exclusive**. For pairs of projects in this category, the expected costs and benefits of one project depend on whether the other one is chosen. For example, Klamath Petroleum may be considering a service station at Fourth Avenue and Main Street as well as one at Twelfth and Main. The costs and benefits from either station will clearly depend on whether the other is built, but it may be possible, and may make sense, to have both stations.

Evaluation of related but not mutually exclusive projects can be simplified by combining them into exhaustive, mutually exclusive sets. For example, the two projects being considered by Klamath can be put into four mutually exclusive sets:

1. Neither station—the "do nothing" option
2. Just the station at Fourth and Main
3. Just the station at Twelfth and Main
4. Both stations

In general, n related projects can be put into 2^n sets including the "do nothing" option. Once the related projects are put into mutually exclusive sets, the analyst treats these sets as the alternatives. We can make 2^n mutually exclusive sets with n related projects by noting that for any single set there are exactly two possibilities for each project. The project may be *in* or *out* of that set. To get the total number of sets, we multiply the n twos to get 2^n. In the Klamath example, there were two possibilities for the station at Fourth and Main—accept or reject. These are combined with the two possibilities for the station at Twelfth and Main, to give the four sets that we listed.

A special case of related projects is where one project is *contingent* on another. Consider the case where project A could be done alone or A and B could be done together, but B could not be done by itself. Project B is then contingent on project A because it cannot be taken unless A is taken first. For example, the Athens and Manchester Development Company is considering building a shopping mall on the outskirts of town. They are also considering building a parking garage to avoid long outdoor walks by patrons. Clearly, they would not build the parking garage unless they were also building the mall.

Another special case of related projects is due to resource constraints. Usually the constraints are financial. For example, Bismuth may be considering two office buildings at different sites, where the expected costs and benefits of the two are unrelated, but Bismuth may be able to finance only one building. The two office-building projects would then be mutually exclusive because of financial constraints. If there are more than two projects, then all of the sets of projects that meet the budget form a mutually exclusive set of alternatives.

When there are several related projects, the number of logically possible combinations becomes quite large. If there are four related projects, there are $2^4 = 16$ mutually exclusive sets, including the "do nothing" alternative. If there are five related projects, the number of alternatives doubles to 32. A good way to keep track of these alternatives is to construct a table with all possible combinations of projects. Example 4.1 demonstrates the use of a table.

EXAMPLE 4.1

The Small Street Residential Association wants to improve the district. Four ideas for renovation projects have been proposed: (1) converting part of the roadway to gardens, (2) adding old-fashioned light standards, (3) replacing the pavement with cobblestones, and (4) making the street one way. However, there are a number of restrictions. The association can afford to do only two of the first three projects together. Also, gardens are possible only if the street is one way. Finally, old-fashioned light standards would look out of place unless the pavement was replaced with cobblestones. The residential association feels it must do something. They do not want simply to leave things the way they are. What mutually exclusive alternatives are possible?

Since the association does not want to "do nothing," only $15 = 2^4 - 1$ alternatives will be considered. These are shown in Table 4.1. The potential projects are listed in rows. The alternatives, which are sets of projects, are in columns. An "x" in a cell indicates that a project is in the alternative represented by that column. Not all logical combinations of projects represent feasible alternatives, as seen in the special cases of contingent alternatives or budget constraints. A last row, below the potential-project rows, indicates whether the sets are feasible alternatives.

The result is that there are seven feasible mutually exclusive alternatives:

1. Cobblestones (alternative 3)
2. One-way street (alternative 4)
3. One-way street with gardens (alternative 7)

4. Cobblestones with lights (alternative 8)

5. One-way street with cobblestones (alternative 10)

6. One-way street with cobblestones and gardens (alternative 13)

7. One-way street with cobblestones and lights (alternative 14) ∎

Table 4.1   Potential Alternatives for the Small Street Renovation

Potential Alternative	1	2	3	4	5	6	7	8	9	10	11	12	13	14	15
Gardens	x				x	x	x				x	x	x		x
Lights		x			x			x	x		x	x		x	x
Cobblestones			x			x		x		x	x		x	x	x
One-way				x			x		x	x		x	x	x	x
Feasible?	No	No	Yes	Yes	No	No	Yes	Yes	No	Yes	No	No	Yes	Yes	No

To summarize our investigation of possible relations among projects, we have a three-fold classification system: (1) independent projects, (2) mutually exclusive projects, and (3) related but not mutually exclusive projects. We can, however, arrange related projects into mutually exclusive sets and treat the sets as mutually exclusive alternatives. This reduces the system to two categories, independent and mutually exclusive. (See Figure 4.1.) Therefore, in the remainder of this chapter we consider only independent and mutually exclusive projects.

Figure 4.1   Possible Relations Among Projects and How to Treat Them

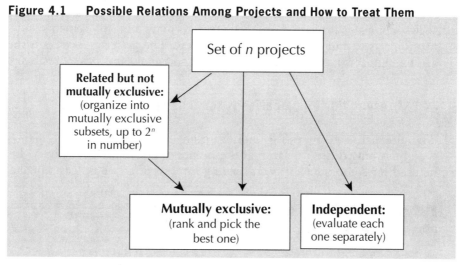

4.3 | Minimum Acceptable Rate of Return (MARR)

A company evaluating projects will set for itself a lower limit for investment acceptability known as the **minimum acceptable rate of return (MARR)**. The MARR is an interest rate that must be earned for any project to be accepted. Projects that earn at least the MARR are desirable, since this means that the money is earning at least as much as can be earned elsewhere. Projects that earn less than the MARR are not desirable, since investing money in these projects denies the opportunity to use the money more profitably elsewhere.

The MARR can also be viewed as the rate of return required to get investors to invest in a business. If a company accepts projects that earn less than the MARR, investors will not be willing to put money into the company. This minimum return required to induce investors to invest in the company is the company's **cost of capital**. Methods for determining the cost of capital are presented in Appendix 4A.

The MARR is thus an opportunity cost in two senses. First, investors have investment opportunities outside any given company. Investing in a given company implies forgoing the opportunity of investing elsewhere. Second, once a company sets a MARR, investing in a given project implies giving up the opportunity of using company funds to invest in other projects that pay at least the MARR.

We shall show in this chapter and in Chapter 5 how the MARR is used in calculations involving the present worth, annual worth, or internal rate of return to evaluate projects. Henceforth, it is assumed that a value for the MARR has been supplied.

4.4 | Present Worth (PW) and Annual Worth (AW) Comparisons

The present worth (PW) comparison method and the annual worth (AW) comparison method are based on finding a comparable basis to evaluate projects in monetary units. With the present worth method, the analyst compares project A and project B by computing the present worths of the two projects at the MARR. The preferred project is the one with the greater present worth. The value of any company can be considered to be the present worth of all of its projects. Therefore, choosing projects with the greatest present worth maximizes the value of the company. With the annual worth method, the analyst compares projects A and B by transforming all disbursements and receipts of the two projects to a uniform series at the MARR. The preferred project is the one with the greater annual worth. One can also speak of *present cost* and *annual cost*. See Close-Up 4.1.

4.4.1 Present Worth Comparisons for Independent Projects

The alternative to investing money in an independent project is to "do nothing." Doing nothing doesn't mean that the money is not used productively. In fact, it would be used for some other project, earning interest at a rate at least equal to the MARR.

CLOSE-UP 4.1	Present Cost and Annual Cost

Sometimes mutually exclusive projects are compared in terms of present cost or annual cost. That is, the best project is the one with minimum present worth of cost as opposed to the maximum present worth. Two conditions should hold for this to be valid: (1) all projects have the same major benefit, and (2) the estimated value of the major benefit clearly outweighs the projects' costs, even if that estimate is imprecise. Therefore, the "do nothing" option is rejected. The value of the major benefit is ignored in further calculations since it is the same for all projects. We choose the project with the lowest cost, considering secondary benefits as offsets to costs.

However, the present worth of any money invested at the MARR is zero, since the present worth of future receipts would exactly offset the current disbursement. Consequently, if an independent project has a present worth greater than zero, it is acceptable. If an independent project has a present worth less than zero, it is unacceptable. If an independent project has a present worth of exactly zero, it is considered *marginally* acceptable.

EXAMPLE 4.2

Steve Chen, a third-year electrical engineering student, has noticed that the networked personal computers provided by his university for its students are frequently fully utilized, so that students often have to wait to get on a machine. The university has a building plan that will create more space for network computers, but the new facilities won't be available for five years. In the meantime, Steve sees the opportunity to create an alternative network in a mall near the campus. The first cost for equipment, furniture, and software is expected to be $70 000. Students would be able to rent time on computers by the hour and to use the printers at a charge per page. Annual net cash flow from computer rentals and other charges, after paying for labour, supplies, and other costs, is expected to be $30 000 a year for five years. When the university opens new facilities at the end of five years, business at Steve's network would fall off and net cash flow would turn negative. Therefore, the plan is to dismantle the network after five years. The five-year-old equipment and furniture are expected to have zero value. If investors in this type of service enterprise demand a return of 20% per year, is this a good investment?

The present worth of the project is

$$PW = -70\ 000 + 30\ 000(P/A,20\%,5)$$

$$= -70\ 000 + 30\ 000(2.9906)$$

$$= 19\ 718$$

$$\cong 20\ 000$$

The project is acceptable since the present worth of about $20 000 is greater than zero.

Another way to look at the project is to suppose that, once Steve has set up the network off campus, he tries to sell it. If he can convince potential investors, who demand a return of 20% a year, that the expectation of a $30 000 per year cash flow for five years is accurate, how much would they be willing to pay for the network? Investors would calculate the present worth of a 20% annuity paying $30 000 for five years. This is given by

$$PW = 30\ 000(P/A,20\%,5)$$

$$= 30\ 000(2.9906)$$

$$= 89\ 718$$

$$\cong 90\ 000$$

Investors would be willing to pay approximately $90 000. Steve will have taken $70 000, the first cost, and used it to create an asset worth almost $90 000. As illustrated in Figure 4.2, the $20 000 difference may be viewed as profit.■

Figure 4.2   Present Worth as a Measure of Profit

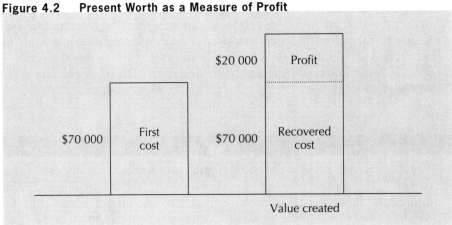

Let us now consider an example in which the benefit of an investment is a reduction in cost.

EXAMPLE 4.3

A mechanical engineer is considering building automated materials-handling equipment for a production line. On the one hand, the equipment would substantially reduce the manual labour currently required to move items from one part of the production process to the next. On the other hand, the equipment would consume energy, require insurance, and need periodic maintenance.

Alternative 1: Continue to use the current method. Yearly labour costs are $9200.

Alternative 2: Build automated materials-handling equipment with an expected service life of 10 years.

First cost	$15 000	
Labour	$3300	per year
Power	$400	per year
Maintenance	$2400	per year
Taxes and insurance	$300	per year

If the MARR is 9%, which alternative is better? Use a present worth comparison.

The investment of $15 000 can be viewed as yielding a positive cash flow of 2800 = 9200 − (3300 + 400 + 2400 + 300) per year in the form of a reduction in cost.

$$PW = -15\ 000 + [9200 - (3300 + 400 + 2400 + 300)](P/A,9\%,10)$$

$$= -15\ 000 + 2800(P/A,9\%,10)$$

$$= -15\ 000 + 2800(6.4176)$$

$$= 2969.44$$

The present worth of the cost savings is approximately $3000 greater than the $15 000 first cost. Therefore, Alternative 2 is worth implementing.■

4.4.2 Present Worth Comparisons for Mutually Exclusive Projects

It is very easy to use the present worth method to choose the best project among a set of mutually exclusive projects *when the service lives are the same*. One just computes the present worth of each project using the MARR. The project with the greatest present worth is the preferred project because it is the one with the greatest profit.

EXAMPLE 4.4

Fly-by-Night Aircraft must purchase a new lathe. It is considering four lathes, each of which has a life of 10 years with no scrap value.

Lathe	1	2	3	4
First cost	$100 000	$150 000	$200 000	$255 000
Annual savings	25 000	34 000	46 000	55 000

Given a MARR of 15%, which alternative should be taken?

The present worths are:

Lathe 1: $PW = -100\ 000 + 25\ 000(P/A,15\%,10)$

$\qquad\qquad = -100\ 000 + 25\ 000(5.0187) \cong 25\ 468$

Lathe 2: $PW = -150\ 000 + 34\ 000(P/A,15\%,10)$

$\qquad\qquad = -150\ 000 + 34\ 000(5.0187) \cong 20\ 636$

Lathe 3: $PW = -200\ 000 + 46\ 000(P/A,15\%,10)$

$\qquad\qquad = -200\ 000 + 46\ 000(5.0187) \cong 30\ 860$

Lathe 4: $PW = -255\ 000 + 55\ 000(P/A,15\%,10)$

$\qquad\qquad = -155\ 000 + 55\ 000(5.0187) \cong 21\ 029$

Lathe 3 has the greatest present worth, and is therefore the preferred alternative.∎

4.4.3 Annual Worth Comparisons

Annual worth comparisons are essentially the same as present worth comparisons, except that all disbursements and receipts are transformed to a uniform series at the MARR, rather than to the present worth. Any present worth *P* can be converted to an annuity *A* by the capital recovery factor $(A/P,i,N)$. Therefore, a comparison of two projects *that have the same life* by the present worth and annual worth methods will always indicate the same preferred alternative. Note that, although the method is called annual worth, the uniform series is not necessarily on a yearly basis.

Present worth comparisons make sense because they compare the worth today of each alternative, but annual worth comparisons can sometimes be more easily grasped mentally. For example, to say that operating an automobile over five years has a present cost of $20 000 is less meaningful than saying that it will cost about $5300 per year for each of the following five years.

NET VALUE 4.1

Car Payment Calculators

The internet offers websites that are useful when you are thinking of buying a car—you can learn more about different makes and models, optional features, prices, what's available (used or new) at which dealer, and financing information if a car is to be purchased and not leased. Major car manufacturers, financial services companies, and car information websites make it easy for customers to figure out their financing plans by offering web-based car payment calculators.

A typical monthly payment calculator determines how much a customer pays every month on the basis of the purchase price, down payment,

interest rate, and loan term. This is essentially an annuity calculation. An affordability calculator, on the other hand, gives a present worth (what price of a car you could afford to buy now) or a future worth (total amount of money that would be spent on a car after all payments are made including interest) based on down payment, monthly payment, interest rate, and loan term. The calculators are useful in making instant comparisons among different cars or car companies, studying what-if scenarios with various payment amounts or lengths of the loan, and determining budget limitations. Similar calculators are available for house mortgage payments.

Sometimes there is no clear justification for preferring either the present worth method or the annual worth method. Then it is reasonable to use the one that requires less conversion. For example, if most receipts or costs are given as annuities or gradients, one can more easily perform an annual worth comparison. Sometimes it can be useful to compare projects on the basis of future worths. See Close-Up 4.2.

CLOSE-UP 4.2   Future Worth

Sometimes it may be desirable to compare projects with the **future worth method**, on the basis of the future worth of each project. This is most likely to be true for cases where money is being saved for some future expense.

For example, two investment plans are being compared to see which accumulates more money for retirement. Plan A consists of a payment of €10 000 today and then €2000 per year over 20 years. Plan B is €3000 per year over 20 years. Interest for both plans is 10%. Rather than convert these cash flows to either present worth or annual worth, it is sensible to compare the future worths of the plans, since the actual euro value in 20 years has particular meaning.

$$FW_A = 10\,000(F/P,10\%,20) + 2000(F/A,10\%,20)$$
$$= 10\,000(6.7275) + 2000(57.275)$$
$$= 181\,825$$
$$FW_B = 3000(F/A,10\%,20)$$
$$= 3000(57.275)$$
$$= 171\,825$$

Plan A is the better choice. It will accumulate to €181 825 over the next 20 years.

EXAMPLE 4.5

Sweat University is considering two alternative types of bleachers for a new athletic stadium.

Alternative 1: Concrete bleachers. The first cost is $350 000. The expected life of the concrete bleachers is 90 years and the annual upkeep costs are $2500.

Alternative 2: Wooden bleachers on earth fill. The first cost of $200 000 consists of $100 000 for earth fill and $100 000 for the wooden bleachers. The annual painting costs are $5000. The wooden bleachers must be replaced every 30 years at a cost of $100 000. The earth fill will last the entire 90 years.

One of the two alternatives will be chosen. It is assumed that the receipts and other benefits of the stadium are the same for both construction methods. Therefore, the greatest net benefit is obtained by choosing the alternative with the lower cost. The university uses a MARR of 7%. Which of the two alternatives is better?

For this example, let us base the analysis on annual worth. Since both alternatives have a life of 90 years, we shall get the equivalent annual costs over 90 years for both at an interest rate of 7%.

Alternative 1: Concrete bleachers

The equivalent annual cost over the 90-year life span of the concrete bleachers is

$$AW = 350\ 000(A/P,7\%,90) + 2500$$

$$= 350\ 000(0.07016) + 2500$$

$$= 27\ 056 \text{ per year}$$

Alternative 2: Wooden bleachers on earth fill

The total annual costs can be broken into three components: AW_1 (for the earth fill), AW_2 (for the bleachers), and AW_3 (for the painting). The equivalent annual cost of the earth fill is

$$AW_1 = 100\ 000(A/P,7\%,90)$$

The equivalent annual cost of the bleachers is easy to determine. The first set of bleachers is put in at the start of the project, the second set at the end of 30 years, and the third set at the end of 60 years, but the cost of the bleachers is the same at each installation. Therefore, we need to get only the cost of the first installation.

$$AW_2 = 100\ 000(A/P,7\%,30)$$

The last expense is for annual painting:

$$AW_3 = 5000$$

The total equivalent annual cost for alternative 2, wooden bleachers on earth fill, is the sum of AW_1, AW_2, and AW_3:

$$AW = AW_1 + AW_2 + AW_3$$

$$= 100\ 000[(A/P,7\%,90) + (A/P,7\%,30)] + 5000$$

$$= 100\ 000(0.07016 + 0.08059) + 5000$$

$$= 20\ 075$$

The concrete bleachers have an equivalent annual cost of about $7000 more than the wooden ones. Therefore, the wooden bleachers are the better choice.■

4.4.4 Comparison of Alternatives With Unequal Lives

When making present worth comparisons, we must always use the same time period in order to take into account the full benefits and costs of each alternative. If the lives of the alternatives are not the same, we can transform them to equal lives with one of the following two methods:

1. Repeat the *service life* of each alternative to arrive at a common time period for all alternatives. Here we assume that each alternative can be repeated with the same costs and benefits in the future—an assumption known as **repeated lives**. Usually we use the *least common multiple* of the lives of the various alternatives. Sometimes it is convenient to assume that the lives of the various alternatives are repeated indefinitely. Note that the assumption of repeated lives may not be valid where it is reasonable to expect technological improvements.

2. Adopt a specified **study period**—a time period that is given for the analysis. To set an appropriate study period, a company will usually take into account the time of required service, or the length of time they can be relatively certain of their forecasts. The study period method necessitates an additional assumption about *salvage value* whenever the life of one of the alternatives exceeds that of the given study period. Arriving at a reliable estimate of salvage value may be difficult sometimes.

Because they rest on different assumptions, the repeated lives and the study period methods can lead to different conclusions when applied to a particular project choice.

EXAMPLE 4.6 (MODIFICATION OF EXAMPLE 4.3)

A mechanical engineer has decided to introduce automated materials-handling equipment for a production line. She must choose between two alternatives: building the equipment, or buying the equipment off the shelf. Each alternative has a different service life and a different set of costs.

Alternative 1: Build custom automated materials-handling equipment.

First cost	$15 000	
Labour	$3300	per year
Power	$400	per year
Maintenance	$2400	per year
Taxes and insurance	$300	per year
Service life	10	years

Alternative 2: Buy off-the-shelf standard automated materials-handling equipment.

First cost	$25 000	
Labour	$1450	per year
Power	$600	per year
Maintenance	$3075	per year
Taxes and insurance	$500	per year
Service life	15	years

If the MARR is 9%, which alternative is better?

The present worth of the custom system over its 10-year life is

$$PW(1) = -15\ 000 - (3300 + 400 + 2400 + 300)(P/A,9\%,10)$$

$$= -15\ 000 - 6400(6.4176)$$

$$\cong -56\ 073$$

The present worth of the off-the-shelf system over its 15-year life is:

$$PW(2) = -25\ 000 - (1450 + 600 + 3075 + 500)(P/A,9\%,15)$$

$$= -25\ 000 - 5625(8.0606)$$

$$\cong -70\ 341$$

The custom system has a lower cost for its 10-year life than the off-the-shelf system for its 15-year life, but it would be *wrong* to conclude from these calculations that the custom system should be preferred. The custom system yields benefits for only 10 years, whereas the off-the-shelf system lasts 15 years. It would be surprising if the cost of 15 years of benefits were not higher than the cost of 10 years of benefits. A fair comparison of the costs can be made only if equal lives are compared.

Let us apply the repeated lives method. If each alternative is repeated enough times, there will be a point in time where their service lives are simultaneously completed. This will happen first at the time equal to the least common multiple of the service lives. The least common multiple of 10 years and 15 years is 30 years. Alternative 1 will be repeated twice (after 10 years and after 20 years), while alternative 2 will be repeated once (after 15 years) during the 30-year period. At the end of 30 years, both alternatives will be completed simultaneously. See Figure 4.3.

Figure 4.3   Least Common Multiple of the Service Lives

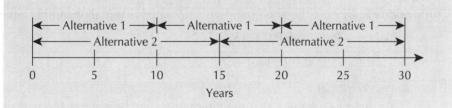

With the same time period of 30 years for both alternatives, we can now compare present worths.

Alternative 1: Build custom automated materials-handling equipment and repeat twice

$$PW(1) = -15\ 000 - 15\ 000(P/F,9\%,10) - 15\ 000(P/F,9\%,20)$$

$$- (3300 + 400 + 2400 + 300)(P/A,9\%,30)$$

$$= -15\ 000 - 15\ 000(0.42241) - 15\ 000(0.17843) - 6400\ (10.273)$$

$$\cong -89\ 760$$

Alternative 2: Buy off-the-shelf standard automated materials-handling equipment and repeat once

$$PW(2) = -25\ 000 - 25\ 000(P/F,9\%,15)$$

$$- (1450 + 600 + 3075 + 500)(P/A,9\%,30)$$

$$= -25\ 000 - 25\ 000(0.27454) - 5625(10.273)$$

$$\cong -89\ 649$$

Using the repeated lives method, we find little difference between the alternatives. An annual worth comparison can also be done over a period of time equal to the least common multiple of the service lives by multiplying each of these present worths by the capital recovery factor for 30 years.

$$AW(1) = -89\ 760(A/P,9\%,30)$$

$$= -89\ 760(0.09734)$$

$$\cong -8737$$

$$AW(2) = -89\ 649(A/P,9\%,30)$$

$$= -89\ 649(0.09734)$$

$$\cong -8726$$

As we would expect, there is again little difference in the annual cost between the alternatives. However, there is a more convenient approach for an annual worth comparison if it can be assumed that the alternatives are repeated indefinitely. Since the annual costs of an alternative remain the same no matter how many times it is repeated, it is not necessary to determine the least common multiple of the service lives. The annual worth of each alternative can be assessed for whatever time period is most convenient for each alternative.

Alternative 1: Build custom automated materials-handling equipment

$$AW(1) = -15\ 000(A/P,9\%,10) - 6400$$

$$= -15\ 000(0.15582) - 6400$$

$$\cong -8737$$

Alternative 2: Buy off-the-shelf standard automated materials-handling equipment

$$AW(2) = -25\ 000(A/P,9\%,15) - 5625$$

$$= -25\ 000(0.12406) - 5625$$

$$\cong -8726$$

If it cannot be assumed that the alternatives can be repeated to permit a calculation over the least common multiple of their service lives, then it is necessary to use the study period method.

Suppose that the given study period is 10 years, because the engineer is uncertain about costs past that time. The service life of the off-the-shelf system (15 years) is greater than the study period (10 years). Therefore, we have to make an assumption about the salvage value of the off-the-shelf system after 10 years. Suppose the engineer judges that its salvage value will be $5000. We can now proceed with the comparison.

Alternative 1: Build custom automated materials-handling equipment (10-year study period)

$$PW(1) = -15\ 000 - (3300 + 400 + 2400 + 300)(P/A,9\%,10)$$

$$= -15\ 000 - 6400(6.4176)$$

$$\cong -56\ 073$$

Alternative 2: Buy off-the-shelf standard automated materials-handling equipment (10-year study period)

$$PW(2) = -25\ 000 - (1450 + 600 + 3075 + 500)(P/A,9\%,10)$$

$$+ 5000(P/F,9\%,10)$$

$$= -25\,000 - 5625(6.4176) + 5000(0.42241)$$

$$\cong -58\,987$$

Using the study period method of comparison, alternative 1 has the smaller present worth of costs at $56 073 and is, therefore, preferred.

Note that here the study period method gives a different answer than the repeated lives method gives. The study period method is often sensitive to the chosen salvage value. A larger salvage value tends to make an alternative with a life longer than the study period more attractive, and a smaller value tends to make it less attractive.

In some instances, it may be difficult to arrive at a reliable estimate of salvage value. Given the sensitivity of the study period method to the salvage value estimate, the analyst may be uncertain about the validity of the results. One way of circumventing this problem is to avoid estimating the salvage value at the outset. Instead we calculate what salvage value would make the alternatives equal in value. Then we decide whether the actual salvage value will be above or below the break-even value found. Applying this approach to our example, we set $PW(1) = PW(2)$ so that

$$PW(1) = PW(2)$$

$$56.073 = -25\,000 - 5625(6.4176) + S(0.42241)$$

where S is the salvage value.

Solving for S, we find $S = 11\,834$. Is a reasonable estimate of the salvage value above or below $11 834? If it is above $11 834, then we conclude that the off-the-shelf system is the preferred choice. If it is below $11 834, then we conclude that the custom system is preferable.■

The study period can also be used for the annual worth method if the assumption of being able to indefinitely repeat the choice of alternatives is not justified.

EXAMPLE 4.7

Joan is renting a flat while on a one-year assignment in England. The flat does not have a refrigerator. She can rent one for a £100 deposit (returned in a year) and £15 per month (paid at the end of each month). Alternatively, she can buy a refrigerator for £300, which she would sell in a year when she leaves. For how much would Joan have to be able to sell the refrigerator in one year when she leaves, in order to be better off buying the refrigerator than renting one? Interest is at 6% nominal, compounded monthly.

Let S stand for the unknown salvage value (i.e., the amount Joan will be able to sell the refrigerator for in a year). We then equate the present worth of the rental alternative with the present worth of the purchase alternative for the one-year study period:

$PW(\text{rental}) = PW(\text{purchase})$

$-100 - 15(P/A,0.5\%,12) + 100(P/F,0.5\%,12) = -300 + S(P/F,0.5\%,12)$

$-100 - 15(11.616) + 100(0.94192) = -300 + S(0.94192)$

$S = 127.35$

If Joan can sell the used refrigerator for more than about £127 after one year's use, she is better off buying it rather than renting one.■

4.5 | Payback Period

The simplest method for judging the economic viability of projects is the payback period method. It is a rough measure of the time it takes for an investment to pay for itself. More precisely, the **payback period** is the number of years it takes for an investment to be recouped when the interest rate is assumed to be zero. When annual savings are constant, the payback period is usually calculated as follows:

$$\text{Payback period} = \frac{\text{First cost}}{\text{Annual savings}}$$

For example, if a first cost of $20 000 yielded a return of $8000 per year, then the payback period would be 20 000/8000 = 2.5 years.

If the annual savings are not constant, we can calculate the payback period by deducting each year of savings from the first cost until the first cost is recovered. The number of years required to pay back the initial investment is the payback period. For example, suppose the saving from a $20 000 first cost is $5000 the first year, increasing by $1000 each year thereafter. By adding the annual savings one year at a time, we see that it would take a just over three years to pay back the first cost (5000 + 6000 + 7000 + 8000 = 26 000). The payback period would then be stated as either four years (if we assume that the $8000 is received at the end of the fourth year) or 3.25 years (if we assume that the $8000 is received uniformly over the fourth year).

According to the payback period method of comparison, the project with the shorter payback period is the preferred investment. A company may have a policy of rejecting projects for which the payback period exceeds some preset number of years. The length of the maximum payback period depends on the type of project and the company's financial situation. If the company expects a cash constraint in the near future, or if a project's returns are highly uncertain after more than a few periods, the company will set a maximum payback period that is relatively short. As a common rule, a payback period of two years is often considered acceptable, while one of more than four years is unacceptable. Accordingly, government grant programs often target projects with payback periods of between two and four years on the rationale that in this range the grant can justify economically feasible projects that a company with limited cash flow would otherwise be unwilling to undertake.

The payback period need not, and perhaps should not, be used as the sole criterion for evaluating projects. It is a rough method of comparison and possesses some glaring weaknesses (as we shall discuss after Examples 4.8 and 4.9). Nevertheless, the payback period method can be used effectively as a preliminary filter. All projects with paybacks within the minimum would then be evaluated, using either rate of return methods (Chapter 5) or present/annual worth methods.

EXAMPLE 4.8

Elyse runs a second-hand book business out of her home where she advertises and sells the books over the internet. Her small business is becoming quite successful and she is considering purchasing an upgrade to her computer system that will give her more reliable uptime. The cost is $5000. She expects that the investment will bring about an annual savings of $2000, due to the fact that her system will no longer suffer long failures and thus she will be able to sell more books. What is the payback period on her investment, assuming that the savings accrue over the whole year?

$$\text{Payback period} = \frac{\text{First cost}}{\text{Annual savings}} = \frac{5000}{2000} = 2.5 \text{ years} \blacksquare$$

EXAMPLE 4.9

Pizza-in-a-Hurry operates a pizza delivery service to its customers with two eight-year-old vehicles, both of which are large, consume a great deal of gas and are starting to cost a lot to repair. The owner, Ray, is thinking of replacing one of the cars with a smaller, three-year-old car that his sister-in-law is selling for $8000. Ray figures he can save $3000, $2000, and $1500 per year for the next three years and $1000 per year for the following two years by purchasing the smaller car. What is the payback period for this decision?

The payback period is the number of years of savings required to pay back the initial cost. After three years, $3000 + $2000 + $1500 = $6500 has been paid back, and this amount is $7500 after four years and $8500 after five years. The payback period would be stated as five years if the savings are assumed to occur at the end of each year, or 4.5 years if the savings accrue continuously throughout the year.∎

The payback period method has four main advantages:

1. It is very easy to understand. One of the goals of engineering decision making is to communicate the reasons for a decision to managers or clients with a variety of backgrounds. The reasons behind the payback period and its conclusions are very easy to explain.

2. The payback period is very easy to calculate. It can usually be done without even using a calculator, so projects can be very quickly assessed.

3. It accounts for the need to recover capital quickly. Cash flow is almost always a problem for small- to medium-sized companies. Even large companies sometimes can't tie up their money in long-term projects.

4. The future is unknown. The future benefits from an investment may be estimated imprecisely. It may not make much sense to use precise methods like present worth on numbers that are imprecise to start with. A simple method like the payback period may be good enough for most purposes.

But the payback period method has three important disadvantages:

1. It discriminates against long-term projects. No houses or highways would ever be built if they had to pay themselves off in two years.

2. It ignores the effect of the timing of cash flows within the payback period. It disregards interest rates and takes no account of the time value of money. (Occasionally, a discounted payback period is used to overcome this disadvantage. See Close-Up 4.3.)

3. It ignores the expected service life. It disregards the benefits that accrue after the end of the payback period.

CLOSE-UP 4.3   Discounted Payback Period

In a discounted payback period calculation, the present worth of each year's savings is subtracted from the first cost until the first cost is diminished to zero. The number of years of savings required to do this is the discounted payback period. The main disadvantages of using a discounted payback period include the more complicated calculations and the need for an interest rate.

For instance, in Example 4.8, Elyse had an investment of $5000 recouped by annual savings of $2000. If interest were at 10%, the present worth of savings would be:

Year	Present Worth	Cumulative
Year 1	$2000(P/F,10\%,1) = 2000(0.90909) = 1818$	1818
Year 2	$2000(P/F,10\%,2) = 2000(0.82645) = 1653$	3471
Year 3	$2000(P/F,10\%,3) = 2000(0.75131) = 1503$	4974
Year 4	$2000(P/F,10\%,4) = 2000(0.68301) = 1366$	6340

Thus the discounted payback period is over 3 years, compared with 2.5 years calculated for the standard payback period.

Example 4.10 illustrates how the payback period method can ignore future cash flows.

EXAMPLE 4.10

Self Defence Systems of Cape Town is going to upgrade its paper-shredding facility. The company has a choice between two models. Model 007, with a first cost of R500 000 and a service life of seven years, would save R100 000 per year. Model MX, with a first cost of R100 000 and an expected service life of 20 years, would save R15 000 per year. If the company's MARR is 8%, which model is the better buy?

Using payback period as the sole criterion:

    Model 007: Payback period $= 500\ 000/100\ 000 = 5$ years

    Model MX: Payback period $= 100\ 000/15\ 000 = 6.6$ years

    It appears that the 007 model is better.

Using annual worth:

    Model 007: AW $= -500\ 000(A/P,8\%,7) + 100\ 000 = 3965$

    Model MX: AW $= -100\ 000(A/P,8\%,20) + 15\ 000 = 4815$

    Here, Model MX is substantially better.

The difference in the results from the two comparison methods is that the payback period method has ignored the benefits of the models that occur after the models have paid themselves off. This is illustrated in Figure 4.4. For Model MX, about 14 years of benefits have been omitted, whereas for model 007, only two years of benefits have been left out.■

Figure 4.4   Flows Ignored by the Payback Period

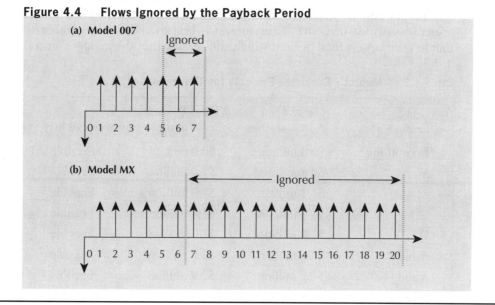

REVIEW PROBLEM 4.1

Tilson Dairies operates several cheese plants. The plants are all old and in need of renovation. Tilson's engineers have developed plans to renovate all the plants. Each project would have a positive present worth at the company's MARR. Tilson has $3.5 million available to invest in these projects. The following facts about the potential renovation projects are available:

Project	First Cost	Present Worth
A: Renovate plant 1	$0.8 million	$1.1 million
B: Renovate plant 2	$1.2 million	$1.7 million
C: Renovate plant 3	$1.4 million	$1.8 million
D: Renovate plant 4	$2.0 million	$2.7 million

Which projects should Tilson accept?

ANSWER

Table 4.2 shows the possible mutually exclusive projects that Tilson can consider.

Tilson should accept projects A, B, and C. They have a combined present worth of $4.6 million. Other feasible combinations that come close to using all available funds are B and D with a total present worth of $4.4 million, and C and D with a total present worth of $4.5 million.

Note that it is not necessary to consider explicitly the "leftovers" of the $3.5 million budget when comparing the present worths. The assumption is that any leftover part of the budget will be invested and provide interest at the MARR, resulting in a zero present worth for that part. Therefore, it is best to choose the combination of projects that has the largest total present worth and stays within the budget constraint.∎

Table 4.2 **Mutually Exclusive Projects for Tilson Dairies**

Project	Total First Cost	Total Present Worth	Feasibility
Do nothing	$0.0 million	$0.0 million	Feasible
A	$0.8 million	$1.1 million	Feasible
B	$1.2 million	$1.7 million	Feasible
C	$1.4 million	$1.8 million	Feasible
D	$2.0 million	$2.7 million	Feasible
A and B	$2.0 million	$2.8 million	Feasible
A and C	$2.2 million	$2.9 million	Feasible
A and D	$2.8 million	$3.8 million	Feasible
B and C	$2.6 million	$3.5 million	Feasible
B and D	$3.2 million	$4.4 million	Feasible
C and D	$3.4 million	$4.5 million	Feasible
A, B, and C	$3.4 million	$4.6 million	Feasible
A, B, and D	$4.0 million	$5.5 million	Not feasible
A, C, and D	$4.2 million	$5.6 million	Not feasible
B, C, and D	$4.6 million	$6.2 million	Not feasible
A, B, C, and D	$5.4 million	$7.3 million	Not feasible

REVIEW PROBLEM 4.2

City engineers are considering two plans for municipal aqueduct tunnels. They are to decide between the two, using an interest rate of 8%.

Plan A is a full-capacity tunnel that will meet the needs of the city forever. Its cost is $3 000 000 now, and $100 000 every 10 years for lining repairs.

Plan B involves building a half-capacity tunnel now and a second half-capacity tunnel in 20 years, when the extra capacity will be needed. Each of the half-capacity tunnels costs $2 000 000. Maintenance costs for each tunnel are $80 000 every 10 years. There is also an additional $15 000 per tunnel per year required to pay for extra pumping costs caused by greater friction in the smaller tunnels.

(a) Which alternative is preferred? Use a present worth comparison.

(b) Which alternative is preferred? Use an annual worth comparison.

ANSWER

(a) *Plan A: Full-Capacity Tunnel*

First, the $100 000 paid at the end of 10 years can be thought of as a future amount which has an equivalent annuity.

$$AW = 100\ 000(A/F,8\%,10) = 100\ 000(0.06903) = 6903$$

Thus, at 8% interest, $100 000 every 10 years is equivalent to $6903 every year.

Since the tunnel will have (approximately) an infinite life, the present cost of the lining repairs can be found using the capitalized cost formula, giving a total cost of

PW(Plan A) = 3 000 000 + 6903/0.08 = 3 086 288

Plan B: Half-Capacity Tunnels

For the first tunnel, the equivalent annuity for the maintenance and pumping costs is

AW = 15 000 + 80 000(0.06903) = 20 522

The present cost is then found with the capitalized cost formula, giving a total cost of

PW_1 = 2 000 000 + 20 522/0.08 = 2 256 525

Now, for the second tunnel, basically the same calculation is used, except that the present worth calculated must be discounted by 20 years at 8%, since the second tunnel will be built 20 years in the future.

PW_2 = {2 000 000 + [15 000 + 80 000(0.06903)]/0.08}(*P/F*,8%,20)

= 2 256 525(0.21455) ≅ 484 137

PW(Plan B) = PW_1 + PW_2 = 2 740 662

Consequently, the two half-capacity aqueducts with a present worth of costs of $2 740 662 are economically preferable.

(b) *Plan A: Full-Capacity Tunnel*

First, the $100 000 paid at the end of 10 years can be thought of as a future amount that has an equivalent annuity of

AW = 100 000(*A/F*,8%,10) = 100 000(0.06903) = 6903

Thus, at 8% interest, $100 000 every 10 years is equivalent to $6903 every year.

Since the tunnel will have (approximately) an infinite life, an annuity equivalent to the initial cost can be found using the capitalized cost formula, giving a total annual cost of

AW(Plan A) = 3 000 000(0.08) + 6903 = 246 903

Plan B: Half-Capacity Tunnels

For the first tunnel, the equivalent annuity for the maintenance and pumping costs is

AW = 15 000 + 80 000(0.06903) ≅ 20 522

The annual equivalent of the initial cost is then found with the capitalized cost formula, giving a total cost of

AW_1 = 2 000 000(0.08) + 20 522 = 180 522

Now, for the second tunnel, basically the same calculation is used, except that the annuity must be discounted by 20 years at 8%, since the second tunnel will be built 20 years in the future.

AW_2 = AW_1(*P/F*,8%,20)

= 180 522(0.21455) ≅ 38 731

$$AW(\text{Plan B}) = AW_1 + AW_2$$

$$= 180\ 522 + 38\ 731 = 219\ 253$$

Consequently, the two half-capacity aqueducts with an annual worth of costs of $219 253 are economically preferable.■

REVIEW PROBLEM 4.3

Fernando Constantia, an engineer at Brandy River Vineyards, has a $100 000 budget for winery improvements. He has identified four mutually exclusive investments, all of five years' duration, which have the cash flows shown in Table 4.3. For each alternative, he wants to determine the payback period and the present worth. For his recommendation report, he will order the alternatives from most preferred to least preferred in each case. Brandy River uses an 8% MARR for such decisions.

Table 4.3   Cash Flows for Review Problem 4.3

| Alternative | Cash Flow at the End of Each Year | | | | | |
	0	1	2	3	4	5
A	−$100 000	$25 000	$25 000	$25 000	$25 000	$ 25 000
B	−100 000	5000	10 000	20 000	40 000	80 000
C	−100 000	50 000	50 000	10 000	0	0
D	−100 000	0	0	0	0	1 000 000

ANSWER

The payback period can be found by decrementing yearly. The payback periods for the alternatives are then

A: 4 years
B: 4.3125 or 5 years
C: 2 years
D: 4.1 or 5 years

The order of the alternatives from most preferred to least preferred using the payback period method with yearly decrementing is: C, A, D, B. The present worth computations for each alternative are:

A: $PW = -100\ 000 + 25\ 000(P/A,8\%,5)$

$= -100\ 000 + 25\ 000(3.9926)$

$= -185$

B: $PW = -100\ 000 + 5000(P/F,8\%,1) + 10\ 000(P/F,8\%,2)$

$+ 20\ 000(P/F,8\%,3) + 40\ 000(P/F,8\%,4) + 80\ 000(P/F,8\%,5)$

$= -100\ 000 + 5000(0.92593) + 10\ 000(0.85734)$

$+ 20\ 000(0.79383) + 40\ 000(0.73503) + 80\ 000(0.68059)$

$= 12\ 982$

C: $PW = -100\ 000 + 50\ 000(P/F,8\%,1) + 50\ 000(P/F,8\%,2)$

$+\ 10\ 000(P/F,8\%,3)$

$= -100\ 000 + 50\ 000(0.92593) + 50\ 000(0.85734)$

$+\ 10\ 000(0.79283)$

$= -2908$

D: $PW = -100\ 000 + 1\ 000\ 000(P/F,8\%,5)$

$= -100\ 000 + 1\ 000\ 000(0.68059)$

$= 580\ 590$

The order of the alternatives from most preferred to least preferred using the present worth method is: D, B, A, C.■

SUMMARY

This chapter discussed relations among projects, and the present worth, annual worth, and payback period methods for evaluating projects. There are three classes of relations among projects, (1) independent, (2) mutually exclusive, and (3) related but not mutually exclusive. We then showed how the third class of projects, those that are related but not mutually exclusive, could be combined into sets of mutually exclusive projects. This enabled us to limit the discussion to the first two classes, independent and mutually exclusive. Independent projects are considered one at a time and are either accepted or rejected. Only the best of a set of mutually exclusive projects is chosen.

The present worth method compares projects on the basis of converting all cash flows for the project to a present worth. An independent project is acceptable if its present worth is greater than zero. The mutually exclusive project with the highest present worth should be taken. Projects with unequal lives must be compared by assuming that the projects are repeated or by specifying a study period. Annual worth is similar to present worth, except that the cash flows are converted to a uniform series. The annual worth method may be more meaningful, and also does not require more complicated calculations when the projects have different service lives.

The payback period is a simple method that calculates the length of time it takes to pay back an initial investment. It is inaccurate but very easy to calculate.

Engineering Economics in Action, Part 4B:
Doing It Right

Naomi stopped for coffee on her way back from Clem's office. She needed time to think about how to decide which potential forge shop investments were best. She wasn't sure that she knew what "best" meant. She got down her engineering economics text and looked at the table of contents. There were a couple of chapters on comparison methods that seemed to be what she wanted. She sat down with the coffee in her right hand and the text on her lap, and hoped for an uninterrupted hour.

One read through the chapters was enough to remind Naomi of the main relevant ideas that she had learned in school. The first thing she had to do was decide whether the investments were independent or not. They clearly were not independent. It would not make sense to refurbish the current forging hammer and replace it with a

→

mechanical press. Where potential investments were not independent, it was easiest to form mutually exclusive combinations as investment options. Naomi came up with seven options. She ranked the options by first cost, starting with the one with the lowest cost:

1. Refurbish the current machine.
2. Refurbish the current machine plus replace the materials-handling equipment.
3. Buy a manually operated mechanical press.
4. Buy a manual mechanical press plus replace the materials-handling equipment.
5. Buy an automated mechanical press.
6. Buy an automated mechanical press plus replace the materials-handling equipment.
7. Buy an automated mechanical press plus integrate it with the materials-handling equipment.

At this point, Naomi wasn't sure what to do next. There were different ways of comparing the options.

Naomi wanted a break from thinking about theory. She decided to take a look at Dave Sullivan's work. She started up her computer and opened up Dave's email. In it Dave apologized for dumping the work on her and invited Naomi to call him in Florida if she needed help. Naomi decided to call him. The phone was answered by Dave's wife, Helena. After telling Naomi that her father was out of intensive care and was in good spirits, Helena turned the phone over to Dave.

"Hi, Naomi. How's it going?"

"Well, I'm trying to finish off the forge project that you started. And I'm taking you up on your offer to consult."

"You have my attention. What's the problem?"

"Well, I've gotten started. I have formed seven mutually exclusive combinations of potential investments." Naomi went on to explain her selection of alternatives.

"That sounds right, Naomi. I like the way you've organized that. Now, how are you going to make the choice?"

"I've just reread the present worth, annual worth, and payback period stuff, and of those three, present worth makes the most sense to me. I can just compare the present worths of the cash flows for each alternative, and the one whose present worth is highest is the best one. Annual worth is the same, but I don't see any good reason in this case to look at the costs on an annual basis."

"What about internal rate of return?"

"Well, actually, Dave, I haven't reviewed IRR yet. I'll need it, will I?"

"You will. Have a look at it, and also remember that your recommendation is for Burns and Kulkowski. Think about how they will be looking at your information."

"Thanks, Dave. I appreciate your help."

"No problem. This first one is important for you; let's make sure we do it right."

PROBLEMS

For additional practice, please see the problems (with selected solutions) provided on the Student CD-ROM that accompanies this book.

4.1 IQ Computer assembles Unix workstations at its plant. The current product line is nearing the end of its marketing life, and it is time to start production of one or more new products. The data for several candidates are shown below.

The maximum budget for research and development is $300 000. A minimum of $200 000 should be spent on these projects. It is desirable to spread out the introduction of new products, so if two products are to be developed together, they should have different lead times. Resource draw refers to the labour and space that are available to the new products; it cannot exceed 100%.

	Potential Product			
	A	**B**	**C**	**D**
Research and development costs	$120 000	$60 000	$150 000	$75 000
Lead time	1 year	2 years	1 year	2 years
Resource draw	60%	50%	40%	30%

On the basis of the above information, determine the set of feasible mutually exclusive alternative projects that IQ Computers should consider.

4.2 The Alabaster Marble Company (AM) is considering opening three new quarries. One, designated T, is in Tusksarelooser County; a second, L, is in Lefant County; the third, M, is in Marxbro County. Marble is shipped mainly within a 500-kilometre range of its quarry because of its weight. The market within this range is limited. The returns that AM can expect from any of the quarries depends on how many quarries AM opens. Therefore, these potential projects are related.

(a) Construct a set of mutually exclusive alternatives from these three potential projects.

(b) The Lefant County quarry has very rich deposits of marble. This makes the purchase of mechanized cutter-loaders a reasonable investment at this quarry. Such loaders would not be considered at the other quarries. Construct a set of mutually exclusive alternatives from the set of quarry projects augmented by the potential mechanized cutter-loader project.

(c) AM has decided to invest no more than $2.5 million in the potential quarries. The first costs are as follows:

Project	First Cost
T quarry	$0.9 million
L quarry	1.4 million
M quarry	1.0 million
Cutter-loader	0.4 million

Construct a set of mutually exclusive alternatives that are feasible, given the investment limitation.

4.3 Angus Automotive has $100 000 to invest in internal projects. The choices are:

Project	Cost
1. Line improvements	$20 000
2. New manual tester	30 000
3. New automatic tester	60 000
4. Overhauling press	50 000

Only one tester may be bought and the press will not need overhauling if the line improvements are not made. What mutually exclusive project combinations are available if Angus Auto will invest in at least one project?

4.4 The intersection of King and Main Streets needs widening and improvement. The possibilities include

1. Widen King

2. Widen Main

3. Add a left-turn lane on King

4. Add a left-turn lane on Main

5. Add traffic lights at the intersection

6. Add traffic lights at the intersection with advanced green for Main

7. Add traffic lights at the intersection with advanced green for King

A left-turn lane can be installed only if the street in question is widened. A left-turn lane is necessary if the street has traffic lights with an advanced green. The city cannot afford to widen both streets. How many mutually exclusive projects are there?

4.5 Yun is deciding among a number of business opportunities. She can

(a) Sell the X division of her company, Yunco

(b) Buy Barzoo's company, Barco

(c) Get new financing

(d) Expand into Tasmania

There is no sense in getting new financing unless she is either buying Barco or expanding into Tasmania. She can only buy Barco if she gets financing or sells the X division. She can only expand into Tasmania if she has purchased Barco. The X division is necessary to compete in the Tasmania. What are the feasible projects she should consider?

4.6 Nottawasaga Printing has four printing lines, each of which consists of three printing stations, A, B, and C. They have allocated $20 000 for upgrading the printing stations. Station A costs $7000 and takes 10 days to upgrade. Station B costs $5000 and takes 5 days, and station C costs $3000 and takes 3 days. Due to the limited number of technicians, Nottawasaga can only upgrade one printing station at a time. That is, if they decide to upgrade two Bs, the total downtime will be 10 days. During the upgrading period, the downtime should not exceed 14 days in total. Also, at least two printing lines must be available at all times to satisfy the current customer demand. The entire line will not be available if any of the printing stations is turned off for upgrading. Nottawasaga Printing wants to know which line and which printing station to upgrade. Determine the feasible mutually exclusive combinations of lines and stations for Nottawasaga Printing.

4.7 Margaret has a project with a £28 000 first cost that returns £5000 per year over its 10-year life. It has a salvage value of £3000 at the end of 10 years. If the MARR is 15%,

(a) What is the present worth of this project?

(b) What is the annual worth of this project?

(c) What is the future worth of this project after 10 years?

(d) What is the payback period of this project?

(e) What is the discounted payback period for this project?

4.8 Appledale Dairy is considering upgrading an old ice-cream maker. Upgrading is available at two levels: moderate and extensive. Moderate upgrading costs $6500 now and yields annual savings of $3300 in the first year, $3000 in the second year, $2700 in the third year, and so on. Extensive upgrading costs $10 550 and saves $7600 in the first year. The savings then decrease by 20% each year thereafter. If the upgraded ice-cream maker will last for seven years, which upgrading option is better? Use a present worth comparison. Appledale's MARR is 8%.

4.9 Kiwidale Dairy is considering purchasing a new ice-cream maker. Two models, Smoothie and Creamy, are available and their information is given below.

(a) What is Kiwidale's MARR that makes the two alternatives equivalent? Use a present worth comparison.

	Smoothie	**Creamy**
First cost	$15 000	$36 000
Service life	12 years	12 years
Annual profit	$4200	$10 800
Annual operating cost	$1200	$3520
Salvage value	$2250	$5000

(b) It turned out that the service life of Smoothie was 14 years. Which alternative is better on the basis of the MARR computed in part (a)? Assume that each alternative can be repeated indefinitely.

4.10 Nabil is considering buying a house while he is at university. The house costs €100 000 today. Renting out part of the house and living in the rest over his five years at school will net, after expenses, €1000 per month. He estimates that he will sell the house after five years for €105 000. If Nabil's MARR is 18%, compounded monthly, should he buy the house?

4.11 A young software genius is selling the rights to a new video game he has developed. Two companies have offered him contracts. The first contract offers $10 000 at the end of each year for the next five years, and then $20 000 per year for the following 10 years. The second offers 10 payments, starting with $10 000 at the end of the first year, $13 000 at the end of the second, and so forth, increasing by $3000 each year (i.e., the tenth payment will be $10 000 + 9 × $3000). Assume the genius uses a MARR of 9%. Which contract should the young genius choose? Use a present worth comparison.

4.12 Sam is considering buying a new lawnmower. He has a choice between a "Lawn Guy" mower and a Bargain Joe's "Clip Job" mower. Sam has a MARR of 5%. The salvage value of each mower at the end of its service life is zero.

(a) Using the information on the next page, determine which alternative is preferable. Use a present worth comparison and the least common multiple of the service lives.

(b) For a four-year study period, what salvage value for the Lawn Guy mower would result in its being the preferred choice? What salvage value for the Lawn Guy would result in the Clip Job being the preferred choice?

	Lawn Guy	Clip Job
First cost	$350	$120
Life	10 years	4 years
Annual gas	$60	$40
Annual maintenance	$30	$60

4.13 Water supply for an irrigation system can be obtained from a stream in some nearby mountains. Two alternatives are being considered, both of which have essentially infinite lives, provided proper maintenance is performed. The first is a concrete reservoir with a steel pipe system and the second is an earthen dam with a wooden aqueduct. Below are the costs associated with each.

Compare the present worths of the two alternatives, using an interest rate of 8%. Which alternative should be chosen?

	Concrete Reservoir	Earthen Dam
First cost	$500 000	$200 000
Annual maintenance costs	$2000	$12 000
Replacing the wood portion of the aqueduct each 15 years	N/A	$100 000

4.14 CB Electronix needs to expand its capacity. It has two feasible alternatives under consideration. Both alternatives will have essentially infinite lives.

Alternative 1: Construct a new building of 20 000 square metres now. The first cost will be $2 000 000. Annual maintenance costs will be $10 000. In addition, the building will need to be painted every 15 years (starting in 15 years) at a cost of $15 000.

Alternative 2: Construct a new building of 12 500 square metres now and an additional 7500 square metres in 10 years. The first cost of the 12 500-square-metre building will be $1 250 000. The annual maintenance costs will be $5000 for the first 10 years (i.e., until the addition is built). The 7500-square-metre addition will have a first cost of $1 000 000. Annual maintenance costs of the renovated building (the original building and the addition) will be $11 000. The renovated building will cost $15 000 to repaint every 15 years (starting 15 years after the addition is done).

Carry out an annual worth comparison of the two alternatives. Which is preferred if the MARR is 15%?

4.15 Katie's project has a five-year term, a first cost, no salvage value, and annual savings of $20 000 per year. After doing present worth and annual worth calculations with a 15% interest rate, Katie notices that the calculated annual worth for the project is exactly three times the present worth. What is the project's present worth and annual worth? Should Katie undertake the project?

4.16 Nighhigh Newsagent wants to replace its cash register and is currently evaluating two models that seem reasonable. The information on the two alternatives, CR1000 and CRX, is shown in the table.

	CR1000	**CRX**
First cost	£680	£1100
Annual savings	£245	£440
Annual maintenance cost	£35 in year 1, increasing by £10 each year thereafter	£60
Service life	4 years	6 years
Scrap value	£100	£250

(a) If Nighhigh Newsagent's MARR is 10%, which type of cash register should they choose? Use the present worth method.

(b) For the less preferred type of cash register found in part (a), what scrap value would make it the preferred choice?

4.17 Midland Metalworking is examining a 750-tonne hydraulic press and a 600-tonne moulding press for purchase. Midland has only enough budget for one of them. If Midland's MARR is 12% and the relevant information is as given below, which press should they purchase? Use an annual worth comparison.

	Hydraulic Press	**Moulding Press**
Initial cost	$275 000	$185 000
Annual savings	$33 000	$24 500
Annual maintenance cost	$2000, increasing by 15% each year thereafter	$1000, increasing by $350 each year thereafter
Life	15 years	10 years
Salvage value	$19 250	$14 800

4.18 Westmount Waxworks is considering buying a new wax melter for its line of replicas of statues of government leaders. There are two choices of supplier, Finedetail and Simplicity. Their proposals are as follows:

	Finedetail	**Simplicity**
Expected life	7 years	10 years
First cost	$200 000	$350 000
Maintenance	$10 000/year + $0.05/unit	$20 000/year + $0.01/unit
Labour	$1.25/unit	$0.50/unit
Other costs	$6500/year + $0.95/unit	$15 500/year + $0.55/unit
Salvage value	$5000	$20 000

Management thinks they will sell about 30 000 replicas per year if there is stability in world governments. If the world becomes very unsettled so that there are frequent overturns of governments, sales may be as high as 200 000 units a year. Westmount Waxworks uses a MARR of 15% for equipment projects.

(a) Who is the preferred supplier if sales are 30 000 units per year? Use an annual worth comparison.

(b) Who is the preferred supplier if sales are 200 000 units per year? Use an annual worth comparison.

(c) How sensitive is the choice of supplier to sales level? Experiment with sales levels between 30 000 and 200 000 units per year. At what sales level will the costs of the two melters be equal?

4.19 The City of Brussels is installing a new swimming pool in the municipal recreation centre. Two designs are under consideration, both of which are to be permanent (i.e., lasting forever). The first design is for a reinforced concrete pool which has a first cost of €1 500 000. Every 10 years the inner surface of the pool would have to be refinished and painted at a cost of €200 000.

The second design consists of a metal frame and a plastic liner, which would have an initial cost of €500 000. For this alternative, the plastic liner must be replaced every 5 years at a cost of €100 000, and every 15 years the metal frame would need replacement at a cost of €150 000. Extra insurance of €5000 per year is required for the plastic liner (to cover repair costs if the liner leaks). The city's cost of long-term funds is 5%.

Determine which swimming pool design has the lower present cost.

4.20 Sam is buying a refrigerator. He has two choices. A used one, at $475, should last him about three years. A new one, at $1250, would likely last eight years. Both have a scrap value of $0. The interest rate is 8%.

(a) Which refrigerator has a lower cost? (Use a present worth analysis with repeated lives. Assume operating costs are the same.)

(b) If Sam knew that he could resell the new refrigerator after three years for $1000, would this change the answer in part (a)? (Use a present worth analysis with a three-year study period. Assume operating costs are the same.)

4.21 Val is considering purchasing a new video plasma display panel to use with her notebook computer. One model, the XJ3, costs $4500 new, while another, the Y19, sells for $3200. Val figures that the XJ3 will last about three years, at which point it could be sold for $1000, while the Y19 will last for only two years and will also sell for $1000. Both panels give similar service, except that the Y19 is not suitable for client presentations. If she buys the Y19, about four times a year she will have to rent one similar to the XJ3, at a total year-end cost of about $300. Using present worth and the least common multiple of the service lives, determine which display panel Val should buy. Val's MARR is 10%.

4.22 For Problem 4.21, Val has determined that the salvage value of the XJ3 after two years of service is $1900. Which display panel is the better choice, on the basis of present worth with a two-year study period?

4.23 Tom is considering purchasing a £24 000 car. After five years, he will be able to sell the vehicle for £8000. Petrol costs will be £2000 per year, insurance £600 per year, and parking £600 per year. Maintenance costs for the first year will be £1000, rising by £400 per year thereafter.

The alternative is for Tom to take taxis everywhere. This will cost an estimated £6000 per year. Tom will rent a vehicle each year at a total cost (to year-end) of £600 for the family vacation, if he has no car. If Tom values money at 11% annual interest, should he buy the car? Use an annual worth comparison method.

4.24 A new gizmo costs R10 000. Maintenance costs R2000 per year, and labour savings are R6567 per year. What is the gizmo's payback period?

4.25 Building a bridge will cost $65 million. A round-trip toll of $12 will be charged to all vehicles. Traffic projections are estimated to be 5000 per day. The operating and maintenance costs will be 20% of the toll revenue. Find the payback period (in years) for this project.

4.26 A new packaging machine will save Greene Cheese Pty. Ltd. $3000 per year in reduced spoilage, $2500 per year in labour, and $1000 per year in packaging material. The new machine will have additional expenses of $700 per year in maintenance and $200 per year in energy. If it costs $20 000 to purchase, what is its payback period? Assume that the savings are earned throughout the year, not just at year-end.

4.27 Diana usually uses a three-year payback period to determine if a project is acceptable. A recent project with uniform yearly savings over a five-year life had a payback period of almost exactly three years, so Diana decided to find the project's present worth to help determine if the project was truly justifiable. However, that calculation didn't help either since the present worth was exactly 0. What interest rate was Diana using to calculate the present worth? The project has no salvage value at the end of its five-year life.

4.28 The Biltmore Garage has lights in places that are difficult to reach. Management estimates that it costs about $2 to change a bulb. Standard 100-watt bulbs with an expected life of 1000 hours are now used. Standard bulbs cost $1. A long-life bulb that requires 90 watts for the same effective level of light is available. Long-life bulbs cost $3. The bulbs that are difficult to reach are in use for about 500 hours a month. Electricity costs $0.08/kilowatt-hour payable at the end of each month. Biltmore uses a 12% MARR (1% per month) for projects involving supplies.

  (a) What minimum life for the long-life bulb would make its cost lower?

  (b) If the cost of changing bulbs is ignored, what is the minimum life for the long-life bulb for them to have a lower cost?

  (c) If the solutions are obtained by linear interpolation of the capital recovery factor, will the approximations understate or overstate the required life?

4.29 A chemical recovery system costs 300 000 yuan and saves 52 800 yuan each year of its seven-year life. The salvage value is estimated at 75 000 yuan. The MARR is 9%. What is the net annual benefit or cost of purchasing the chemical recovery system? Use the capital recovery formula.

4.30 Savings of $5600 per year can be achieved through either a $14 000 machine (A) with a seven-year service life and a $2000 salvage value, or a $25 000 machine (B) with a ten-year service life and a $10 000 salvage value. If the MARR is 9%, which machine is a better choice, and for what annual benefit or cost? Use annual worth and the capital recovery formula.

4.31 Ridgley Custom Metal Products (RCMP) must purchase a new tube bender. RCMP's MARR is 11%. They are considering two models:

Model	First Cost	Economic Life	Yearly Net Savings	Salvage Value
T	$100 000	5 years	$50 000	$20 000
A	150 000	5 years	60 000	30 000

(a) Using the *present worth* method, which tube bender should they buy?

(b) RCMP has discovered a third alternative, which has been added to the table below. Now which tube bender should they buy?

Model	First Cost	Economic Life	Yearly Net Savings	Salvage Value
T	$100 000	5 years	$50 000	$ 20 000
A	150 000	5 years	60 000	30 000
X	200 000	3 years	75 000	100 000

4.32 RCMP (see Problem 4.31, part (b)) can forecast demand for its products for only three years in advance. The salvage value after three years is $40 000 for model T and $80 000 for model A. Using the study period method, which of the three alternatives is best?

4.33 Using the annual worth method, which of the three tube benders should RCMP buy? The MARR is 11%. Use the data from Problem 4.31, part (b).

4.34 What is the payback period for each of the three alternatives from the RCMP problem? Use the data from Problem 4.31, part (b).

4.35 Data for two independent investment opportunities are shown below.

	Machine A	Machine B
Initial cost	¥1 500 000	¥2 000 000
Revenues (annual)	¥ 900 000	¥1 100 000
Costs (annual)	¥ 600 000	¥ 800 000
Scrap value	¥ 100 000	¥ 200 000
Service life	5 years	10 years

(a) For a MARR of 8%, should either, both, or neither machine be purchased? Use the annual worth method.

(b) For a MARR of 8%, should either, both, or neither machine be purchased? Use the present worth method.

(c) What are the payback periods for these machines? Should either, both, or neither machine be purchased, based on the payback periods? The required payback period for investments of this type is three years.

4.36 Xaviera is comparing two mutually exclusive projects, A and B, that have the same initial investment and the same present worth over their service lives. Wolfgang points out that, using the annual worth method, A is clearly better than B. What can be said about the service lives for the two projects?

4.37 Xaviera noticed that two mutually exclusive projects, A and B, have the same payback period and the same economic life, but A has a larger present worth than B does. What can be said about the size of the annual savings for the two projects?

4.38 Two plans have been proposed for accumulating money for capital projects at Bobbin Bay Lighting. One idea is to put aside 100 000 rupees per year, independent of growth. The second is to start with a smaller amount, 80 000 rupees per year, but to increase this in proportion to the expected company growth. The money will accumulate interest at 10%, and the company is expected to grow about 5% per year. Which plan will accumulate more money in 10 years?

4.39 Cleanville Environmental Services is evaluating two alternative methods of disposing of municipal waste. The first involves developing a landfill site near the city. Costs of the site include $1 000 000 start-up costs, $100 000 closedown costs 30 years from now, and operating costs of $20 000 per year. Starting in 10 years, it is expected that there will be revenues from user fees of $30 000 per year. The alternative is to ship the waste out of the region. An area firm will agree to a long-term contract to dispose of the waste for $130 000 per year. Using the *annual worth* method, which alternative is economically preferred for a MARR of 11%? Would this likely be the actual preferred choice?

4.40 Alfredo Auto Parts is considering investing in a new forming line for grille assemblies. For a five-year study period, the cash flows for two separate designs are shown below. Create a spreadsheet that will calculate the present worths for each project for a variable MARR. Through trial and error, establish the MARR at which the present worths of the two projects are exactly the same.

	Cash Flows for Grille Assembly Project					
	Automated Line			**Manual Line**		
Year	**Disburse-ments**	**Receipts**	**Net Cash Flow**	**Disburse-ments**	**Receipts**	**Net Cash Flow**
0	€1 500 000	€ 0	– €1 500 000	€1 000 000	€ 0	– €1 000 000
1	50 000	300 000	250 000	20 000	200 000	180 000
2	60 000	300 000	240 000	25 000	200 000	175 000
3	70 000	300 000	230 000	30 000	200 000	170 000
4	80 000	300 000	220 000	35 000	200 000	165 000
5	90 000	800 000	710 000	40 000	200 000	160 000

4.41 Stayner Catering is considering setting up a temporary division to handle demand created by their city's special tourist promotion during the coming year. They will invest in tables, serving equipment and trucks for a one-year period. Labour is employed on a monthly basis. Warehouse space is rented monthly, and revenue is generated monthly. The items purchased will be sold at the end of the year, but the salvage values are somewhat uncertain. Given below are the known or expected cash flows for the project.

Month	Purchase	Labour Expenses	Warehouse Expenses	Revenue
January (beginning)	$200 000			
January (end)		$ 2000	$3000	$ 2000
February		2000	3000	2000
March		2000	3000	2000
April		2000	3000	2000
May		4000	3000	10 000
June		10 000	6000	40 000
July		10 000	6000	110 000
August		10 000	6000	60 000
September		4000	3000	30 000
October		2000	3000	10 000
November		2000	3000	5000
December	Salvage?	2000	3000	2000

For an interest rate of 12% compounded monthly, create a spreadsheet that calculates and graphs the present worth of the project for a range of salvage values of the purchased items from 0% to 100% of the purchase price. Should Stayner Catering go ahead with this project?

4.42 Alfredo Auto Parts has two options for increasing efficiency. They can expand the current building or keep the same building but remodel the inside layout. For a five-year study period, the cash flows for the two options are shown below. Construct a spreadsheet that will calculate the present worth for each option for a variable MARR. By trial and error, determine the MARR at which the present worths of the two options are equivalent.

	Expansion Option			Remodelling Option		
Year	Disbursements	Receipts	Net Cash Flow	Disbursements	Receipts	Net Cash Flow
0	€850 000	€ 0	– €850 000	€230 000	€ 0	– €230 000
1	25 000	200 000	175 000	9000	80 000	71 000
2	30 000	225 000	195 000	11 700	80 000	68 300
3	35 000	250 000	215 000	15 210	80 000	64 790
4	40 000	275 000	235 000	19 773	80 000	60 227
5	45 000	300 000	255 000	25 705	80 000	54 295

4.43 Derek has two choices for a heat-loss prevention system for the shipping doors at Kirkland Manufacturing. He can isolate the shipping department from the rest of the plant, or he can curtain off each shipping door separately. Isolation consists of building a permanent wall around the shipping area. It will cost $60 000 and will save $10 000 in heating costs per year. Plastic curtains around each shipping door will have a total cost of about $5000, but will have to be replaced about once every two years. Savings in heating costs for installing the curtains will be about $3000 per year. Use the payback period method to determine which alternative is better. Comment on the use of the payback period for making this decision.

4.44 Assuming that the wall built to isolate the shipping department in Problem 4.43 will last forever, and that the curtains have zero salvage value, compare the annual worths of the two alternatives. The MARR for Kirkland Manufacturing is 11%. Which alternative is better?

4.45 Cleanville Environmental Services is considering investing in a new water treatment system. On the basis of the information given below for two alternatives, a fully automated and a partially automated system, construct a spreadsheet for computing the annual worths for each alternative with a variable MARR. Through trial and error, determine the MARR at which the annual worths of the two alternatives are equivalent.

	Fully Automated System			Partially Automated System		
Year	Disburse-ments	Receipts	Net Cash Flow	Disburse-ments	Receipts	Net Cash Flow
0	$1 000 000	$ 0	– $1 000 000	$650 000	$ 0	– $650 000
1	30 000	300 000	270 000	30 000	220 000	190 000
2	30 000	300 000	270 000	30 000	220 000	190 000
3	80 000	300 000	220 000	35 000	220 000	185 000
4	30 000	300 000	270 000	35 000	220 000	185 000
5	30 000	300 000	270 000	40 000	220 000	180 000
6	80 000	300 000	220 000	40 000	220 000	180 000
7	30 000	300 000	270 000	45 000	220 000	175 000
8	30 000	300 000	270 000	45 000	220 000	175 000
9	80 000	300 000	220 000	50 000	220 000	170 000
10	30 000	300 000	270 000	50 000	220 000	170 000

More Challenging Problem

4.46 Fred has projects to consider for economic feasibility. All of his projects consist of a first cost P and annual savings A. His rule of thumb is to accept all projects for which the series present worth factor (for the appropriate MARR and service life) is equal to or greater than the payback period. Is this a sensible rule?

MINI-CASE 4.1

Rockwell International

The Light Vehicle Division of Rockwell International makes seat-slide assemblies for the automotive industry. They have two major classifications for investment opportunities: developing new products to be manufactured and sold, and developing new machines to improve production. The overall approach to assessing whether an investment should be made depends on the nature of the project.

In evaluating a new product, they consider the following:

1. *Marketing strategy:* Does it fit the business plan for the company?
2. *Work force:* How will it affect human resources?
3. *Margins:* The product should generate appropriate profits.
4. *Cash flow:* Positive cash flow is expected within two years.

In evaluating a new machine, they consider the following:

1. *Cash flow:* Positive cash flow is expected within a limited time period.
2. *Quality issues:* For issues of quality, justification is based on cost avoidance rather than positive cash flow.
3. *Cost avoidance:* Savings should pay back an investment within one year.

Discussion

All companies consider more than just the economics of a decision. Most take into account the other issues—often called *intangibles*—by using managerial judgment in an informal process. Others, like Rockwell International, explicitly consider a selection of intangible issues.

The trend today is to carefully consider several intangible issues, either implicitly or explicitly. Human resource issues are particularly important since employee enthusiasm and commitment have significant repercussions. Environmental impacts of a decision can affect the image of the company. Health and safety is another intangible with significant effects.

However, the economics of the decision is usually (but not always) the single most important factor in a decision. Also, economics is the factor that is usually the easiest to measure.

Questions

1. Why do you think Rockwell International has different issues to consider depending on whether an investment was a new product or a new machine?
2. For each of the issues mentioned, describe how it would be measured. How would you determine if it was worth investing in a new product or new machine with respect to that issue?
3. There are two kinds of errors that can be made. The first is that an investment is made when it should not be, and the second is that an investment is not made when it should be. Describe examples of both kinds of errors for both products and machines (four examples in total) if the issues listed for Rockwell International are strictly followed. What are some sensible ways to prevent such errors?

Appendix 4A The MARR and the Cost of Capital

For a business to survive, it must be able to earn a high enough return to induce investors to put money into the company. The minimum rate of return required to get investors to invest in a business is that business's **cost of capital**. A company's cost of capital is also its minimum acceptable rate of return for projects, its MARR. This appendix reviews how the cost of capital is determined. We first look at the relation between risk and the cost of capital. Then, we discuss sources of capital for large businesses and small businesses.

4A.1 | Risk and the Cost of Capital

There are two main forms of investment in a company, *debt* and *equity*. Investors in a company's debt are lending money to the company. The loans are contracts that give lenders rights to repayment of their loans, and to interest at predetermined interest rates. Investors in a company's equity are the owners of the company. They hold rights to the residual after all contractual payments, including those to lenders, are made.

Investing in equity is more risky than investing in debt. Equity owners are paid only if the company first meets its contractual obligations to lenders. This higher risk means that equity owners require an expectation of a greater return on average than the interest rate paid to debt holders. Consider a simple case in which a company has three possible performance levels—weak results, normal results, and strong results. Investors do not know which level will actually occur. Each level is equally probable. To keep the example simple, we assume that all after-tax income is paid to equity holders as dividends so that there is no growth. The data are shown in Table 4A.1.

We see that, no matter what happens, lenders will get a return of 10%:

$$0.1 = \frac{10\,000}{100\,000}$$

Table 4A.1   Cost of Capital Example

	Possible Performance Levels		
	Weak Results	Normal Results	Strong Results
Net operating income ($/year)[1]	40 000	100 000	160 000
Interest payment ($/year)	10 000	10 000	10 000
Net income before tax ($/year)	30 000	90 000	150 000
Tax at 40% ($/year)	12 000	36 000	60 000
After-tax income = Dividends ($/year)	18 000	54 000	90 000
Debt ($)	100 000	100 000	100 000
Value of shares ($)	327 273	327 273	327 273

[1]Net operating income per year is revenue per year minus cost per year.

Owners get one of three possible returns:

$$5.5\% \left(0.055 = \frac{18\,000}{327\,273}\right),$$

$$16.5\% \left(0.165 = \frac{54\,000}{327\,273}\right), \text{ or}$$

$$27.5\% \left(0.275 = \frac{90\,000}{327\,273}\right)$$

These three possibilities average out to 16.5%. If things are good, owners do better than lenders. If things are bad, owners do worse. But their average return is greater than returns to lenders.

The lower rate of return to lenders means that companies would like to get their capital with debt. However, reliance on debt is limited for two reasons.

1. If a company increases the share of capital from debt, it increases the chance that it will not be able to meet the contractual obligations to lenders. This means the company will be bankrupt. Bankruptcy may lead to reorganizing the company or possibly closing the company. In either case, bankruptcy costs may be high.

2. Lenders are aware of the dangers of high reliance on debt and will, therefore, limit the amount they lend to the company.

4A.2 | Company Size and Sources of Capital

Large, well-known companies can secure capital both by borrowing and by selling ownership shares with relative ease because there will be ready markets for their shares as well as any debt instruments, like bonds, they may issue. These companies will seek ratios of debt to equity that balance the marginal advantages and disadvantages of debt financing. Hence, the cost of capital for large, well-known companies is a weighted average of the costs of borrowing and of selling shares, which is referred to as the **weighted average cost of capital**. The weights are the fractions of total capital that come from the different sources. If market conditions do not change, a large company that seeks to raise a moderate amount of additional capital can do so at a stable cost of capital. This cost of capital is the company's MARR.

We can compute the after-tax cost of capital for the example shown in Table 4A.1 as follows.

Weighted average cost of capital

$$= 0.1\left(\frac{100\,000}{427\,273}\right) + 0.165\left(\frac{327\,273}{427\,273}\right) = 0.150$$

This company has a cost of capital of about 15%.

For smaller, less well-known companies, raising capital may be more difficult. Most investors in large companies are not willing to invest in unknown small companies. At start-up, a small company may rely entirely on the capital of the owners and their friends and relatives. Here the cost of capital is the opportunity cost for the investors.

If a new company seeks to grow more rapidly than the owners' investment plus cash flow permits, the next source of capital with the lowest cost is usually a bank loan. Bank loans are limited because banks are usually not willing to lend more than some fraction of the amount an owner(s) puts into a business.

If bank loans do not add up to sufficient funds, then the company usually has two options. One option is the sale of financial securities such as stocks and bonds through stock exchanges that specialize in small, speculative companies. Another option is venture capitalists. Venture capitalists are investors who specialize in investing in new companies. The cost of evaluating new companies is usually high and so is the risk of investing in them. Together, these factors usually lead venture capitalists to want to put enough money into a small company so that they will have enough control over the company.

In general, new equity investment is very expensive for small companies. Studies have shown that venture capitalists typically require the expectation of at least a 35% rate of return after tax. Raising funds on a stock exchange is usually even more expensive than getting funding from a venture capitalist.

CHAPTER **5**

Comparison Methods Part 2

Engineering Economics in Action, Part 5A: What's Best? Revisited

Review Problems
Summary

Engineering Economics in Action, Part 5B: The Invisible Hand

Problems

Mini-Case 5.1

Appendix 5A: Tests for Multiple IRRs

Clem had said, "I have to make a recommendation to Ed Burns and Anna Kulkowski for their Wednesday meeting on this forging hammer in the South Shop. I'd like you to handle it." Dave Sullivan, who had started the project, had gone to Florida to see his sick father-in-law. Naomi welcomed the opportunity, but she still had to figure out exactly what to recommend.

Naomi looked carefully at the list of seven mutually exclusive alternatives for replacing or refurbishing the machine. Present worth could tell her which of the seven was "best," but present worth was just one of several comparison methods. Which method should she use?

Dave would help more, if she asked him. In fact, he could no doubt tell her exactly what to do, if she wanted. But this one she knew she could handle, and it was a matter of pride to do it herself. Opening her engineering economics textbook, she read on.

5.1 | Introduction

In Chapter 4, we showed how to structure projects so that they were either independent or mutually exclusive. The present worth, annual worth, and payback period methods for evaluating projects were also introduced. This chapter continues on the theme of comparing projects by presenting a commonly used but somewhat more complicated comparison method called the *internal rate of return*, or IRR.

Although the IRR method is widely used, all of the comparison methods have value in particular circumstances. Selecting which method to use is also covered in this chapter. It is also shown that the present worth, annual worth, and IRR methods all result in the same recommendations for the same problem. We close this chapter with a chart summarizing the strengths and weaknesses of the four comparison methods presented in Chapters 4 and 5.

5.2 | The Internal Rate of Return

Investments are undertaken with the expectation of a return in the form of future earnings. One way to measure the return from an investment is as a rate of return per dollar invested, or in other words as an interest rate. The rate of return usually calculated for a project is known as the *internal rate of return (IRR)*. The adjective *internal* refers to the fact that the internal rate of return depends only on the cash flows due to the investment. The internal rate of return is that interest rate at which a project just breaks even. The meaning of the IRR is most easily seen with a simple example.

EXAMPLE 5.1

Suppose $100 is invested today in a project that returns $110 in one year. We can calculate the IRR by finding the interest rate at which $100 now is equivalent to $110 at the end of one year:

$$P = F(P/F, i^*, 1)$$

$$100 = 110(P/F, i^*, 1)$$

$$100 = \frac{110}{1 + i^*}$$

where i^* is the internal rate of return.

Solving this equation gives a rate of return of 10%. In a simple example like this, the process of finding an internal rate of return is finding the interest rate that makes the present worth of benefits equal to the first cost. This interest rate is the IRR.■

Of course, cash flows associated with a project will usually be more complicated than in the example above. A more formal definition of the IRR is stated as follows. The **internal rate of return (IRR)** on an investment is the interest rate, i^*, such that, when all cash flows associated with the project are discounted at i^*, the present worth of the cash inflows equals the present worth of the cash outflows. That is, the project just breaks even. An equation that expresses this is

$$\sum_{t=0}^{T} \frac{(R_t - D_t)}{(1 + i^*)^t} = 0 \tag{5.1}$$

where

R_t = the cash inflow (receipts) in period t
D_t = the cash outflow (disbursements) in period t
T = the number of time periods
i^* = the internal rate of return

Since Equation (5.1) can also be expressed as

$$\sum_{t=0}^{T} R_t(1 + i^*)^{-t} = \sum_{t=0}^{T} D_t(1 + i^*)^{-t}$$

it can be seen that, in order to calculate the IRR, one sets the disbursements equal to the receipts and solves for the unknown interest rate. For this to be done, the disbursements and receipts must be comparable, as a present worth, a uniform series, or a future worth. That is, use

PW(disbursements) = PW(receipts) and solve for the unknown i^*,
AW(disbursements) = AW(receipts) and solve for the unknown i^*, or
FW(disbursements) = FW(receipts) and solve for the unknown i^*.

The IRR is usually positive, but can be negative as well. A negative IRR means that the project is losing money rather than earning it.

We usually solve the equations for the IRR by trial and error, as there is no explicit means of solving Equation (5.1) for projects where the number of periods is large. A spreadsheet provides a quick way to perform trial-and-error calculations; most spreadsheet programs also include a built-in IRR function.

EXAMPLE 5.2

Clem is considering buying a tuxedo. It would cost $500, but would save him $160 per year in rental charges over its five-year life. What is the IRR for this investment?

As illustrated in Figure 5.1, Clem's initial cash outflow for the purchase would be $500. This is an up-front outlay relative to continuing to rent tuxedos. The investment would create a saving of $160 per year over the five-year life of the tuxedo. These savings can be viewed as a series of receipts relative to rentals. The IRR of Clem's investment can be found by determining what interest rate makes the present worth of the disbursements equal to the present worth of the cash inflows.

Present worth of disbursements = 500

Present worth of receipts $= 160(P/A,i^*,5)$

Setting the two equal,

$$500 = 160(P/A,i^*,5)$$

$$(P/A,i^*,5) = 500/160$$

$$= 3.125$$

Figure 5.1   Clem's Tuxedo

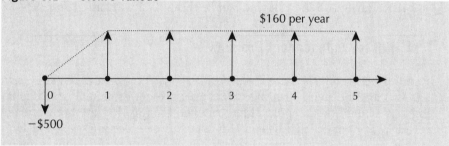

From the interest factor tables, we find that

$$(P/A,15\%,5) = 3.3521$$

$$(P/A,20\%,5) = 2.9906$$

Interpolating between $(P/A,15\%,5)$ and $(P/A,20\%,5)$ gives

$$i^* = 15\% + (5\%)[(3.125 - 3.3521)/(2.9906 - 3.3521)]$$

$$= 18.14\%$$

An alternative way to get the IRR for this problem is to convert all cash outflows and inflows to equivalent annuities over the five-year period. This will yield the same result as when present worth was used.

Annuity equivalent to the disbursements $= 500(A/P,i^*,5)$

Annuity equivalent to the receipts = 160

Again, setting the two equal,

$$500(A/P,i^*,5) = 160$$

$$(A/P,i^*,5) = 160/500$$

$$= 0.32$$

From the interest factor tables,

$$(A/P,15\%,5) = 0.29832$$

$$(A/P,20\%,5) = 0.33438$$

An interpolation gives

$$i^* = 15\% + 5\%[(0.32 - 0.29832)/(0.33438 - 0.29832)]$$

$$\cong 18.0\%$$

Note that there is a slight difference in the answers, depending on whether the disbursements and receipts were compared as present worths or as annuities. This difference is due to the small error induced by the linear interpolation.■

5.3 | Internal Rate of Return Comparisons

In this section, we show how the internal rate of return can be used to decide whether a project should be accepted. We first show how to use the IRR to evaluate independent projects. Then we show how to use the IRR to decide which of a group of mutually exclusive alternatives to accept. We then show that it is possible for a project to have more than one IRR. Finally, we show how to handle this difficulty by using an *external rate of return*.

5.3.1 IRR for Independent Projects

Recall from Chapter 4 that projects under consideration are evaluated using the MARR, and that any independent project that has a present or annual worth equal to or exceeding zero should be accepted. The principle for the IRR method is analogous. We will invest in any project that has an IRR equal to or exceeding the MARR. Just as projects with a zero present or annual worth are marginally acceptable, projects with IRR = MARR have a marginally acceptable rate of return (by definition of the MARR).

Also analogous to Chapter 4, when we perform a rate of return comparison on several independent projects, the projects must have equal lives. If this is not the case, then the approaches covered in Section 4.4.4 (Comparison of Alternatives With Unequal Lives) must be used.

EXAMPLE 5.3

The High Society Baked Bean Co. is considering a new canner. The canner costs $120 000, and will have a scrap value of $5000 after its 10-year life. Given the expected increases in sales, the total savings due to the new canner, compared with continuing with the current operation, will be $15 000 the first year, increasing by $5000 each year thereafter. Total extra costs due to the more complex equipment will be $10 000 per year. The MARR for High Society is 12%. Should they invest in the new canner?

The cash inflows and outflows for this problem are summarized in Figure 5.2. We need to compute the internal rate of return in order to decide if High Society should buy the canner. There are several ways we can do this. In this problem, equating annual outflows and receipts appears to be the easiest approach, because most of the cash flows are already stated on a yearly basis.

$$5000(A/F,i^*,10) + 15\,000 + 5000(A/G,i^*,10)$$
$$- 120\,000(A/P,i^*,10) - 10\,000 = 0$$

Dividing by 5000,

$$(A/F,i^*,10) + 1 + (A/G,i^*,10) - 24(A/P,i^*,10) = 0$$

Figure 5.2   High Society Baked Bean Canner

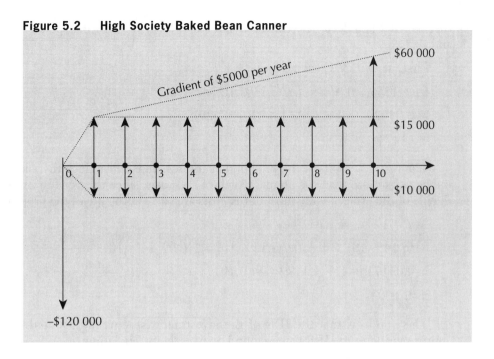

The IRR can be found by trial and error alone, by trial and error and linear interpolation, or by a spreadsheet IRR function. A trial-and-error process is particularly easy using a spreadsheet, so this is often the best approach. A good starting point for the process is at zero interest. A graph (Figure 5.3) derived from the spreadsheet indicates that the IRR is between 13% and 14%. This may be good enough for a decision, since it exceeds the MARR of 12%.

If finer precision is required, there are two ways to proceed. One way is to use a finer grid on the graph, for example, one that covers 13% to 14%. The other way is to interpolate between 13% and 14%. We shall first use the interest factor tables to show

Figure 5.3   Estimating the IRR for Example 5.3

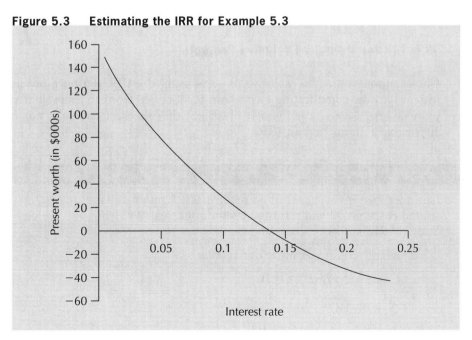

that the IRR is indeed between 13% and 14%. Next we shall interpolate between 13% and 14%.

First, at $i = 13\%$, we have

$(A/F,13\%,10) + 1 + (A/G,13\%,10) - 24(A/P,13\%,10)$

$= 0.05429 + 1 + 3.5161 - 24(0.18429)$

$\cong 0.1474$

The result is a bit too high. A higher interest rate will reduce the annual worth of the benefits more than the annual worth of the costs, since the benefits are spread over the life of the project while most of the costs are early in the life of the project.

At $i = 14\%$, we have

$(A/F,14\%,10) + 1 + (A/G,14\%,10) - 24(A/P,14\%,10)$

$= 0.05171 + 1 + 3.4489 - 24(0.19171)$

$\cong -0.1004$

This confirms that the IRR of the investment is between 13% and 14%. A good approximation to the IRR can be found by linearly interpolating:

$i^* = 13\% + (0 - 0.1474)/(0.1004 - 0.1474)$

$\cong 13.6\%$

The IRR for the investment is approximately 13.6%. Since this is greater than the MARR of 12%, the company should buy the new canner. Note again that it was not actually necessary to determine where in the range of 13% to 14% the IRR fell. It was enough to demonstrate that it was 12% or more.■

In summary, if there are several independent projects, the IRR for each is calculated separately, and those having an IRR equal to or exceeding the MARR should be chosen.

5.3.2 IRR for Mutually Exclusive Projects

Choice among mutually exclusive projects using the IRR is a bit more involved. Some insight into the complicating factors can be obtained from an example that involves two mutually exclusive alternatives. It illustrates that the best choice is not necessarily the alternative with the highest IRR.

EXAMPLE 5.4

Consider two investments. The first costs $1 today and returns $2 in one year. The second costs $1000 and returns $1900 in one year. Which is the preferred investment? Your MARR is 70%.

The first project has an IRR of 100%:

$-1 + 2(P/F,i^*,1) = 0$

$(P/F,i^*,1) = 1/2 = 0.5$

$i^* = 100\%$

The second has an IRR of 90%:

$$-1000 + 1900(P/F,i^*,1) = 0$$

$$(P/F,i^*,1) = 1000/1900 = 0.52631$$

$$i^* = 90\%$$

If these were independent projects, both would be acceptable since their IRRs exceed the MARR. If one of the two projects must be chosen, it might be tempting to choose the first project, the alternative with the larger rate of return. However, this approach is incorrect because it can overlook projects that have a rate of return equal to or greater than the MARR, but don't have the maximum IRR. In the example, the correct approach is to first observe that the least cost investment provides a rate of return that exceeds the MARR. The next step is to find the rate of return on the more expensive investment to see if the *incremental* investment has a rate of return equal to or exceeding the MARR. The incremental investment is the additional $999 that would be invested if the second investment was taken instead of the first:

$$-(1000 - 1) + (1900 - 2)(P/F,i^*,1) = 0$$

$$(P/F,i^*,1) = 999/1898 = 0.52634$$

$$i^* = 89.98\%$$

Indeed, the incremental investment has an IRR exceeding 70% and thus the second investment should be chosen.∎

The next example illustrates the process again, showing this time how an incremental investment is not justified.

EXAMPLE 5.5

Monster Meats can buy a new meat slicer system for €50 000. They estimate it will save them €11 000 per year in labour and operating costs. The same system with an automatic loader is €68 000, and will save approximately €14 000 per year. The life of either system is thought to be eight years. Monster Meats has three feasible alternatives:

Alternative	First Cost	Annual Savings
"Do nothing"	€ 0	€ 0
Meat slicer alone	50 000	11 000
Meat slicer with automatic loader	68 000	14 000

Monster Meats uses a MARR of 12% for this type of project. Which alternative is better?

We first consider the system without the loader. Its IRR is 14.5%, which exceeds the MARR of 12%. This can be seen by solving for i^* in

$$-50\ 000 + 11\ 000(P/A,i^*,8) = 0$$

$$(P/A,i^*,8) = 50\ 000/11\ 000$$

$$(P/A,i^*,8) = 4.545$$

From the interest factor tables, or by trial and error with a spreadsheet,

$$(P/A,14\%,8) = 4.6388$$
$$(P/A,15\%,8) = 4.4873$$

By interpolation or further trial and error,

$$i^* \cong 14.5\%$$

The slicer alone is thus economically justified and is better than the "do nothing" alternative.

We now consider the system with the slicer and loader. Its IRR is 12.5%, which may be seen by solving for i^* in

$$-68\ 000 + 14\ 000(P/A,i^*,8) = 0$$
$$(P/A,i^*,8) = 68\ 000/14\ 000$$
$$(P/A,i^*,8) = 4.857$$
$$(P/A,12\%,8) = 4.9676$$
$$(P/A,13\%,8) = 4.7987$$
$$i^* \cong 12.5\%$$

The IRR of the meat slicer and automatic loader is about 12.5%, which on the surface appears to meet the 12% MARR requirement. But, on the incremental investment, Monster Meats would be earning only 7%. This may be seen by looking at the IRR on the *extra*, or *incremental*, €18 000 spent on the loader.

$$-(68\ 000 - 50\ 000) + (14\ 000 - 11\ 000)(P/A,i^*,8) = 0$$
$$-18\ 000 + 3000(P/A,i^*,8) = 0$$
$$(P/A,i^*,8) = 18\ 000/3000$$
$$(P/A,i^*,8) = 6$$
$$i^* \cong 7\%$$

This is less than the MARR; therefore, Monster Meats should not buy the automated loader.

When the IRR was calculated for the system including the loader, the surplus return on investment earned by the slicer alone essentially subsidized the loader. The slicer investment made enough of a return so that, even when it was coupled with the money-losing loader, the whole machine still seemed to be a good buy. In fact, the extra €18 000 would be better spent on some other project at the MARR or higher. The relation between the potential projects is shown in Figure 5.4.■

Figure 5.4   Monster Meats

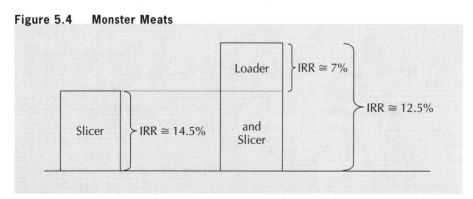

The fundamental principle illustrated by the two examples is that, to use the IRR to compare two or more mutually exclusive alternatives properly, we cannot make the decision on the basis of the IRRs of individual alternatives alone; we must take the IRRs of the *incremental* investments into consideration. In order to properly assess the worthiness of the incremental investments, it is necessary to have a systematic way to conduct pair-wise comparisons of projects. Note that before undertaking a systematic analysis of mutually exclusive alternatives with the IRR method, you should ensure that the alternatives have equal lives. If they do not have equal lives, then the methods of Section 4.4.4 (study period or repeated lives methods) must be applied first to set up comparable cash flows.

The first step in the process of comparing several mutually exclusive alternatives using the IRR is to order the alternatives from the smallest first cost to the largest first cost. Since one alternative must be chosen, accept the alternative with the smallest first cost (which may be the "do nothing" alternative with $0 first cost) as the *current best alternative* regardless of its IRR exceeding the MARR. This means that the current best alternative may have an IRR *less* than the MARR. Even if that's the case, a proper analysis of the IRRs of the incremental investments will lead us to the *correct* best overall alternative. For this reason, we don't have to check the IRR of any of the individual alternatives.

The second step of the analysis consists of looking at the incremental investments of alternatives that have a higher first cost than the current best alternative. Assume that there are *n* projects and they are ranked from 1 (the current best) to *n*, in increasing order of first costs. The current best is "challenged" by the project ranked second. One of two things occurs:

1. The incremental investment to implement the challenger does not have an IRR at least equal to the MARR. In this case, the challenger is excluded from further consideration and the current best is challenged by the project ranked third.

2. The incremental investment to implement the challenger has an IRR at least as high as the MARR. In this case, the challenger replaces the current best. It then is challenged by the alternative ranked third.

The process then continues with the next alternative challenging the current best until all alternatives have been compared. The current best alternative remaining at the end of the process is then selected as the best overall alternative. Figure 5.5 summarizes the incremental investment analysis for the mutually exclusive projects.

EXAMPLE 5.6 (REPRISE OF EXAMPLE 4.4)

Fly-by-Night Aircraft must purchase a new lathe. It is considering one of four new lathes, each of which has a life of 10 years with no scrap value. Given a MARR of 15%, which alternative should be chosen?

Lathe	1	2	3	4
First cost	$100 000	$150 000	$200 000	$255 000
Annual savings	25 000	34 000	46 000	55 000

The alternatives have already been ordered from lathe 1, which has the smallest first cost, to lathe 4, which has the greatest first cost. Since one lathe must be purchased, accept lathe 1 as the current best alternative. Calculating the IRR for lathe 1, although not necessary, is shown as follows:

$$100\ 000 = 25\ 000(P/A,i^*,10)$$

$$(P/A,i^*,10) = 4$$

An approximate IRR is obtained by trial and error with a spreadsheet.

$$i^* \cong 21.4\%$$

The current best alternative is then challenged by the first challenger, lathe 2, which has the next-highest first cost. The IRR of the incremental investment from lathe 1 to lathe 2 is calculated as follows:

$$(150\ 000 - 100\ 000) - (34\ 000 - 25\ 000)(P/A,i^*,10) = 0$$

or

$$[150\ 000 - 34\ 000(P/A,i^*,10)] - [100\ 000 - 25\ 000(P/A,i^*,10)] = 0$$

$$(P/A,i^*,10) = 50\ 000/9000 = 5.556$$

An approximate IRR is obtained by trial and error.

$$i^* \cong 12.4\%$$

Since the IRR of the incremental investment falls below the MARR, lathe 2 fails the challenge to become the current best alternative. The reader can verify that lathe 2 alone has an IRR of approximately 18.7%. Even so, lathe 2 is not considered a viable alternative. In other words, the incremental investment of $50 000 could be put to better use elsewhere. Lathe 1 remains the current best and the next challenger is lathe 3.

As before, the incremental IRR is the interest rate at which the present worth of lathe 3 less the present worth of lathe 1 is 0:

$$[200\ 000 - 46\ 000(P/A,i^*,10)] - [100\ 000 - 25\ 000(P/A,i^*,10)] = 0$$

$$(P/A,i^*,10) = 100\ 000/21\ 000 = 4.762$$

An approximate IRR is obtained by trial and error.

$$i^* \cong 16.4\%$$

The IRR on the incremental investment exceeds the MARR, and therefore lathe 3 is preferred to lathe 1. Lathe 3 now becomes the current best. The new challenger is lathe 4. The IRR on the incremental investment is

$$[255\ 000 - 55\ 000(P/A,i^*,10)] - [200\ 000 - 46\ 000(P/A,i^*,10)] = 0$$

$$(P/A,i^*,10) = 55\ 000/9000 = 6.11$$

$$i^* \cong 10.1\%$$

The additional investment from lathe 3 to lathe 4 is not justified. The reader can verify that the IRR of lathe 4 alone is about 17%. Once again, we have a challenger with an IRR greater than the MARR, but it fails as a challenger because the incremental investment from the current best does not have an IRR at least equal to the MARR. The current best remains lathe 3. There are no more challengers, and so the best overall alternative is lathe 3.■

In the next section, the issue of multiple IRRs is discussed, and methods for identifying and eliminating them are given. Note that the process described in Figure 5.5 requires that

Figure 5.5   Flowchart for Comparing Mutually Exclusive Alternatives

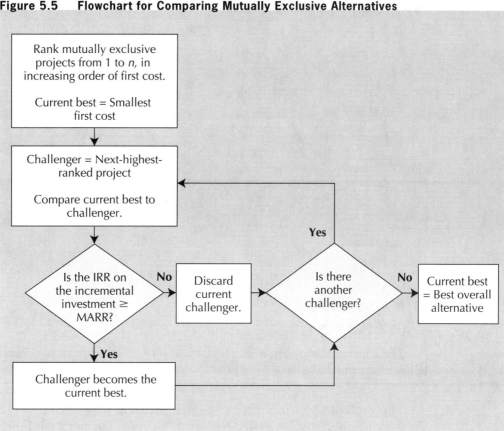

a single IRR (or ERR, as discussed later) be determined for each incremental investment. If there are multiple IRRs, they must be dealt with for *each* increment of investment.

5.3.3 Multiple IRRs

A problem with implementing the internal rate of return method is that there may be more than one internal rate of return. Consider the following example.

EXAMPLE 5.7

A project pays $1000 today, costs $5000 a year from now, and pays $6000 in two years. (See Figure 5.6.) What is its IRR?

Equating the present worths of disbursements and receipts and solving for the IRR gives the following:

$$1000 - 5000(P/F,i^*,1) + 6000(P/F,i^*,2) = 0$$

Recalling that $(P/F,i^*,N)$ stands for $1/(1 + i^*)^N$, we have

$$1 - \frac{5}{1 + i^*} + \frac{6}{(1 + i^*)^2} = 0$$

$$(1 + i^*)^2 - 5(1 + i^*) + 6 = 0$$

Figure 5.6   Multiple IRR Example

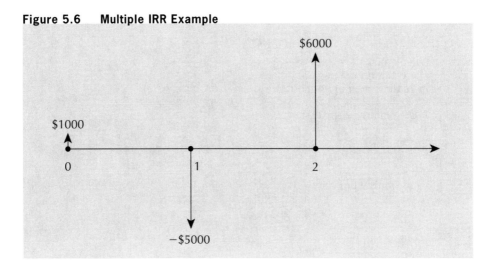

$$(1 + 2i^* + i^{*2}) - 5i^* + 1 = 0$$

$$i^{*2} - 3i^* + 2 = 0$$

$$(i^* - 1)(i^* - 2) = 0$$

The roots of this equation are $i^*_1 = 1$ and $i^*_2 = 2$. In other words, this project has two IRRs: 100% and 200%! ■

The multiple internal rates of return problem may be stated more generally. Consider a project that has cash flows over T periods. The **net cash flow**, A_t, associated with period t is the difference between cash inflows and outflows for the period (i.e., $A_t = R_t - D_t$ where R_t is cash inflow in period t and D_t is cash outflow in period t). We set the present worth of the net cash flows over the entire life of the project equal to zero to find the IRR(s). We have

$$A_0 + A_1(1 + i)^{-1} + A_2(1 + i)^{-2} + \ldots + A_T(1 + i)^{-T} = 0 \qquad (5.2)$$

Any value of i that solves Equation (5.2) is an internal rate of return for that project. That there may be multiple solutions to Equation (5.2) can be seen if we rewrite the equation as

$$A_0 + A_1 x + A_2 x^2 + \ldots + A_T x^T = 0 \qquad (5.3)$$

where $x = (1 + i)^{-1}$.

Solving the Tth degree polynomial of Equation (5.3) is the same as solving for the internal rates of return in Equation (5.2). In general, when finding the roots of Equation (5.3), there may be as many positive real solutions for x as there are sign changes in the coefficients, the A's. Thus, there may be as many IRRs as there are sign changes in the A's.

We can see the meaning of multiple roots most easily with the concept of **project balance**. If a project has a sequence of net cash flows $A_0, A_1, A_2, \ldots, A_T$, and the interest rate is i', there are $T + 1$ project balances, $B_0, B_1, B_2, \ldots, B_T$, one at the end of each period t, $t = 0, 1, \ldots, T$. A project balance, B_t, is the accumulated future value of all cash flows, up to the end of period t, compounded at the rate, i'. That is,

$$B_0 = A_0$$

$$B_1 = A_0(1 + i') + A_1$$

$$B_2 = A_0(1 + i')^2 + A_1(1 + i') + A_2$$
$$B_T = A_0(1 + i')^T + A_1(1 + i')^{T-1} + \ldots + A_T$$

Table 5.1 shows the project balances at the end of each year for both 100% and 200% interest rates for the project in Example 5.7. The project starts with a cash inflow of $1000. At a 100% interest rate, the $1000 increases to $2000 over the first year. At the end of the first year, there is a $5000 disbursement, leaving a negative project balance of $3000. At 100% interest, this negative balance increases to $6000 over the second year. This negative $6000 is offset exactly by the $6000 inflow. This makes the project balance zero at the end of the second year. The project balance at the end of the project is the future worth of all the cash flows in the project. When the future worth at the end of the project life is zero, the present worth is also zero. This verifies that the 100% is an IRR.

Table 5.1   Project Balances for Example 5.7

End of Year	At $i' = 100\%$	At $i' = 200\%$
0	1000	1000
1	$1000(1 + 1) - 5000 = -3000$	$1000(1 + 2) - 5000 = -2000$
2	$-3000(1 + 1) + 6000 = 0$	$-2000(1 + 2) + 6000 = 0$

Now consider the 200% interest rate. Over the first year, the $1000 inflow increases to $3000. At the end of the first year, $5000 is paid out, leaving a negative project balance of $2000. This negative balance grows at 200% to $6000 over the second year. This is offset exactly by the $6000 inflow so that the project balance is zero at the end of the second year. This verifies that the 200% is also an IRR!

NET VALUE 5.1

Capital Budgeting and Financial Management Resources

The internet can be an excellent source of information about the project comparison methods presented in Chapters 4 and 5. A broad search for materials on the PW, AW, and IRR comparison methods might use key words such as *financial management* or *capital budgeting*. Such a search can yield very useful supports for practising engineers or financial managers responsible for making investment decisions. For example, one can find short courses on the process of project evaluation using present worth or IRR methods and how to determine the cost of capital in the Education Centre at http://www.globeadvisor.com. Investopedia (www.investopedia.com) provides online investment resources, including definitions for a wide number of commonly used financial terms.

With more focused key words such as *IRR* one might find examples of how the internal rate of return is used in practice in a wide range of engineering and other applications. Some consulting companies publish promotional white papers on their websites that provide examples of project evaluation methods. For example, cautionary words on the use of the IRR method and advocacy for the use of ERR (also known as the *modified internal rate of return* or *MIRR*) can be found at McKinsey's online newsletter, *McKinsey on Finance* (www.corporatefinance.mckinsey.com).

Source: J. Kelleher and J. MacCormack, "Internal Rate of Return: A Cautionary Tale," *McKinsey on Finance*, Issue 12, accessed May 8, 2008, as follows: http://corporatefinance.mckinsey.com/_downloads/knowledge/mckinsey_on_finance/MoF_Issue_12.pdf

Looking at Table 5.1, it's actually fairly obvious that an important assumption is being made about the initial $1000 received. The IRR computation implicitly assumes that the $1000 is *invested* during the first period at either 100% or 200%, one of the two IRRs. However, during the first period, the project is not an investment. The project balance is positive. The project is *providing* money, not using it. This money cannot be reinvested immediately in the project. It is simply cash on hand. The $1000 must be invested elsewhere for one year if it is to earn any return. It is unlikely, however, that the $1000 provided by the project in this example would be invested in something else at 100% or 200%. More likely, it would be invested at a rate at or near the company's MARR.

5.3.4 External Rate of Return Methods

To resolve the multiple IRR difficulty, we need to consider what return is earned by money associated with a project that is not invested in the project. The usual assumption is that the funds are invested elsewhere and earn an *explicit rate of return* equal to the MARR. This makes sense, because when there is cash on hand that is not invested in the project under study, it will be used elsewhere. These funds would, by definition, gain interest at a rate at least equal to the MARR. The **external rate of return (ERR)**, denoted by i^*_e, is the rate of return on a project where any cash flows that are not invested in the project are assumed to earn interest at a predetermined explicit rate (usually the MARR). For a given explicit rate of return, a project can have only one value for its ERR.

It is possible to calculate a precise ERR that is comparable to the IRRs of other projects using an explicit interest rate when necessary. Because the cash flows of Example 5.7 are fairly simple, let us use them to illustrate how to calculate the ERR precisely.

EXAMPLE 5.8 (EXAMPLE 5.7 REVISITED: ERR)

A project pays $1000 today, costs $5000 a year from now, and pays $6000 in two years. What is its rate of return? Assume that the MARR is 25%.

The first $1000 is not invested immediately in the project. Therefore, we assume that it is invested outside the project for one year at the MARR. Thus, the cumulative cash flow for year 1 is

$$1000(F/P,25\%,1) - 5000 = 1250 - 5000 = -\$3750$$

With this calculation, we transform the cash flow diagram representing this problem from that in Figure 5.7(a) to that in Figure 5.7(b). These cash flows provide a single (precise) ERR, as follows:

$$-3750 + 6000(P/F,i^*_e,1) = 0$$

$$(P/F,i^*_e,1) = 3750/6000 = 0.625$$

$$\frac{1}{1+i^*_e} = 0.625$$

$$1+i^*_e = \frac{1}{0.625} = 1.6$$

$$i^*_e = 0.6$$

$$ERR = 60\% \ \blacksquare$$

Figure 5.7   Multiple IRR Solved

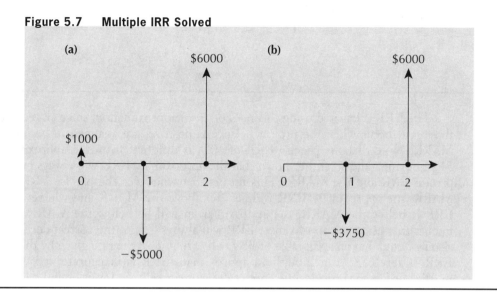

In general, computing a precise ERR can be a complex procedure because of the difficulty in determining exactly when the explicit interest rate should be applied. In order to do such a calculation, project balances have to be computed for trial ERRs. In periods in which project balances are positive for the trial ERR, the project is a source of funds. These funds would have to be invested outside the project at the MARR. During periods when the project balance is negative for the trial ERR, any receipts would be invested in the project and will typically yield more than the MARR. Whether the project balances are negative or positive will depend on the trial ERRs. This implies that the calculation process requires much experimenting with trial ERRs before an ERR is found that makes the future worth zero. A more convenient, but approximate, method is to use the following procedure:

1. Take all *net* receipts forward at the MARR to the time of the last cash flow.
2. Take all *net* disbursements forward at an unknown interest rate, i^*_{ea}, also to the time of the last cash flow.
3. Equate the future value of the receipts from Step 1 to the future value of the disbursements from Step 2 and solve for i^*_{ea}.
4. The value for i^*_{ea} is the *approximate ERR* for the project.

EXAMPLE 5.9 (EXAMPLE 5.7 REVISITED AGAIN: AN APPROXIMATE ERR)

To approximate the ERR, we compute the interest rate that gives a zero future worth at the end of the project when all receipts are brought forward at the MARR. In Example 5.7, the $1000 is thus assumed to be reinvested at the MARR for two years, the life of the project. The disbursements are taken forward to the end of the two years at an unknown interest rate, i^*_{ea}. With a MARR of 25%, the revised calculation is

$$1000(F/P,25\%,2) + 6000 = 5000(F/P,i^*_{ea},1)$$

$$(F/P,i^*_{ea},1) = [1000(1.5625) + 6000]/5000$$

$$(F/P,i^*_{ea},1) = 1.5125$$

$$1 + i_{ea}^* = 1.5125$$

$$i_{ea}^* = 0.5125 \text{ or } 51.25\%$$

$$ERR \cong 51\% \ \blacksquare$$

The ERR calculated using this method is an approximation, since all receipts, not just those that occur when the project balance is positive, are assumed to be invested at the MARR. Note that the precise ERR of 60% is different from the approximate ERR of 51%. Fortunately, it can be shown that the approximate ERR will always be between the precise ERR and the MARR. This means that whenever the precise ERR is above the MARR, the approximate ERR will also be above the MARR and whenever the precise ERR is below the MARR, the approximation will be below the MARR as well. This implies that using the approximate ERR will always lead to the correct decision. It should also be noted that an acceptable project will earn *at least* the rate given by the approximate ERR. Therefore, even though an approximate ERR is inaccurate, it is often used in practice because it provides the correct decision as well as a lower bound on the return on an investment, while being easy to calculate.

5.3.5 When to Use the ERR

The ERR (approximate or precise) must be used whenever there are multiple IRRs possible. Unfortunately, it can be difficult to know in advance whether there will be multiple IRRs. On the other hand, it is fortunate that most ordinary projects have a structure that precludes multiple IRRs.

Most projects consist of one or more periods of outflows at the start, followed only by one or more periods of inflows. Such projects are called **simple investments**. The cash flow diagram for a simple investment takes the general form shown in Figure 5.8. In terms of Equations (5.2) and (5.3), there is only one change of sign, from negative to positive in the A's, the sequence of coefficients. Hence, a simple investment always has a unique IRR.

If a project is not a simple investment, there may or may not be multiple IRRs—there is no way of knowing for sure without further analysis. In practice, it may be reasonable to use an approximate ERR whenever the project is not a simple investment. Recall from

Figure 5.8   The General Form of Simple Investments

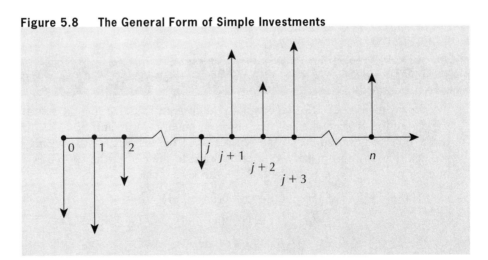

Section 5.3.4 that the approximate ERR will always provide a correct decision, whether its use is required or not, since it will understate the true rate of return.

However, it is generally desirable to compute an IRR whenever it is possible to do so, and to use an approximate ERR only when there may be multiple IRRs. In this way, the computations will be as accurate as possible. When it is desirable to know for sure whether there will be only one IRR, there are several steps of analysis that can be undertaken. These are covered in detail in Appendix 5A.

To reiterate, the approximate ERR can be used to evaluate any project, whether it is a simple investment or not. However, the approximate ERR will tend to be a less accurate rate than the IRR. The inaccuracy will tend to be similar for projects with cash flows of a similar structure, and either method will result in the same decision in the end.

5.4 | Rate of Return and Present/Annual Worth Methods Compared

A comparison of rate of return and present/annual worth methods leads to two important conclusions:

1. The two sets of methods, when properly used, give the same decisions.
2. Each set of methods has its own advantages and disadvantages.

Let us consider each of these conclusions in more detail.

5.4.1 Equivalence of Rate of Return and Present/Annual Worth Methods

If an independent project has a unique IRR, the IRR method and the present worth method give the same decision. Consider Figure 5.9. It shows the present worth as a function of the interest rate for a project with a unique IRR. The maximum of the curve lies at the vertical axis (where the interest rate = 0) at the point given by the sum of all

Figure 5.9   Present Worth (PW) as a Function of Interest Rate (i) for a Simple Investment

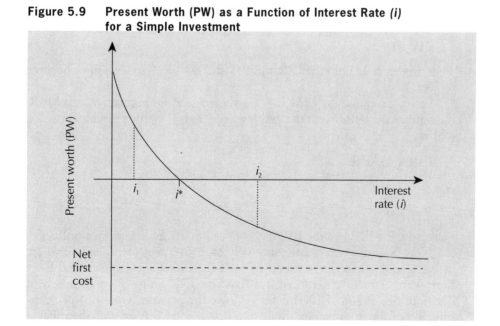

undiscounted net cash flows. (We assume that the sum of all the undiscounted net cash flows is positive.) As the interest rate increases, the present worth of all cash flows after the first cost decreases. Therefore, the present worth curve slopes down to the right. To determine what happens as the interest rate increases indefinitely, let us recall the general equation for present worth

$$PW = \sum_{t=0}^{T} A_t (1 + i)^{-t} \tag{5.4}$$

where

i = the interest rate
A_t = the net cash flow in period t
T = the number of periods

Letting $i \to \infty$, we have

$$\lim_{i \to \infty} \frac{1}{(1+i)^t} = 0 \text{ for } t = 1, 2, \dots, T$$

Therefore, as the interest rate becomes indefinitely large, all terms in Equation (5.4) approach zero except the first term (where $t = 0$), which remains at A_0. In Figure 5.9, this is shown by the asymptotic approach of the curve to the first cost, which, being negative, is below the horizontal axis.

The interest rate at which the curve crosses the horizontal axis (i^* in Figure 5.9), where the present worth is zero, is, by definition, the IRR.

To demonstrate the equivalence of the rate of return and the present/annual worth methods for decision making, let us consider possible values for the MARR. First, suppose the MARR = i_1, where $i_1 < i^*$. In Figure 5.9, this MARR would lie to the left of the IRR. From the graph we see that the present worth is positive at i_1. In other words, we have

IRR > MARR

and

PW > 0

Thus, in this case, both the IRR and PW methods lead to the same conclusion: Accept the project.

Second, suppose the MARR = i_2, where $i_2 > i^*$. In Figure 5.9, this MARR would lie to the right of the IRR. From the graph we see that, at i_2, the present worth is negative. Thus we have

IRR < MARR

and

PW < 0

Here, too, the IRR and the PW method lead to the same conclusion: Reject the project.

Now consider two simple, mutually exclusive projects, A and B, where the first cost of B is greater than the first cost of A. If the increment from A to B has a unique IRR, then we can readily demonstrate that the IRR and PW methods lead to the same decision. See Figure 5.10(a), which shows the present worths of projects A and B as a function of the interest rate. Since the first cost of B is greater than that of A, the curve

Figure 5.10   Present Worth as a Function of Interest Rate (i) for Two Simple, Mutually Exclusive Projects

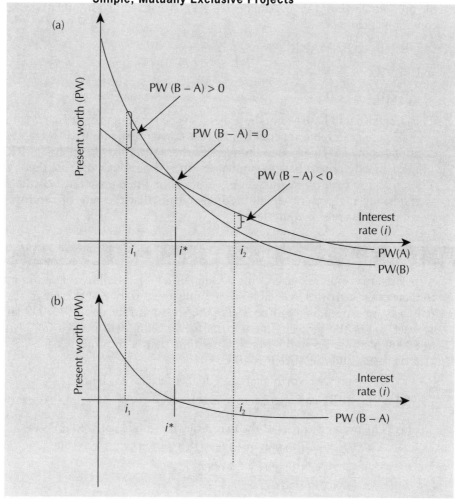

for project B asymptotically approaches a lower present worth than does the curve for project A as the interest rate becomes indefinitely large, and thus the two curves must cross at some point.

To apply the IRR method, we must consider the increment (denoted by B – A). The present worth of the increment (B – A) will be zero where the two curves cross. This point of intersection is marked by the interest rate, i^*. We have plotted the curve for the increment (B – A) in Figure 5.10(b) to clarify the relationships.

Let us again deal with possible values of the MARR. First, suppose the MARR (i_1) is less than i^*. Then, as we see in Figure 5.10(b), the present worth of (B – A) is positive at i_1. That is, the following conditions hold:

IRR(B – A) > MARR

and

PW(B – A) > 0

Thus, according to both the IRR method and the PW method, project B is better than project A.

Second, suppose the MARR $= i_2$, where $i_2 > i^*$. Then we see from Figure 5.10(b) that the present worth of the increment (B – A) is negative at i_2. In other words, the following conditions hold:

IRR(B – A) < MARR

and

PW(B – A) < 0

Thus, according to both methods, project A is better than project B.

In a similar fashion, we could show that the approximate ERR method gives the same decisions as the PW method in those cases where there may be multiple IRRs.

We already noted that the annual worth and present worth methods are equivalent. Therefore, by extension, our demonstration of the equivalence of the rate of return methods and the present worth methods means that the rate of return and the annual worth methods are also equivalent.

EXAMPLE 5.10

A tourist-area resort is considering adding either a parasailing operation or kayak rentals to their other activities. Available space limits them to one of these two choices. The initial costs for parasailing will be $100 000, with net returns of $15 000 annually for the 15-year life of the project. Initial costs for kayaking will be $10 000, with net returns of $2000 per year for its 15-year life. Assume that both projects have a $0 salvage value after 15 years, and the MARR is 10%.

(a) Using present worth analysis, which alternative is better?

(b) Using IRR, which alternative is better?

(a) The present worths of the two projects are calculated as follows:

$$\begin{aligned}
\text{PW}_{\text{para}} &= -100\ 000 + 15\ 000(P/A,10\%,15) \\
&= -100\ 000 + 15\ 000(7.6061) \\
&= 14\ 091.50
\end{aligned}$$

$$\begin{aligned}
\text{PW}_{\text{kayak}} &= -10\ 000 + 2000(P/A,10\%,15) \\
&= -10\ 000 + 2000(7.6061) \\
&= 5212.20
\end{aligned}$$

The parasailing venture has a higher present worth at about $14 000 and is thus preferred.

(b) The IRRs of the two projects are calculated as follows:

Parasailing

$$100\ 000 = 15\ 000(P/A,i^*,15)$$

$$(P/A,i^*,15) = 100\ 000/15\ 000 = 6.67 \rightarrow i^*_{\text{para}} = 12.4\%$$

Kayaking

$$10\ 000 = 2000(P/A,i^*,15)$$

$$(P/A,i^*,15) = 5 \rightarrow i^*_{\text{kayak}} = 18.4\%$$

One might conclude that, because IRR$_{\text{kayak}}$ is larger, the resort should invest in the kayaking project, but this is *wrong*. When done correctly, a present worth analysis and an IRR analysis will always agree. The error here is that the parasailing project was assessed without consideration of the increment from the kayaking project. Checking the IRR of the increment (denoted by the subscript "kayak-para"):

$$(100\ 000 - 10\ 000) = (15\ 000 - 2000)(P/A,i^*,15)$$

$$(P/A,i^*,15) = 90\ 000/13\ 000 = 6.923 \rightarrow i^*_{kayak-para} = 11.7\%$$

Since the increment from the kayaking project also exceeds the MARR, the larger parasailing project should be taken.■

5.4.2 Why Choose One Method Over the Other?

Although rate of return methods and present worth/annual worth methods give the same decisions, each set of methods has its own advantages and disadvantages. The choice of method may depend on the way the results are to be used and the sort of data the decision makers prefer to consider. In fact, many companies, by policy, require that several methods be applied so that a more complete picture of the situation is presented. A summary of the advantages and disadvantages of each method is given in Table 5.2.

Rate of return methods state results in terms of *rates*, while present/annual worth methods state results in absolute figures. Many managers prefer rates to absolute figures because rates facilitate direct comparisons of projects whose sizes are quite different. For example, a petroleum company comparing performances of a refining division and a distribution division would not look at the typical values of present or annual worth for projects in the two divisions. A refining project may have first costs in the range of hundreds of *millions*, while distribution projects may have first costs in the range of *thousands*. It would not be meaningful to compare the absolute profits between a refining project and a distribution project. The absolute profits of refining projects will almost certainly be larger than those of distribution projects. Expressing project performance in terms of rates of return permits understandable comparisons. A disadvantage of rate of return methods, however, is the possible complication that there may be more than one rate of return. Under such circumstances, it is necessary to calculate an ERR.

Table 5.2   Advantages and Disadvantages of Comparison Methods

Method	Advantages	Disadvantages
IRR	Facilitates comparisons of projects of different sizes Commonly used	Relatively difficult to calculate Multiple IRRs may exist
Present worth	Gives explicit measure of profit contribution	Difficult to compare projects of different sizes
Annual worth	Annual cash flows may have familiar meanings to decision makers	Difficult to compare projects of different sizes
Payback period	Very easy to calculate Commonly used Takes into account the need to have capital recovered quickly	Discriminates against long-term projects Ignores time value of money Ignores the expected service life

In contrast to a rate of return, a present or annual worth computation gives a direct measure of the profit provided by a project. A company's main goal is likely to earn profits for its owners. The present and annual worth methods state the contribution of a project toward that goal. Another reason that managers prefer these methods is that present worth and annual worth methods are typically easier to apply than rate of return methods.

For completeness of coverage, we note that the payback period method may not give results consistent with rate of return or present/annual worth methods as it ignores the time value of money and the service life of projects. It is, however, a method commonly used in practice due to its ease of use and intuitive appeal.

EXAMPLE 5.11

Each of the following scenarios suggests a best choice of comparison method.

1. Edward has his own small firm that will lease injection-moulding equipment to make polyethylene containers. He must decide on the specific model to lease. He has estimates of future monthly sales.

   The annual worth method makes sense here, because Edward's cash flows, including sales receipts and leasing expenses, will probably all be on a monthly basis. As a sole proprietor, Edward need not report his conclusions to others.

2. Ramesh works for a large power company and must assess the viability of locating a transformer station at various sites in the city. He is looking at the cost of the building lot, power lines, and power losses for the various locations. He has fairly accurate data about costs and future demand for electricity.

   As part of a large firm, Ramesh will probably be obliged to use a specific comparison method. This would probably be IRR. A power company makes many large and small investments, and the IRR method allows them to be compared fairly. Ramesh has the data necessary for the IRR calculations.

3. Sehdev must buy a relatively inexpensive log splitter for his agricultural firm. There are several different types that require a higher or lower degree of manual assistance. He has only rough estimates of how this machine will affect future cash flows.

   This relatively inexpensive purchase is a good candidate for the payback period method. The fact that it is inexpensive means that extensive data gathering and analysis are probably not warranted. Also, since future cash flows are relatively uncertain, there is no justification for using a particularly precise comparison method.

4. Ziva will be living in the Arctic for six months, testing her company's equipment under hostile weather conditions. She needs a field office and must determine which of the following choices is economically best: (1) renting space in an industrial building, (2) buying and outfitting a trailer, (3) renting a hotel room for the purpose.

   For this decision, a present worth analysis would be appropriate. The cash flows for each of the alternatives are of different types, and bringing them to present worth would be a fair way to compare them. It would also provide an accurate estimate to Ziva's firm of the expected cost of the remote office for planning purposes.■

REVIEW PROBLEMS

REVIEW PROBLEM 5.1

Wei-Ping's consulting firm needs new quarters. A downtown office building is ideal. The company can either buy or lease it. To buy the office building will cost 60 000 000 yuan. If the building is leased, the lease fee is 4 000 000 yuan payable at the beginning of each year. In either case, the company must pay city taxes, maintenance, and utilities.

Wei-Ping figures that the company needs the office space for only 15 years. Therefore, they will either sign a 15-year lease or buy the building. If they buy the building, they will then sell it after 15 years. The value of the building at that time is estimated to be 150 000 000 yuan.

What rate of return will Wei-Ping's firm receive by buying the office building instead of leasing it?

ANSWER

The rate of return can be calculated as the IRR on the incremental investment necessary to buy the building rather than lease it.

The IRR on the incremental investment is found by solving for i^* in

$$(60\ 000\ 000 - 4\ 000\ 000) - 150\ 000\ 000(P/F,i^*,15) = 4\ 000\ 000(P/A,i^*,14)$$

$$4(P/A,i^*,14) + 150(P/F,i^*,15) = 56$$

For $i^* = 11\%$, the result is

$$4(P/A,11\%,14) + 150(P/F,11\%,15)$$

$$= 4(6.9819) + 150(0.20900)$$

$$= 59.2781$$

For $i^* = 12\%$,

$$4(P/A,12\%,14) + 150(P/F,12\%,15)$$

$$= 4(6.6282) + 150(0.1827)$$

$$= 53.9171$$

A linear interpolation between 11% and 12% gives the IRR

$$i^* = 11\% + (59.2781 - 56)/(59.2781 - 53.9171) = 11.6115\%$$

By investing their money in buying the building rather than leasing, Wei-Ping's firm is earning an IRR of about 11.6%.∎

REVIEW PROBLEM 5.2

The Real S. Tate Company is considering investing in one of four rental properties. Real S. Tate will rent out whatever property they buy for four years and then sell it at the end of that period. The data concerning the properties is shown on the next page.

Rental Property	Purchase Price	Net Annual Rental Income	Sale Price at the End of Four Years
1	$100 000	$ 7200	$100 000
2	120 000	9600	130 000
3	150 000	10 800	160 000
4	200 000	12 000	230 000

On the basis of the purchase prices, rental incomes, and sale prices at the end of the four years, answer the following questions.

(a) Which property, if any, should Tate invest in? Real S. Tate uses a MARR of 8% for projects of this type.

(b) Construct a graph that depicts the present worth of each alternative as a function of interest rates ranging from 0% to 20%. (A spreadsheet would be helpful in answering this part of the problem.)

(c) From your graph, determine the range of interest rates for which your choice in part (a) is the best investment. If the MARR were 9%, which rental property would be the best investment? Comment on the sensitivity of your choice to the MARR used by the Real S. Tate Company.

ANSWER

(a) Since the "do nothing" alternative is feasible and it has the least first cost, it becomes the current best alternative. The IRR on the incremental investment for property 1 is given by:

$$-100\ 000 + 100\ 000(P/F,i^*,4) + 7200(P/A,i^*,4) = 0$$

The IRR on the incremental investment is 7.2%. Because this is less than the MARR of 8%, property 1 is discarded from further consideration.

Next, the IRR for the incremental investment for property 2, the alternative with the next-highest first cost, is found by solving for i^* in

$$-120\ 000 + 130\ 000(P/F,i^*,4) + 9600(P/A,i^*,4) = 0$$

The interest rate that solves the above equation is 9.8%. Since an IRR of 9.8% exceeds the MARR, property 2 becomes the current best alternative. Now the incremental investments over and above the first cost of property 2 are analyzed.

Next, property 3 challenges the current best. The IRR in the incremental investment to property 3 is

$$(-150\ 000 + 120\ 000) + (160\ 000 - 130\ 000)(P/F,i^*,4)$$
$$+ (10\ 800 - 9600)(P/A,i^*,4) = 0$$
$$-30\ 000 + 30\ 000(P/F,i^*,4) + 1200(P/A,i^*,4) = 0$$

This gives an IRR of only 4%, which is below the MARR. Property 2 remains the current best alternative and property 3 is discarded.

Finally, property 4 challenges the current best. The IRR on the incremental investment from property 2 to property 4 is

$$(-200\ 000 + 120\ 000) + (230\ 000 - 130\ 000)(P/F,i^*,4)$$
$$+ (12\ 000 - 9600)(P/A,i^*,4) = 0$$
$$-80\ 000 + 100\ 000(P/F,i^*,4) + 2400(P/A,i^*,4) = 0$$

Figure 5.11   Present Worth for Review Problem 5.2

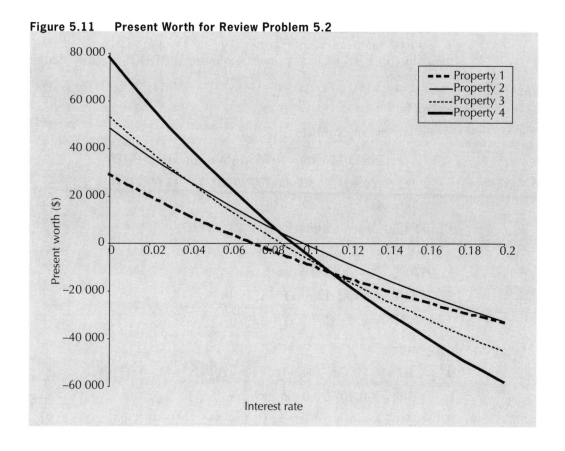

The IRR on the incremental investment is 8.5%, which is above the MARR. Property 4 becomes the current best choice. Since there are no further challengers, the choice based on IRR is the current best, property 4.

(b) The graph for part (b) is shown in Figure 5.11.

(c) From the graph, one can see that property 4 is the best alternative provided that the MARR is between 0% and 8.5%. This is the range of interest rates over which property 4 has the largest present worth.

   If the MARR is 9%, the best alternative is property 2. This can be seen by going back to the original IRR computations and observing that the results of the analysis are essentially the same, except that the incremental investment from property 2 to property 4 no longer has a return exceeding the MARR. This can be confirmed from the diagram (Figure 5.11) as well, since the property with the largest present worth at 9% is property 2.

   With respect to sensitivity analysis, the graph shows that, for a MARR between 0% and 8.5%, property 4 is the best choice and, for a MARR between 8.5% and 9.8%, property 2 is the best choice. If the MARR is above 9.8%, no property has an acceptable return on investment, and the "do nothing" alternative would be chosen.■

REVIEW PROBLEM 5.3

You are in the process of arranging a marketing contract for a new Java applet you are writing. It still needs more development, so your contract will pay you £5000 today to finish the prototype. You will then get royalties of £10 000 at the end of each of the second and third years. At the end of each of the first and fourth years, you will be required to spend £20 000 and £10 000 in upgrades, respectively. What is the (approximate) ERR on this project, assuming a MARR of 20%? Should you accept the contract?

ANSWER

To calculate the approximate ERR, set

FW(receipts @ MARR) = FW(disbursements @ ERR)

$5000(F/P,20\%,4) + 10\ 000(F/P,20\%,2) + 10\ 000(F/P,20\%,1)$
$= 20\ 000(F/P,i^*_{ea},3) + 10\ 000$

$5000(2.0736) + 10\ 000(1.44) + 10\ 000(1.2)$
$= 20\ 000(F/P,i^*_{ea},3) + 10\ 000$

$(F/P,i^*_{ea},3) = 1.3384$

At ERR = 10%, $(F/P,i,3) = 1.3310$

At ERR = 11%, $(F/P,i,3) = 1.3676$

Interpolating:

$i^*_{ea} = 10\% + (1.3384 - 1.3310)(11 - 10)/(1.3676 - 1.3310)$

$= 10\% + 0.0074/0.0366 \cong 10.2\%$

The (approximate) ERR is 10.2%. Since this is below the MARR of 20%, the contract should not be accepted.∎

SUMMARY

This chapter presented the IRR method for evaluating projects and also discussed the relationship among the present worth, annual worth, payback period, and IRR methods.

The IRR method consists of determining the rate of return for a sequence of cash flows. For an independent project, the calculated IRR is compared with a MARR, and if it is equal to or exceeds the MARR it is an acceptable project. To determine the best project of several mutually exclusive ones, it is necessary to determine the IRR of each increment of investment.

The IRR selection procedure is complicated by the possibility of having more than one rate of return because of a cash flow structure that, over the course of a project, requires that capital, eventually invested in the project at some point, be invested externally. Under such circumstances, it is necessary to calculate an ERR.

The present worth and annual worth methods are closely related, and both give results identical to those of the IRR method. Rate of return measures are readily understandable, especially when comparing projects of unequal sizes, whereas present/annual worth measures give an explicit expression of the profit contribution of a project. The main advantage of the payback period method is that it is easy to implement and understand, and takes into account the need to have capital recovered quickly.

Engineering Economics in Action, Part 5B:
The Invisible Hand

"Hello." Dave's voice was clear enough over the phone that he could have been in his office down the hall, but Naomi could tell from his relaxed tone that the office was not on his mind.

"Hi, Dave, it's Naomi. Can I bend your ear about that drop forge project again?"

"Oh, hi, Naomi. Sure, what have you got?"

"Well, as I see it, IRR has got to be the way to go. Of course, present worth or annual worth will give the same answer, but I'm sure Ed Burns and Anna Kulkowski would prefer IRR. They have to compare potential investments across different parts of the organization. It's kind of hard to compare net present worths of investments in information systems, where you rarely get above a first cost of $100 000, with forge investments where you can easily get up to a few hundred thousand first cost. And, as I said before, the drop forge operation isn't one in which the annual cost has any particular significance."

There was a short pause. Naomi suddenly regretted speaking as if she was so sure of herself—but, darn it, she was sure on this one.

"Exactly right," Dave replied. Naomi could feel an invisible hand pat her on the back. "So how exactly would you proceed?"

"Well, I have the options ranked by first cost. The first one is just refurbishing the existing machine. There is no test on that one unless we are willing to stop making our own parts, and Clem told me that was out . . . "

Dave interjected with "You don't mean that you're automatically going to refurbish the existing machine, do you?"

"No, no. The simple refurbishing option is the base. I then go to the next option, which is to refurbish the drop forging hammer and replace the materials-handling system. I compare this with the just-refurbish option by looking at the incremental first cost. I will check to see if the additional first cost has an IRR of at least 15% after tax, which, Clem tells me, is the minimum acceptable rate of return. If the incremental first cost has an IRR of at least 15%, the combination of refurbishing and replacing the materials-handling system is better than just refurbishing. I then consider the next option, which is to buy the manually operated mechanical press with no change in materials handling. I look at the incremental investment here and see if its IRR is at least 15%. To go back a step, if the IRR of replacing materials handling plus refurbishing the old machine did not pay off at 15%, I would have rejected that and compared the manually operated mechanical press with the first option, just refurbishing the old machine. I then work my way, option by option, up to the seventh. How does that sound?"

"Well, that sounds great, as far as it goes. Have you checked for problems with multiple IRRs?"

"Well, so far each set of cash flows has been a simple investment, but I will be careful."

"I would also compute payback periods for them in case we are having cash flow problems. If the payback is too long, they may not necessarily take an option even with the incremental IRR being above their 15% MARR."

Naomi considered this for a second. "One other question, Dave. What should I do about intangibles?"

"You mean the noise from the forging hammer?"

"Yes. It's important, but you can't evaluate it in dollars and cents."

"Just remind them of it in your report. If they want a more formal analysis, they'll come back to you."

"Thanks, Dave. You've been a big help."

As Naomi hung up the phone, she couldn't help smiling ruefully to herself. She had ignored the payback period altogether—after all, it didn't take either interest or service life into account. "I guess that's what they call practical experience," she said to herself as she got out her laptop.

PROBLEMS

For additional practice, please see the problems (with selected solutions) provided on the Student CD-ROM that accompanies this book.

5.1 What is the IRR for a $1000 investment that returns $200 at the end of each of the next

(a) 7 years?

(b) 6 years?

(c) 100 years?

(d) 2 years?

5.2 New windows are expected to save £400 per year in energy costs over their 30-year life for Fab Fabricating. At an initial cost of £8000 and zero salvage value, are they a good investment? Fab's MARR is 8%.

5.3 An advertising campaign will cost ¥200 000 000 for planning and ¥40 000 000 in each of the next six years. It is expected to increase revenues permanently by ¥40 000 000 per year. Additional revenues will be gained in the pattern of an arithmetic gradient with ¥20 000 000 in the first year, declining by ¥5 000 000 per year to zero in the fifth year. What is the IRR of this investment? If the company's MARR is 12%, is this a good investment?

5.4 Aline has three contracts from which to choose. The first contract will require an outlay of $100 000 but will return $150 000 one year from now. The second contract requires an outlay of $200 000 and will return $300 000 one year from now. The third contract requires an outlay of $250 000 and will return $355 000 one year from now. Only one contract can be accepted. If her MARR is 20%, which one should she choose?

5.5 Refer to Review Problem 4.3. Assuming the four investments are independent, use the IRR method to select which, if any, should be chosen. Use a MARR of 8%.

5.6 Fantastic Footwear can invest in one of two different automated clicker cutters. The first, A, has a 100 000 yuan first cost. A similar one with many extra features, B, has a 400 000 yuan first cost. A will save 50 000 yuan per year over the cutter now in use. B will save 150 000 yuan per year. Each clicker cutter will last five years. If the MARR is 10%, which alternative is better? Use an IRR comparison.

5.7 CB Electronix must buy a piece of equipment to place electronic components on the printed circuit boards it assembles. The proposed equipment has a 10-year life with no scrap value.

The supplier has given CB several purchase alternatives. The first is to purchase the equipment for $850 000. The second is to pay for the equipment in 10 equal installments of $135 000 each, starting one year from now. The third is to pay $200 000 now and $95 000 at the end of each year for the next 10 years.

(a) Which alternative should CB choose if their MARR is 11% per year? Use an IRR comparison approach.

(b) Below what MARR does it make sense for CB to buy the equipment now for $850 000?

5.8 The following table summarizes information for four projects:

Project	First Cost	IRR on Overall Investment	IRR on Increments of Investment Compared With Project		
			1	**2**	**3**
1	$100 000	19%			
2	175 000	15%	9%		
3	200 000	18%	17%	23%	
4	250 000	16%	12%	17%	13%

The data can be interpreted in the following way: The IRR on the incremental investment between project 4 and project 3 is 13%.

(a) If the projects are independent, which projects should be undertaken if the MARR is 16%?

(b) If the projects are mutually exclusive, which project should be undertaken if the MARR is 15%? Indicate what logic you have used.

(c) If the projects are mutually exclusive, which project should be undertaken if the MARR is 17%? Indicate what logic you have used.

 5.9 There are several mutually exclusive ways Grazemont Dairy can meet a requirement for a filling machine for their creamer line. One choice is to buy a machine. This would cost €65 000 and last for six years with a salvage value of €10 000. Alternatively, they could contract with a packaging supplier to get a machine free. In this case, the extra costs for packaging supplies would amount to €15 000 per year over the six-year life (after which the supplier gets the machine back with no salvage value for Grazemont). The third alternative is to buy a used machine for €30 000 with zero salvage value after six years. The used machine has extra maintenance costs of €3000 in the first year, increasing by €2500 per year. In all cases, there are installation costs of €6000 and revenues of €20 000 per year. Using the IRR method, determine which is the best alternative. The MARR is 10%.

5.10 Project X has an IRR of 16% and a first cost of $20 000. Project Y has an IRR of 17% and a first cost of $18 000. The MARR is 15%. What can be said about which (if either) of the two projects should be undertaken if (a) the projects are independent and (b) the projects are mutually exclusive?

5.11 Charlie has a project for which he had determined a present worth of R56 740. He now has to calculate the IRR for the project, but unfortunately he has lost complete information about the cash flows. He knows only that the project has a five-year service life and a first cost of R180 000, that a set of equal cash flows occurred at the end of each year, and that the MARR used was 10%. What is the IRR for this project?

5.12 Lucy's project has a first cost P, annual savings A, and a salvage value of $1000 at the end of the 10-year service life. She has calculated the present worth as $20 000, the annual worth as $4000, and the payback period as three years. What is the IRR for this project?

5.13 Patti's project has an IRR of 15%, first cost P, and annual savings A. She observed that the salvage value S at the end of the five-year life of the project was exactly half of the purchase price, and that the present worth of the project was exactly double the annual savings. What was Patti's MARR?

5.14 Jerry has an opportunity to buy a bond with a face value of $10 000 and a coupon rate of 14%, payable semiannually.

(a) If the bond matures in five years and Jerry can buy one now for $3500, what is his IRR for this investment?

(b) If his MARR for this type of investment is 20%, should he buy the bond?

5.15 The following cash flows result from a potential construction contract for Erstwhile Engineering.

1. Receipts of £500 000 at the start of the contract and £1 200 000 at the end of the fourth year

2. Expenditures at the end of the first year of £400 000 and at the end of the second year of £900 000

3. A net cash flow of £0 at the end of the third year

   Using an appropriate rate of return method, for a MARR of 25%, should Erstwhile Engineering accept this project?

5.16 Samiran has entered into an agreement to develop and maintain a computer program for symbolic mathematics. Under the terms of the agreement, he will pay 9 000 000 rupees in royalties to the investor at the end of the fifth, tenth, and fifteenth years, with the investor paying Samiran 4 500 000 rupees now, and then 6 500 000 rupees at the end of the twelfth year.

   Samiran's MARR for this type of investment is 20%. Calculate the ERR of this project. Should he accept this agreement, on the basis of these disbursements and receipts alone? Are you sure that the ERR you calculated is the only ERR? Why? Are you sure that your recommendation to Samiran is correct? Justify your answer.

5.17 Refer to Problem 4.12. Find which alternative is preferable using the IRR method and a MARR of 5%. Assume that one of the alternatives must be chosen. Answer the following questions by using present worth computations to find the IRRs. Use the least common multiple of service lives.

(a) What are the cash flows for each year of the comparison period (i.e., the least common multiple of service lives)?

(b) Are you able to conclude that there is a single IRR on the incremental investment? Why or why not?

(c) Which of the two alternatives should be chosen? Use the ERR method if necessary.

5.18 Refer to Example 4.6 in which a mechanical engineer has decided to introduce automated materials-handling equipment to a production line. Use a present worth approach with an IRR analysis to determine which of the two alternatives is best. The MARR is 9%. Use the repeated lives method to deal with the fact that the service lives of the two alternatives are not equal.

5.19 Refer to Problem 4.20. Use an IRR analysis to determine which of the two alternatives is best. The MARR is 8%. Use the repeated lives method to deal with the unequal service lives of the two alternatives.

5.20 Refer to Problem 4.21. Val has determined that the salvage value of the XJ3 after two years of service is $1900. Using the IRR method, which display panel is the better choice? Use a two-year study period. She must choose one of the alternatives.

5.21 Yee Swian has received an advance of 200 000 yuan on a software program she is writing. She will spend 1 200 000 yuan this year writing it (consider the money to have been spent at the end of year 1), and then receive 1 000 000 yuan at the end of the second year. The MARR is 12%.

(a) What is the IRR for this project? Does the result make sense?

(b) What is the precise ERR?

(c) What is the approximate ERR?

 5.22 Zhe develops truss analysis software for civil engineers. He has the opportunity to contract with at most one of two clients who have approached him with development proposals. One contract pays him $15 000 immediately, and then $22 000 at the end of the project three years from now. The other possibility pays $20 000 now and $5000 at the end of each of the three years. In either case, his expenses will be $10 000 per year. For a MARR of 10%, which project should Zhe accept? Use an appropriate rate of return method.

5.23 The following table summarizes cash flows for a project:

Year	Cash Flow at End of Year
0	−€5000
1	3000
2	4000
3	−1000

(a) Write out the expression you need to solve to find the IRR(s) for this set of cash flows. Do not solve.

(b) What is the maximum number of solutions for the IRR that could be found in part (a)? Explain your answer in one sentence.

(c) You have found that an IRR of 14.58% solves the expression in part (a). Compute the project balances for each year.

(d) Can you tell (without further computations) if there is a unique IRR from this set of cash flows? Explain in one sentence.

5.24 Pepper Properties screens various projects using the payback period method. For renovation decisions, the minimum acceptable payback period is five years. Renovation projects are characterized by an immediate investment of P dollars which is recouped as an annuity of A dollars per year over 20 years. They are considering changing to the IRR method for such decisions. If they changed to the IRR method, what MARR would result in exactly the same decisions as their current policy using payback period?

5.25 Six mutually exclusive projects, A, B, C, D, E, and F, are being considered. They have been ordered by first costs so that project A has the smallest first cost, F the largest. The data in the table on the next page applies to these projects. The data can be interpreted as follows: the IRR on the incremental investment between project D and project C is 6%. Which project should be chosen using a MARR of 15%?

Project	IRR on Overall Investment	IRR on Increments of Investment Compared With Project				
		A	B	C	D	E
A	20%					
B	15%	12%				
C	24%	30%	35%			
D	16%	18%	22%	6%		
E	17%	16%	19%	15%	16%	
F	21%	20%	21%	19%	18%	11%

5.26 Three mutually exclusive designs for a bypass are under consideration. The bypass has a 10-year life. The first design incurs a cost of $1.2 million for a net savings of $300 000 per annum. The second design would cost $1.5 million for a net savings of $400 000 per annum. The third has a cost of $2.1 million for a net savings of $500 000 per annum. For each of the alternatives, what range of values for the MARR results in its being chosen? It is not necessary that any be chosen.

5.27 Linus's project has cash flows at times 0, 1, and 2. He notices that for a MARR of 12%, the ERR falls exactly halfway between the MARR and the IRR, while for a MARR of 18%, the ERR falls exactly one-quarter of the way between the MARR and the IRR. If the cash flow is $2000 at time 2 and negative at time 0, what are the possible values of the cash flow at time 1?

5.28 Three construction jobs are being considered by Clam City Construction (see the following table). Each is characterized by an initial deposit paid by the client to CCC, a yearly cost incurred by CCC at the end of each of three years, and a final payment to CCC by the client at the end of three years. CCC has the capacity to do only one of these contracts. Use an appropriate rate of return method to determine which they should do. Their MARR is 10%.

Job	Deposit ($)	Cost Per Year ($)	Final Payment ($)
1	100 000	75 000	200 000
2	150 000	100 000	230 000
3	175 000	150 000	300 000

5.29 Kool Karavans is considering three investment proposals. Each of them is characterized by an initial cost, annual savings over four years, and no salvage value, as illustrated in the following table. They can only invest in two of these proposals. If their MARR is 12%, which two should they choose?

Proposal	First Cost ($)	Annual Savings ($)
A	40 000	20 000
B	110 000	30 000
C	130 000	45 000

5.30 Development projects done by Standalone Products are subsidized by a government grant program. The program pays 30% of the total cost of the project (costs summed without discounting, i.e., the interest rate is zero), half at the beginning of the project and half at the end, up to a maximum of €100 000. There are two projects being considered. One is a customized checkweigher for cheese products, and the other is an automated production scheduling system. Each project has a service life of five years. Costs and benefits for both projects, not including grant income, are shown below. Only one can be done, and the grant money is certain. PTR has a MARR of 15% for projects of this type. Using an appropriate rate of return method, which project should be chosen?

	Checkweigher	Scheduler
First cost	€30 000	€10 000
Annual costs	5 000	12 000
Annual benefits	14 000	17 000
Salvage value	8 000	0

5.31 Jacob is considering the replacement of the heating system for his building. There are three alternatives. All are natural-gas–fired furnaces, but they vary in energy efficiency. Model A is leased at a cost of $500 per year over a 10-year study period. There are installation charges of $500 and no salvage value. It is expected to provide energy savings of $200 per year. Model B is purchased for a total cost of $3600, including installation. It has a salvage value of $1000 after 10 years of service, and is expected to provide energy savings of $500 per year. Model C is also purchased, for a total cost of $8000, including installation. However, half of this cost is paid now, and the other half is paid at the end of two years. It has a salvage value of $1000 after 10 years, and is expected to provide energy savings of $1000 per year. For a MARR of 12% and using a rate of return method, which heating system should be installed? One model must be chosen.

5.32 Corral Cartage leases trucks to service its shipping contracts. Larger trucks have cheaper operating costs if there is sufficient business, but are more expensive if they are not full. CC has estimates of monthly shipping demand. What comparison method(s) would be appropriate for choosing which trucks to lease?

5.33 The bottom flaps of shipping cartons for Yonge Auto Parts are fastened with industrial staples. Yonge needs to buy a new stapler. What comparison method(s) would be appropriate for choosing which stapler to buy?

5.34 Joan runs a dog kennel. She is considering installing a heating system for the interior runs which will allow her to operate all year. What comparison method(s) would be appropriate for choosing which heating system to buy?

5.35 A large food company is considering replacing a scale on its packaging line with a more accurate one. What comparison method(s) would be appropriate for choosing which scale to buy?

5.36 Mona runs a one-person company producing custom paints for hobbyists. She is considering buying printing equipment to produce her own labels. What comparison method(s) would be appropriate for choosing which equipment to buy?

5.37 Peter is the president of a rapidly growing company. There are dozens of important things to do, and cash flow is tight. What comparison method(s) would be appropriate for Peter to make acquisition decisions?

5.38 Lemuel is an engineer working for the electric company. He must compare several routes for transmission lines from a distant nuclear plant to new industrial parks north of the city. What comparison method(s) is he likely to use?

5.39 Vicky runs a music store that has been suffering from thefts. She is considering installing a magnetic tag system. What comparison method(s) would be best for her to use to choose among competing leased systems?

5.40 Thanh's company is growing very fast and has a hard time meeting its orders. An opportunity to purchase additional production equipment has arisen. What comparison method(s) would Thanh use to justify to her manager that the equipment purchase was prudent?

More Challenging Problem

5.41 Charro Environmental is considering taking over a contaminated building site on a former military base. In return for $10 000 000 from the government, it will invest $8 000 000 per year for the following three years to clean up the site. Once the site is clean (end of year 3), it will receive a further $15 000 000 from the government. Over the following two years it will invest $5 000 000 per year constructing a new commercial building on the site. The new building will last forever and net $5 000 000 per year for Charro from tenant leases (starting at the end of year 6). If Charro's MARR is 10%, what is the exact ERR for this project? Should Charro proceed with the project? (*Note:* All disbursements can be assumed to occur at the end of the year.)

MINI-CASE 5.1

The Galore Creek Project

NovaGold Resources is a former gold exploration company that has recently been transforming itself into a gold producer. Its first independent development is the Galore Creek Project. It is also involved as a partner with Placer Dome in another project, and with Rio Tinto in a third. Galore Creek is expected to produce an average of 7650 kilograms of gold, 51 030 kilograms of silver, and 5 670 000 kilograms of copper over its first five years.

In a news release, NovaGold reported that an independent engineering services company calculated that the project would pay back the US$500 million mine capital costs in 3.4 years of a 23-year life. They also calculated a pre-tax rate of return of 12.6% and an undiscounted after-tax NPV of US$329 million. All of these calculations were done at long-term average metal prices. At then-current metal prices the pre-tax rate of return almost doubles to 24.3% and the NPV (net present value = present worth) increases to US$1.065 billion.

Source: "Higher Grades and Expanded Tonnage Indicated by Drilling at Galore Creek Gold–Silver–Copper Project," news release, August 18, 2004, NovaGold Resources Inc. site, www.novagold.net, accessed May 11, 2008.

Discussion

Companies have a choice of how to calculate the benefits of a project in order to determine if it is worth doing. They also have a choice of how to report the benefits of a project to others.

NovaGold is a publicly traded company. Because of this, when a large and very important project is being planned, not only does NovaGold want to make good business decisions, but it also must maintain strong investor confidence and interest.

In this news release, payback period, IRR, and NPV were used to communicate the value of the Galore Creek project. However, you need to look carefully at the wording to ensure that you can correctly interpret the claims about the economic viability of the project.

Questions

1. "[A]n independent engineering services company calculated that the project would pay back the US$500 million mine capital costs in 3.4 years of a 23-year life." There are a variety of costs associated with any project. The payback period here is calculated with respect to "mine capital costs." This suggests that there might be "non-mine" capital costs—for example, administrative infrastructure, transportation system, etc. It also means that operating costs are not included in this calculation. What do you think is the effect of calculating the payback period on "mine capital costs" alone?

2. "They also calculated a pre-tax rate of return of 12.6%. . . ." Taxes reduce the profit from an enterprise, and correspondingly reduce the rate of return. As will be seen in Chapter 8, a 50% corporate tax rate is fairly common. Thus if the pre-tax rate is 12.6%, the after-tax rate would be about 6.3%. Does 6.3% seem to you a sufficient return for a capital-intensive, risky project of this nature, given other investment opportunities available?

3. "[A]nd an 'undiscounted' after-tax NPV of US$329 million." The term *undiscounted* means that the present worth of the project was calculated with an interest rate of 0%. Using a spreadsheet, construct a graph showing the present worth of the project for a range of interest rates from 0% to 20%, assuming the annual returns for the project are evenly distributed over the 23-year life of the project. Does the reported value of $US329 million fairly represent a meaningful NPV for the project?

4. The returns for the Galore Creek Project are much more attractive at then-current metal prices, which were significantly higher than long-term average metal prices. Which metal prices are more sensible to use when evaluating the worth of the project?

5. Did NovaGold report its economic evaluation of the Galore Creek Project in an ethical manner?

Appendix 5A Tests for Multiple IRRs

When the IRR method is used to evaluate projects, we have to test for multiple IRRs. If there are undetected multiple IRRs, an IRR might be calculated that seems correct, but is in error. We consider three tests for multiple IRRs, forming essentially a three-step procedure. In the first test, the signs of the cash flows are examined to see if the project is a simple investment. In the second test, the present worth of the project is plotted against the interest rate to search for interest rates at which the present value is zero. In the third test, project balances are calculated. Each of these tests has three possible outcomes.

1. There is definitely a unique IRR and there is no possibility for multiple IRRs.

2. There are definitely multiple IRRs because two or more IRRs have been found.

3. The test outcome is inconclusive; a unique IRR or multiple IRRs are both possible.

The tests are applied sequentially. The second test is applied only if the outcome of the first test is not clear. The third test is applied only if the outcomes of the first two are not clear. Keep in mind that, even after all three tests have been applied, the test outcomes may remain inconclusive.

The first test examines whether the project is simple. Recall that most projects consist of one or more periods of outflows at the start, followed only by one or more periods of inflows; these are called simple investments. Although simple investments guarantee a single IRR, a project that is not simple may have a single IRR or multiple IRRs. Some investment projects have large cash outflows during their lives or at the ends of their lives that cause net cash flows to be negative after years of positive net cash flows. For example, a project that involves the construction of a manufacturing plant may involve a planned expansion of the plant requiring a large expenditure some years after its initial operation. As another example, a nuclear electricity plant may have planned large cash outflows for disposal of spent fuel at the end of its life. Such a project may have a unique IRR, but it may also have multiple IRRs. Consequently we must examine such projects further.

Where a project is not simple, we go to the second test. The second test consists of making a graph plotting present worth against interest rate. Points at which the present worth crosses or just touches the interest-rate axis (i.e., where present worth = 0) are IRRs. (We assume that there is at least one IRR.) If there is more than one such point, we know that there is more than one IRR. A convenient way to produce such a graph is using a spreadsheet. See Example 5A.1.

EXAMPLE 5A.1 (EXAMPLE 5.7 RESTATED)

A project pays $1000 today, costs $5000 a year from now, and pays $6000 in two years. Are there multiple IRRs?

Table 5A.1 was obtained by computing the present worth of the cash flows in Example 5.7 for a variety of interest rates. Figure 5A.1 shows the graph of the values in Table 5A.1.

While finding multiple IRRs in a plot ensures that the project does indeed have multiple IRRs, failure to find multiple IRRs does not necessarily mean that multiple IRRs do not exist. *Any* plot will cover only a finite set of points. There may be values of the interest rate for which the present worth of the project is zero that are not in the range of interest rates used.

Table 5A.1   Spreadsheet Cells Used to Construct Figure 5A.1

Interest Rate, i	Present Worth ($)
0.6	218.8
0.8	74.1
1.0	0.0
1.2	−33.1
1.4	−41.7
1.6	−35.5
1.8	−20.4
2.0	0.0
2.2	23.4
2.4	48.4

Figure 5A.1   Illustration of Two IRRs for Example 5A.1

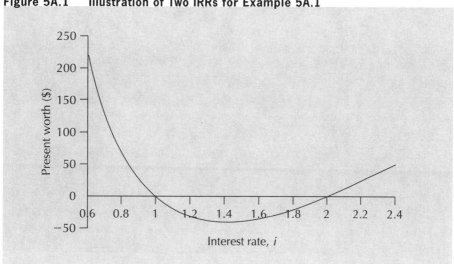

Where the project is not simple and a plot does not show multiple IRRs, we apply the third test. The third test entails calculating the project balances. As we mentioned earlier, project balances refer to the cumulative net cash flows at the end of each time period. For an IRR to be unique, there should be no time when the project balances, computed using that IRR, are positive. This means that there is no extra cash not reinvested in the project. This is a sufficient condition for there to be a unique IRR. (Recall that it is the cash generated by a project, but not reinvested in the project, that creates the possibility of multiple IRRs.)

We now present three examples. All three examples involve projects that are not simple investments. In the first, a plot shows multiple IRRs. In the second, the plot shows only a single IRR. This is inconclusive, so project balances are computed. None of the project balances is positive, so we know that there is a single IRR. In the third example, the plot shows only one IRR, so the project balances are computed. One of these is positive, so the results of all tests are inconclusive.■

EXAMPLE 5A.2

Wellington Woods is considering buying land that they will log for three years. In the second year, they expect to develop the area that they clear as a residential subdivision that will entail considerable costs. Thus, in the second year, the net cash flow will be negative. In the third year, they expect to sell the developed land at a profit. The net cash flows that are expected for the project are:

End of Year	Cash Flow
0	−100 000
1	440 000
2	−639 000
3	306 000

The negative net cash flow in the second period implies that this is not a simple project. Therefore, we apply the second test. We plot the present worth against interest rates to

Figure 5A.2 **Wellington Woods Present Worth**

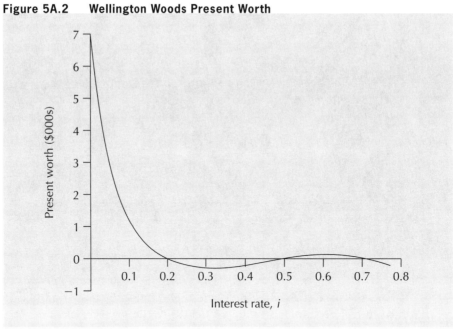

search for IRRs. (See Figure 5A.2.) At 0% interest, the present worth is a small positive amount, $7000. The present worth is then 0 at 20%, 50%, and 70%. Each of these values is an IRR. The spreadsheet cells that were used for the plot are shown in Table 5A.2.

Table 5A.2 **Spreadsheet Cells Used to Generate Figure 5A.2**

Interest Rate	Present Worth	Interest Rate	Present Worth	Interest Rate	Present Worth
0%	$7000.00	28%	− 352.48	56%	$ 79.65
2%	5536.33	30%	− 364.13	58%	92.49
4%	4318.39	32%	− 356.87	60%	97.66
6%	3310.12	34%	− 335.15	62%	94.84
8%	2480.57	36%	− 302.77	64%	83.79
10%	1803.16	38%	− 263.01	66%	64.36
12%	1255.01	40%	− 218.66	68%	36.44
14%	816.45	42%	− 172.11	70%	0.00
16%	470.50	44%	− 125.39	72%	− 44.96
18%	202.55	46%	− 80.20	74%	− 98.41
20%	0.00	48%	− 38.00	76%	− 160.24
22%	− 148.03	50%	0.00	78%	− 230.36
24%	− 250.91	52%	32.80		
26%	− 316.74	54%	59.58		

In this example, a moderately fine grid of two percentage points was used. Depending on the problem, the analyst may wish to use a finer or coarser grid.

The correct decision in this case can be obtained by the approximate ERR method. Suppose the MARR is 15%; then the approximate ERR is the interest rate that makes the future worth of outlays equal to the future worth of receipts when the receipts earn 15%. In other words, the approximate ERR is the value of i that solves

$$100\,000(1 + i)^3 + 639\,000(1 + i) = 440\,000(1.15)^2 + 306\,000$$

$$100\,000(1 + i)^3 + 639\,000(1 + i) = 887\,900$$

Try 15% for i^*_{ea}. Using the tables for the left-hand side of the above equation, we have

$$100\,000(F/P,15\%,3) + 639\,000(F/P,15\%,1)$$

$$= 100\,000(1.5208) + 639\,000(1.15)$$

$$= 887\,000 < 887\,900$$

Thus, the approximate ERR is slightly above 15%. The project is (marginally) acceptable by this calculation because the approximate ERR, which is a conservative estimate of the correct ERR, is above the MARR.■

EXAMPLE 5A.3

Investment in a new office coffeemaker has the following effects:

1. There is a three-month rental fee of $40 for the equipment, payable immediately and in three months.
2. A rebate of $30 from the supplier is given immediately for an exclusive six-month contract.
3. Supplies will cost $20 per month, payable at the beginning of each month.
4. Income from sales will be $30 per month, received at the end of each month.

Will there be more than one IRR for this problem?

We apply the first test by calculating net cash flows for each time period. The net cash flows for this project are as follows:

End of Month	Receipts	Disbursements	Net Cash Flow
0	+ $30	− $40 + (− $20)	− $30
1	+ 30	− 20	+ 10
2	+ 30	− 20	+ 10
3	+ 30	− 40 + (− 20)	− 30
4	+ 30	− 20	+ 10
5	+ 30	− 20	+ 10
6	+ 30	0	+ 30

As illustrated in Figure 5A.3, the net cash flows for this problem do not follow the pattern of a simple investment. Therefore, there may be more than one IRR for this problem. Accordingly, we apply the second test.

Figure 5A.3   Net Cash Flows for the Coffeemaker

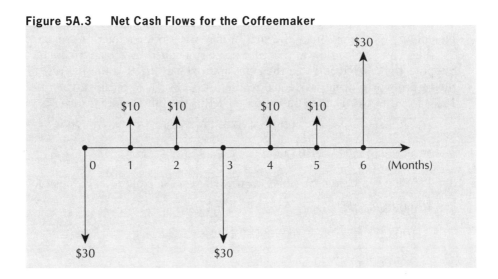

A plot of present worth against the interest rate is shown in Figure 5A.4. The plot starts with a zero interest rate where the present worth is just the arithmetic sum of all the net cash flows over the project's life. This is a positive $10. The present worth as a function of the interest rate decreases as the interest rate increases from zero. The plot continues down, and passes through the interest-rate axis at about 5.8%. There is only one IRR in the range plotted. We need to apply the third test by computing project balances at the 5.8% interest rate.

Figure 5A.4   IRR for New Coffeemaker

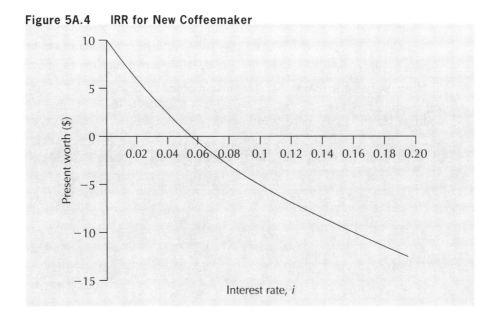

The project balances at the 5.8% interest rate are shown on the next page.

Since all project balances are negative or zero, we know that this investment has only one IRR. It is 5.8% per month or about 69.3% per year.∎

Month	Project Balance ($)
0	$B_0 = -30$
1	$B_1 = -30.0(1.058) + 10 = -21.7$
2	$B_2 = -21.7(1.058) + 10 = -13.0$
3	$B_3 = -13.0(1.058) - 30 = -43.7$
4	$B_4 = -43.7(1.058) + 10 = -36.3$
5	$B_5 = -36.3(1.058) + 10 = -28.4$
6	$B_6 = -28.4(1.058) + 30 = 0$

EXAMPLE 5A.4

Green Woods, like Wellington Woods, is considering buying land that they will log for three years. In the second year, they also expect to develop the area that they have logged as a residential subdivision, which again will entail considerable costs. Thus, in the second year, the net cash flow will be negative. In the third year, they expect to sell the developed land at a profit. But Green Woods expects to do much better than Wellington Woods in the sale of the land. The net cash flows that are expected for the project are:

Year	Cash Flow
0	−$100 000
1	455 000
2	−667 500
3	650 000

The negative net cash flow in the second period implies that this is not a simple project. We now plot the present worth against interest rate. See Figure 5A.5. At zero interest rate, the present worth is a positive $337 500.

The present worth falls as the interest rate rises. It crosses the interest-rate axis at about 206.4%. There are no further crossings of the interest-rate axis in the range plotted, but since this is not conclusive we compute project balances.

Year	Project Balance ($)
0	$B_0 = -100\ 000$
1	$B_1 = -100\ 000(3.064) + 455\ 000 = 148\ 600$
2	$B_2 = -148\ 600(3.064) - 667\ 500 = -2\ 121\ 900$
3	$B_3 = -2\ 121\ 900(3.064) + 650\ 000 = -1500$

We note that the project balance is positive at the end of the first period. This means that a unique IRR is *not* guaranteed. We have gone as far as the three tests can

Figure 5A.5   Present Worth for Example 5A.4

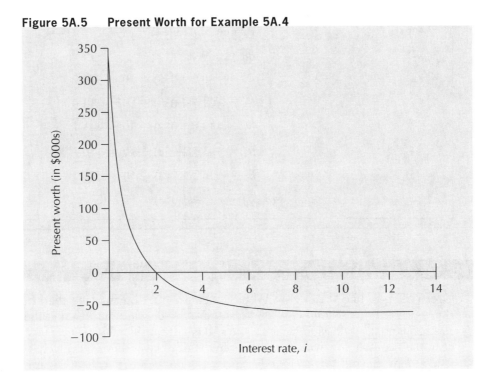

take us. On the basis of the three tests, there may be only the single IRR that we have found, 206.4%, or there may be multiple IRRs.

In this case, we use the approximate ERR to get a decision. Suppose the MARR is 30%. The approximate ERR, then, is the interest rate that makes the future worth of outlays equal to the future worth of receipts when the receipts earn 30%. That is, the approximate ERR is the value of i that solves the following:

$$100\ 000(1 + i)^3 + 667\ 500(1 + i) = 455\ 000(1.3)^2 + 650\ 000$$

Trial and error with a spreadsheet gives the approximate ERR as $i^*_{ea} \cong 57\%$. This is above the MARR of 30%. Therefore, the investment is acceptable.

It is possible, using the precise ERR, to determine that the IRR that we got with the plot of present worth against the interest rate, 206.4%, is, in fact, unique. The precise ERR equals the IRR in this case. Computation of the precise ERR may be cumbersome, and we do not cover this computation in this book. Note, however, that we got the same decision using the approximate ERR as we would have obtained with the precise ERR. Also note that the approximate ERR is a conservative estimate of the precise ERR, which is equal to the unique IRR.■

To summarize, we have discussed three tests that are to be applied sequentially, as shown in Figure 5A.6. The first and easiest test to apply is to see if a project is a simple investment. If it is a simple investment, there is a single IRR, and the correct decision can be obtained by the IRR method. If the project is not a simple investment, we apply the second test, which is to plot the project's present worth against the interest rate. If the plot shows at least two IRRs, we know that there is not a unique IRR, and the correct decision can be obtained with the approximate ERR method. If a plot does not show multiple IRRs, we next compute project balances using the IRR found from the plot. If none of the project balances is positive, the IRR is unique, and the correct decision can be obtained

Figure 5A.6   Tests for Multiple IRRs

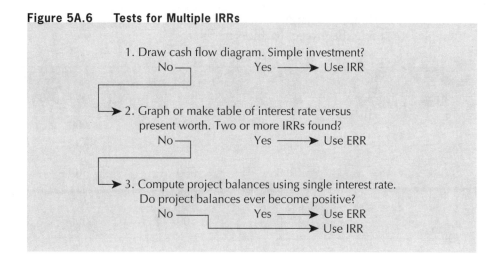

with the IRR method. If one or more of the project balances are positive, we don't know whether there is a unique IRR. Accordingly, we use the approximate ERR method which always will yield a correct decision.

PROBLEMS FOR APPENDIX 5A

5A.1 A five-year construction project for Wawa Engineering receives staged payments in years 2 and 5. The resulting net cash flows are as follows:

Year	Cash Flow
0	−$300 000
1	− 500 000
2	700 000
3	− 400 000
4	− 100 000
5	900 000

The MARR for Wawa Engineering is 15%.

(a) Is this a simple project?

(b) Plot the present worth of the project against interest rates from 0% to 100%. How many times is the interest-rate axis crossed? How many IRRs are there?

(c) Calculate project balances over the five-year life of the project. Can we conclude that the IRR(s) observed in part (b) is (are) the only IRR(s)? If so, should the project be accepted?

(d) Calculate the approximate ERR for this project. Should the project be accepted?

5A.2 For the cash flows associated with the projects below, determine whether there is a unique IRR, using the project balances method.

	Project		
End of Period	**1**	**2**	**3**
0	−$3000	−$1500	$ 600
1	900	7000	− 2000
2	900	− 9000	500
3	900	2900	500
4	900	500	1000

5A.3 A mining opportunity in a third-world country has the following cash flows.

1. $10 000 000 is received at time 0 as an advance against expenses.

2. Costs in the first year are $20 000 000, and in the second year $10 000 000.

3. Over years 3 to 10, annual revenues of $5 000 000 are expected.

After 10 years, the site reverts to government ownership. MARR is 30%.

(a) Is this a simple project?

(b) Plot the present worth of the project against interest rates from 0% to 100%. How many times is the interest-rate axis crossed? How many IRRs are there?

(c) Calculate project balances over the 10-year life of the project. Can we conclude that the IRR(s) observed in part (b) is (are) the only IRR(s)? If so, should the project be accepted?

(d) Calculate the approximate ERR for this project. Should the project be accepted?

(e) Calculate the exact ERR for this project. Should the project be accepted?

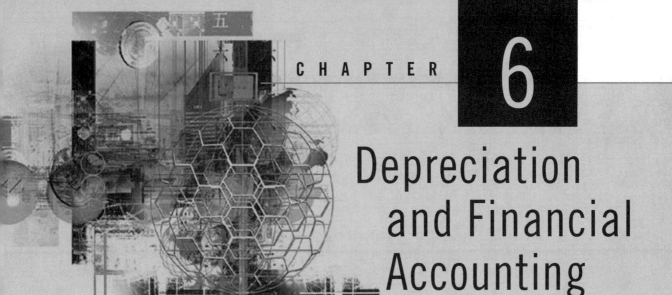

CHAPTER 6

Depreciation and Financial Accounting

Engineering Economics in Action, Part 6A:
The Pit Bull

Naomi liked to think of Terry as a pit bull. Terry had this endearing habit of finding some detail that irked him, and not letting go of it until he was satisfied that things were done properly. Naomi had seen this several times in the months they had worked together. Terry would sink his teeth into some quirk of Global Widgets' operating procedures and, just like a fighting dog, not let go until the fight was over.

This time, it was about the disposal of some computing equipment. Papers in hand, he quietly approached Naomi and earnestly started to explain his concern. "Naomi, I don't know what Bill Fisher is doing, but something's definitely not right here. Look at this."

Terry displayed two documents to Naomi. One was an accounting statement showing the book value of various equipment, including some CAD/CAM computers that had been sold for scrap the previous week. The other was a copy of a sales receipt from a local salvage firm for that same equipment.

"I don't like criticizing my fellow workers, but I really am afraid that Bill might be doing something wrong." Bill Fisher was the buyer responsible for capital equipment at Global Widgets, and he also disposed of surplus assets. "You know the CAD/CAM workstations they had in engineering design? Well, they were replaced recently and sold. Here is the problem. They were only three years old, and our own accounting department estimated their value as about $5000 each." Terry's finger pointed to the evidence on the accounting statement. "But here," his finger moving to the guilty figure on the sales receipt, "they were actually sold for $300 each!" Terry sat back in his chair. "How about that!"

Naomi smiled. Unfortunately, she would have to pry his teeth out of this one. "Interesting observation, Terry. But you know, I think it's probably OK. Let me explain."

6.1 | Introduction

Engineering projects often involve an investment in equipment, buildings, or other assets that are put to productive use. As time passes, these assets lose value, or *depreciate*. The first part of this chapter is concerned with the concept of depreciation and several methods that are commonly used to model depreciation. Depreciation is taken into account when a firm states the value of its assets in its financial statements, as seen in the second half of this chapter. It also forms an important part of the decision of when to replace an aging asset and when to make cyclic replacements, as will be seen in Chapter 7, and has an important impact on taxation, as we will see in Chapter 8.

With the growth in importance of small technology-based enterprises, many engineers have taken on broad managerial responsibilities that include financial accounting. Financial accounting is concerned with recording and organizing the financial data of businesses. The data cover both *flows over time*, like revenues and expenses, and *levels*, like an enterprise's resources and the claims on those resources at a given date. Even engineers who do not have broad managerial responsibilities need to know the elements of financial accounting to understand the enterprises with which they work.

In the second part of this chapter, we explain two basic financial statements used to summarize the financial dimensions of a business. We then explain how these statements can be used to make inferences about the financial health of the firm.

6.2 | Depreciation and Depreciation Accounting

6.2.1 Reasons for Depreciation

An asset starts to lose value as soon as it is purchased. For example, a car bought for $20 000 today may be worth $18 000 next week, $15 000 next year, and $1000 in 10 years. This loss in value, called **depreciation**, occurs for several reasons.

Use-related physical loss: As something is used, parts wear out. An automobile engine has a limited life span because the metal parts within it wear out. This is one reason why a car diminishes in value over time. Often, use-related physical loss is measured with respect to *units of production*, such as thousands of kilometres for a car, hours of use for a light bulb, or thousands of cycles for a punch press.

Time-related physical loss: Even if something is not used, there can be a physical loss over time. This can be due to environmental factors affecting the asset or to endogenous physical factors. For example, an unused car can rust and thus lose value over time. Time-related physical loss is expressed in units of time, such as a 10-year-old car or a 40-year-old sewage treatment plant.

Functional loss: Losses can occur without any physical changes. For example, a car can lose value over time because styles change so that it is no longer fashionable. Other examples of causes of loss of value include legislative changes, such as for pollution control or safety devices, and technical changes. Functional loss is usually expressed simply in terms of the particular unsatisfied function.

6.2.2 Value of an Asset

Models of depreciation can be used to estimate the loss in value of an asset over time, and also to determine the remaining value of the asset at any point in time. This remaining value has several names, depending on the circumstances.

Market value is usually taken as the actual value an asset can be sold for in an open market. Of course, the only way to determine the actual market value for an asset is to sell it. Consequently, the term *market value* usually means an *estimate* of the market value. One way to make such an estimation is by using a depreciation model that reasonably captures the true loss in value of an asset.

Book value is the depreciated value of an asset for accounting purposes, as calculated with a depreciation model. The book value may be more or less than market value. The depreciation model used to arrive at a book value might be controlled by regulation for some purposes, such as taxation, or simply by the desirability of an easy calculation scheme. There might be several different book values for the same asset, depending on the purpose and depreciation model applied. We shall see how book values are reported in financial statements later in this chapter.

Scrap value can be either the actual value of an asset at the end of its physical life (when it is broken up for the material value of its parts) or an estimate of the scrap value calculated using a depreciation model.

Salvage value can be either the actual value of an asset at the end of its useful life (when it is sold) or an estimate of the salvage value calculated using a depreciation model.

It is desirable to be able to construct a good model of depreciation in order to state a book value of an asset for a variety of reasons:

1. In order to make many managerial decisions, it is necessary to know the value of owned assets. For example, money may be borrowed using the firm's assets as collateral. In order to demonstrate to the lender that there is security for the

loan, a credible estimate of the assets' value must be made. A depreciation model permits this to be done. The use of depreciation for this purpose is explored more thoroughly in the second part of this chapter.

2. One needs an estimate of the value of owned assets for planning purposes. In order to decide whether to keep an asset or replace it, you have to be able to judge how much it is worth. More than that, you have to be able to assess how much it will be worth at some time in the future. The impact of depreciation in replacement studies is covered in Chapter 7.

3. Government tax legislation requires that taxes be paid on company profits. Because there can be many ways of calculating profits, strict rules are made concerning how to calculate income and expenses. These rules include a particular scheme for determining depreciation expenses. This use of depreciation is discussed more thoroughly in Chapter 8.

To match the way in which certain assets depreciate and to meet regulatory or accuracy requirements, many different depreciation models have been developed over time. Of the large number of depreciation schemes available (see Close-Up 6.1), straight-line and declining-balance are certainly the most commonly used. Straight-line depreciation is popular primarily because it is particularly easy to calculate. The declining-balance method is required by tax law in many jurisdictions for determining corporate taxes, as is discussed in Chapter 8. In particular, straight-line and declining-balance depreciation are the only ones necessary for corporate tax calculations in Canada, the UK, Australia, and the United States. Consequently, these are the only depreciation methods presented in detail in this book.

6.2.3 Straight-Line Depreciation

The **straight-line method of depreciation** assumes that the rate of loss in value of an asset is constant over its useful life. This is illustrated in Figure 6.1 for an asset worth

CLOSE-UP 6.1	Depreciation Methods
Method	**Description**
Straight-line	The book value of an asset diminishes by an equal *amount* each year.
Declining-balance	The book value of an asset diminishes by an equal *proportion* each year.
Sum-of-the-years'-digits	An accelerated method, like declining-balance, in which the depreciation rate is calculated as the ratio of the remaining years of life to the sum of the digits corresponding to the years of life.
Double-declining-balance	A declining-balance method in which the depreciation rate is calculated as $2/N$ for an asset with a service life of N years.
150%-declining-balance	A declining-balance method in which the depreciation rate is calculated as $1.5/N$ for an asset with a service life of N years.
Units-of-production	Depreciation rate is calculated per unit of production as the ratio of the units produced in a particular year to the total estimated units produced over the asset's lifetime.

**Figure 6.1   Book Value Under Straight-Line Depreciation
($1000 Purchase and $200 Salvage Value After Eight Years)**

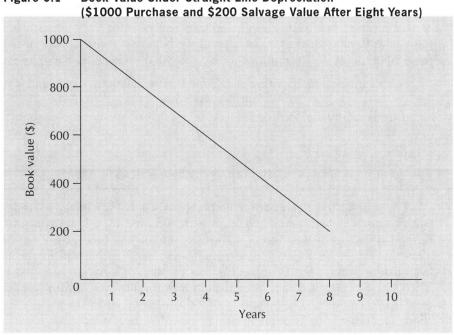

$1000 at the time of purchase and $200 eight years later. Graphically, the book value of the asset is determined by drawing a straight line between its first cost and its salvage or scrap value.

Algebraically, the assumption is that the rate of loss in asset value is constant and is based on its original cost and salvage value. This gives rise to a simple expression for the depreciation charge per period. We determine the depreciation per period from the asset's current value and its estimated salvage value at the end of its useful life, N periods from now, by

$$D_{sl}(n) = \frac{P - S}{N} \tag{6.1}$$

where

$D_{sl}(n)$ = the depreciation charge for period n using the straight-line method

P = the purchase price or current market value

S = the salvage value after N periods

N = the useful life of the asset, in periods

Similarly, the book value at the end of any particular period is easy to calculate:

$$BV_{sl}(n) = P - n\left[\frac{P - S}{N}\right] \tag{6.2}$$

where

$BV_{sl}(n)$ = the book value at the end of period n using straight-line depreciation

EXAMPLE 6.1

A laser cutting machine was purchased four years ago for £380 000. It will have a salvage value of £30 000 two years from now. If we believe a constant rate of depreciation is a reasonable means of determining book value, what is its current book value?

From Equation (6.2), with P = £380 000, S = £30 000, N = 6, and n = 4,

$$BV_{sl}(4) = 380\,000 - 4\left[\frac{380\,000 - 30\,000}{6}\right]$$

$$BV_{sl}(4) = 146\,667$$

The current book value for the cutting machine is £146 667. ∎

The straight-line depreciation method has the great advantage of being easy to calculate. It also is easy to understand and is in common use. The main problem with the method is that its assumption of a constant rate of loss in asset value is often not valid. Thus, book values calculated using straight-line depreciation will frequently be different from market values. For example, the loss in value of a car over its first year (say, from $20 000 to $15 000) is clearly more than its loss in value over its fifth year (say, from $6000 to $5000). The declining-balance method of depreciation covered in the next section allows for "faster" depreciation in earlier years of an asset's life.

6.2.4 Declining-Balance Depreciation

Declining-balance depreciation (also known as *reducing-balance depreciation*) models the loss in value of an asset over a period as a constant fraction of the asset's current value. In other words, the depreciation charge in a particular period is a constant proportion (called the depreciation rate) of its closing book value from the previous period. The effect of various depreciation rates on book values is illustrated in Figure 6.2.

Figure 6.2   Book Value Under Declining-Balance Depreciation ($1000 Purchase With Various Depreciation Rates)

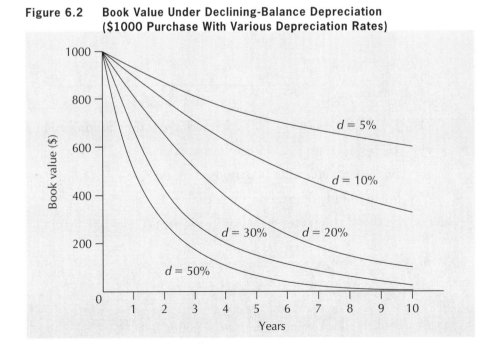

Algebraically, the depreciation charge for period n is simply the depreciation rate multiplied by the book value from the end of period $(n-1)$. Noting that $BV_{db}(0) = P$,

$$D_{db}(n) = BV_{db}(n-1) \times d \qquad (6.3)$$

where

$D_{db}(n)$ = the depreciation charge in period n using the declining-balance method

$BV_{db}(n)$ = the book value at the end of period n using the declining-balance method

d = the depreciation rate

Similarly, the book value at the end of any particular period is easy to calculate, by noting that the remaining value after each period is $(1-d)$ times the value at the end of the previous period.

$$BV_{db}(n) = P(1-d)^n \qquad (6.4)$$

where

P = the purchase price or current market value

In order to use the declining-balance method of depreciation, we must determine a reasonable depreciation rate. By using an asset's current value, P, and a salvage value, S, n periods from now, we can use Equation (6.4) to find the declining balance rate that relates P and S.

$$BV_{db}(n) = S = P(1-d)^n$$

$$(1-d) = \sqrt[n]{\frac{S}{P}}$$

$$d = 1 - \sqrt[n]{\frac{S}{P}} \qquad (6.5)$$

EXAMPLE 6.2

Paquita wants to estimate the scrap value of a smokehouse 20 years after purchase. She feels that the depreciation is best represented using the declining-balance method, but she doesn't know what depreciation rate to use. She observes that the purchase price of the smokehouse was €245 000 three years ago, and an estimate of its current salvage value is €180 000. What is a good estimate of the value of the smokehouse after 20 years?

From Equation (6.5),

$$d = 1 - \sqrt[n]{\frac{S}{P}}$$

$$= 1 - \sqrt[3]{\frac{180\ 000}{245\ 000}}$$

$$= 0.097663$$

Then, by using Equation (6.4), we have

$$BV_{db}(20) = 245\ 000(1 - 0.097663)^{20} = 31\ 372$$

An estimate of the salvage value of the smokehouse after 20 years using the declining-balance method of depreciation is €31 372. ∎

The declining-balance method has a number of useful features. For one thing, it matches the observable loss in value that many assets have over time. The rate of loss is expressed in one parameter, the depreciation rate. It is relatively easy to calculate, although perhaps not quite as easy as the straight-line method. In particular, it is often required to be used for taxation purposes, as discussed in detail in Chapter 8.

EXAMPLE 6.3

Sherbrooke Data Services has purchased a new mass storage system for $250 000. It is expected to last six years, with a $10 000 salvage value. Using both the straight-line and declining-balance methods, determine the following:

(a) The depreciation charge in year 1
(b) The depreciation charge in year 6
(c) The book value at the end of year 4
(d) The accumulated depreciation at the end of year 4

This is an ideal application for a spreadsheet solution. Table 6.1 illustrates a spreadsheet that calculates the book value, depreciation charge, and accumulated depreciation for both depreciation methods over the six-year life of the system.

Table 6.1   Spreadsheet for Example 6.3

	Straight-Line Depreciation		
Year	Depreciation Charge	Accumulated Depreciation	Book Value
0			$250 000
1	$40 000	$ 40 000	210 000
2	40 000	80 000	170 000
3	40 000	120 000	130 000
4	40 000	160 000	90 000
5	40 000	200 000	50 000
6	40 000	240 000	10 000

	Declining-Balance Depreciation		
Year	Depreciation Charge	Accumulated Depreciation	Book Value
0			$250 000
1	$103 799	$103 799	146 201
2	60 702	164 501	85 499
3	35 499	200 000	50 000
4	20 760	220 760	29 240
5	12 140	232 900	17 100
6	7 100	240 000	10 000

The depreciation charge for each year with the *straight-line* method is $40 000:

$$D_{sl}(n) = (250\,000 - 10\,000)/6 = 40\,000$$

The depreciation rate for the *declining-balance* method is

$$d = 1 - \sqrt[n]{\frac{S}{P}} = 1 - \sqrt[6]{\frac{10\,000}{250\,000}} = 0.4152$$

The detailed calculation for each of the questions is as follows:

(a) Depreciation charge in year 1

$$D_{sl}(1) = (250\,000 - 10\,000)/6 = 40\,000$$
$$D_{db}(1) = BV_{db}(0)d = 250\,000(0.4152) = 103\,799.11$$

(b) Depreciation charge in year 6

$$D_{sl}(6) = D_{sl}(1) = 40\,000$$
$$D_{db}(6) = BV_{db}(5)d = 250\,000(0.5848)^5(0.4152) = 7099.82$$

(c) Book value at the end of year 4

$$BV_{sl}(4) = 250\,000 - 4(250\,000 - 10\,000)/6 = 90\,000$$

$$BV_{db}(4) = 250\,000(1 - 0.4152)^4 = 29\,240.17$$

(d) Accumulated depreciation at the end of year 4
Using the straight-line method: $P - BV_{sl}(4) = 160\,000$
Using the declining-balance method: $P - BV_{db}(4) = 220\,759.83$ ■

In summary, depreciation affects economic analyses in several ways. First, it allows us to estimate the value of an owned asset, as illustrated in the above examples. We shall see in the next part of this chapter how these values are reported in a firm's financial statements. Next, the capability of estimating the value of an asset is particularly useful in replacement studies, which is the topic of Chapter 7. Finally, in Chapter 8, we cover aspects of tax systems that affect decision making; in particular we look at the effect of depreciation expenses.

6.3 | Elements of Financial Accounting

How well is a business doing? Can it survive an unforeseen temporary drop in cash flows? How does a business compare with others of its size in the industry? Answering these questions and others like them is part of the accounting function. The accounting function has two parts, financial accounting and management accounting. **Financial accounting** is concerned with recording and organizing the financial data of a business, which include revenues and expenses, and an enterprise's resources and the claims on those resources. **Management accounting** is concerned with the costs and benefits of the various activities of an enterprise. The goal of management accounting is to provide managers with information to help in decision making.

Engineers have always played a major role in management accounting, especially in a part of management accounting called *cost* accounting. They have not, for the most part, had significant responsibility for financial accounting until recently. With the growth in importance of small technology-based enterprises, many engineers have taken on broad managerial responsibilities that include financial accounting. Even engineers who do not have broad managerial responsibilities need to know the elements of financial accounting to understand the enterprises in which they work. Management accounting is not covered in this text because it is difficult to provide useful information without taking more than a single chapter. Instead, we focus on financial accounting.

The object of financial accounting is to provide information to internal management and interested external parties. Internally, management uses financial accounting information for processes such as budgeting, cash management, and management of long-term debt. External users include actual and potential investors and creditors who wish to make rational decisions about an enterprise. External users also include government agencies concerned with taxes and regulation.

Areas of interest to all these groups include an enterprise's revenues and expenses, and assets (resources held by the enterprise) and liabilities (claims on those resources).

In the next few sections, we discuss two basic summary financial statements that give information about these matters: the *balance sheet* and the *income statement*. These statements form the basis of a financial report, which is usually produced on a monthly, quarterly, semiannual, or yearly basis. Following the discussion of financial statements, we shall consider the use of information in these statements when making inferences about an enterprise's performance compared with industry standards and with its own performance over time.

6.3.1 Measuring the Performance of a Firm

The flow of money in a company is much like the flow of water in a network of pipes or the flow of electricity in an electrical circuit, as illustrated in Figure 6.3. In order to measure the performance of a water system, we need to determine the flow through the system and the pressure in the system. For an electrical circuit, the analogous parameters are current and voltage. Flow and current are referred to as *through variables*, and are measured with respect to time (flow is litres per second and current is amperes, which are coulombs per second). Pressure and voltage are referred to as *across variables*, and are measured at a point in time.

The flow of money in an organization is measured in a similar way with the income statement and balance sheet. The income statement represents a *through variable* because it summarizes revenues and expenses over a period of time. It is prepared by listing the revenues earned during a period and the expenses incurred during the same period, and by subtracting total expenses from total revenues, arriving at a net income. An income statement is always associated with a particular period of time, be it a month, quarter, or year.

The balance sheet, in contrast to the income statement, is a snapshot of the financial position of a firm at a particular point in time, and so represents an *across variable*. The financial position is summarized by listing the assets of the firm, its liabilities (debts), and the equity of the owner or owners.

6.3.2 The Balance Sheet

A **balance sheet** (also called a *position statement*) is a snapshot of an enterprise's financial position at a particular point in time, normally the last day of an accounting period. A firm's financial position is summarized in a balance sheet by listing its assets, liabilities, and owners' (or shareholders') equity. The heading of the balance sheet gives the name of the enterprise and the date.

Figure 6.3   Through and Across Variables

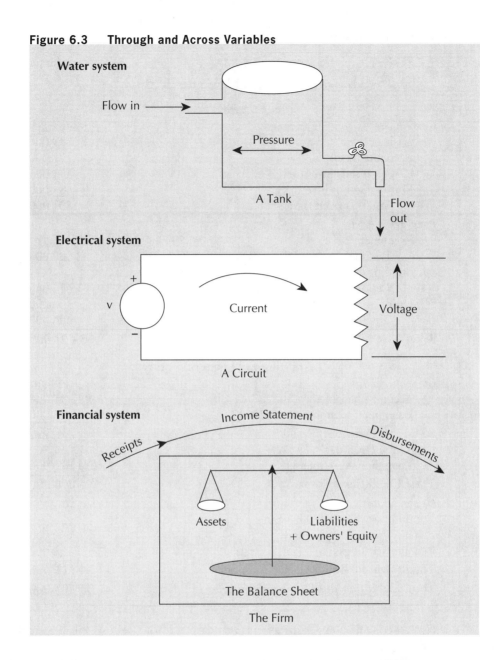

EXAMPLE 6.4

Table 6.2 shows a balance sheet for the Major Electric Company, a manufacturer of small electrical appliances.

The first category of financial information in a balance sheet reports the **assets** of the enterprise. These are the economic resources owned by the enterprise, or more simply, everything that the enterprise owns. Assets are classified on a balance sheet as current assets and long-term assets. **Current assets** are cash and other assets that could be converted to cash within a relatively short period of time, usually a year or less. Inventory and accounts receivable are examples of non-cash current assets. **Long-term assets** (also called fixed assets or non-current assets) are assets that are not expected to be converted to cash in the short term, usually taken to be one year. Indeed, it may be

Table 6.2   **Balance Sheet for the Major Electric Company**

Major Electric Company Balance Sheet as of November 30, 2010		
Assets		
Current Assets		
Cash		$ 39 000
Accounts receivable		27 000
Raw materials inventory		52 000
Finished goods inventory		683 000
Total Current Assets		$ 801 000
Long-Term Assets		
Equipment	$6 500 000	
Less accumulated depreciation	4 000 000	2 500 000
Buildings	1 750 000	
Less accumulated depreciation	150 000	1 600 000
Land		500 000
Total Long-Term Assets		$ 4 600 000
Total Assets		**$5 401 000**
Liabilities and Owners' Equity		
Current Liabilities		
Accounts payable		$ 15 000
Loan due December 31, 2010		75 000
Total Current Liabilities		90 000
Long-Term Liabilities		
Loan due December 31, 2013		1 000 000
Total Liabilities		**$1 090 000**
Owners' Equity		
Common stock: 1 000 000 shares at $3 par value per share		$ 3 000 000
Retained earnings		1 311 000
Total Owners' Equity		**$4 311 000**
Total Liabilities and Owners' Equity		**$5 401 000**

difficult to convert long-term assets into cash without selling the business as a going concern. Equipment, land, and buildings are examples of long-term assets.

An enterprise's **liabilities** are claims, other than those of the owners, on a business's assets, or simply put, everything that the enterprise owes. Debts are usually the most important liabilities on a balance sheet. There may be other forms of liabilities as well. A commitment to the employees' pension plan is an example of a non-debt liability. As with assets, liabilities may be classified as current or long-term. **Current liabilities** are liabilities that are due within some short period of time, usually a year or less. Examples of current liabilities are debts that are close to maturity, accounts payable to suppliers, and taxes due. **Long-term liabilities** are liabilities that are not expected to draw on the business's current assets. Long-term loans and bonds issued by the business are examples of long-term liabilities.

The difference between a business's assets and its liabilities is the amount due to the owners—their equity in the business. That is, owners' equity is what is left over from assets after claims of others are deducted. We have, therefore,

Owners' equity = Assets − Liabilities

or

Assets = Liabilities + Owners' equity

Owners' equity is the interest of the owner or owners of a firm in its assets. For the basic types of ownership structure in business organizations, see Close-Up 6.2. Owners' equity usually appears as two components on a balance sheet of a corporation. The first is the par value of the owners' shares. When an enterprise is first organized, it is authorized to issue a certain number of shares. **Par value** is the price per share set by the corporation at the time the shares are originally issued. At any time after the first sale, the shares may be traded at prices that are greater than or less than the par value, depending on investors' expectations of the return that will be earned by the business in the future. There is no reason to expect the market price to equal the par value for very long after the shares are first sold. Nonetheless, the amount recorded in the balance sheet is the original par value. New shares sold anytime after the first issue may have a par value of their own, distinct from those of the original issue.

The second part of owners' equity usually shown on the balance sheet is retained earnings. **Retained earnings** includes the cumulative sum of earnings from normal operations, in addition to gains (or losses) from transactions like the sale of plant assets or

CLOSE-UP 6.2	Types of Business Ownerships

There are three basic ways to structure a business organization.

A **sole proprietorship** (sole trader, single proprietor) is a business owned by one person. It is the simplest and least regulated form of business to start (in essence, all you need is a business name), and accounts for the largest number of businesses. Under a sole proprietorship, the owner keeps all the profits, but at the same time, has *unlimited liability*; that is, the owner is personally responsible for all business debts and the creditors can come after even his or her personal assets in order to recoup debts.

A **partnership** is a business owned by two or more owners (partners). In a *general partnership*, the partners run the business together and share all profits and losses (unlimited liability) according to a partnership agreement. In a *limited partnership*, some partners are involved only as investors (limited partners) and they let one or more general partners take charge of day-to-day operation. The limited partners have *limited liability*; that is, they are only liable for up to the amount of their investment, and their personal assets are protected from the creditors of business debts.

A **corporation** is owned by shareholders. The shareholders elect the board of directors, and the board is responsible for selecting the managers to run the business in the interest of shareholders. A corporation is set up as a business entity, with its own rights and responsibilities, separate from the owners. This means that the corporation is responsible for its own debts, and the owners have limited liability (up to the amount of their investment). In the United Kingdom, a corporation is called either a *limited company* (Limited or Ltd.) or *public limited company* (PLC). Other terms include *Aktiengesellschaft* (AG; Germany, Austria, Switzerland), *Société anonyme* (SA; France), *Naamloze Vennootschap* (NV; The Netherlands).

Generally speaking, the main advantage of incorporating a business is its relative ease in raising capital for a growth opportunity. Sole proprietorships and partnerships, although easier to start up than corporations, are both disadvantaged by personal liability issues, limited availability of equity, and difficulty of ownership transfer, which all contribute to potential difficulties in raising sufficient funds for growth.

investments the proceeds of which have been reinvested in the business (i.e., not paid out as dividends). Firms retain earnings mainly to expand operations through the purchase of additional assets. Contrary to what one may think, retained earnings do not represent cash. They may be invested in assets such as equipment and inventory.

The balance sheet gets its name from the idea that the total assets are equal in value to or *balanced by* the sum of the total liabilities and the owners' equity. A simple way of thinking about it is that the capital used to buy each asset has to come from debt (liabilities) and/or equity (owners' equity). At a company's start-up, the original shareholders may provide capital in the form of equity, and there will also likely be debt capital. The general term used to describe the markets in which short or long-term debt and equity are exchanged is the **financial market**. The capital provided through a financial market finances the assets and working capital for production. As the company undertakes its business activities, it will make or lose money. As it does so, assets will rise or fall, and equity and/or debt will rise or fall correspondingly. For example, if the company makes a profit, the profits will either pay off debts, be invested in new assets, or be paid as dividends to shareholders. Figure 6.4 provides an overview of the sources and uses of capital in an organization.■

**Figure 6.4   Cash Flow Relationship Between the Company's Assets
and the Financial Markets**

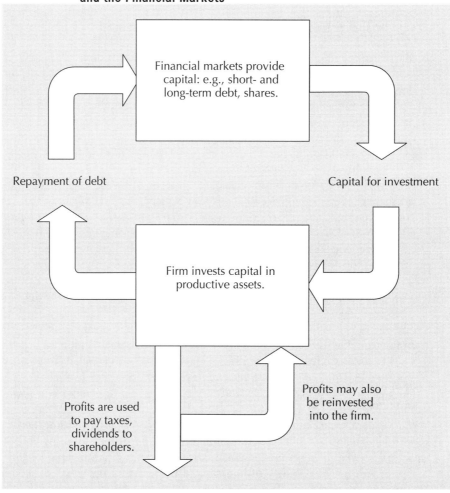

EXAMPLE 6.5

Ian Claymore is the accountant at Major Electric. He has just realized that he forgot to include in the balance sheet for November 30, 2010, a government loan of $10 000 to help in the purchase of a $25 000 test stand (which he also forgot to include). The loan is due to be repaid in two years. When he revises the statement, what changes should he make?

The government loan is a long-term liability because it is due in more than one year. Consequently, an extra $10 000 would appear in long-term liabilities. This extra $10 000 must also appear as an asset for the balance to be maintained. The $25 000 value of the test stand would appear as equipment, increasing the equipment amount to $6 525 000. The $15 000 extra must come from a decrease in cash from $39 000 to $24 000.

Depending on the timing of the purchase, depreciation for the test stand might also be recognized in the balance sheet. Depreciation would reduce the net value of the equipment by the depreciation charge. The same amount would be balanced in the liabilities section by a reduction in retained earnings.■

6.3.3 The Income Statement

An **income statement** summarizes an enterprise's revenues and expenses over a specified accounting period. Months, quarters, and years are commonly used as reporting periods. As with the balance sheet, the heading of an income statement gives the name of the enterprise and the reporting period. The income statement first lists revenues, by type, followed by expenses. Expenses are then subtracted from revenues to give income (or profit) before taxes. Income taxes are then deducted to obtain net income. See Close-Up 6.3 for a measure used for operating profit.

The income statement summarizes the revenues and expenses of a business over a period of time. However, it does not directly give information about the generation of cash. For this reason, it may be useful to augment the income statement with a **statement of changes in financial position** (also called a cash flow statement, a statement of sources and uses of funds, or a funds statement), which shows the amounts of cash generated by a company's operation and by other sources, and the amounts of cash used for investments and other non-operating disbursements.

CLOSE-UP 6.3	Earnings Before Interest and Income Tax (EBIT)

Income before taxes *and* before interest payments, that is, total revenue minus operating expenses (all expenses except for income tax and interest), is commonly referred to as the **earnings before interest and taxes** (EBIT). EBIT measures the company's operating profit, which results from making sales and controlling operating expenses. Due to its focus on operating profit, EBIT is often used to judge whether there is enough profit to recoup the cost of capital (see Appendix 4A).

EXAMPLE 6.6

The income statement for the Major Electric Company for the year ended November 30, 2010, is shown in Table 6.3.

We see that Major Electric's largest source of revenue was the sale of goods. The firm also earned revenue from the sale of management services to other companies. The largest

Table 6.3   Income Statement for the Major Electric Company

Major Electric Company Income Statement for the Year Ended November 30, 2010		
Revenues		
Sales	$7 536 000	
Management fees earned	106 000	
Total Revenues		**$7 642 000**
Expenses		
Cost of goods sold	$6 007 000	
Selling costs	285 000	
Administrative expenses	757 000	
Interest paid	86 000	
Total Expenses		**$7 135 000**
Income before taxes		$507 000
Taxes (at 40%)		202 800
Net Income		**$304 200**

expense was the cost of the goods sold. This includes the cost of raw materials, production costs, and other costs incurred to produce the items sold. Sometimes firms choose to list cost of goods sold as a *negative revenue*. The cost of goods sold will be subtracted from the sales to give a net sales figure. The net sales amount is the listed revenue, and the cost of goods sold does not appear as an expense.

The particular entries listed as either revenues or expenses will vary, depending on the nature of the business and the activities carried out. All revenues and expenses appear in one of the listed categories. For Major Electric, for example, the next item on the list of expenses, selling costs, includes delivery cost and other expenses such as salespersons' salaries. Administrative expenses include those costs incurred in running the company that are not directly associated with manufacturing. Payroll administration is an example of an administrative expense.

Subtracting the expenses from the revenues gives a measure of profit for the company over the accounting period, one year in the example. However, this profit is taxed at a rate that depends on the company's particular characteristics. For Major Electric, the tax rate is 40%, so the company's profit is reduced by that amount.■

EXAMPLE 6.7

Refer back to Example 6.5. Ian Claymore also forgot to include the effects of the loan and test stand purchase in the income statement shown in Example 6.6. When he revises the statement, what changes should he make?

Neither the loan nor the purchase of the asset appears directly in the income statement. The loan itself is neither income nor expense; if interest is paid on it, this is an expense. The test stand is a depreciable asset, which means that only the depreciation for the test stand appears as an expense.■

6.3.4 Estimated Values in Financial Statements

The values in financial statements appear to be authoritative. However, many of the values in financial statements are estimates based on the **cost principle of accounting**. The cost principle of accounting states that assets are to be valued on the basis of their cost as opposed to market or other values. For example, the $500 000 given as the value of the land held by Major Electric is what Major Electric paid for the land. The market value of the land may now be greater or less than $500 000.

The value of plant and equipment is also based on cost. The value reported in the balance sheet is given by the initial cost minus accumulated depreciation. If Major Electric tried to sell the equipment, they might get more or less than this because depreciation models only approximate market value. For example, if there were a significant improvement in new equipment offered now by equipment suppliers compared with when Major Electric bought their equipment, Major Electric might get less than the $2 500 000 shown on the balance sheet.

Consider the finished goods inventory as another example of the cost principle of accounting. The value reported is Major Electric's manufacturing cost for producing the items. The implicit assumption being made is that Major Electric will be able to sell these goods for at least the cost of producing them. Their value may be reduced in later balance sheets if it appears that Major Electric cannot sell the goods easily.

The value shown for accounts receivable is clearly an estimate. Some fraction of accounts receivable may never be collected by Major. The value in the balance sheet reflects what the accountant believes to be a conservative estimate based on experience.

In summary, when examining financial statement data, it is important to remember that many reported values are estimates. Most firms include their accounting methods and assumptions within their periodic reports to assist in the interpretation of the statements.

6.3.5 Financial Ratio Analysis

Performance measures are calculated values that allow conclusions to be drawn from data. Performance measures drawn from financial statements can be used to answer such questions as:

1. Is the firm able to meet its short-term financial obligations?
2. Are sufficient profits being generated from the firm's assets?
3. How dependent is the firm on its creditors?

Financial ratios are one kind of performance measure that can answer these questions. They give an analyst a framework for asking questions about the firm's liquidity, asset management, leverage, and profitability. Financial ratios are ratios between key amounts taken from the financial statements of the firm. While financial ratios are simple to compute, they do require some skill to interpret, and they may be used for different purposes. For example, internal management may be concerned with the firm's ability to pay its current liabilities or the effect of long-term borrowing for a plant expansion. An external investor may be interested in past and current earnings to judge the wisdom of investing in the firm's stock. A bank will assess the riskiness of lending money to a firm before extending credit.

To properly interpret financial ratios, analysts commonly make comparisons with ratios computed for the same company from previous financial statements (a **trend analysis**) and with industry standard ratios. This is referred to as **financial ratio analysis**.

Industry standards can be obtained from various commercial and government websites and publications. In Canada, Statistics Canada (www.statscan.ca) publishes *Financial and*

NET VALUE 6.1

Securities Regulators

Countries that trade in corporate stocks and bonds (collectively called *securities*) generally have a regulatory body to protect investors, ensure that trading is fair and orderly, and facilitate the acquisition of capital by businesses.

In Canada, the Canadian Securities Administrators (CSA) coordinates regulators from each of Canada's provinces and territories, and also educates Canadians about the securities industry, the stock markets, and how to protect investors from scams by providing a variety of educational materials on securities and investing.

In the United States, the Securities Exchange Commission (SEC) regulates securities for the country. The SEC has been particularly active in enforcement in recent years. In the United Kingdom, the Financial Services Authority (FSA) regulates the securities industry as well as other financial services such as banks. In Australia, the regulator is the Australian Securities and Investments Commission (ASIC).

Australia: **www.asic.gov.au**

Canada: **www.csa-acvm.ca**

United Kingdom: **www.fsa.gov.uk**

United States: **www.sec.gov**

Taxation Statistics for Enterprises, which lists financial data from the balance sheets and income statements as well as selected financial ratios for numerous industries. In the United States, Standard & Poor's *Industry Surveys* or Dun & Bradstreet's *Industry Handbook* are classic commercial sources of information, and less recent information for some industries can be found at the US Census Bureau site (www.census.gov). In the United Kingdom, the Centre for Interfirm Comparison (www.cifc.co.uk) is a commercial source for industry norms. Finally, in Australia, industry norms are available on a tax-year basis from the Australian Taxation Office website (www.ato.gov.au). Examples of a past industry-total balance sheet and income statement (in millions of dollars) for the electronic products manufacturing industry are shown in Tables 6.4 and 6.5. These statistics allow an analyst to compare an individual firm's financial statements with those of the appropriate industry.

We shall see in the next section how the financial ratios derived from industry-total financial data can be used to assess the health of a firm.

6.3.6 Financial Ratios

Numerous financial ratios are used in the financial analysis of a firm. Here we shall introduce six commonly used ratios to illustrate their use in a comparison with ratios from the industry-total data and in trend analysis. To facilitate the discussion, we shall use Tables 6.6 and 6.7, which show the balance sheets and income statements for Electco Electronics, a small electronics equipment manufacturer.

The first two financial ratios we address are referred to as **liquidity ratios**. Liquidity ratios evaluate the ability of a business to meet its current liability obligations. In other words, they help us evaluate its ability to weather unforeseen fluctuations in cash flows. A company that does not have a reserve of cash, or other assets that can be converted into cash easily, may not be able to fulfill its short-term obligations.

A company's net reserve of cash and assets easily converted to cash is called its working capital. **Working capital** is simply the difference between total current assets and total current liabilities:

Working capital = Current assets – Current liabilities

Table 6.4   Example of Industry-Total Balance Sheet (in Millions of $)

Electronic Product Manufacturing Balance Sheet	
Assets	
Assets	
Cash and deposits	$ 2 547
Accounts receivable and accrued revenue	9 169
Inventories	4 344
Investments and accounts with affiliates	10 066
Portfolio investments	1 299
Loans	839
Capital assets, net	4 966
Other assets	2 898
Total assets	$36 128
Liabilities and Owners' Equity	
Liabilities	
Accounts payable and accrued liabilities	$ 8 487
Loans and accounts with affiliates	3 645
Borrowings:	
Loans and overdrafts from banks	1 408
Loans and overdrafts from others	1 345
Bankers' acceptances and paper	232
Bonds and debentures	2 487
Mortgages	137
Deferred income tax	(247)
Other liabilities	1 931
Total liabilities	$19 426
Owners' Equity	
Share capital	9 177
Contributed surplus and other	415
Retained earnings	7 110
Total owners' equity	$16 702
Current assets	$17 663
Current liabilities	11 692

The adequacy of working capital is commonly measured with two ratios. The first, the **current ratio**, is the ratio of all current assets relative to all current liabilities. The current ratio may also be referred to as the **working capital ratio**.

$$\text{Current ratio} = \frac{\text{Current assets}}{\text{Current liabilities}}$$

Electco Electronics had a current ratio of 4314/2489 = 1.73 in 2008 (Table 6.6). Ordinarily, a current ratio of 2 is considered adequate, although this determination may depend a great deal on the composition of current assets. It also may depend on industry standards. In the case of Electco, the industry standard is 1.51, which is listed in Table 6.8. It would appear that Electco had a reasonable amount of liquidity in 2008 from the industry's perspective.

Table 6.5 Example of Industry-Total Income Statement (Millions of $)

Electronic Product Manufacturing Income Statement	
Operating revenues	**$50 681**
Sales of goods and services	49 357
Other operating revenues	1324
Operating expenses	**48 091**
Depreciation, depletion, and amortization	1501
Other operating expenses	46 590
Operating profit	**2590**
Other revenue	**309**
Interest and dividends	309
Other expenses	**646**
Interest on short-term debt	168
Interest on long-term debt	478
Gains (Losses)	**(122)**
On sale of assets	25
Others	(148)
Profit before income tax	**2130**
Income tax	715
Equity in affiliates' earnings	123
Profit before extraordinary gains	**1538**
Extraordinary gains/losses	—
Net profit	**1538**

A second ratio, the acid-test ratio, is more conservative than the current ratio. The **acid-test ratio** (also known as the **quick ratio**) is the ratio of quick assets to current liabilities:

$$\text{Acid-test ratio} = \frac{\text{Quick assets}}{\text{Current liabilities}}$$

The acid-test ratio recognizes that some current assets, for example, inventory and prepaid expenses, may be more difficult to turn into cash than others. *Quick assets* are cash, accounts receivable, notes receivable, and temporary investments in marketable securities—those current assets considered to be highly *liquid*.

The acid-test ratio for Electco for 2008 was (431 + 2489)/2489 = 1.17. Normally, an acid-test ratio of 1 is considered adequate, as this indicates that a firm could meet all its current liabilities with the use of its *quick* current assets if it were necessary. Electco appears to meet this requirement.

The current ratio and the acid-test ratio provide important information about how liquid a firm is, or how well it is able to meet its current financial obligations. The extent to which a firm relies on debt for its operations can be captured by what are called **leverage** or **debt-management ratios**. An example of such a ratio is the **equity ratio**. It is the ratio of total owners' equity to total assets. The smaller this ratio, the more dependent the firm is on debt for its operations and the higher are the risks the company faces.

Table 6.6   Balance Sheets for Electco Electronics for Years Ended 2008, 2009, and 2010

Electco Electronics Year-End Balance Sheets *(in thousands of dollars)*			
	2008	**2009**	**2010**
Assets			
Current assets			
Cash	431	340	320
Accounts receivable	2489	2723	2756
Inventories	1244	2034	2965
Prepaid services	150	145	149
Total current assets	4314	5242	6190
Long-term assets			
Buildings and equipment (net of depreciation)	3461	2907	2464
Land	521	521	521
Total long-term assets	3982	3428	2985
Total Assets	8296	8670	9175
Liabilities			
Current liabilities			
Accounts payable	1493	1780	2245
Bank overdraft	971	984	992
Accrued taxes	25	27	27
Total current liabilities	2489	2791	3264
Long-term liabilities			
Mortgage	2489	2455	2417
Total Liabilities	4978	5246	5681
Owners' Equity			
Share capital	1825	1825	1825
Retained earnings	1493	1599	1669
Total Owners' Equity	3318	3424	3494
Total Liabilities and Owners' Equity	8296	8670	9175

$$\text{Equity ratio} = \frac{\text{Total owners' equity}}{\text{Total liabilities} + \text{Total equity}}$$

$$= \frac{\text{Total owners' equity}}{\text{Total assets}}$$

The equity ratio for Electco in 2008 was 3318/8296 = 0.40 and for the industry was 0.46 as shown in Table 6.8. Electco has paid for roughly 60% of its assets with debt; the remaining 40% represents equity. This is close to the industry practice as a whole and would appear acceptable.

Table 6.7   Income Statements for Electco Electronics for Years Ended 2008, 2009, 2010

Electco Electronics Income Statements *(in thousands of dollars)*			
	2008	**2009**	**2010**
Revenues			
Sales	12 440	11 934	12 100
Total Revenues	12 440	11 934	12 100
Expenses			
Cost of goods sold (excluding depreciation)	10 100	10 879	11 200
Depreciation	692	554	443
Interest paid	346	344	341
Total Expenses	11 138	11 777	11 984
Profit before taxes	1302	157	116
Taxes (at 40%)	521	63	46
Profit before extraordinary items	781	94	70
Extraordinary gains/losses	70		
Profit after taxes	**851**	**94**	**70**

Table 6.8   Industry-Standard Ratios and Financial Ratios for Electco Electronics

Financial Ratio	Industry Standard	Electco Electronics		
		2008	**2009**	**2010**
Current ratio	1.51	1.73	1.88	1.90
Quick ratio	—	1.17	1.10	0.94
Equity ratio	0.46	0.40	0.39	0.38
Inventory-turnover ratio	11.36	10.00	5.87	4.08
Return-on-total-assets ratio	4.26%	9.41%	1.09%	0.76%
Return-on-equity ratio	9.21%	23.53%	2.75%	2.00%

Another group of ratios is called the **asset-management ratios** or **efficiency ratios**. They assess how efficiently a firm is using its assets. The inventory-turnover ratio is an example. The **inventory-turnover ratio** specifically looks at how efficiently a firm is using its resources to manage its inventories. This is reflected in the number of times that its inventories are replaced (or turned over) per year. The inventory-turnover ratio provides a measure of whether the firm has more or less inventory than normal.

$$\text{Inventory-turnover ratio} = \frac{\text{Sales}}{\text{Inventories}}$$

Electco's turnover ratio for 2008 was 12 440/1244 = 10 turns per year. This is reasonably close to the industry standard of 11.36 turns per year as shown in Table 6.8.

In 2008, Electco invested roughly the same amount in inventory per dollar of sales as the industry did, on average.

Two points should be observed about the inventory-turnover ratio. First, the sales amount in the numerator has been generated over a period of time, while the inventory amount in the denominator is for one point in time. A more accurate measure of inventory turns would be to approximate the average inventory over the period in which sales were generated.

A second point is that sales refer to market prices, while inventories are listed at cost. The result is that the inventory-turnover ratio as computed above will be an overstatement of the true turnover. It may be more reasonable to compute inventory turnover based on the ratio of cost of goods sold to inventories. Despite this observation, traditional financial analysis uses the sales-to-inventories ratio.

The next two ratios give evidence of how productively assets have been employed in producing a profit. The **return-on-assets (ROA) ratio** or **net-profit ratio** is the first example of a **profitability ratio**:

$$\text{Return-on-assets ratio} = \frac{\text{Net income (before extraordinary items)}}{\text{Total assets}}$$

Electco had a return on assets of $781/8296 = 0.0941$ or 9.41% in 2008. Table 6.8 shows that the industry-total ROA for 2008 was 4.26%. Note the comments on extraordinary items in Close-Up 6.4.

The second example of a profitability ratio is the **return-on-equity (ROE) ratio**:

$$\text{Return-on-equity ratio} = \frac{\text{Net income (before extraordinary items)}}{\text{Total equity}}$$

The return-on-equity ratio looks at how much profit a company has earned in comparison to the amount of capital that the owners have tied up in the company. It is often compared to how much the owners could have earned from an alternative investment and used as a measure of investment performance. Electco had a ROE of $781/3318 = 0.2354$ or 23.54% in 2008, whereas the industry-standard ROE was 9.21% as shown in Table 6.8. The year 2008 was an excellent one for the owners at Electco from their investment point of view.

Overall, Electco's performance in 2008 was similar to that of the electronic product manufacturing industry as a whole. One exception is that Electco generated higher profits than the norm; it may have been extremely efficient in its operations that year.

The rosy profit picture painted for Electco in 2008 does not appear to extend into 2009 and 2010, as a trend analysis shows. Table 6.8 shows the financial ratios computed for 2009 and 2010 with those of 2008 and the industry standard.

For more convenient reference, we have summarized the six financial ratios we have dealt with and their definitions in Table 6.9.

CLOSE-UP 6.4   Extraordinary Items

Extraordinary items are gains and losses that do not typically result from a company's normal business activities, are not expected to occur regularly, and are not recurring factors in any evaluations of the ordinary operations of the business. For example, cost or loss of income caused by natural disasters (floods, tornados, ice storms, etc.) would be extraordinary loss. Revenue created by the sale of a division of a firm is an example of extraordinary gain. Extraordinary items are reported separately from regular items and are listed net of applicable taxes.

Table 6.9   A Summary of Financial Ratios and Definitions

Ratio	Definition	Comments
Current ratio (Working capital ratio)	$\dfrac{\text{Current assets}}{\text{Current liabilities}}$	A liquidity ratio
Acid-test ratio (Quick ratio)	$\dfrac{\text{Quick assets}}{\text{Current liabilities}}$	A liquidity ratio (Quick assets = Current assets − Inventories − Prepaid items)
Equity ratio	$\dfrac{\text{Total equity}}{\text{Total assets}}$	A leverage or debt-management ratio
Inventory-turnover ratio	$\dfrac{\text{Sales}}{\text{Inventories}}$	An asset management or efficiency ratio
Return-on-assets ratio (Net-profit ratio)	$\dfrac{\text{Net income}}{\text{Total assets}}$	A profitability ratio (excludes extraordinary items)
Return-on-equity ratio	$\dfrac{\text{Net income}}{\text{Total equity}}$	A profitability ratio (measure of investment performance; excludes extraordinary items)

Electco's return on assets has dropped significantly over the three-year period. Though the current and quick ratios indicate that Electco should be able to meet its short-term liabilities, there has been a significant buildup of inventories over the period. Electco is not selling what it is manufacturing. This would explain the drop in Electco's inventory turns.

Coupled with rising inventory levels is a slight increase in the cost of goods sold over the three years. From the building and equipment entries in the balance sheet, we know that no major capital expenditures on new equipment have occurred during the period. Electco's equipment may be aging and in need of replacement, though further analysis of what is happening is necessary before any conclusions on this issue can be drawn.

A final observation is that Electco's accounts receivable seems to be growing over the three-year period. Since there may be risks associated with the possibility of bad debt, Electco should probably investigate the matter.

In summary, Electco's main problem appears to be a mismatch between production levels and sales levels. Other areas deserving attention are the increasing cost of goods sold and possible problems with accounts receivable collection. These matters need investigation if Electco is to recover from its current slump in profitability.

We close the section on financial ratios with some cautionary notes on their use. First, since financial statement values are often approximations, we need to interpret the financial ratios accordingly. In addition, accounting practices vary from firm to firm and may lead to differences in ratio values. Wherever possible, look for explanations of how values are derived in financial reports.

A second problem encountered in comparing a firm's financial ratios with the industry-standard ratios is that it may be difficult to determine what industry the firm best fits into. Furthermore, within every industry, large variations exist. In some cases, an analyst may construct a relevant "average" by searching out a small number of similar firms (in size and business type) that may be used to form a customized industry average.

Finally, it is important to recognize the effect of seasonality on the financial ratios calculated. Many firms have highly seasonal operations with natural high and low periods of activity. An analyst needs to judge these fluctuations in context. One solution to this problem is to adjust the data seasonally through the use of averages. Another is to collect financial data from several seasons so that any deviations from the normal pattern of activity can be picked up.

Despite our cautionary words on the use of financial ratios, they do provide a useful framework for analyzing the financial health of a firm and for answering many questions about its operations.

REVIEW PROBLEMS

REVIEW PROBLEM 6.1

Joan is the sole proprietor of a small lawn-care service. Last year, she purchased an eight-horsepower chipper–shredder to make mulch out of small tree branches and leaves. At the time it cost $760. She expects that the value of the chipper–shredder will decline by a constant amount each year over the next six years. At the end of six years, she thinks that it will have a salvage value of $100.

Construct a table that gives the book value of the chipper–shredder at the end of each year, for six years. Also indicate the accumulated depreciation at the end of each year. A spreadsheet may be helpful.

ANSWER

The depreciation charge for each year is

$$D_{sl}(n) = \frac{(P-S)}{N} = \frac{760 - 100}{6} = 110 \qquad n = 1, \ldots, 6$$

This is the requested table:

Year	Depreciation Charge	Book Value	Accumulated Depreciation
0		$760	
1	$110	650	$110
2	110	540	220
3	110	430	330
4	110	320	440
5	110	210	550
6	110	100	660

REVIEW PROBLEM 6.2

A three-year-old extruder used in making plastic yogurt cups has a current book value of £168 750. The declining-balance method of depreciation with a rate $d = 0.25$ is used to determine depreciation charges. What was its original price? What will its book value be two years from now?

ANSWER

Let the original price of the extruder be P. The book value three years after purchase is £168 750. This means that the original price was

$$BV_{db}(3) = P(1 - d)^3$$

$$168\ 750 = P(1 - 0.25)^3$$

$$P = 400\ 000$$

The original price was £400 000.

The book value two years from now can be determined either from the original purchase price (depreciated for five years) or the current value (depreciated for two years):

$$BV_{db}(5) = 400\ 000(1 - 0.25)^5 = 94\ 921.88$$

or

$$BV_{db}(2) = 168\ 750(1 - 0.25)^2 = 94\ 921.88$$

The book value two years from now will be £94 921.88. ∎

REVIEW PROBLEM 6.3

You have been given the following data from the Fine Fishing Factory for the year ending December 31, 2010. Construct an income statement and a balance sheet from the data.

Accounts payable	$ 27 500
Accounts receivable	32 000
Advertising expense	2500
Bad debts expense	1100
Buildings, net	14 000
Cash	45 250
Common stock	125 000
Cost of goods sold	311 250
Depreciation expense, buildings	900
Government bonds	25 000
Income taxes	9350
Insurance expense	600
Interest expense	500
Inventory, December 31, 2010	42 000
Land	25 000
Machinery, net	3400
Mortgage due May 30, 2012	5000
Office equipment, net	5250
Office supplies expense	2025
Other expenses	7000

Prepaid expenses	3000
Retained earnings	?
Salaries expense	69 025
Sales	421 400
Taxes payable	2500
Wages payable	600

ANSWER

Solving this problem consists of sorting through the listed items and identifying which are balance sheet entries and which are income statement entries. Then, assets can be separated from liabilities and owners' equity, and revenues from expenses.

Fine Fishing Factory **Balance Sheet** **as of December 31, 2010**	
Assets	
Current assets	
Cash	$ 45 250
Accounts receivable	32 000
Inventory, December 31, 2010	42 000
Prepaid expenses	3000
Total current assets	$ 122 250
Long-term assets	
Land	25 000
Government bonds	25 000
Machinery, net	3400
Office equipment, net	5250
Buildings, net	14 000
Total long-term assets	$ 72 650
Total assets	**$194 900**
Liabilities	
Current liabilities	
Accounts payable	$ 27 500
Taxes payable	2500
Wages payable	600
Total current liabilities	$ 30 600
Long-term liabilities	
Mortgage due May 30, 2012	5000
Total long-term liabilities	$ 5000
Total liabilities	**$ 35 600**
Owners' Equity	
Common stock	$ 125 000
Retained earnings	34300
Total owners' equity	**$ 159300**
Total liabilities and owners' equity	**$194 900**

Fine Fishing Factory
Income Statement
for the Year Ending December 31, 2010

Revenues

Sales	$421 400
Cost of goods sold	311 250
Net revenue from sales	$110 150

Expenses

Salaries expense	69 025
Bad debts expense	1100
Advertising expense	2500
Interest expense	500
Insurance expense	600
Office supplies expense	2025
Other expenses	7000
Depreciation expense, buildings	900
Depreciation expense, office equipment	850
Total Expenses	$ 84 500

Income before taxes	$ 25 650
Income taxes	9350
Income after taxes	**$ 16 300**

REVIEW PROBLEM 6.4

Perform a financial ratio analysis for the Major Electric Company using the balance sheet and income statement from Sections 6.3.2 and 6.3.3. Industry standards for the ratios are as follows:

Ratio	Industry Standard
Current ratio	1.80
Acid-test ratio	0.92
Equity ratio	0.71
Inventory-turnover ratio	14.21
Return-on-assets ratio	7.91%
Return-on-equity ratio	11.14%

ANSWER

The ratio computations for Major Electric are:

$$\text{Current ratio} = \frac{\text{Current assets}}{\text{Current liabilities}} = \frac{801\ 000}{90\ 000} = 8.9$$

$$\text{Acid-test ratio} = \frac{\text{Quick assets}}{\text{Current liabilities}} = \frac{66\ 000}{90\ 000} = 0.73$$

$$\text{Equity ratio} = \frac{\text{Total equity}}{\text{Total assets}} = \frac{4\ 311\ 000}{5\ 401\ 000} = 0.7982 \cong 0.80$$

$$\text{Inventory-turnover ratio} = \frac{\text{Sales}}{\text{Inventories}} = \frac{7\ 536\ 000}{683\ 000}$$

$$= 11.03 \text{ turns per year}$$

$$\text{Return-on-total-assets ratio} = \frac{\text{Profits after taxes}}{\text{Total assets}} = \frac{304\ 200}{5\ 401\ 000}$$

$$= 0.0563 \text{ or } 5.63\% \text{ per year}$$

$$\text{Return-on-equity ratio} = \frac{\text{Net income}}{\text{Total equity}} = \frac{304\ 200}{4\ 311\ 000}$$

$$= 0.0706 \text{ or } 7.06\% \text{ per year}$$

A summary of the ratio analysis results follows:

Ratio	Industry Standard	Major Electric
Current ratio	1.80	8.90
Acid-test ratio	0.92	0.73
Equity ratio	0.71	0.80
Inventory-turnover ratio	14.21	11.03
Return-on-assets ratio	7.91%	5.63%
Return-on-equity ratio	11.14%	7.06%

Major Electric's current ratio is well above the industry standard and well above the general guideline of 2. The firm appears to be quite liquid. However, the acid-test ratio, with a value of 0.73, gives a slightly different view of Major Electric's liquidity. Most of Major Electric's current assets are inventory; thus, the current ratio is somewhat misleading. If we look at the acid-test ratio, Major Electric's quick assets are only 73% of their current liabilities. The firm may have a difficult time meeting their current debt obligations if they have unforeseen difficulties with their cash flow.

Major Electric's equity ratio of 0.80 indicates that it is not heavily reliant on debt and therefore is not at high risk of going bankrupt. Major Electric's inventory turns are lower than the industry norm, as are its ROA and ROE.

Taken together, Major Electric appears to be in reasonable financial shape. One matter they should probably investigate is why their inventories are so high. With lower inventories, they could improve their inventory turns and liquidity, as well as their return on assets. ■

SUMMARY

This chapter opened with a discussion of the concept of depreciation and various reasons why assets lose value. Two popular depreciation models, straight-line and declining-balance, were then presented as methods commonly used to determine book value of capital assets and depreciation charges.

The second part of the chapter dealt with the elements of financial accounting. We first presented the two main financial statements: the balance sheet and the income statement. Next, we showed how these statements can be used to assess the financial health of a firm through the use of ratios. Comparisons with industry norms and trend analysis are normally part of financial analysis. We closed with cautionary notes on the interpretation of the ratios.

The significance of the material in this chapter is twofold. First, it sets the groundwork for material in Chapters 7 and 8, replacement analysis and taxation. Second, and perhaps more importantly, it is increasingly necessary for all engineers to have an understanding of depreciation and financial accounting as they become more and more involved in decisions that affect the financial positions of the organizations in which they work.

Engineering Economics in Action, Part 6B:
Usually the Truth

Terry showed Naomi what he thought was evidence of wrongdoing by a fellow employee. Naomi said, "Interesting observation, Terry. But you know, I think it's probably OK. Let me explain. The main problem is that you are looking at two kinds of evaluation here, book value and market value. The book value is what an asset is worth from an accounting point of view, while the market value is what it sells for."

Terry nodded. "Yes, I know that. That's true about anything you sell. But this is different. We've got a $5000 estimate against a $300 sale. You can't tell me that our guess about the sales price can be that far out!"

"Yes, it can, and I'll tell you why. That book value is an estimate of the market value that has been calculated according to very particular rules. Corporate tax rules require us to use a particular depreciation scheme for all of our assets. Now, the reality is that things decline in value at different rates, and computers lose value really quickly. We could, for our own purposes, determine a book value for any asset that is a better estimate of its market value, but sometimes it's too much trouble to keep one set of figures for tax reasons and another for other purposes. So often everything is given a book value according to the tax rules, and consequently sometimes the difference between the book value and the market value can be a lot."

"But surely the government can see that computers in particular don't match the depreciation scheme they mandate. Or are they just ripping us off?"

"Well, they can see that. They are always tweaking the rules, and sometimes they make changes to help make book values more realistic. But it's hard to keep up with technology when it changes so fast."

Terry smiled ruefully, "So our accounting statements don't really show the truth?"

Naomi smiled back, "I guess not, if by 'truth' you mean market value. But usually they're close. Usually."

PROBLEMS

For additional practice, please see the problems (with selected solutions) provided on the Student CD-ROM that accompanies this book.

6.1 For each of the following, state whether the loss in value is due to use-related physical loss, time-related physical loss, or functional loss:

(a) Albert sold his two-year-old computer for $500, but he paid $4000 for it new. It wasn't fast enough for the new software he wanted.

(b) Beatrice threw out her old tennis shoes because the soles had worn thin.

(c) Claudia threw out her old tennis shoes because she is jogging now instead.

(d) Day-old bread is sold at half-price at the neighbourhood bakery.

(e) Egbert sold his old lawnmower to his neighbour for £20.

(f) Fred picked up a used overcoat at the thrift store for less than 10% of the new price.

(g) Gunther notices that newspapers cost €0.50 on the day of purchase, but are worth less than €0.01 each as recyclable newsprint.

(h) Harold couldn't get the price he wanted for his house because the exterior paint was faded and flaking.

6.2 For each of the following, state whether the value is a market value, book value, scrap value, or salvage value:

(a) Inta can buy a new stove for $800 at Joe's Appliances.

(b) Jacques can sell his used stove to Inta for €200.

(c) Kitty can sell her used stove to the recyclers for $20.

(d) Liam can buy Jacques' used stove for €200.

(e) Noriko is adding up the value of the things she owns. She figures her stove is worth at least ¥20 000.

6.3 A new industrial sewing machine costs in the range of $5000 to $10 000. Technological change in sewing machines does not occur very quickly, nor is there much change in the functional requirements of a sewing machine. A machine can operate well for many years with proper care and maintenance. Discuss the different reasons for depreciation and which you think would be most appropriate for a sewing machine.

6.4 Communications network switches are changing dramatically in price and functionality as changes in technology occur in the communications industry. Prices drop frequently as more functionality and capacity are achieved. A switch only six months old will have depreciated since it was installed. Discuss the different reasons for depreciation and which you think would be most appropriate for this switch.

6.5 Ryan owns a five-hectare plot of land in the countryside. He has been planning to build a cottage on the site for some time, but has not been able to afford it yet. However, five years ago, he dug a pond to collect rainwater as a water supply for the cottage. It has never been used, and is beginning to fill in with plant life and garbage that has been dumped there. Ryan realizes that his investment in the pond has depreciated in value since he dug it. Discuss the different reasons for depreciation and which you think would be most appropriate for the pond.

6.6 A company that sells a particular type of web-indexing software has had two larger firms approach it for a possible buyout. The current value of the company, based on recent financial statements, is $4.5 million. The two bids were for $4 million and $7 million, respectively. Both bids were bona fide, meaning they were real offers. What is the market value of the company? What is its book value?

6.7 An asset costs $14 000 and has a scrap value of $3000 after seven years. Calculate its book value using straight-line depreciation

(a) After one year

(b) After four years

(c) After seven years

6.8 An asset costs $14 000. At a depreciation rate of 20%, calculate its book value using the declining-balance method

(a) After one year

(b) After four years

(c) After seven years

6.9 (a) An asset costs £14 000. What declining-balance depreciation rate would result in the scrap value of £3000 after seven years?

(b) Using the depreciation rate from part (a), what is the book value of the asset after four years?

 6.10 Using a spreadsheet program, chart the book value of a $14 000 asset over a seven-year life using declining-balance depreciation ($d = 0.2$). On the same chart, show the book value of the $14 000 asset using straight-line depreciation with a scrap value of $3000 after seven years.

 6.11 Using a spreadsheet program, chart the book value of a $150 000 asset for the first 10 years of its life at declining-balance depreciation rates of 5%, 20%, and 30%.

6.12 A machine has a life of 30 years, costs €245 000, and has a salvage value of €10 000 using straight-line depreciation. What depreciation rate will result in the same book value for both the declining-balance and straight-line methods at the end of year 20?

6.13 A new press brake costs York Steel £780 000. It is expected to last 20 years, with a £60 000 salvage value. What rate of depreciation for the declining-balance method will produce a book value after 20 years that equals the salvage value of the press?

6.14 (a) Using straight-line depreciation, what is the book value after four years for an asset costing $150 000 that has a salvage value of $25 000 after 10 years? What is the depreciation charge in the fifth year?

(b) Using declining-balance depreciation with $d = 20\%$, what is the book value after four years for an asset costing $150 000? What is the depreciation charge in the fifth year?

(c) What is the depreciation rate using declining-balance for an asset costing $150 000 that has a salvage value of $25 000 after 10 years?

6.15 Julia must choose between two different designs for a safety enclosure, which will be in use indefinitely. Model A has a life of three years, a first cost of $8000, and maintenance of $1000 per year. Model B will last four years, has a first cost of $10 000, and has maintenance of $800 per year. A salvage value can be estimated for model A using a depreciation rate of 40% and declining-balance depreciation, while a salvage value for model B can be estimated using straight-line depreciation and the knowledge that after one year its salvage value will be $7500. Interest is at 14%. Using a present worth analysis, which design is better?

6.16 Adventure Airline's new baggage handling conveyor costs $250 000 and has a service life of 10 years. For the first six years, depreciation of the conveyor is calculated using the declining-balance method at the rate of 30%. During the last four years, the straight-line method is used for accounting purposes in order to have a book value of 0 at the end of the service life. What is the book value of the conveyor after 7, 8, 9, and 10 years?

6.17 Molly inherited $5000 and decided to start a lawn-mowing service. With her inheritance and a bank loan of $5000, she bought a used ride-on lawnmower and a used truck. For five years, Molly had a gross income of $30 000, which covered the annual operating costs and the loan payment, both of which totalled $4500. She spent the rest of her income personally. At the end of five years, Molly found that her loan was paid off but the equipment was wearing out.

(a) If the equipment (lawnmower and truck) was depreciating at the rate of 50% according to the declining-balance method, what is its book value after five years?

(b) If Molly wanted to avoid being left with a worthless lawnmower and truck and with no money for renewing them at the end of five years, how much should she have saved annually toward the second set of used lawnmower and used truck of the same initial value as the first set? Assume that Molly's first lawnmower and truck could be sold at their book value. Use an interest rate of 7%.

6.18 Enrique is planning a trip around the world in three years. He will sell all of his possessions at that time to fund the trip. Two years ago, he bought a used car for €12 500. He observes that the market value for the car now is about €8300. He needs to know how much money his car will add to his stash for his trip when he sells it three years from now. Use the declining-balance depreciation method to tell him.

6.19 Ben is choosing between two different industrial fryers using an annual worth calculation. Fryer 1 has a five-year service life and a first cost of $400 000. It will generate net year-end savings of $128 000 per year. Fryer 2 has an eight-year service life and a first cost of $600 000. It will generate net year-end savings of $135 000 per year. If the salvage value is estimated using declining-balance depreciation with a 20% depreciation rate, and the MARR is 12%, which fryer should Ben buy?

6.20 Dick noticed that the book value of an asset he owned was exactly $500 higher if the value was calculated by straight-line depreciation over declining-balance depreciation, exactly halfway through the asset's service life, and the scrap value at the end of the service life was the same by either method. If the scrap value was $100, what was the purchase price of the asset?

6.21 A company had net sales of $20 000 last month. From the balance sheet at the end of last month, and an income statement for the month, you have determined that the current ratio was 2.0, the acid-test ratio was 1.2, and the inventory turnover was two per month. What was the value of the company's current assets?

6.22 In the last quarter, the financial-analysis report for XYZ Company revealed that the current, quick, and equity ratios were 1.9, 0.8, and 0.37, respectively. In order to improve the firm's financial health based on these financial ratios, the following strategies are considered by XYZ for the current quarter:

(i) Reduce inventory

(ii) Pay back short-term loans

(iii) Increase retained earnings

(a) Which strategy (or strategies) is effective for improving each of the three financial ratios?

(b) If only one strategy is considered by XYZ, which one seems to be most effective? Assume no other information is available for analysis.

6.23 The end-of-quarter balance sheet for XYZ Company indicated that the current ratio was 1.8 and the equity ratio was 0.45. It also indicated that the long-term assets were twice as much as the current assets, and half of the current assets were highly liquid. The total equity was $68 000. Since the current ratio was close to 2, XYZ feels that the company had a reasonable amount of liquidity in the last quarter. However, if XYZ wants more assurance, which financial ratio would provide further information? Using the information provided, compute the appropriate ratio and comment on XYZ's concern.

6.24 A potentially very large customer for Milano Metals wants to fully assess the financial health of the company in order to decide whether to commit to a long-term, high-volume relationship. You have been asked by the company president, Roch, to review the company's financial performance over the last three years and make a complete report to him. He will then select from your report information to present to the customer. Consequently, your report should be as thorough and honest as possible.

Research has revealed that in your industry (sheet metal products), the average value of the current ratio is 2.79, the equity ratio is 0.54, the inventory turnover is 4.9, and the net-profit ratio is 3.87. Beyond that information, you have access to only the balance sheet (on the next page) and the income statement shown here, and should make no further assumptions. Your report should be limited to the equivalent of about 300 words.

Milano Metals Income Statement for the Years Ending December 31, 2008, 2009, and 2010 *(in thousands of dollars)*			
	2008	**2009**	**2010**
Total revenue	9355	9961	8470
Less: Costs	8281	9632	7654
Net revenue	1074	329	816
Less:			
Depreciation	447	431	398
Interest	412	334	426
Income taxes	117	21	156
Net income from operations	98	(457)	(164)
Add: Extraordinary item	770		(1832)
Net income	868	(457)	(1996)

6.25 The Milano Metals income statement and balance sheets shown for Problem 6.24 were in error. A piece of production equipment was sold for $100 000 cash in 2010 and was not accounted for. Which items on these statements must be changed, and (if known) by how much?

6.26 The Milano Metals income statement and balance sheet shown for Problem 6.24 were in error. An extra $100 000 in sales was made in 2010 and not accounted for. Only half of the payments for these sales have been received. Which items on these statements must be changed, and (if known) by how much?

6.27 At the end of last month, Paarl Manufacturing had 45 954 rand (R45 954) in the bank. It owed the bank, because of the mortgage, R224 000. It also had a working capital loan of R30 000. Its customers owed R22 943, and it owed its suppliers R12 992. The company owned property worth R250 000. It had R123 000 in finished goods, R102 000 in raw materials, and R40 000 in work in progress. Its production equipment was worth R450 000 when new (partially paid for by a large government loan due to be paid back in three years) but had accumulated a total of R240 000 in depreciation, R34 000 worth last month.

Milano Metals Consolidated Balance Sheets December 31, 2008, 2009, and 2010 *(in thousands of dollars)*			

Assets

	2008	2009	2010
Current Assets			
Cash	19	19	24
Accounts receivable	779	884	1176
Inventories	3563	3155	2722
	4361	4058	3922
Fixed Assets			
Land	1136	1064	243
Buildings and equipment	2386	4682	2801
	3552	5746	3044
Other Assets	3413	....	3
Total Assets	8296	9804	6969

Liabilities and Owners' Equity

	2008	2009	2010
Current Liabilities			
Due to bank	1431	1 929	2040
Accounts payable	1644	1 349	455
Wages payable	341	312	333
Income tax payable	562	362	147
Long-Term Debt	2338	4743	2528
Total Liabilities	6316	8695	5503
Owners' Equity			
Capital stock	1194	1191	1091
Retained earnings	786	(82)	375
Total Owners' Equity	1980	1109	1466
Total Liability and Owners' Equity	8296	9804	6969

The company has investors who put up R100 000 for their ownership. It has been reasonably profitable; this month the gross income from sales was R220 000, and the cost of the sales was only R40 000. Expenses were also relatively low; salaries were R45 000 last month, while the other expenses were depreciation, maintenance at R1500, advertising at R3400, and insurance at R300. In spite of R32 909 in accrued taxes (Paarl pays taxes at 55%), the company had retained earnings of R135 000.

Construct a balance sheet (as at the end of this month) and income statement (for this month) for Paarl Manufacturing. Should the company release some of its retained earnings through dividends at this time?

6.28 Salvador Industries bought land and built its plant 20 years ago. The depreciation on the building is calculated using the straight-line method, with a life of 30 years and a salvage value of $50 000. Land is not depreciated. The depreciation for the equipment, all of which was purchased at the same time the plant was constructed, is calculated using declining-balance at 20%. Salvador currently has two outstanding loans: one for $50 000 due December 31, 2010, and another one for which the next payment is due in four years.

Salvador Industries
Balance Sheet as of June 30, 2010

Assets

Cash	$ 350 000
Accounts receivable	2 820 000
Inventories	2 003 000
Prepaid services	160 000
Total Current Assets	☐

Long-Term Assets

Building	$200 000	
Less accumulated depreciation	☐	☐
Equipment	$480 000	
Less accumulated depreciation	☐	☐
Land		540 000
Total Long-Term Assets		☐

Total Assets ☐

Liabilities and Owners' Equity

Current Liabilities

Accounts payable ☐	$ 921 534
Accrued taxes	29 000
Total ☐	☐

Long-Term Liabilities

Mortgage ☐	$1 200 000
	318 000
Total Long-Term Liabilities	☐

Total ☐ ☐

Owners' Equity

Capital stock ☐	$1 920 000

Total Owners' Equity ☐
Total Liabilities and Owners' Equity ☐

During April 2010, there was a flood in the building because a nearby river overflowed its banks after unusually heavy rain. Pumping out the water and cleaning up the basement and the first floor of the building took a week. Manufacturing was suspended during this period, and some inventory was damaged. Because of lack of adequate insurance, this unusual and unexpected event cost the company $100 000 net.

(a) Fill in the blanks and complete a copy of the balance sheet and income statement here, using any of the above information you feel is necessary.

Salvador Industries		
Income Statement for the Year Ended June 30, 2010		
Income		
Gross income from sales	$8 635 000	
Less [ ]	7 490 000	[ ]
Total income		[ ]
[ ]		
Depreciation		70 000
Interest paid		240 000
Other expenses		100 000
Total expenses		[ ]
Income before taxes		[ ]
Taxes at 40%		[ ]
[ ]		[ ]
[ ]		[ ]
Net income		[ ]

(b) Show how information from financial ratios can indicate whether Salvador Industries can manage an unusual and unexpected event such as the flood without threatening its existence as a viable business.

6.29 Movit Manufacturing has the following alphabetized income statement and balance sheet entries from the year 2010. Construct an income statement and a balance sheet from the information given.

Accounts payable	$ 7500
Accounts receivable	15 000
Accrued wages	2850
Cash	2100
Common shares	150
Contributed capital	3000
Cost of goods sold	57 000
Current assets	
Current liabilities	
Deferred income taxes	2250
Depreciation expense	750
General expense	8100
GICs	450
Income taxes	1800
Interest expense	1500
Inventories	18 000
Land	3000

Less: Accumulated depreciation	10 950
Long-term assets	
Long-term bonds	4350
Long-term liabilities	
Mortgage	9450
Net income after taxes	2700
Net income before taxes	4500
Net plant and equipment	7500
Net sales	76 500
Operating expenses	
Owners' equity	
Prepaid expenses	450
Selling expenses	4650
Total assets	46 500
Total current assets	36 000
Total current liabilities	15 000
Total expenses	15 000
Total liabilities and owners' equity	46 500
Total long-term assets	10 500
Total long-term liabilities	16 050
Total owners' equity	15 450
Working capital loan	4650

6.30 Calculate for Movit Manufacturing in Problem 6.29 the financial ratios listed in the table below. Using these ratios and those provided for 2008 and 2009, conduct a short analysis of Movit's financial health.

Movit Manufacturing Financial Ratios			
Ratio	**2010**	**2009**	**2008**
Current ratio		1.90	1.60
Acid-test ratio		0.90	0.75
Equity ratio		0.40	0.55
Inventory-turnover ratio		7.00	12.00
Return-on-assets ratio		0.08	0.10
Return-on-equity ratio		0.20	0.18

6.31 Fraser Phraser operates a small publishing company. He is interested in getting a loan for expanding his computer systems. The bank has asked Phraser to supply them with his financial statements from the past two years. His statements appear below. Comment on Phraser's financial position with regard to the loan based on the results of financial ratio analysis.

Fraser Phraser Company
Comparative Balance Sheets
for the Years Ending in 2009 and 2010
(in thousands of dollars)

	2009	2010
Assets		
Current Assets		
Cash	22 500	1250
Accounts receivable	31 250	40 000
Inventories	72 500	113 750
Total Current Assets	126 250	155 000
Long-Term Assets		
Land	50 000	65 000
Plant and equipment	175 000	250 000
Less: Accumulated depreciation	70 000	95 000
Net plant and equipment	105 000	155 000
Total Long-Term Assets	155 000	220 000
Total Assets	281 250	375 000
Liabilities and Owners' Equity		
Current Liabilities		
Accounts payable	26 250	55 000
Working capital loan	42 500	117 500
Total Current Liabilities	68 750	172 500
Long-Term Liabilities		
Mortgage	71 875	57 375
Total Long-Term Liabilities	71 875	57 375
Owners' Equity		
Common shares	78 750	78 750
Retained earnings	61 875	66 375
Total Owners' Equity	140 625	145 125
Total Liabilities and Owners' Equity	281 250	375 000

Fraser Phraser Company
Income Statements
for Years Ending 2009 and 2010
(in thousands of dollars)

	2009	2010
Revenues		
Sales	156 250	200 000
Cost of goods sold	93 750	120 000
Net revenue from sales	62 500	80 000
Expenses		
Operating expenses	41 875	46 250
Depreciation expense	5625	12 500
Interest expense	3750	7625
Total expenses	51 250	66 375
Income before taxes	11 250	13 625
Income taxes	5625	6813
Net income	5625	6813

6.32 A friend of yours is thinking of purchasing shares in Petit Ourson SA in the near future, and decided that it would be prudent to examine its financial statements for the past two years before making a phone call to his stockbroker. The statements are shown below.

Your friend has asked you to help him conduct a financial ratio analysis. Fill out the financial ratio information on a copy of the third table on the next page. After comparison with industry standards, what advice would you give your friend?

Petit Ourson SA
Comparative Balance Sheets
for the Years Ending 2009 and 2010
(in thousands of euros)

	2009	2010
Assets		
Current Assets		
Cash	500	375
Accounts receivable	1125	1063
Inventories	1375	1563
Total Current Assets	3000	3000
Long-Term Assets		
Plant and equipment	5500	6500
Less: Accumulated depreciation	2500	3000
Net plant and equipment	3000	3500
Total Long-Term Assets	3000	3500
Total Assets	6000	6500

Liabilities and Owners' Equity

Current Liabilities

Accounts payable	500	375
Working capital loan	000	375
Total Current Liabilities	500	750

Long-Term Liabilities

Bonds	1500	1500
Total Long-Term Liabilities	1500	1500

Owners' Equity

Common shares	750	750
Contributed capital	1500	1500
Retained earnings	1750	2000
Total Owners' Equity	4000	4250
Total Liabilities and Owners' Equity	6000	6500

Petit Ourson SA
Income Statements
for the Years Ending 2009 and 2010
(in thousands of euros)

	2009	2010
Revenues		
Sales	3000	3625
Cost of goods sold	1750	2125
Net revenue from sales	1250	1500
Expenses		
Operating expenses	75	100
Depreciation expense	550	500
Interest expense	125	150
Total expenses	750	750
Income Before Taxes	500	750
Income taxes	200	300
Net Income	300	450

Petit Ourson SA
Financial Ratios

	Industry Norm	2009	2010
Current ratio	4.50		
Acid-test ratio	2.75		
Equity ratio	0.60		
Inventory-turnover ratio	2.20		
Return-on-assets ratio	0.09		
Return-on-equity ratio	0.15		

6.33 Construct an income statement and a balance sheet from the scrambled entries for Paradise Pond Company from the years 2009 and 2010 shown in the table below.

Paradise Pond Company	2009	2010
Accounts receivable	$ 675	$ 638
Less: Accumulated depreciation	1500	1800
Accounts payable	300	225
Bonds	900	900
Cash	300	225
Common shares	450	450
Contributed capital	900	900
Cost of goods sold	1750	2125
Depreciation expense	550	500
Income taxes	200	300
Interest expense	125	150
Inventories	825	938
Net plant and equipment	1800	2100
Net revenue from sales	1250	1500
Operating expenses	075	100
Plant and equipment	3300	3900
Profit after taxes	300	450
Profit before taxes	500	750
Retained earnings	1050	1200
Sales	3000	3625
Total assets	3600	3900
Total current assets	1800	1800
Total current liabilities	300	450
Total expenses	750	750
Total liabilities and owners' equity	3600	3900
Total long-term assets	1800	2100
Total long-term liabilities	900	900
Total owners' equity	2400	2550
Working capital loan	000	225

More Challenging Problem

6.34 Like all companies, Exco has to pay taxes on its profits. However, for each additional dollar in depreciation Exco claims, Exco's taxes are reduced by t, since the depreciation expense offsets income, thus reducing profits subject to tax. Exco is required by tax law to use declining-balance depreciation to calculate the depreciation of assets on a year-by-year basis, with a depreciation rate of d. This depreciation calculation continues forever, even if the asset is sold or scrapped.

The actual depreciation for a particular asset is best represented as straight-line over y years, with zero salvage value. If the asset costs P dollars, and interest is i, what is the present worth of all tax savings Exco will incur by purchasing the asset? (*Hint:* The depreciation savings in the first year is Ptd.)

MINI-CASE 6.1

Business Expense or Capital Expenditure?

Courtesy of *Stabroek News* (Guyana), September 15, 2007:

Commissioner-General of the Guyana Revenue Authority (GRA), Khurshid Sattaur, says that the GRA has developed an extra-statutory ruling to allow for the depreciation of software over a two-year period.

A release from the Office of the Commissioner-General yesterday quoted Sattaur as saying that "the Revenue Authority for the purposes of Section 17 of the *Income Tax Act* Chapter 81:01 and the *Income Tax (Depreciate Rates) Regulation 1992* as amended by the *Income (Depreciate Rates) (Amendments) 1999* allow wear and tear at a rate of 50% per annum."

He noted that while Guyana's income tax laws adequately provide for the depreciation of computer equipment (hardware) for 50% write-off over two years, the law does not make provision for software.

According to Sattaur, the GRA feels that businesses have been taking advantage of the inadequate provisions in the law and have been treating the software as an expense to be written off in the first year or in the year of acquisition.

Therefore, the release stated, the GRA is allowing wear and tear on website development costs and software, whether or not they form part of the installed software over a two-year period.

Discussion

Calculating depreciation is made difficult by many factors. First, the value of an asset can change over time in many complicated ways. Age, wear and tear, and functional changes all have their effects, and are often unpredictable. A 30-year old VW Beetle, for example, can suddenly increase in value because a new Beetle is introduced by the manufacturer.

A second complication is created by tax laws that make it desirable for companies to depreciate things quickly, while at the same time restricting the way they calculate depreciation. Tax laws require that companies, at least for tax purposes, use specific methods of calculating depreciation that may bear little relevance to market value or remaining life.

A third complication is that, in real life, it is sometimes hard to determine what is an asset and what is not. If some costly item is recognized as an expense, a company can claim the costs early and offset profits to reduce taxes. However, as in the case of the software in Guyana, new rules can suddenly change an expense into an asset, which then has to be depreciated over time at higher cost to the company. Companies and tax departments often conflict because of ambiguity about assets and expenses.

In the end, depreciation calculations are simply estimates that are useful only with a clear understanding of the assumptions underlying them.

Questions

1. For each of the following, indicate whether the straight-line method or the declining-balance method would likely give the most accurate estimate of how the asset's value changes over time, or explain why neither method is suitable.

   (a) A $20 bill

   (b) A $2 bill

    (c) A pair of shoes

    (d) A haircut

    (e) An engineering degree

    (f) A Van Gogh painting

2. What differences occur in the balance sheet and income statement of a company in Guyana now that a software acquisition is considered to be a capital expenditure as opposed to an expense? Is the company's profit greater or less in the first year? How does the total profit over the life of the asset compare?

A.1 Introduction

Clem looked up from his computer as Naomi walked into his office. "Hi, Naomi. Sit down. Just let me save this stuff."

After a few seconds Clem turned around, showing a grin. "I'm working on our report for the last quarter's operations. Things went pretty well. We exceeded our targets on defect reductions and on reducing overtime. And we shipped everything required—over 90% on time."

Naomi caught a bit of Clem's exuberance. "Sounds like a report you don't mind writing."

"Yeah, well, it was a team job. Everyone did good work. Talking about doing good work, I should have told you this before, but I didn't think about it at the right time. Ed Burns and Anna Kulkowski were really impressed with the report you did on the forge project."

Naomi leaned forward. "But they didn't follow my recommendation to get a new manual forging press. I assumed there was something wrong with what I did."

"I read your report carefully before I sent it over to them. If there had been something wrong with it, you would have heard right away. Trust me. I'm not shy. It's just that we were a little short of cash at the time. We could stay in business with just fixing up the guides on the old forging hammer. And Burns and Kulkowski decided there were more important things to do with our money."

"If I didn't have confidence in you, you wouldn't be here this morning. I'm going to ask you and Dave Sullivan to look into an important strategic issue concerning whether we continue to buy small aluminum parts or whether we make them ourselves. We're just waiting for Dave to show up."

"OK. Thanks, Clem. But please tell me next time if what I do is all right. I'm still finding my way around here."

"You're right. I guess that I'm still more of an engineer than a manager."

Voices carried into Clem's office from the corridor. "That sounds like Dave in the hall saying hello to Carole," Naomi observed. "It looks like we can get started."

Dave Sullivan came in with long strides and dropped into a chair. "Good morning, everybody. It is still morning, barely. Sorry to be late. What's up?"

Clem looked at Dave and started talking. "What's up is this. I want you and Naomi to look into our policy about buying or making small aluminum parts. We now use about 200 000 pieces a month. Most of these, like bolts and sleeves, are cold-formed.

"Prabha Vaidyanathan has just done a market projection for us. If she's right, our demand for these parts will continue to grow. Unfortunately, she wasn't very precise about the *rate* of growth. Her estimate was for anything between 5% and 15% a year. We now contract this work out. But even if growth is only 5%, we may be at the level where it pays for us to start doing this work ourselves.

"You remember we had a couple of engineers from Hamilton Tools looking over our processes last week? Well, they've come back to us with an offer to sell us a cold-former. They have two possibilities. One is a high-volume job that is a version of their Model E2. The other is a low-volume machine based on their Model E1.

"The E2 will do about 2000 pieces an hour, depending on the sizes of the parts and the number of changeovers we make. The E1 will do about 1000 pieces an hour."

Naomi asked, "About how many hours per year will these formers run?"

"Well, with our two shifts, I think we're talking about 3600 hours a year for either model."

Dave came in. "Hold it. If my third-grade arithmetic still works, that sounds like either 3.6 million or 7.2 million pieces a year. You say that we are using only 2.4 million pieces a year."

Clem answered, "That's right. Ms. Vaidyanathan has an answer to that one. She says we can sell excess capacity until we need it ourselves. Again, unfortunately, she isn't very precise about what this means. We now pay about five cents a piece. Metal cost is in addition to that. We pay for that by weight. She says that we won't get as much as five cents because we don't have the market connections. But she says we should be able to find a broker so that we net somewhere between three cents and four cents a piece, again plus metal."

Naomi spoke. "That's a pretty wide range, Clem."

"I know. Prabha says that she couldn't do any better with the budget Burns and Kulkowski gave her. For another $5000, she says that she can narrow the range on *either* the growth rate or the potential prices for selling pieces from any excess capacity. Or, for about $7500, she could do both. I spoke to Anna Kulkowski about this. Anna says that they won't approve anything over $5000. One of the things I want you two to look at is whether or not it's necessary to get more information. If you do recommend spending on market research, it has to be for just one of either the selling price range or the growth rate.

"I have the proposal from Hamilton Tools here. It has information on the two formers. This is Wednesday. I'd like a report from you by Friday afternoon so that I can look at it over the weekend.

"Did I leave anything out?"

Naomi asked, "Are we still working with a 15% after-tax MARR?"

Clem hesitated. "This is just a first cut. Don't worry about details on taxes. We can do a more precise calculation before we actually make a decision. Just bump up the MARR to 25% before tax. That will about cover our 40% marginal tax rate."[1]

Dave asked, "What time frame should we use in our calculations?"

"Right. Use 10 years. Either of these models should last at least that long. But I wouldn't want to stretch Prabha's market projections beyond 10 years."

Dave stood up and announced, "It's about a quarter to one." He turned to Naomi. "Do you want to start on this over at the Grand China Restaurant? It's past the lunch rush, and we'll be able to talk while we eat. I think Clem will buy us lunch out of his budget."

Clem interjected, "All right, Dave. Just don't order the most expensive thing on the menu."

Naomi laughed. "I'm glad we have one big spender around here."

A.2 Problem Definition

About 40 minutes later, Dave and Naomi were most of the way through their main courses. Dave suggested that they get started. He took a pad from his briefcase and said that he would take notes. Naomi agreed to let him do that.

Dave started. "OK. What are our options?"

"Well, I did a bit of arithmetic on my calculator while you were on the phone before lunch. It looks as though, even if the demand growth rate is only 5%, a single small former will not have enough capacity to see us through 10 years. This means that there are four options. The first is a sequence of two low-capacity formers. The second is just a high-capacity former. The third, which would kick in only if the growth rate is high, would be a sequence of three low-capacity formers. The fourth is a low-capacity and a high-capacity. I'm not sure of the sequence for that."

Dave thought for a bit. "I don't think so. I assume that, even if we put in our own former, we could contract out requirements that our own shop couldn't handle. That might be the way to go. That is, you wouldn't want to add to capacity if there was only one year left in the 10-year horizon. There probably would not be enough unsatisfied output requirement to amortize the first cost of a second small former."

"That sounds as though there are a whole bunch of options. It looks like we're in for a couple of long nights." Naomi sounded dejected.

"Well, maybe not."

"Maybe not what?"

"Maybe we won't have to spend those long nights. I think we can look at just three options at the start. We could look at a sequence of two small formers. We could look at a small former followed by outsourcing when capacity is exhausted. And we could look at a large former, possibly followed by outsourcing if capacity is exhausted. If we can rule out the two-small-formers option, we can certainly rule out three small formers or a big former combined with a small one."

"Smart, Dave. How should we proceed?"

"Well, there are a couple of ways of doing this. But, given that we have only 10 years to look at,

---

1. Businesses are required to pay tax on their earnings. Determining the effect of taxes on before-tax earnings may involve extensive computation. Managers frequently approximate the effect of taxes on decisions by increasing the MARR. However, it is good practice to do precise calculations before actually making a decision.

it's probably easiest to use a spreadsheet to develop cash flow sequences for the three options. The two options in which only a single machine is bought at time zero are pretty straightforward. The one with a sequence of two small formers is a bit more complicated. One of us can do the two easy ones. The other can do the sequence. For each option we need to look at, say, nine outcomes: three possible demand growth rates—5%, 10%, and 15%—times three possible prices—3¢, 3.5¢, and 4¢—for selling excess capacity. That should be enough to show us what's happening. What part do you want to do?"

"I'll take the hard one, the sequence of two. I'd like the practice. I'll let you check my analysis when I'm finished."

"OK. That's fine. But notice that the two-low-capacity-formers sequence is not a simple investment, so let's stick with present worths at this stage. Also, we are going to have to make some decision about how we record the timing of the purchase of the second former if we run out of capacity during a year. I suggest that we assume the second former is bought at the end of the year before we run out of capacity."

"OK."

Dave continued, "We need to put together a simple table on the specifications for these two machines. I'll do that, since you are doing the hard job with cash flows. Why don't we go back to the plant. I'll make up the table. I'll get you a copy later this afternoon."

A.3 Crunch Time

Naomi was sitting in her office thinking about structuring the cash flows for the two-small-machines sequence. Dave knocked and came in.

He handed Naomi a sheet of paper.

"Here's the table. Shall we meet tomorrow morning to compare results?" (See Table A.1.)

Naomi glanced at the sheet of paper, and motioned for Dave to stay. "This is a bit new to me. Would it be okay if we get all of our assumptions down before we go too far?"

"Sure, we can do that," replied Dave as he grabbed a seat beside Naomi's desk. Naomi already had a pad and pencil out.

Dave went on, "OK, first, what's our goal?"

"Well, right now we are faced with a 'make or buy' decision," suggested Naomi. "We need to find out if it is cheaper to make these parts or to continue to buy them, and if making them is cheaper we need to choose which machine or machines we should buy."

"Pretty close, Naomi, but I think that may be more than we need at this point. Remember, Clem is primarily interested in whether this project is worth pursuing as a full-blown proposal to top management. So we can use a greatly simplified model now to answer that question, and get into the details if Clem decides it's worth pursuing."

"Right. If we look at the present worths of the three options we came up with at lunch, we should see quickly enough if investing in our own machines is feasible. Positive present worths

Table A.1   Specifications for the Two Cold Formers

Characteristics		Model E1 (Small)	Model E2 (Large)
First cost ($)		125 000	225 000
Hourly operating cost ($)		35.00	61.25
Average number of pieces/hour		1000	2000
Hours/year		3600	3600
Depreciation	20%/year declining balance, for both machines		
Market facts	Buying price: $0.05/piece plus cost of metal Selling price: $0.03 to $0.04 per piece plus metal Demand growth: 5% to 15%		

indicate buying machines is the best course; negative values would suggest it's probably not worth pursuing further."

Dave nodded. "OK," continued Naomi, "so we want to do a 'first approximation' calculation of present worths of each of our three options. Now, what assumptions should we use?"

"Let's see. We want to consider only a 10-year study period for the moment. I think we can assume the machines will have no salvage value at the end of the study period, so we don't have to worry about estimating depreciation and salvage values. This will be a conservative estimate in any case.

"Next," Dave continued, "we will ignore tax implications, and use a before-tax MARR of 25%. The 'sequence of machines' option has a complex cash flow, but by only considering present worths we can avoid dealing with possible multiple IRRs."

"You know," interjected Naomi, "it just occurred to me that the small machine makes 1000 pieces per hour at an operating cost of $35."

"And this is news to you?" said a bemused Dave.

"No, Dave. What I mean is that the operating cost is $0.035 per part on this machine, while one scenario calls for our selling excess capacity at only $0.03. We'll be losing half a cent on every part made for sale."

"Hmmm. You know, you're right. And look at this. Even the bigger machine's operating cost is $0.030625 per piece. We'd be losing money on that one too."

"When we run the different cases, it looks like we'll have to consider whether excess capacity will be sold or whether the machine will just be shut down. Just when I thought this was going to be easy . . . ," trailed Naomi.

"Maybe, but maybe not. Look at this another way. If the small machine sells its excess capacity at a loss of half a cent per piece, that comes to five bucks per hour. On the other hand, if the machine is shut down, we'll have an operator standing around at a cost of a lot more than that."

"But can't an idle operator be given other work to do?"

"Again, maybe, but maybe not. With the job security clause the union has now, if we lay the operator off for any period of time, the company has to pay most of the wages anyway, so there isn't much savings that way. On the other hand, the operator would probably be idled at unpredictable times, so other departments would have a difficult job of scheduling work for him or her to do. I'm not saying it's impossible, but it's a job in itself to figure out if we can place the operator in productive work when trying to idle the machine."

"Boy, they didn't talk about these problems in engineering school!"

"Welcome to the real world, Naomi, where nothing is simple."

"Then how about if we do this?" Naomi continued. "We assume that since an idle operator is probably more costly to the company than $5 per hour, the machines will run a full 3600 hours per year each regardless of whether the excess capacity is being sold at a price above operating costs or not."

"Sounds good, Naomi. In fact, we could claim an indirect benefit of this otherwise unprofitable operation since our operators will gain more experience with doing quick setups and statistical process control. That should make the brass happy," Dave said with a quick wink.

"I guess we can also assume that a machine is purchased and paid for at the beginning of a year while savings and operating costs accrue at the end of the year," continued Naomi. "And that the demand for parts is constant for each month within a year but grows by a fixed proportion from one year to the next."

Naomi finished scribbling down the assumptions, including the ones they had discussed with Clem earlier that day. Turning to Dave, she said, "So, how does this look?"

"Looks fine to me. Should we get together tomorrow and compare results?"

"OK. What time?"

"Why don't we exchange results first thing in the morning and then meet about nine-thirty in my office?"

"That's fine. See you then."

As Dave left, Naomi looked more closely at the table he had given her (Table A.1). "Time to crunch those numbers," she said to herself.

QUESTIONS

1. Construct spreadsheets for calculating present worths of the three proposals. For each proposal, you need to calculate PWs for each of 5%, 10%, and 15% demand growth and $0.03,

$0.035, and $0.04 selling price (nine combinations in all). Present the results in tabular and/or graphical format to support your analysis. A portion of a sample spreadsheet layout is given in Table A.2.

2. Write a memo to Clem presenting your findings. The goal of the analysis is to determine if bringing production in-house appears feasible, and if so, which machine purchase sequence(s) should be studied in further detail. The memo should contain a tentative recommendation about which option looks best and what additional research, if any, should be done. Keep the memo as concise as possible.

Table A.2   Portion of a Present Worth Spreadsheet

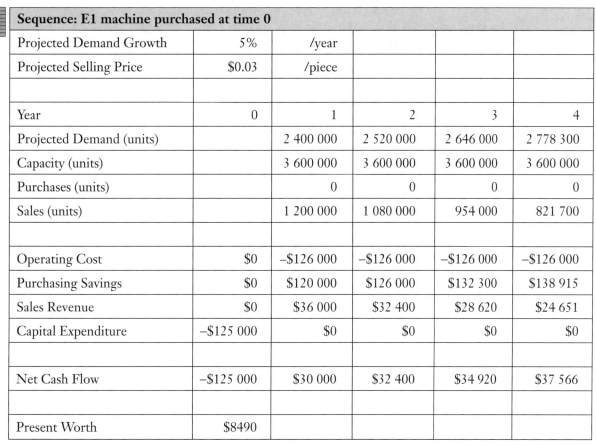

Sequence: E1 machine purchased at time 0					
Projected Demand Growth	5%	/year			
Projected Selling Price	$0.03	/piece			
Year	0	1	2	3	4
Projected Demand (units)		2 400 000	2 520 000	2 646 000	2 778 300
Capacity (units)		3 600 000	3 600 000	3 600 000	3 600 000
Purchases (units)		0	0	0	0
Sales (units)		1 200 000	1 080 000	954 000	821 700
Operating Cost	$0	–$126 000	–$126 000	–$126 000	–$126 000
Purchasing Savings	$0	$120 000	$126 000	$132 300	$138 915
Sales Revenue	$0	$36 000	$32 400	$28 620	$24 651
Capital Expenditure	–$125 000	$0	$0	$0	$0
Net Cash Flow	–$125 000	$30 000	$32 400	$34 920	$37 566
Present Worth	$8490				

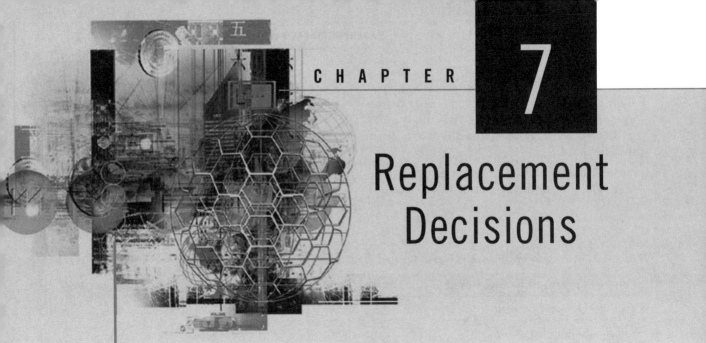

CHAPTER 7

Replacement Decisions

Engineering Economics in Action, Part 7A: You Need the Facts

Review Problems
Summary

Engineering Economics in Action, Part 7B: Decision Time

Problems

Mini-Case 7.1

Engineering Economics in Action, Part 7A:
You Need the Facts

"You know the 5-stage progressive die that we use for the Admiral Motors rocker arm contract?" Naomi was speaking to Terry, her co-op student, one Tuesday afternoon. "Clem asked me to look into replacing it with a 10-stage progressive die that would reduce the hand finishing substantially. It's mostly a matter of labour cost saving, but there is likely to be some quality improvement with the 10-stage die as well. I would like you to come up with a ballpark estimate of the cost of switching to the 10-stage progressive die."

Terry asked, "Don't you have the cost from the supplier?"

"Yes, but not really," said Naomi. "The supplier is Hamilton Tools. They've given us a price for the machine, but there are a lot of other costs involved in replacing one production process with another."

"You mean things like putting the machine in place?" Terry asked.

"Well, there's that," responded Naomi. "But there is also a lot more. For example, we will lose production during the changeover. That's going to cost us something."

"Is that part of the cost of the 10-stage die?"

"It's part of the first cost of switching to the 10-stage die," Naomi said. "If we decide to go ahead with the 10-stage die and incur these costs, we'll never recover them—they are sunk. We have already incurred those costs for the 5-stage die and it's only two years old. It still has lots of life in it. If the first costs of the 10-stage die are large, it's going to be hard to make a cost justification for switching to the 10-stage die at this time."

"OK. How do I go about this?" Terry asked.

Naomi sat back and chewed on her yellow pencil for about 15 seconds. She leaned forward and began. "Let's start with order-of-magnitude estimates of what it's going to cost to get the 10-stage die in place. If it looks as if there is no way that the 10-stage die is going to be cost-effective now, we can just stop there."

"It sounds like a lot of fuzzy work," said Terry.

"Terry, I know you like to be working with mathematical models. I'm also sure that you can read the appropriate sections on replacement models in an engineering economics book. But none of those models is worth anything unless you have data to put in it. You need the models to know what information to look for. And once you have the information, you will make better decisions using the models. But you do need the facts."

7.1 | Introduction

Survival of businesses in a competitive environment requires regular evaluation of the plant and equipment used in production. As these assets age, they may not provide adequate quality, or their costs may become excessive. When a plant or piece of equipment is evaluated, one of four mutually exclusive choices will be made.

1. An existing asset may be kept in its current use without major change.
2. An existing asset may be overhauled so as to improve its performance.
3. An existing asset may be removed from use without replacement by another asset.
4. An existing asset may be replaced with another asset.

This chapter is concerned with methods of making choices about possible replacement of long-lived assets. While the comparison methods developed in Chapters 4 and 5 for choosing between alternatives are often used for making these choices, the issues of replacement deserve a separate chapter for several reasons. First, the relevant costs for making replacement decisions are not always obvious, since there are costs associated with taking the replaced assets out of service that should be considered. This was ignored in the studies in Chapters 4 and 5. Second, the service lives of assets were provided to the reader in Chapters 4 and 5. As seen in this chapter, the principles of replacement allow the calculation of these service lives. Third, assumptions about how an asset might be replaced in the future can affect a decision now. Some of these assumptions are implicit in the methods analysts use to make the choices. It is therefore important to be aware of these assumptions when making replacement decisions.

The chapter starts with an example to introduce some of the basic concepts involved in replacement decisions. Following this is a discussion of the reasons why a long-lived asset may be replaced. We then develop the idea of the *economic life* of an asset—the service life that minimizes the average cost of using the asset. This is followed by a discussion of replacement with an asset that differs from the current asset, in which the built-in cost advantage of existing assets is revealed. Finally, we look at the case of replacement where there may be a series of replacements for a current asset, each of which might be different.

We shall not consider the implications of taxes for replacement decisions in this chapter. This is postponed until Chapter 8. We shall assume in this chapter that no future price changes due to inflation are expected. The effect of expected price changes on replacement decisions will be considered in Chapter 9.

7.2 | A Replacement Example

We introduce some of the basic concepts involved in replacement decisions through an example.

EXAMPLE 7.1

Sergio likes hiring engineering students to work in his landscaping business during the summer because they are such hard workers and have a lot of common sense. The students are always complaining about maintenance problems with the lawnmowers, which are subject to a lot of use and wear out fairly quickly. His routine has been to replace the machines every five years. Clarissa, one of the engineering student workers, has suggested that replacing them more often might make sense, since so much time is lost when there is a breakdown, in addition to the actual repair cost.

"I've checked the records, and have made some estimates that might help us figure out the best time to replace the machines," Clarissa reports. "Every time there is a breakdown, the average cost to fix the machine is $60. In addition to that, there is an average loss of two hours of time at $20 per hour. As far as the number of repairs required goes, the pattern seems to be zero repairs the first season we use a new lawnmower. However, in the second season, you can expect about one repair, two repairs in the third season, four repairs in the fourth, and eight in the fifth season. I can see why you only keep each mower five years!"

"Given that the cost of a new lawnmower has averaged about $600 and that they decline in value at 40% per year, and assuming an interest rate of 5%, I think we have enough information to determine how often the machines should be replaced," Clarissa concludes authoritatively.

How often should Sergio replace his lawnmowers? How much money will he save?

To keep things simple for now, let's assume that Sergio has to have lawns mowed every year, for an indefinite number of years into the future. If he keeps each lawnmower for, say, three years rather than five years, he will end up buying lawnmowers more frequently, and his overall capital costs for owning the lawnmowers will increase. However, his repair costs should decrease at the same time since newer lawnmowers require fewer repairs. We want to find out which replacement period results in the lowest overall costs—this is the replacement period Sergio should adopt and is referred to as the *economic life* of the lawnmowers.

We could take any period of time as a study period to compare the various possible replacement patterns using a *present worth* approach. The least common multiple of the service lives method (see Chapter 3), suggests that we would need to use a $3 \times 4 \times 5 = 60$-year period if we were considering service lives between one and five years. This is an awkward approach in this case, and even worse in situations where there are more possibilities to consider. It makes much more sense to use an *annual worth* approach. Furthermore, since we typically analyze the costs associated with owning and operating an asset, annual worth calculated in the context of a replacement study is commonly referred to as **equivalent annual cost (EAC)**. In the balance of the chapter, we will therefore use EAC rather than annual worth. However, we should not lose sight of the fact that EAC computations are nothing more than the annual worth approach with a different name adopted for use in replacement studies.

Returning to Sergio's replacement problem, if we calculate the EAC for the possibility of a one-year, two-year, three-year (and so on) replacement period, the pattern that has the lowest EAC would indicate which is best for Sergio. It would be the best because he would be spending the least, keeping both the cost of purchase and the cost of repairs in mind.

This can be done in a spreadsheet, as illustrated in Table 7.1. The first column, "Replacement Period," lists the possible replacement periods being considered. In this case, Sergio is looking at periods ranging from one to five years. The second column lists the salvage value of a lawnmower at the end of the replacement period, in this case estimated using the declining-balance method with a rate of 40%. This is used to compute the entries of the third column. "EAC Capital Costs" is the annualized cost of purchasing and salvaging each lawnmower, assuming replacement periods ranging from one to five years. Using the capital recovery formula (refer back to Close-Up 3.1), this annualized cost is calculated as:

$$EAC(\text{capital costs}) = (P - S)(A/P, i, N) + Si$$

where

$EAC(\text{capital costs})$ = annualized capital costs for an N-year replacement period

P = purchase price

S = salvage value of the asset at the end of N years

Table 7.1 Total Equivalent Annual Cost Calculations for Lawnmower Replacement Example

Replacement Period (Years)	Salvage Value	EAC Capital Costs	Annual Repair Costs	EAC Repair Costs	EAC Total
1	$360.00	$270.00	$ 0.00	$ 0.00	$270.00
2	216.00	217.32	100.00	48.78	266.10
3	129.60	179.21	200.00	96.75	275.96
4	77.76	151.17	400.00	167.11	318.27
5	46.66	130.14	800.00	281.64	411.79

S can be calculated in turn as

$$BV_{db}(n) = P(1 - d)^n$$

For example, in the case of a three-year replacement period, the calculation is

$$EAC(\text{capital costs}) = [600 - 600(1 - 0.40)^3](A/P,5\%,3) + [600(1 - 0.40)^3](0.05)$$

$$= (600 - 129.60)(0.36721) + (129.60)(0.05)$$

$$\cong 179.21$$

The "average" annual cost of repairs (under the heading "EAC Repair Costs"), assuming for simplicity that the cash flows occur at the end of the year in each case, can be calculated as:

$$EAC(\text{repairs}) = [(60 + 40)(P/F,5\%,2) + (60 + 40)(2)(P/F,5\%,3)](A/P,5\%,3)$$

$$= [100(0.90703) + 200(0.86584)](0.36721)$$

$$\cong 96.75$$

The total EAC is then:

$$EAC(\text{total}) = EAC(\text{capital costs}) + EAC(\text{repairs}) = 179.22 + 96.75 \cong 275.96$$

Examining the EAC calculations in Table 7.1, it can be seen that the EAC is minimized when the replacement period is two years (and this is only slightly cheaper than replacing the lawnmowers every year). So this is saying that if Sergio keeps his lawnmowers for two years and then replaces them, his average annual costs over time will be minimized. We can also see how much money Clarissa's observation can save him, by subtracting the expected yearly costs from the estimate of what he is currently paying. This shows that on average Sergio saves $411.79 – $266.10 = $145.69 per lawnmower per year by replacing his lawnmowers on a two-year cycle over replacing them every five years.■

The situation illustrated in Example 7.1 is the simplest case possible in replacement studies. We have assumed that the physical asset to be replaced is identical to the one replacing it, and such a sequence continues indefinitely into the future. By *asset*, we mean any machine or resource of value to an enterprise. An existing physical asset is called the **defender**, because it is currently performing the value-generating activity. The potential replacement is called the **challenger**. However, it is not always the case that the challenger and the defender are the same. It is generally more common that the defender is outmoded or less adequate than the challenger, and also that new challengers can be expected in the future.

This gives rise to several cases to consider. Situations like Sergio's lawnmower problem are relatively uncommon—we live in a technological age where it is unlikely that a replacement for an asset will be identical to the asset it is replacing. Even lawnmowers improve in price, capability, or quality in the space of a few years. A second case, then, is that of a defender that is different from the challenger, with the assumption that the replacement will continue in a sequence of identical replacements indefinitely into the future. Finally, there is the case of a defender different from the challenger, which is itself different from another replacing it farther in the future. All three of these cases will be addressed in this chapter.

Before we look at these three cases in detail, let's look at why assets have to be replaced at all and how to incorporate various costs into the replacement decision.

7.3 | Reasons for Replacement or Retirement

If there is an ongoing need for the service an asset provides, at some point it will need replacement. **Replacement** becomes necessary if there is a cheaper way to get the service the asset provides, or if the service provided by the existing asset is no longer adequate.

An existing asset is **retired** if it is removed from use without being replaced. This can happen if the service that the asset provides is no longer needed. Changes in customer demand, changes in production methods, or changes in technology may result in an asset no longer being necessary. For example, the growth in the use of MP3 players for audio recordings has led manufacturers of compact discs to retire some production equipment since the service it provided is no longer needed.

There may be a cheaper way to get the service provided by the existing asset for several reasons. First, productive assets often deteriorate over time because of wearing out in use, or simply because of the effect of time. As a familiar example, an automobile becomes less valuable with age (older cars, unless they are collectors' cars, are worth less than newer cars with the same mileage) or if it has high mileage (the kilometres driven reflect the wear on the vehicle). Similarly, production equipment may become less productive or more costly to operate over time. It is usually more expensive to maintain older assets. They need fixing more often, and parts may be harder to find and cost more.

Technological or organizational change can also bring about cheaper methods of providing service than the method used by an existing asset. For example, the technological changes associated with the use of computers have improved productivity. Organizational changes, both within a company and in markets outside the company, can lead to lower-cost methods of production. A form of organizational change that has become very popular is the specialist company. These companies take on parts of the production activities of other companies. Their specialization may enable them to have lower costs than the companies can attain themselves. See Close-Up 7.1.

The second major reason why a current asset may be replaced is inadequacy. An asset used in production can become inadequate because it has insufficient capacity to meet growing demand or because it no longer produces items of high enough quality. A company may have a choice between adding new capacity parallel to the existing capacity, or replacing the existing asset with a higher capacity asset, perhaps one with more advanced technology. If higher quality is required, there may be a choice between upgrading an

---

CLOSE-UP 7.1   Specialist Companies

Specialist companies concentrate on a limited range of very specialized products. They develop the expertise to manufacture these products at minimal cost. Larger firms often find it more economical to contract out production of low-volume components instead of manufacturing the components themselves.

In some industries, the use of specialist companies is so pervasive that the companies apparently manufacturing a product are simply assembling it; the actual manufacturing takes place at dozens or sometimes hundreds of supplier firms.

A good example of this is the automotive industry. In North America, auto makers support an extremely large network of specialist companies, linked by computer. A single specialist company might supply brake pads, for example, to all three major auto manufacturers. Producing brake pads in huge quantities, the specialist firm can refine its production process to extremes of efficiency and profitability.

existing piece of equipment or replacing it with equipment that will yield the higher quality. In either case, contracting out the work to a specialist is one possibility.

In summary, there are two main reasons for replacing an existing asset. First, an existing asset will be replaced if there is a cheaper way to get the service that the asset provides. This can occur because the asset ages or because of technological or organizational changes. Second, an existing asset will be replaced if the service it provides is inadequate in either quantity or quality.

7.4 | Capital Costs and Other Costs

When a decision is made to acquire a new asset, it is essentially a decision to purchase the capacity to perform tasks or produce output. **Capacity** is the ability to produce, often measured in units of production per time period. Although production requires capacity, it is also important to understand that just acquiring the capacity entails costs that are incurred whether or not there is actual production. Furthermore, a large portion of the capacity cost is incurred early in the life of the capacity. There are two main reasons for this:

1. Part of the cost of acquiring capacity is the expense incurred over time because the assets required for that capacity gradually lose their value. This expense is often called the **capital cost** of the asset. It is incurred by the difference between what is paid for the assets required for a particular capacity and what the assets could be resold for some time after purchase. The largest portion of the capital costs typically occur early in the lives of the assets.

2. Installing a new piece of equipment or new plant sometimes involves substantial up-front costs, called **installation costs**. These are the costs of acquiring capacity, excluding the purchase cost, which may include disruption of production, training of workers, and perhaps a reorganization of other processes. Installation costs are not reversible once the capacity has been put in place.

For example, if Sergio bought a new lawnmower to accommodate new landscaping clients, rather than for replacement, he would be increasing the capacity of his lawn-mowing service. The capital cost of the lawnmower in the first year would be its associated loss in market value over that year. The installation cost would probably be negligible.

It is worth noting that, in general, the total cost of a new asset includes both the installation costs and the cost of purchasing the asset. When we compute the capital costs of an asset over a period of time, the first cost (usually denoted by P) includes the installation costs. However, when we compute a salvage value for the asset as it ages, we do *not* include the installation costs as part of the depreciable value of the asset, since these costs are expended upon installation and cannot be recovered at a later time.

The large influence of capital costs associated with acquiring new capacity means that, once the capacity has been installed, the *incremental* cost of continuing to use that capacity during its planned life is relatively low. This gives a defender a cost advantage during its planned life over a challenger.

In addition to the capital and installation costs, the purchase of an asset carries with it future **operating and maintenance costs**. Operating costs might include electricity, gasoline, or other consumables, and maintenance might include periodic servicing and repairs. Also, it is worth noting that a challenger may also give rise to changes in revenues as well as changes in costs that should not be neglected.

The different kinds of costs discussed in this section can be related to the more general ideas of fixed and variable costs. **Fixed costs** are those that remain the same, regardless of actual units of production. For example, capital costs are usually fixed—once an asset is purchased, the cost is incurred even if there is zero production. **Variable costs** are costs that change depending on the number of units produced. The costs of the raw materials used in production are certainly variable costs, and to a certain degree operating and maintenance costs are as well.

With this background, we now look at the three different replacement cases:

1. Defender and challenger are identical, and repeat indefinitely
2. Challenger repeats indefinitely, but is different from defender
3. Challenger is different from defender, but does not repeat

---

N E T   V A L U E   7 . 1

Estimating Salvage Values and Scrap Values

An operating asset can have considerable value as long as it continues to perform the function for which it is intended. However, if it is replaced, its salvage value greatly depends on what is done with it. It is likely that the salvage value is different from whatever depreciation value is calculated for accounting or taxation purposes. However, to make a good replacement decision, it is desirable to have an accurate estimate of the actual salvage value of the asset.

The internet can be very helpful in estimating salvage values. A search for the asset by type, year, and model may reveal a similar asset in a used market. Or a more general search may reveal a broker for used assets of this nature who might be contacted to provide an estimate of salvage value based on the broker's experience with similar assets. In both of these cases, not only can a salvage value be determined, but also a channel for disposing of the asset is found. Even if the asset is scrapped, its value can be estimated from, for example, posted values for metal scrap available on the web.

7.5 | Defender and Challenger Are Identical

All long-lived assets eventually require replacement. Consequently, the issue in replacement studies is not *whether* to replace an asset, but *when* to replace it. In this section we consider the case where there is an ongoing need for a service provided by an asset and where the asset technology is not changing rapidly. (This is the case for Sergio's landscaping example at the beginning of this chapter.)

Several assumptions are made.

1. The defender and challenger are assumed to be technologically identical. It is also assumed that this remains true for the company's entire planning horizon.
2. The lives of these identical assets are assumed to be short relative to the time horizon over which the assets are required.
3. Relative prices and interest rates are assumed to be constant over the company's time horizon.

These assumptions are quite restrictive. In fact, there are only a few cases where the assumptions strictly hold (cable used for electric power delivery is an example). Nonetheless, the idea of economic life of an asset is still useful to our understanding of replacement analysis.

Assumptions 1 and 2 imply that we may model the replacement decision as being repeated an indefinitely large number of times. The objective is then to determine a minimum-cost lifetime for the assets, a lifetime that will be the same for all the assets in the sequence of replacements over the company's time horizon.

We have seen that the relevant costs associated with acquiring a new asset are the capital costs, installation costs (which are often pooled with the capital costs), and operating and maintenance costs. It is usually true that operating and maintenance costs of assets—plant or equipment—rise with the age of the asset. Offsetting increases in operating and maintenance costs is the fall in capital costs per year that usually occurs as the asset life is extended and the capital costs are spread over a greater number of years. The rise in operating and maintenance costs per year, and the fall in capital costs per year as the life of an asset is extended, work in opposite directions. In the early years of an asset's life, the capital costs per year (although decreasing) usually, but not always, dominate total yearly costs. As the asset ages, increasing operating and maintenance costs usually overtake the declining annual capital costs. This means that there is a lifetime that will minimize the *average* cost (adjusting for the time value of money) per year of owning and using long-lived assets. This is referred to as the **economic life** of the asset.

These ideas are illustrated in Figure 7.1. Here we see the capital costs per period decrease as the number of periods the asset is kept increases because assets tend to depreciate in value by a smaller amount each period of use. On the other hand, the operating and maintenance costs per period increase because older assets tend to need more repairs and have other increasing costs with age. The economic life of an asset is found at the point where the rate of increase in operating and maintenance costs per period equals the rate of decrease in capital costs per period, or equivalently where the total cost per period is minimized.

Figure 7.1   Cost Components for Replacement Studies

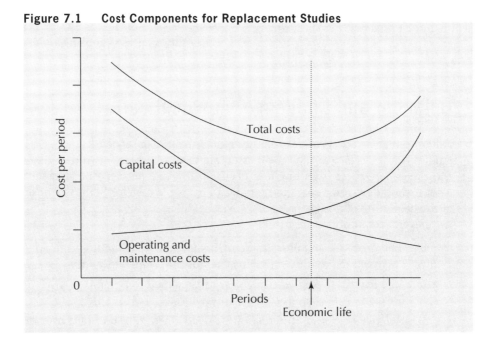

EXAMPLE 7.2

The Jiffy Printer Company produces printers for home use. Jiffy is considering installing an automated plastic moulding system to produce parts for the printers. The moulder itself costs €20 000 and the installation costs are estimated to be €5000. Operating and maintenance costs are expected to be €30 000 in the first year and to rise at the rate of 5% per year. Jiffy estimates depreciation with a declining-balance model using a rate of 40%, and uses a MARR of 15% for capital investments. Assuming that there will be an ongoing need for the moulder, and assuming that the technology does not change (i.e., no cheaper or better method will arise), how long should Jiffy keep a moulder before replacing it with a new model? In other words, what is the economic life of the automated moulding system?

Determining the economic life of an asset is most easily done with a spreadsheet. Table 7.2 shows the development of the equivalent annual costs for the automated plastic moulding system of Example 7.2.

In general, the EAC for an asset has two components:

EAC = EAC(capital costs) + EAC(operating and maintenance)

If

P = the current value of an asset = (for a new asset) purchase price + installation costs

S = the salvage value of an asset N years in the future

then

EAC(capital costs) = $[P - (P/F, i, N)S] (A/P, i, N)$

which in Close-Up 3.1 was shown to be equivalent to

EAC(capital costs) = $(P - S)(A/P, i, N) + Si$

In the first column of Table 7.2 is the life of the asset, in years. The second column shows the salvage value of the moulding system as it ages. The equipment costs €20 000 originally, and as the system ages the value declines by 40% of current value each year,

Table 7.2   Computation of Total Equivalent Annual Costs of the Moulding System With a MARR = 15%

Life in Years	Salvage Value	EAC Capital Costs	EAC Operating and Maintenance Costs	EAC Total
0	€20 000.00			
1	12 000.00	€16 750.00	€30 000.00	€46 750.00
2	7200.00	12 029.07	30 697.67	42 726.74
3	4320.00	9705.36	31 382.29	41 087.65
4	2592.00	8237.55	32 052.47	40 290.02
5	1555.20	7227.23	32 706.94	39 934.17
6	933.12	6499.33	33 344.56	39 843.88
7	559.87	5958.42	33 964.28	39 922.70
8	335.92	5546.78	34 565.20	40 111.98
9	201.55	5227.34	35 146.55	40 373.89
10	120.93	4975.35	35 707.69	40 683.04

giving the estimated salvage values listed in Table 7.2. For example, the salvage value at the end of the fourth year is:

$$BV_{db}(4) = 20\,000(1 - 0.4)^4$$

$$\cong 2592$$

The next column gives the equivalent annual capital costs if the asset is kept for N years, $N = 1, \ldots, 10$. This captures the loss in value of the asset over the time it is kept in service. As an example of the computations, the equivalent annual capital cost of keeping the moulding system for four years is

$$EAC(\text{capital costs}) = (P - S)(A/P,15\%,4) + Si$$

$$= (20\,000 + 5000 - 2592)(0.35027) + 2592(0.15)$$

$$\cong 8238$$

Note that the installation costs have been included in the capital costs, as these are expenses incurred at the time the asset is originally put into service. Table 7.2 illustrates that the equivalent annual capital costs decline as the asset's life is extended. Next, the equivalent annual operating and maintenance costs are found by converting the stream of operating and maintenance costs (which are increasing by 5% per year) into a stream of equal-sized annual amounts. Continuing with our sample calculations, the EAC of operating and maintenance costs when the moulding system is kept for four years is

EAC(operating and maintenance costs)

$$= 30\,000\,[(P/F,15\%,1) + (1.05)(P/F,15\%,2) + (1.05)^2(P/F,15\%,3) \\ + (1.05)^3(P/F,15\%,4)]\,(A/P,15\%,4)$$

$$\cong 32\,052$$

Notice that the equivalent annual operating and maintenance costs increase as the life of the moulding system increases.

Finally, we obtain the total equivalent annual cost of the moulding system by adding the equivalent annual capital costs and the equivalent annual operating and maintenance costs. This is shown in the last column of Table 7.2. We see that at a six-year life the declining equivalent annual installation and capital costs offset the increasing operating and maintenance costs. In other words, the economic life of the moulder is six years, with a total EAC of €39 844.■

While it is *usually* true that capital cost per year falls with increasing life, it is not always true. Capital costs per year can rise at some point in the life of an asset if the decline in value of the asset is not smooth or if the asset requires a major overhaul.

If there is a large drop in the value of the asset in some year during the asset's life, the cost of holding the asset over that year will be high. Consider the following example.

EXAMPLE 7.3

An asset costs $50 000 to buy and install. The asset has a resale value of $40 000 after installation. It then declines in value by 20% per year until the fourth year when its value drops from over $20 000 to $5000 because of a predictable wearing-out of a major component. Determine the equivalent annual capital cost of this asset for lives ranging from one to four years. The MARR is 15%.

The computations are summarized in Table 7.3. The first column gives the life of the asset in years. The second gives the salvage value of the asset as it ages. The asset loses 20% of its previous year's value each year except in the fourth, when its value drops to $5000. The last column summarizes the equivalent annual capital cost of the asset. Sample computations for the third and fourth years are:

EAC(capital costs, three-year life)

$$= (P - S)(A/P, 15\%, 3) + Si$$

$$= (40\ 000 + 10\ 000 - 20\ 480)(0.43798) + 20\ 480(0.15)$$

$$\cong 16\ 001$$

EAC(capital costs, four-year life)

$$= (P - S)(A/P, 15\%, 4) + Si$$

$$= (40\ 000 + 10\ 000 - 5000)(0.35027) + 5000(0.15)$$

$$\cong 16\ 512$$

Table 7.3   EAC of Capital Costs for Example 7.3

Life in Years	Salvage Value	EAC Capital Costs
0	$40 000	
1	32 000	$25 500
2	25 600	18 849
3	20 480	16 001
4	5000	16 512

The large drop in value in the fourth year means that there is a high cost of holding the asset in the fourth year. This is enough to raise the average capital cost per year.■

In summary, when we replace one asset with another with an identical technology, it makes sense to speak of its economic life. This is the lifetime of an individual asset that will minimize the average cost per year of owning and using it. In the next section, we deal with the case where the challenger is different from the defender.

7.6 | Challenger Is Different From Defender; Challenger Repeats Indefinitely

The decision rule that minimizes cost in the case where a defender is faced by a challenger that is expected to be followed by a sequence of identical challengers is as follows.

1. Determine the economic life of the challenger and its associated EAC.
2. Determine the remaining economic life of the defender and its associated EAC.
3. If the EAC of the defender is greater than the EAC of the challenger, replace the defender now. Otherwise, do not replace now.

In many cases, the computations in step 2 can be reduced somewhat. For assets that have been in use for several years, the yearly capital costs will typically be low compared to the yearly operating costs—the asset's salvage value won't change much from year to year but the operating costs will continue to increase. If this is true, as it often is for assets being replaced, the **one year principle** can be used. This principle states that if the capital costs for the defender are small compared to the operating costs, and the yearly operating costs are monotonically increasing, the economic life of the defender is one year and its total EAC is the cost of using the defender for one more year.

The principle thus says that if the EAC of keeping the defender one more year exceeds the EAC of the challenger at its economic life, the defender should be replaced immediately. The advantage of the one year principle is that there is no need to find the service life for the defender that minimizes the EAC—it is known in advance that the EAC is minimized in a one-year period. The principle is particularly useful because for most assets the operating costs increase smoothly and monotonically as the asset is kept for longer and longer periods of time, while the capital costs decrease smoothly and monotonically. For a defender that has been in use for some time, the EAC(operating costs) will typically dominate the EAC(capital costs), and thus the total EAC will increase over any additional years that the asset is kept. This is illustrated in Figure 7.2, which can be compared to Figure 7.1 earlier in this chapter.

Figure 7.2   Cost Components for Certain Older Assets

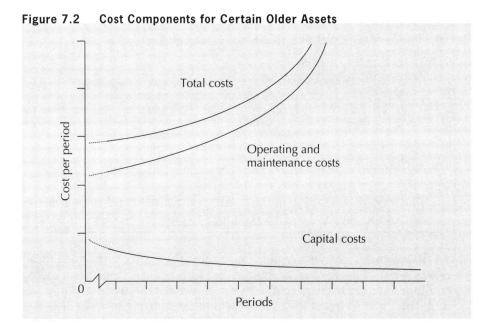

For older assets that conform to this pattern of costs, it is only necessary to check whether the defender needs replacing due to costs over the next year, because in subsequent years the case for the defender will only get worse, not better. If there is an uneven yearly pattern of operating costs, the one year principle cannot be used, because the current year might give an unrealistic value due to the particular expenses in that year.

EXAMPLE 7.4

An asset is three years old. Yearly operating and maintenance costs are currently 5000 rand per year, increasing by 10% per year. The salvage value for the asset is currently 60 000 rand and is expected to be 50 000 rand one year from now. Can the one year principle be used?

The capital costs are not low (and thus the EAC(capital costs) for any given service life is not low) compared to the operating and maintenance costs. Even though costs have a regular pattern, the one year principle cannot be used.■

EXAMPLE 7.5

An asset is 10 years old. Yearly operating and maintenance costs are currently $5000 per year, increasing by 10% per year. The salvage value of the asset is currently $8000 and is expected to be $7000 one year from now. Can the one year principle be used?

The capital costs are low compared to the operating and maintenance costs, and costs have a regular pattern. The one year principle can be used.■

EXAMPLE 7.6

An asset is 10 years old. Operating and maintenance costs average £5000 per year, increasing by 10% per year. However, most of the operating and maintenance costs consist of a periodic overhaul costing £15 000 that occurs every three years. The salvage value of the asset is currently £8000 and is expected to be £7000 one year from now. Can the one year principle be used?

The capital costs are low compared to the operating and maintenance costs but the operating and maintenance costs do not have an even pattern from year to year. The one year principle cannot be used.■

The one year principle can be used when it is clear that the key conditions—low capital costs and a regular year-to-year pattern of monotonically increasing operating and maintenance costs—are satisfied. Where a situation is ambiguous, it is always prudent to fully assess the EAC of the defender.

To fully explore the case of a defender being replaced by a challenger that repeats, as well as to explore some other ideas useful in replacement analysis, the next three subsections cover examples to illustrate the analysis steps. In the first, we examine the situation of replacing subcontracted capacity with in-house production. This is an example of replacing a service with an asset. In the second example, the issue of sunk costs is examined by considering the replacement of an existing productive asset with a different challenger. In the final example, we look at the situation of making replacement decisions when there are irregular cash flows.

7.6.1 Converting From Subcontracted to In-House Production

EXAMPLE 7.7

Currently, the Jiffy Printer Company of Example 7.2 pays a custom moulder €0.25 per piece (excluding material costs) to produce parts for its printers. Demand is forecast to be 200 000 parts per year. Jiffy is considering installing the automated plastic moulding system described in Example 7.2 to produce the parts. Should it do so now?

In Jiffy's situation, the *defender* is the current technology: a subcontractor. The *challenger* is the automated plastic moulding system. In order to decide whether Jiffy is better off with the defender or the challenger, we need to compute the unit cost (cost per piece) of production with the automated moulder and compare it to the unit cost for the subcontracted parts. If the automated system is better, the challenger wins and should replace the defender.

From Example 7.2:

EAC(moulder) = 39 844

Dividing the EAC by the expected number of parts needed per year allows us to calculate the unit cost as 39 838/200 000 = 0.1992.

When the unit cost of in-house production is compared with the €0.25 unit cost of subcontracting the work, in-house production is cheaper, so Jiffy should replace the subcontracting with an in-house automated plastic moulding system.∎

This example has illustrated the basic idea behind a replacement analysis when we are considering the purchase of a new asset as a replacement to current technology. The cost of the replacement must take into account the capital costs (including installation) and the operating and maintenance costs over the life of the new asset.

In the next subsection, we see how some costs are no longer relevant in the decision to replace an *existing* asset.

7.6.2 The Irrelevance of Sunk Costs

Once an asset has been installed and has been operating for some time, the costs of installation and all other costs incurred up to that time are no longer relevant to any decision to replace the current asset. These costs are called **sunk costs**. Only those costs that will be incurred in keeping and operating the asset from this time on are relevant. This is best illustrated with an example.

EXAMPLE 7.8

Two years have passed since the Jiffy Printer Company from Example 7.7 installed an automated moulding system to produce parts for its printers. At the time of installation, they expected to be producing about 200 000 pieces per year, which justified the investment. However, their actual production has turned out to be only about 150 000 pieces per year. Their unit cost is €39 844/150 000 = €0.2656 rather than the €0.1992 they had expected. They estimate the current market value of the moulder at €7200. In this case, maintenance costs do not depend on the actual production rate. Should Jiffy sell the moulding system and go back to buying from the custom moulder at €0.25 per piece?

In the context of a replacement problem, Jiffy is looking at replacing the existing system (the defender) with a different technology (the challenger). Since Jiffy already has the moulder and has already expended considerable capital on putting it into place, it may be better for Jiffy to keep the current moulder for some time longer. Let us calculate the cost to Jiffy of keeping the moulding system for one more year. This may not be the optimal length of time to continue to use the system, but if the cost is less than €0.25 per piece it is cheaper than retiring or replacing it now.

The reason that the cost of keeping the moulder an additional year may be low is that the capital costs for the two-year-old system are now low compared with the costs of putting the capacity in place. The capital cost for the third year is simply the loss in value of the moulder over the third year. This is the difference between what Jiffy can get for the system now, at the end of the second year, and what it can get a year from now when the system will be three years old. Jiffy can get €7200 for the

$30,000 \ (P/F)$

system now. Using the declining-balance depreciation rate of 40% to calculate a salvage value, we can determine the capital cost associated with keeping the moulder over the third year.

Applying the capital recovery formula from Chapter 3, the EAC for capital costs is

$$\text{EAC(capital costs, third year)} = (P - S)(A/P,15\%,1) + Si$$
$$= [7200 - 0.6(7200)](1.15) + 0.6(7200)(0.15)$$
$$\cong 3960$$

Recall that the operating and maintenance costs started at €30 000 and rose at 5% each year. The operating and maintenance costs for the third year are

$$\text{EAC(operating and maintenance, third year)} = 30\ 000(1.05)^2$$
$$\cong 33\ 075$$

The total cost of keeping the moulder for the third year is the sum of the capital costs and the operating and maintenance costs:

$$\text{EAC(third year)} = \text{EAC(capital costs, third year)} +$$
$$\text{EAC (operating and maintenance, third year)}$$
$$= 3960 + 33\ 075$$
$$\cong 37\ 035$$

Dividing the annual costs for the third year by 150 000 units gives us a unit cost of €0.247 for moulding in-house during the third year. Not only is this lower than the average cost over a six-year life of the system, it is also lower than what the subcontracted custom moulder charges. Similar computations would show that Jiffy could go two more years after the third year with in-house moulding. Only then would the increase in operating and maintenance costs cause the total unit cost to rise above €0.25.

Given the lower demand, we see that installing the automated moulding system was a mistake for Jiffy. The average lifetime costs for in-house moulding were greater than the cost of subcontracting, but once the system was installed, it was not optimal to go back to subcontracting immediately. This is because the capital cost associated with the purchase and installation of an asset (which becomes sunk after its installation) is disproportionately large as compared with the cost of using the asset once it is in place.■

That a defender has a cost advantage over a challenger, or over contracting out during its planned life, is important. It means that if a defender is to be removed from service during its life for cost reasons, the average lifetime costs for the challenger or the costs of contracting out must be considerably lower than the average lifetime costs of the defender.

Just because well-functioning defenders are not often retired for cost reasons does not mean that they will not be retired at all. Changes in markets or technology may make the quantity or quality of output produced by a defender inadequate. This can lead to its retirement or replacement even when it is functioning well.

7.6.3 When Capital or Operating Costs Are Non-Monotonic

Sometimes operating costs do not increase smoothly and monotonically over time. The same can happen to capital costs. When the operating or capital costs are not smooth and monotonic, the one year principle does not apply. The reason that the principle does not apply is that there may be periodic or one-time costs that occur over the course of the next year (as in the case where periodic overhauls are required). These costs may make the cost of keeping the defender for one more year greater than the cost of installing and using the challenger over its economic life. However, there may be a life longer than one year over which the cost of using the defender is less than the cost of installing and using a challenger. Consider this example concerning the potential replacement of a generator.

EXAMPLE 7.9

The Colossal Construction Company uses a generator to produce power at remote sites. The existing generator is now three years old. It cost $11 000 when purchased. Its current salvage value of $2400 is expected to fall to $1400 next year and $980 the year after, and to continue declining at 30% of current value per year. Its ordinary operating and maintenance costs are now $1000 per year and are expected to rise by $500 per year. There is also a requirement to do an overhaul costing $1000 this year and every third year thereafter.

New fuel-efficient generators have been developed, and Colossal is thinking of replacing its existing generator. It is expected that the new generator technology will be the best available for the foreseeable future. The new generator sells for $9500. Installation costs are negligible. Other data for the new generator are summarized in Table 7.4.

Table 7.4   Salvage Values and Operating Costs for New Generator

End of Year	Salvage Value	Operating Cost
0	$9500	
1	8000	$1000
2	7000	1000
3	6000	1200
4	5000	1500
5	4000	2000
6	3000	2000
7	2000	2000
8	1000	3000

Should Colossal replace the existing generator with the new type? The MARR is 12%.

We first determine the economic life for the challenger. The calculations are shown in Table 7.5.

Sample calculations for the EAC of keeping the challenger for one, two, and three years are as follows:

$$EAC(1 \text{ year}) = (P - S)(A/P,12\%,1) + Si + 1000$$
$$= (9500 - 8000)(1.12) + 8000(0.12) + 1000$$
$$\cong 3640$$

Table 7.5   Economic Life of the New Generator

End of Year	Salvage Value	Operating Costs	EAC
1	$8000	$1000	$3640.00
2	7000	1000	3319.25
3	6000	1200	3236.50
4	5000	1500	3233.07*
5	4000	2000	3290.81
6	3000	2000	3314.16
7	2000	2000	3318.68
8	1000	3000	3393.52

*Lowest equivalent annual cost.

$$\begin{aligned}
\text{EAC(2 years)} &= (P - S)(A/P,12\%,2) + Si + 1000 \\
&= (9500 - 7000)(0.5917) + 7000(0.12) + 1000 \\
&\cong 3319
\end{aligned}$$

$$\begin{aligned}
\text{EAC(3 years)} &= (P - S)(A/P,12\%,3) + Si + 1000 + 200(A/F,12\%,3) \\
&= (9500 - 6000)(0.41635) + 6000(0.12) + 1000 \\
&\quad + 200(0.29635) \\
&\cong 3237
\end{aligned}$$

As the number of years increases, this approach for calculating the EAC becomes more difficult, especially since in this case the operating costs are neither a standard annuity nor an arithmetic gradient. An alternative is to calculate the present worths of the operating costs for each year. The EAC of the operating costs can be found by applying the capital recovery factor to the sum of the present worths for the particular service period considered. This approach is particularly handy when using spreadsheets.

By either calculation, we see in Table 7.5 that the economic life of the generator is four years.

Next, to see if and when the defender should be replaced, we calculate the costs of keeping the defender for one more year. Using the capital recovery formula:

$$\begin{aligned}
&\text{EAC (keep defender 1 more year)} \\
&\quad = \text{EAC(capital costs)} + \text{EAC(operating costs)} \\
&\quad = (2400 - 1400)(A/P,12\%,1) + 1400(0.12) + 2000 \\
&\quad \cong 3288
\end{aligned}$$

The equivalent annual cost of using the defender one more year is $3288. This is more than the yearly cost of installing and using the challenger over its economic life. Since the operating costs are not smoothly increasing, we need to see if there is a longer life for the defender for which its costs are lower than for the challenger. This can be done with a spreadsheet, as shown in Table 7.6.

We see that, for an additional life of three years, the defender has a lower cost per year than the challenger, when the challenger is kept over its economic life. Therefore, the defender should not be replaced at this time. Next year a new evaluation should be performed.

Table 7.6 **Equivalent Annual Cost of Additional Life for the Defender**

Additional Life in Years	Salvage Value	Operating Costs	EAC
0	$2400		
1	1400	$2000	$3288
2	980	1500	2722
3	686	2000	**2630**
4	480	3500	2872
5	336	3000	2924
6	235	3500	3013

7.7 | Challenger Is Different From Defender; Challenger Does Not Repeat

In this section, we no longer assume that challengers are alike. We recognize that future challengers will be available and we expect them to be better than the current challenger. We must then decide if the defender should be replaced by the current challenger. Furthermore, if it is to be replaced by the current challenger, *when* should the replacement occur? This problem is quite complex. The reason for the complexity is that, if we believe that challengers will be improving, we may be better off skipping the current challenger and waiting until the next improved challenger arrives. The difficulties are outlined in Example 7.10.

EXAMPLE 7.10

Rita is examining the possibility of replacing the kiln controllers at the Burnaby Insulators plant. She has information about the existing controllers and the best replacement on the market. She also has information about a new controller design that will be available in three years. Rita has a five-year time horizon for the problem. What replacement alternatives should Rita consider?

One way to determine the minimum cost over the five-year horizon is to determine the costs of *all* possible combinations of the defender and the two challengers. This is impossible, since the defender and challengers could replace one another at any time. However, it is reasonable to consider only the combinations of the period length of one year. Any period length could be used, but a year is a natural choice because investment decisions tend, in practice, to follow a yearly cycle. These combinations form a mutually exclusive set of investment opportunities (see Section 4.2). If no time horizon were given in the problem, we would have had to assume one, to limit the number of possible alternatives.

The possible decisions that need to be evaluated in this case are shown in Table 7.7.

For example, the first row in Table 7.7 (Alternative 1) means to keep the defender for the whole five-year period. Alternative 2 is to keep the defender for four years, and then purchase the challenger four years from now, and keep it for one year. Alternative 15 is to replace the defender now with the first challenger, keep it three years, then replace it with the second challenger, and keep the second challenger for the remaining two years.

Table 7.7   Possible Decisions for Burnaby Insulators

Decision Alternative	Defender Life in Years	First Challenger Life in Years	Second Challenger Life in Years
1	5	0	0
2	4	1	0
3	4	0	1
4	3	2	0
5	3	1	1
6	3	0	2
7	2	3	0
8	2	2	1
9	2	1	2
10	1	4	0
11	1	3	1
12	1	2	2
13	0	5	0
14	0	4	1
15	0	3	2

To choose among these possible alternatives, we need information about the following for the defender and both challengers:

1. Costs of installing the challengers
2. Salvage values for different possible lives for all three kiln controllers
3. Operating and maintenance costs for all possible ages for all three

With this information, the minimum-cost solution is obtained by computing the costs for all possible decision alternatives. Since these are mutually exclusive projects, any of the comparison methods of Chapters 4 and 5 are appropriate, including present worth, annual worth, or IRR. The effects of sunk costs are already included in the enumeration of the various replacement possibilities, so looking at the benefits of keeping the defender is already automatically taken into account.■

The difficulty with this approach is that the computational burden becomes great if the number of years in the time horizon is large. On the other hand, it is unlikely that information about a future challenger will be available under normal circumstances. In Example 7.10, Rita knew about a controller that wouldn't be available for three years. In real life, even if somehow Rita had inside information on the supplier research and marketing plans, it is unlikely that she would be confident enough of events three years away to use the information with complete assurance. Normally, if the information were available, the challenger itself would be available, too. Consequently, in many cases it is reasonable to assume that challengers in the planning future will be identical to the current challenger, and the decision procedure to use is the simpler one presented in the previous section.

REVIEW PROBLEMS

REVIEW PROBLEM 7.1

Kenwood Limousines runs a fleet of vans that ferry people from several outlying cities to a major international airport. New vans cost £45 000 each and depreciate at a declining-balance rate of 30% per year. Maintenance for each van is quite expensive, because they are in use 24 hours a day, seven days a week. Maintenance costs, which are about £3000 the first year, double each year the vehicle is in use. Given a MARR of 8%, what is the economic life of a van?

ANSWER

Table 7.8 shows the various components of this problem for replacement periods from one to five years. It can be seen that the replacement period with the minimum equivalent annual cost is two years. Therefore, the economic life is two years.

Table 7.8   Summary Computations for Review Problem 7.1

Year	Salvage Value	Maintenance Costs	Equivalent Annual Costs		
			Capital	Maintenance	Total
0	£45 000				
1	31 500	£ 3000	£17 100	£ 3000	£20 100
2	22 050	6000	14 634	4442	**19 076**
3	15 435	12 000	12 707	6770	19 477
4	10 805	24 000	11 189	10 594	21 783
5	7563	48 000	9981	16 970	26 951

As an example, the calculation for a three-year period is:

EAC(capital costs)

$$= (45\ 000 - 15\ 435)(A/P,8\%,3) + 15\ 435(0.08)$$

$$= 29\ 565(0.38803) + 15\ 435(0.08)$$

$$\cong 12\ 707$$

EAC(maintenance costs)

$$= [3000(F/P,8\%,2) + 6000(F/P,8\%,1) + 12\ 000]\ (A/F,8\%,3)$$

$$= [3000(1.1664) + 6000(1.08) + 12\ 000)](0.30804)$$

$$\cong 6770$$

EAC(total)   = EAC(capital costs) + EAC(maintenance costs)

$$= 12\ 707 + 6770 = 19\ 477\blacksquare$$

REVIEW PROBLEM 7.2

Global Widgets makes rocker arms for car engines. The manufacturing process consists of punching blanks from raw stock, forming the rocker arm in a 5-stage progressive die, and finishing in a sequence of operations using hand tools. A recently developed 10-stage die can eliminate many of the finishing operations for high-volume production.

The existing 5-stage die could be used for a different product, and in this case would have a salvage value of $20 000. Maintenance costs of the 5-stage die will total $3500 this year, and are expected to increase by $3500 per year. The 10-stage die will cost $89 000, and will incur maintenance costs of $4000 this year, increasing by $2700 per year thereafter. Both dies depreciate at a declining-balance rate of 20% per year. The net yearly benefit of the automation of the finishing operations is expected to be $16 000 per year. The MARR is 10%. Should the 5-stage die be replaced?

ANSWER

Since there is no information about subsequent challengers, it is reasonable to assume that the 10-stage die would be repeated. The EAC of using the 10-stage die for various periods is shown in Table 7.9.

A sample EAC computation for keeping the 10-stage die for two years is as follows:

EAC(capital costs, two-year life)

$$= (P - S)(A/P, 10\%, 2) + Si$$

$$= (89\,000 - 56\,960)(0.57619) + 56\,960(0.10)$$

$$\cong 24\,157$$

EAC(maintenance costs, two-year life)

$$= [4000(F/P, 10\%, 1) + 6700](A/F, 10\%, 2)$$

$$= [4000(1.1) + 6700](0.47619)$$

$$\cong 5286$$

Table 7.9   EAC Computations for the Challenger in Review Problem 7.2

Life in Years	Salvage Value	Maintenance Costs	Equivalent Annual Costs		
			Capital	Maintenance	Total
0	$89 000				
1	71 200	$ 4000	$26 700	$ 4000	$30 700
2	56 960	6700	24 157	5286	29 443
3	45 568	9400	22 021	6529	28 550
4	36 454	12 100	20 222	7729	27 951
5	29 164	14 800	18 701	8887	27 589
6	23 331	17 500	17 411	10 004	27 415
7	18 665	20 200	16 314	11 079	**27 393**
8	14 932	22 900	15 377	12 113	27 490

EAC(total, two-year life)

$$= 24\,157 + 5286$$

$$= 29\,443$$

Completing similar computations for other lifetimes shows that the economic life of the 10-stage die is seven years and the associated equivalent annual costs are $27 393.

The next step in the replacement analysis is to consider the annual cost of the 5-stage die (the defender) over the next year. This cost is to be compared with the economic life EAC of the 10-stage die, that is, $27 393. Note that the cost analysis of the defender should include the benefits generated by the 10-stage die as an operating cost for the 5-stage die as this $16 000 is a cost of *not* changing to the 10-stage die. Since the capital costs are low, and operating costs are monotonically increasing, the one year principle applies. The EAC of the capital and operating costs of keeping the defender one additional year are found as follows:

Salvage value of 5-stage die after one year = 20 000(1 – 0.2) = $16 000

EAC(capital costs, one additional year)

$$= (P - S)(A/P,10\%,1) + Si$$

$$= (20\,000 - 16\,000)(1.10) + 16\,000(0.10)$$

$$\cong 6000$$

EAC(maintenance and operating costs, one additional year)

$$= 3500 + 16\,000$$

$$\cong 19\,500$$

EAC(total, one additional year)

$$= 19\,500 + 6000$$

$$\cong 25\,500$$

The 5-stage die should not be replaced this year because the EAC of keeping it one additional year ($25 500) is less than the optimal EAC of the 10-stage die ($27 393). The knowledge that the 5-stage die should not be replaced this year is usually sufficient for the immediate replacement decision. However, if a different challenger appears in the future, we would want to reassess the replacement decision.

It may also be desirable to estimate when in the future the defender might be replaced, even if it is not being replaced now. This can be done by calculating the equivalent annual cost of keeping the defender additional years until the time we can determine when it should be replaced. Table 7.10 summarizes those calculations for additional years of operating the 5-stage die.

As an example of the computations, the EAC of keeping the defender for two additional years is calculated as

Salvage value of 5-stage die after two years = 16 000(1 – 0.2) = 12 800

EAC(capital costs, two additional years)

$$= (P - S)(A/P,10\%,2) + Si$$

$$= (20\,000 - 12\,800)(0.57619) + 12\,800(0.10)$$

$$\cong 5429$$

EAC(maintenance and operating costs, two additional years)

$$= [19\,500(F/P,10\%,1) + (16\,000 + 7000)](A/F,10\%,2)$$

$$= [19\,500(1.1) + 23\,000](0.47619)$$

$$\cong 21\,167$$

Table 7.10   EAC Computations for Keeping the Defender Additional Years

Additional Life in Years	Salvage Value	Maintenance and Operating Costs	Equivalent Annual Costs		
			Capital	Operating	Total
0	$20 000				
1	16 000	$19 500	$6000	$19 500	$25 500
2	12 800	23 000	5429	21 167	26 595
3	10 240	26 500	4949	22 778	27 727
4	8192	30 000	4544	24 334	28 878
5	6554	33 500	4202	25 836	30 038
6	5243	37 000	3913	27 283	31 196
7	4194	40 500	3666	28 677	32 343
8	3355	44 000	3455	30 018	33 473

EAC(total, two additional years)

$$= 5429 + 21\ 167 = 26\ 595$$

Further calculations in this manner will predict that the defender should be replaced at the end of the second year, given that the challenger remains the same during this time. This is because the EAC of keeping the defender for two years is less than the optimal EAC of the 10-stage die, but keeping the defender three years or more is more costly.■

REVIEW PROBLEM 7.3

Avril bought a computer three years ago for $3000, which she can now sell on the open market for $300. The local Mr. Computer store will sell her a new HAL computer for $4000, including the new accounting package she wants. Her own computer will probably last another two years, and then would be worthless. The new computer would have a salvage value of $300 at the end of its economic life of five years. The net benefits to Avril of the new accounting package and other features of the new computer amount to $800 per year. An additional feature is that Mr. Computer will give Avril a $500 trade-in on her current computer. Interest is 15%. What should Avril do?

ANSWER

There are a couple of things to note about this problem. First, the cost of the new computer should be taken as $3800 rather than $4000. This is because, although the price was quoted as $4000, the dealer was willing to give Avril a $500 trade-in on a used computer that had a market value of only $300. This amounts to discounting the price of the new computer to $3800. Similarly, the used computer should be taken to be worth $300, and not $500. The $500 figure does not represent the market value of the used computer, but rather the value of the used computer combined with the discount on the new computer. One must sometimes be careful to extract from the available information the best estimates of the values and costs for the various components of a replacement study.

First, we need to determine the EAC of the challenger over its economic life. We are told that the economic life is five years and hence the EAC computations are as follows:

$$\text{EAC(capital costs)} = (3800 - 300)(A/P,15\%,5) + 300(0.15)$$
$$= 3500(0.29832) + 45$$
$$\cong 1089$$

$$\text{EAC(operating costs)} = 0$$

$$\text{EAC(challenger, total)} = 1089$$

Now we need to check the equivalent annual cost of keeping the existing computer one additional year. A salvage value for the computer for one year was not given. However, we can check to see if the EAC for the defender over two years is less than for the challenger. If it is, this is sufficient to retain the old computer.

$$\text{EAC(capital costs)} = (300 - 0)(A/P,15\%,2) + 0(0.15)$$
$$= 300(0.61512) + 0$$
$$\cong 185$$

$$\text{EAC(operating costs)} = 800$$

$$\text{EAC(defender, total over 2 years)} = 985$$

Avril should hang on to her current computer because its EAC over two years is less than the EAC of the challenger over its five-year economic life.■

SUMMARY

This chapter is concerned with replacement and retirement decisions. Replacement can be required because there may be a cheaper way to provide the same service, or the nature of the service may have changed. Retirement can be required if there is no longer a need for the asset.

If an asset is replaced by a stream of identical assets, it is useful to determine the economic life of the asset, which is the replacement interval that provides the minimum annual cost. The asset should then be replaced at the end of its economic life.

If there is a challenger that is different from the defender, and future changes in technology are not known, one can determine the minimum EAC of the challenger and compare this with the cost of keeping the defender. If keeping the defender for any period of time is cheaper than the minimum EAC of the challenger, the defender should be kept. Often it is sufficient to assess the cost of keeping the defender for one more year.

Defenders that are still functioning well have a significant cost advantage over challengers or over obtaining the service performed by the defender from another source. This is because there are installation costs and because the capital cost per year of an asset diminishes over time.

Where future changes in technology are expected, decisions about when and whether to replace defenders are more complex. In this case, possible replacement decisions must be enumerated as a set of mutually exclusive alternatives and subjected to any of the standard comparison methods.

Figure 7.3 provides a summary of the overall procedure for assessing a replacement decision.

Figure 7.3   The Replacement Decision Making Process

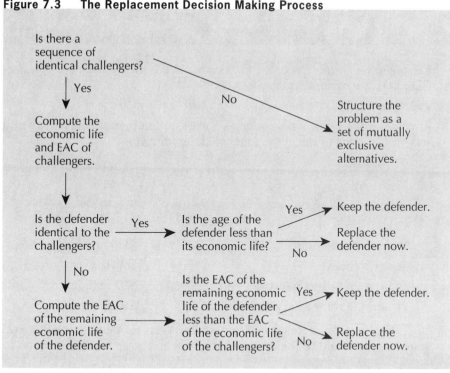

Engineering Economics in Action, Part 7B:
Decision Time

Naomi, Dave, and Clem were meeting in Clem's office. They had just finished a discussion of their steel-ordering policy. Clem turned to Naomi and said, "OK. Let's look at the 10-stage progressive die. Where does that stand?"

Naomi said, "It looks possible. Did you get a chance to read Terry's report?"

"Yes, I did," Clem answered. "Was it his idea to use the 5-stage die for small runs so that we don't have to take a big hit from scrapping it?"

"Actually, it was," Naomi said.

"The kid may be a little intense," Clem said, "but he does good work. So where does that leave us?"

"Well, as I said, it looks possible that the 10-stage die will pay off," Naomi responded.

"We have to decide what the correct time horizon is for making the analysis. Then we need more precise estimates of the costs and salvage value for the 10-stage die."

Clem turned in his chair and asked, "What do you think, Dave?"

Dave straightened himself in his chair and said, "I really don't know. How much experience has Hamilton Tools had at making dies this complicated?"

Naomi answered "Not much. If we took them up on their proposal, we'd be their second or third customer."

"What do you have in mind, Dave?" Clem asked.

Dave said, "Well, if it's only the second or third time they've done something like this, I think we can expect some improvements over the next couple of years. So maybe we ought to wait."

"That makes sense," Clem responded. "I'd like you two to work on this. Give Tan Wang at Hamilton Tools a call. He'll know if anything is in the works. Get him to give you an estimate of what to expect. Then I want you to consider some possibilities. You know: 'Replace now.' 'Wait one year.' 'Wait two years.' And so on. Don't make it too complicated. Then, evaluate the different possibilities. I want a recommendation for next week's meeting. It's getting to be decision time."

PROBLEMS

For additional practice, please see the problems (with selected solutions) provided on the Student CD-ROM that accompanies this book.

7.1 Freeport Brothers have recently purchased a network computer system. Cabling installed in the office walls connects a workstation in each employee's office to a central server. The costs of this system included the following:

5 workstations	$4500 each
New server	$6000
60-metre cable	$11.40 per metre
Cabling hardware	$188
Workstation software	$1190 per workstation
Server software	$1950
10 hours of hardware installer time	$20 per hour
25 hours of software installer time	$60 per hour

If Freeport Brothers wanted to calculate a replacement interval for such a computer system, what would be the capital cost for the first year at a depreciation rate of 30%?

7.2 Last year, Clairbrook Canning Co. bought a fancy colour printer, which cost $20 000, for special printing jobs. Fast changes in colour printing technology have resulted in almost identical printers being available today for about one-quarter of the cost. Should CCC consider selling its printer and buying one of the new ones?

7.3 Maryhill Mines has a pelletizer that it is considering for replacement. Every three years it is overhauled at considerable cost. It is due for an overhaul this year. Evelyn, the company's mining engineer, has calculated that the sum of the operating and capital costs for this year for the pelletizer are significantly more than the EAC for a new pelletizer over its service life. Should the existing pelletizer be replaced?

7.4 Determine the economic life for each of the items listed below. Salvage values can be estimated by the declining-balance method using an annual rate of 20%. The MARR is 8%.

	Purchase	Installation	Operating
Item 1	£10 000	£2000	£300 first year, increasing by £300 per year
Item 2	£20 000	£2000	£200 first year, increasing by £200 per year
Item 3	£30 000	£3000	£2000 first year, increasing by £2000 per year

7.5 A new bottle-capping machine costs $45 000, including $5000 for installation. The machine is expected to have a useful life of eight years with no salvage value at that time (assume straight-line depreciation). Operating and maintenance costs are expected to be $3000 for the first year, increasing by $1000 each year thereafter. Interest is 12%.

(a) Construct a spreadsheet that has the following headings: Year, Salvage Value, Maintenance Costs, EAC(Capital Costs), EAC(Operating Costs), and EAC(Total Costs). Compute the EAC(Total Costs) if the bottle capper is kept for n years, $n = 1, \ldots, 8$.

(b) Construct a chart showing the EAC(Capital Costs), EAC(Operating Costs), and EAC(Total Costs) if the bottle capper were to be kept for n years, $n = 1, \ldots, 8$.

(c) What is the economic life of the bottle capper?

7.6 Gerry likes driving small cars and buys nearly identical ones whenever the old one needs replacing. Typically, he trades in his old car for a new one costing about $15 000. A new car warranty covers all repair costs above standard maintenance (standard maintenance costs are constant over the life of the car) for the first two years. After that, his records show an average repair expense (over standard maintenance) of $2500 in the third year (at the end of the year), increasing by 50% per year thereafter. If a 30% declining-balance depreciation rate is used to estimate salvage values, and interest is 8%, how often should Gerry get a new car?

7.7 Gerry (see Problem 7.6) has observed that the cars he buys are somewhat more reliable now than in the past. A better estimate of the repair costs is $1500 in the third year, increasing by 50% per year thereafter, with all other information in Problem 7.6 being the same. Now how often should Gerry get a new car?

7.8 For each of the following cases, determine whether the one year principle would apply.

(a) A defender has been in use for seven years and has negligible salvage value. Operating costs are €400 per year for electricity. Once every five years it is overhauled at a cost of €1000.

(b) A defender has been in use for seven years and has negligible salvage value. Operating costs are €400 per year for electricity. Once a year it is overhauled at a cost of €1000.

(c) A defender has been in use for two years and has negligible salvage value. Operating costs are €400 per year for electricity. Once a year it is overhauled at a cost of €1000.

(d) A defender has been in use for seven years and has current salvage value of €4000. Its value one year from now is estimated to be €4000. Operating costs are €400 per year for electricity. Once a year it is overhauled at a cost of €1000.

(e) A defender has been in use for seven years and has current salvage value of €4000. Its value one year from now is estimated to be €2000. Operating costs are €400 per year for electricity. Once a year it is overhauled at a cost of €1000.

7.9 If the operating costs for an asset are 500×2^n and the capital costs are $10 000 \times (0.8)^n$, where n is the life in years, what is the economic life of the asset?

7.10 A roller conveyor system used to transport cardboard boxes along an order-filling line costs $100 000 plus $20 000 to install. It is estimated to depreciate at a declining-balance rate of 25% per year over its 15-year useful life. Annual maintenance costs are estimated to be $6000 for the first year, increasing by 20% every year thereafter. In addition, every third year, the rollers must be replaced at a cost of $7000. Interest is at 10%.

(a) Construct a spreadsheet that has the following headings: Year, Salvage Value, Maintenance Costs, EAC(Capital Costs), EAC(Maintenance Costs), and EAC(Total Costs). Compute the EAC(Total Costs) if the conveyor were to be kept for n years, $n = 1, \ldots, 15$.

(b) Construct a chart showing the EAC(Capital Costs), EAC(Maintenance Costs), and EAC(Total Costs) if the conveyor were to be kept for *n* years, *n* = 1, . . . , 15.

(c) What is the economic life of the roller conveyor system?

 7.11 Brockville Brackets (BB) has a three-year-old robot that welds small brackets onto car-frame assemblies. At the time the robot was purchased, it cost $300 000 and an additional $50 000 was spent on installation. BB acquired the robot as part of an eight-year contract to produce the car-frame assemblies. The useful life of the robot is 12 years, and its value is estimated to decline by 20% of current value per year, as shown in the first table below. Operating and maintenance costs estimated when the robot was purchased are also shown in the table.

BB has found that the operating and maintenance costs for the robot have been higher than anticipated. At the end of the third year, new estimates of the operating and maintenance costs are shown in the second table.

	Defender, When New	
Life (Years)	**Salvage Value**	**Operating and Maintenance Costs**
0	$300 000	
1	240 000	$40 000
2	192 000	40 000
3	153 600	40 000
4	122 880	40 000
5	98 304	44 000
6	78 643	48 400
7	62 915	53 240
8	50 332	58 564
9	40 265	64 420
10	32 212	70 862
11	25 770	77 949
12	20 616	85 744

	Costs for Three-Year-Old Defender	
Additional Life (Years)	**Salvage Value**	**Operating and Maintenance Costs**
0	$153 600	
1	122 880	$50 000
2	98 304	55 000
3	78 643	60 500
4	62 915	66 550
5	50 332	73 205

BB has determined that the reason the operating and maintenance costs were in error was that the robot was positioned too close to existing equipment for the mechanics to repair it easily and quickly. BB is considering moving the robot farther away from some adjacent equipment so that mechanics can get easier access for repairs. To move the robot will cause BB to lose valuable production time, which is estimated to have a cost of $25 000. However, once complete, the move will lower maintenance costs to what had originally been expected for the remainder of the contract (e.g., $40 000 for the fourth year, increasing by 10% per year thereafter). Moving the robot will not affect its salvage value.

If BB uses a MARR of 15%, should it move the robot? If so, when? Remember that the contract exists only for a further five years.

7.12 Consider Brockville Brackets from Problem 7.11 but assume that it has a contract to produce the car assemblies for an indefinite period. If it does not move the robot, the operating and maintenance costs will be higher than expected. If it moves the robot (at a cost of $25 000), the operating and maintenance costs are expected to be what were originally expected for the robot. Furthermore, BB expects to be able to obtain new versions of the existing robot for an indefinite period in the future; each is expected to have an installation cost of $50 000.

(a) Construct a spreadsheet table showing the EAC(total costs) if BB keeps the current robot in its current position for n more years, $n = 1, \ldots, 9$.

(b) Construct a spreadsheet table showing the EAC(total costs) if BB moves the current robot and then keeps it for n more years, $n = 1, \ldots, 9$.

(c) Construct a spreadsheet table showing the EAC(total costs) if BB is to buy a new robot and keep it for n years, $n = 1, \ldots, 9$.

(d) On the basis of your answers for parts (a) through (c), what do you advise BB to do?

7.13 Nico has a 20-year-old oil-fired hot air furnace in his house. He is considering replacing it with a new high-efficiency natural gas furnace. The oil-fired furnace has a scrap value of $500, which it will retain indefinitely. A maintenance contract costs $300 per year, plus parts. Nico estimates that parts will cost $200 this year, increasing by $100 per year in subsequent years. The new gas furnace will cost $4500 to buy and $500 to install. It will save $500 per year in energy costs. The maintenance costs for the gas furnace are covered under guarantee for the first five years. The market value of the gas furnace can be estimated from straight-line depreciation with a salvage value of $500 after 10 years. Using a MARR of 10%, should the oil furnace be replaced?

7.14 A certain machine costs £25 000 to purchase and install. It has salvage values and operating costs as shown in the table on the next page. The salvage value of £20 000 listed at time 0 reflects the loss of the installation costs at the time of installation. The MARR is 12%.

(a) What is the economic life of the machine?

(b) What is the equivalent annual cost over that life?

Costs and Salvage Values for Various Lives		
Life in Years	**Salvage Value**	**Operating Cost**
0	£20 000.00	
1	16 000.00	£ 3000.00
2	12 800.00	3225.00
3	10 240.00	3466.88
4	8192.00	3726.89
5	6553.60	4006.41
6	5242.88	4306.89
7	4194.30	4629.90
8	3355.44	4977.15
9	2684.35	5350.43
10	2147.48	5751.72
11	1717.99	6183.09
12	1374.39	6646.83
13	1099.51	7145.34
14	879.61	7681.24
15	703.69	8257.33
16	562.95	8876.63
17	450.36	9542.38
18	360.29	10 258.06

Now assume that the MARR is 5%.

(c) What is the economic life of the machine?

(d) What is the equivalent annual cost over that life?

(e) Explain the effect of decreasing the MARR.

7.15 Jack and Jill live in the suburbs. Jack is a self-employed house painter who works out of their house. Jill works in the city, to which she regularly commutes by car. The car is a four-year-old import. Jill could commute by bus. They are considering selling the car and getting by with the van Jack uses for work.

The car cost $12 000 new. It dropped about 20% in value in the first year. After that it fell by about 15% per year. The car is now worth about $5900. They expect it to continue to decline in value by about 15% of current value every year. Operating and other costs are about $2670 per year. They expect this to rise by about 7.5% per year. A commuter pass costs $112 per month, and is not expected to increase in cost.

Jack and Jill have a MARR of 10%, which is what Jack earns on his business investments. Their time horizon is two years because Jill expects to quit work at that time.

(a) Will commuting by bus save money?

(b) Can you advise Jack and Jill about retiring the car?

 7.16 Ener-G purchases new turbines at a cost of €100 000. Each has a 15-year useful life and must be overhauled periodically at a cost of €10 000. The salvage value of a turbine declines 15% of current value each year, and operating and maintenance costs (including the cost of the overhauls) of a typical turbine are as shown in the table below (the costs for the fifth and tenth years include a €10 000 overhaul, but an overhaul is not done in the fifteenth year since this is the end of the turbine's useful life).

Defender, When New, Overhaul Every Five Years		
Life (Years)	Salvage Value	Operating and Maintenance Costs
0	€100 000	
1	85 000	€15 000
2	72 250	20 000
3	61 413	25 000
4	52 201	30 000
5	44 371	45 000
6	37 715	20 000
7	32 058	25 000
8	27 249	30 000
9	23 162	35 000
10	19 687	50 000
11	16 734	25 000
12	14 224	30 000
13	12 091	35 000
14	10 277	40 000
15	8735	45 000

(a) Construct a spreadsheet that gives, for each year, the EAC(operating and maintenance costs), EAC(capital costs), and EAC(total costs) for the turbines. Interest is 15%. How long should Ener-G keep each turbine before replacing it, given a five-year overhaul schedule? What are the associated equivalent annual costs?

(b) If Ener-G were to overhaul its turbines every six years (at the same cost), the salvage value and operating and maintenance costs would be as shown in the table on the next page. Should Ener-G switch to a six-year overhaul cycle?

Defender, When New, Overhaul Every Six Years		
Life (Years)	**Salvage Value**	**Operating and Maintenance Costs**
0	€100 000	
1	85 000	€15 000
2	72 250	20 000
3	61 413	25 000
4	52 201	30 000
5	44 371	35 000
6	37 715	50 000
7	32 058	20 000
8	27 249	25 000
9	23 162	30 000
10	19 687	35 000
11	16 734	40 000
12	14 224	55 000
13	12 091	25 000
14	10 277	30 000
15	8735	35 000

7.17 The BBBB Machine Company makes a group of metal parts on a turret lathe for a local manufacturer. The current lathe is now six years old. It has a planned further life of three years. The contract with the manufacturer has three more years to run as well. A new, improved lathe has become available. The challenger will have lower operating costs than the defender.

The defender can now be sold for $1200 in the used-equipment market. The challenger will cost $25 000 including installation. Its salvage value after installation, but before use, will be $20 000. Further data for the defender and the challenger are shown in the tables that follow.

Defender		
Additional Life in Years	**Salvage Value**	**Operating Cost**
0	$1200	
1	600	$20 000
2	300	20 500
3	150	21 012.50

Challenger		
Life in Years	Salvage Value	Operating Cost
0	$20 000	
1	14 000	$13 875
2	9800	14 360.63
3	6860	14 863.25

BBBB is not sure if the contract it has with the customer will be renewed. Therefore, BBBB wants to make the decision about replacing the defender with the challenger using a three-year study period. BBBB uses a 12% MARR for this type of investment.

(a) What is the present worth of costs over the next three years for the defender?

(b) What is the present worth of costs over the next three years for the challenger?

(c) Now suppose that BBBB did not have a good estimate of the salvage value of the challenger at the end of three years. What minimum salvage value for the challenger at the end of three years would make the present worth of costs for the challenger equal to that of the defender?

7.18 Suppose, in the situation described in Problem 7.17, BBBB believed that the contract with the manufacturer would be renewed. BBBB also believed that all challengers after the current challenger would be identical to the current challenger. Further data concerning these challengers are given on the next page. Recall that a new challenger costs $25 000 installed.

BBBB was also advised that machines identical to the defender would be available indefinitely. New units of the defender would cost $17 500, including installation. Further data concerning new defenders are shown in the table on the next page. The MARR is 12%.

(a) Find the economic life of the challenger. What is the equivalent annual cost over that life?

(b) Should the defender be replaced with the challenger or with a new defender?

(c) When should this be done?

Challenger		
Life in Years	**Salvage Value**	**Operating Cost**
0	$20 000.00	
1	14 000.00	$13 875.00
2	9800.00	14 369.63
3	6860.00	14 863.25
4	4802.00	15 383.46
5	3361.40	15 921.88
6	2352.98	16 479.15
7	1647.09	17 055.92
8	1152.96	17 652.87
9	807.07	18 270.73
10	564.95	18 910.20
11	395.47	19 572.06
12	276.83	20 257.08

Defender When New		
Life in Years	**Salvage Value**	**Operating Cost**
0	$15 000.00	
1	9846.45	$17 250.00
2	6463.51	17 681.25
3	4242.84	18 123.28
4	2785.13	18 576.36
5	1828.24	19 040.77
6	1200.11	19 516.79
7	600.00	20 004.71
8	300.00	20 504.83
9	150.00	21 017.45
10	150.00	21 542.89
11	150.00	22 081.46
12	150.00	22 633.49
13	150.00	23 199.33

7.19 You own several copiers that are currently valued at R10 000 all together. Annual operating and maintenance costs for all copiers are estimated at R9000 next year, increasing by 10% each year thereafter. Salvage values decrease at a rate of 20% per year.

You are considering replacing your existing copiers with new ones that have a suggested retail price of R25 000. Operating and maintenance costs for the new equipment will be R6000 over the first year, increasing by 10% each year thereafter. The salvage value of the new equipment is well approximated by a 20% drop from the suggested retail price per year. Furthermore, you can get a trade-in allowance of R12 000 for your equipment if you purchase the new equipment at its suggested retail price. Your MARR is 8%. Should you replace your existing equipment now?

7.20 An existing piece of equipment has the following pattern of salvage values and operating and maintenance costs:

Defender					
Additional Life (Years)	Salvage Value	Maintenance Costs	EAC Capital Costs	EAC Operating and Maintenance Costs	EAC Total
0	$10 000				
1	8000	$2000	$3500	$2000	$5500
2	6400	2500	3174	2233	5407
3	5120	3000	2905	2454	5359
4	4096	3500	2682	2663	5345
5	3277	4000	2497	2861	5359
6	2621	4500	2343	3049	5391
7	2097	5000	2214	3225	5439
8	1678	5500	2106	3391	5497
9	1342	6000	2016	3546	5562

A replacement asset is being considered. Its relevant costs over the next nine years are shown on the next page.

There is a need for the asset (either the defender or the challenger) for the next nine years.

(a) What replacement alternatives are there?

(b) What replacement timing do you recommend?

Challenger					
Additional Life (Years)	Salvage Value	Maintenance Costs	EAC Capital Costs	EAC Operating and Maintenance Costs	EAC Total
0	$12 000				
1	9600	$1500	$4200	$1500	$5700
2	7680	1900	3809	1686	5495
3	6144	2300	3486	1863	5349
4	4915	2700	3219	2031	5249
5	3932	3100	2997	2189	5186
6	3146	3500	2811	2339	5150
7	2517	3900	2657	2480	5137
8	2013	4300	2528	2613	5140
9	1611	4700	2419	2737	5156

 7.21 The Brunswick Table Top Company makes tops for tables and desks. The company now owns a seven-year-old planer that is experiencing increasing operating costs. The defender has a maximum additional life of five years. The company is considering replacing the defender with a new planer.

The new planer would cost €30 000 installed. Its value after installation, but before use, would be about €25 000. The company has been told that there will be a new-model planer available in two years. The new model is expected to have the same first costs as the current challenger. However, it is expected to have lower operating costs. Data concerning the defender and the two challengers are shown in the tables that follow. Brunswick Table has a 10-year planning period and uses a MARR of 10%.

Defender		
Additional Life in Years	Salvage Value	Operating Cost
0	€4000	
1	3000	€20 000
2	2000	25 000
3	1000	30 000
4	500	35 000
5	500	40 000

First Challenger		
Life in Years	**Salvage Value**	**Operating Cost**
0	€25 000	
1	20 000	€16 800
2	16 000	17 640
3	12 800	18 522
4	10 240	19 448
5	8192	20 421
6	6554	21 442
7	5243	22 514
8	4194	23 639
9	3355	24 821
10	2684	26 062

Second Challenger		
Life in Years	**Salvage Value**	**Operating Cost**
0	€25 000	
1	20 000	€12 000
2	16 000	12 600
3	12 800	13 230
4	10 240	13 892
5	8192	14 586
6	6554	15 315
7	5243	16 081
8	4194	16 885
9	3355	17 729
10	2684	18 616

(a) What are the combinations of planers that Brunswick can use to cover requirements for the next 10 years? For example, Brunswick may keep the defender one more year, then install the first challenger and keep it for nine years. Notice that the first challenger will not be installed after the second year when the second challenger becomes available. You may ignore combinations that involve installing the first challenger after the second becomes available. Recall also that the maximum additional life for the defender is five years.

(b) What is the best combination?

7.22 You estimate that your two-year-old car is now worth $12 000 and that it will decline in value by 25% of its current value each year of its eight-year remaining useful life. You also estimate that its operating and maintenance costs will be $2100, increasing by 20% per year thereafter. Your MARR is 12%.

(a) Construct a spreadsheet showing (1) additional life in years, (2) salvage value, (3) operating and maintenance costs, (4) EAC(operating and maintenance costs), (5) EAC(capital costs), and (6) EAC(total costs). What additional life minimizes the EAC(total costs)?

(b) Now you are considering the possibility of painting the car in three years' time for $2000. Painting the car will increase its salvage value. By how much will the salvage value have to increase before painting the car is economically justified? Modify the spreadsheet you developed for part (a) to show this salvage value and the EAC (total costs) for each additional year of life. Will painting the car extend its economic life?

7.23 A long-standing principle of computer innovations is that computers double in power for the same price, or, equivalently, halve in cost for the same power, every 18 months. Auckland Data Services (ADS) owns a single computer that is at the end of its third year of service. ADS will continue to buy computers of the same power as its current one. Its current computer would cost $80 000 to buy today, excluding installation. Given that a new model is released every 18 months, what replacement policy should ADS adopt for computers over the next three years? Other facts to be considered are:

1. Installation cost is 15% of purchase price.

2. Salvage values are computed at a declining-balance depreciation rate of 50%.

3. Annual maintenance cost is estimated as 10% of accumulated depreciation or as 15% of accumulated depreciation per 18-month period.

4. ADS uses a MARR of 12%.

 7.24 A water pump to be used by the city's maintenance department costs $10 000 new. A running-in period, costing $1000 immediately, is required for a new pump. Operating and maintenance costs average $500 the first year, increasing by $300 per year thereafter. The salvage value of the pump at any time can be estimated by the declining-balance rate of 20%. Interest is at 10%. Using a spreadsheet, calculate the EAC for replacing the pump after one year, two years, etc. How often should the pump be replaced?

 7.25 The water pump from Problem 7.24 is being considered to replace an existing one. The current one has a salvage value of $1000 and will retain this salvage value indefinitely.

(a) Operating costs are currently $2500 per year and rise by $400 per year. Should the current pump be replaced? When?

(b) Operating costs are currently $3500 per year and rise by $200 per year. Should the current pump be replaced? When?

 7.26 Chatham Automotive purchased new electric forklifts to move steel automobile parts two years ago. These cost $75 000 each, including the charging stand. In practice, it was found that they did not hold a charge as long as claimed by the manufacturer, so operating costs are very high. This also results in their currently having a salvage value of about $10 000.

Chatham is considering replacing them with propane models. The new ones cost $58 000. After one year, they have a salvage value of $40 000, and thereafter decline in value at a declining-balance depreciation rate of 20%, as does the electric model from this time on. The MARR is 8%. Operating costs for the electric model are $20 000 over the first year, rising by 12% per year. Operating costs for the propane model initially will be $10 000 over the first year, rising by 12% per year. Should Chatham Automotive replace the forklifts now?

7.27 Suppose that Chatham Automotive (Problem 7.26) can get a $14 000 trade-in value for their current electric model when they purchase a new propane model. Should they replace the electric forklifts now?

7.28 A joint former cost ¥6 000 000 to purchase and ¥1 000 000 to install seven years ago. The market value now is ¥3 300 000, and this will decline by 12% of current value each year for the next three years. Operating and maintenance costs are estimated to be ¥340 000 this year, and are expected to increase by ¥50 000 per year.

(a) How much should the EAC of a new joint former be over its economic life to justify replacing the old one sometime in the next three years? The MARR is 10%.

(b) The EAC for a new joint former turns out to be ¥1 030 000 for a 10-year life. Should the old joint former be replaced within the next three years? If so, when?

(c) Is it necessary to consider replacing the old joint former more than three years from now, given that a new one has an EAC of ¥1 030 000?

7.29 Northwest Aerocomposite manufactures fibreglass and carbon fibre fairings. Its largest water-jet cutter will have to be replaced some time before the end of four years. The old cutter is currently worth $49 000. Other cost data for the current and replacement cutters can be found in the tables that follow. The MARR is 15%. What is the economic life of the new cutter, and what is the equivalent annual cost for that life? When should the new cutter replace the old?

Challenger		
Life in Years	Salvage Value	Operating and Maintenance Costs
0	$90 000	
1	72 000	$12 000
2	57 600	14 400
3	46 080	17 280
4	36 864	20 736
5	29 491	24 883
6	23 593	29 860
7	18 874	35 832
8	15 099	42 998
9	12 080	51 598

Defender		
Life in Years	Salvage Value	Operating and Maintenance Costs
0	$49 000	
1	36 500	$17 000
2	19 875	21 320
3	15 656	26 806
4	6742	33 774

7.30 The water pump from Problem 7.24 has an option to be overhauled once. It costs $1000 to overhaul a three-year-old pump and $2000 to overhaul a five-year-old pump. The major advantage of an overhaul is that it reduces the operating and maintenance costs to $500, which will increase again by $300 per year thereafter. Should the pump be overhauled? If so, should it be overhauled in three years or five years?

7.31 Northwest Aerocomposite in Problem 7.29 found out that its old water-jet cutter may be overhauled at a cost of $14 000 now. The cost information for the old cutter after an overhaul is as shown in the table below.

Defender With an Overhaul		
Life (Years)	Salvage Value	Operating and Maintenance Costs
0	$55 000	
1	40 970	$16 500
2	22 310	20 690
3	17 574	26 013
4	7568	32 775

Should Northwest overhaul the old cutter? If an overhaul takes place, when should the new cutter replace the old? Assume that the cost information for the replacement cutter is as given in Problem 7.29.

7.32 Tiny Bay Freight Company (TBFC) wants to begin business with one delivery truck. After two years of operation, the company plans to increase the number of trucks to two, and after four years, plans to increase the number to three. TBFC currently has no trucks. The company is considering purchasing one type of truck that costs $30 000. The operating and maintenance costs are estimated to be $7200 per year. The resale value of the truck will decline each year by 40% of the current value. The company will consider replacing a truck every two years. That is, the company may keep a truck for two years, four years, six years, and so on. TBFC's MARR is 12%.

(a) What are the possible combinations for purchasing and replacing trucks over the next five years so that TBFC will meet its expansion goals and will have three trucks in hand at the end of five years?

(b) Which purchase/replacement combination is the best?

Problems 7.33 through 7.36 are concerned with the economic life of assets where there is a sequence of identical assets. The problems explore the sensitivity of the economic life to four parameters: the MARR, the level of operating cost, the rate of increase in operating cost, and the level of first cost. In each problem there is a pair of assets. The assets differ in only a single parameter. The problem asks you to determine the effect of this difference on the economic life and to explain the result. All assets decline in value by 20% of current value each year. Installation costs are zero for all assets. Further data concerning the four pairs of assets are given in the table that follows.

Asset Number	First Cost	Initial Operating Cost	Rate of Operating Cost Increase	MARR
A1	$125 000	$30 000	12.5%/year	5%
B1	125 000	30 000	12.5%/year	25%
A2	100 000	30 000	$2000/year	15%
B2	100 000	40 000	$2000/year	15%
A3	100 000	30 000	5%/year	15%
B3	100 000	30 000	12.5%/year	15%
A4	75 000	30 000	5%/year	15%
B4	150 000	30 000	5%/year	15%

7.33 Consider Assets A1 and B1. They differ only in the MARR.

(a) Determine the economic lives for Assets A1 and B1.

(b) Create a diagram showing the EAC(capital), the EAC(operating), and the EAC(total) for Assets A1 and B1.

(c) Explain the difference in economic life between A1 and B1.

7.34 Consider Assets A2 and B2. They differ only in the level of initial operating cost.

(a) Determine the economic lives for Assets A2 and B2.

(b) Create a diagram showing the EAC(capital), the EAC(operating), and the EAC(total) for Assets A2 and B2.

(c) Explain the difference in economic life between A2 and B2.

7.35 Consider Assets A3 and B3. They differ only in the rate of increase of operating cost.

(a) Determine the economic lives for Assets A3 and B3.

(b) Create a diagram showing the EAC(capital), the EAC(operating), and the EAC(total) for Assets A3 and B3.

(c) Explain the difference in economic life between A3 and B3.

7.36 Consider Assets A4 and B4. They differ only in the level of first cost.

(a) Determine the economic lives for Assets A4 and B4.

(b) Create a diagram showing the EAC(capital), the EAC(operating), and the EAC(total) for Assets A4 and B4.

(c) Explain the difference in economic life between A4 and B4.

7.37 This problem concerns the economic life of assets where there is a sequence of identical assets. In this case there is an opportunity to overhaul equipment. Two issues are explored. The first concerns the optimal life of equipment. The second concerns the decision of whether to replace equipment that is past its economic life. Consider a piece of equipment that costs $40 000 to buy and install. The equipment has a maximum life of 15 years. Overhaul is required in the fourth, eighth, and twelfth years. The company uses a MARR of 20%. Further information is given in the table below.

(a) Show that the economic life for this equipment is seven years.

(b) Suppose that the equipment is overhauled in the eighth year rather than replaced. Show that keeping the equipment for three more years (after the eighth year), until it next comes up for overhaul, has lower cost than replacing the equipment immediately.

Hint for part (b): The comparison must be done fairly and carefully. Assume that under either plan the replacement is kept for its optimal life of seven years. It is easier to compare the plans if they cover the same number of years. One way to do this is to consider an 11-year period as shown on the next page.

Year	Salvage Value	Operating Cost	Overhaul Cost
0	$15 000		
1	12 000	$ 2000	
2	9600	2200	
3	7680	2420	
4	7500	2662	$ 2500
5	6000	2000	
6	4800	2200	
7	3840	2420	
8	4500	2662	32 500
9	3600	2000	
10	2880	2800	
11	2304	3920	
12	2000	5488	17 500
13	1200	4000	
14	720	8000	
15	432	16 000	

Year	Plan A	Plan B
0		
1	Defender	Replacement #1
2	Defender	Replacement #1
3	Defender	Replacement #1
4	Replacement #1	Replacement #1
5	Replacement #1	Replacement #1
6	Replacement #1	Replacement #1
7	Replacement #1	Replacement #1
8	Replacement #1	Replacement #2
9	Replacement #1	Replacement #2
10	Replacement #1	Replacement #2
11	Replacement #2	Replacement #2

First, show that the present worth of costs over the 11 years is lower under plan A than under plan B. Second, point out that the equipment that is in place at the end of the eleventh year is newer under plan A than under plan B.

(c) Why is it necessary to take into account the age of the equipment at the end of the 11-year period?

 7.38 Northfield Metal Works is a household appliance parts manufacturer that has just won a contract with a major appliance company to supply replacement parts to service shops. The contract is for five years. Northfield is considering using three existing manual punch presses or a new automatic press for part of the work. The new press would cost £225 000 installed. Northfield is using a five-year time horizon for the project. The MARR is 25% for projects of this type. Further data concerning the two options are shown in the tables that follow.

Automatic Punch Press		
Life in Years	Salvage Value	Operating Cost
0	£125 000	
1	100 000	£25 000
2	80 000	23 750
3	64 000	22 563
4	51 200	21 434
5	40 960	20 363

Hand-Fed Press		
Additional Life in Years	Salvage Value	Operating Cost
0	£10 000	
1	9000	£25 000
2	8000	25 000
3	7000	25 000
4	6000	25 000
5	5000	25 000

Note that the hand-fed press values are for each of the three presses. Costs must be multiplied by three to get the costs for three presses. Northfield is not sure of the salvage values for the new press. What salvage value at the end of five years would make the two options equal?

More Challenging Problem

7.39 (a) Referring to Figure 7.1, why is the economic life different from the intersection of the reducing capital costs line and the increasing operating and maintenance costs line? Under what circumstances would the economic life correspond with the intersection of the lines?

(b) Referring to Figure 7.2, the point at which total costs are lowest is not shown on the graph, but rather would have appeared on a section of the graph to the left of the part illustrated. As seen in Figure 7.1, the asset should have been replaced when total costs were lowest. Even if this wasn't done, every year of additional life the total costs of an asset in Figure 7.2 are increasing. However, as pointed out in the text, the situation illustrated in Figure 7.2 is very common, justifying the one year rule for such assets. How can it be that asset costs can commonly have the structure seen in Figure 7.2?

MINI-CASE 7.1

Paying for Replacement

From the *MindaNews* (www.mindanews.com), September 29, 2007:

Union to Davao City Water District: Use your savings, don't borrow!
Davao City, Philippines—The workers' union at the Davao City Water District (DCWD) has urged management to finance the repair of old and damaged pipes using its savings, instead of availing of a P126-M loan.

By financing its Comprehensive Mainline Replacement Program (CMRP), a project to replace old and damaged pipes, through internal savings, the DCWD could spare the city's more than a million residents from another likely round of taxes, said the critics of the loan program, among them the workers' union, a consumer group, and a militant group.

The Nagkahiusang Mamumuo sa DCWD (NMDCWD) labour union, the Consumer Alert and the Bayan Muna pressed this point before a public hearing conducted by the City Council's Committee on Government Enterprises Friday.

They said the DCWD "has money to fund the project by its own."

The NMDCWD labour union described the loan program as "unjustifiable and unnecessary."

"The DCWD Board and Management could review its 2007 budget and push for the allocation of funds for the CMRP from its own income and or savings," the labour union said in a position paper presented by Consumer Alert in the hearing. "In this way the water district would not be paying the interest of a loan, which, in the first place is not needed."

The union has been holding daily protest rallies outside their office compounds along J.P. Laurel Avenue, MacArthur Highway and Ulas, demanding that the agency generate the fund from its savings, or stop the project.

Rhodora Gamboa, general manager of DCWD, told the public hearing that they could not rely on their savings alone to finance repair or expansion programs, stressing they would "never be able to catch up with the city's rate of increase in demand for water."

She assured the public that the agency would be able to cover the cost of the loan, including interest payments, from savings that it would generate from "estimated losses it will no longer incur if the leaking pipes are fixed."

Discussion

None of the antagonists in this issue are questioning whether water pipeline replacement is necessary in Davao City. Instead, the concern is how the responsible agency—the DCWD—is going to pay for the pipe replacement. The workers' union wants the DCWD to use its savings, while the DCWD wants to borrow the capital, with the intention of repaying it out of money saved by the improved performance of the replacement pipes.

Questions

1. Does it make sense that the union representing the workers at DCWD should spearhead the opposition to the loan? Do you feel that their concern for the possibility of raising taxes is the real reason for the opposition, or is there an ulterior motive?

2. Does it ever make sense to borrow money to replace something, even if you have enough savings to cover the cost? What possible reasons could there be for the DCWD to borrow the money in this case?

3. How exactly does the DCWD earn money to pay off the loan when the pipes don't leak as much as they used to? Is paying off the loan in this way a realistic expectation?

4. Outline the various costs and anticipated benefits of either the DCWD using its savings, or borrowing the capital.

C H A P T E R 8

Taxes

Engineering Economics in Action, Part 8A: It's in the Details

Review Problems
Summary

Engineering Economics in Action, Part 8B: The Work Report

Problems
Mini-Case 8.1
Appendix 8A: Deriving the Tax Benefit Factor

Engineering Economics in Action, Part 8A:
It's in the Details

"Details, Terry. Sometimes it's all in the details." Naomi pursed her lips and nodded sagely. Terry and Naomi were sitting in the coffee room together. The main break periods for the line workers had passed, so they were alone except for a maintenance person on the other side of the room who was enjoying either a late breakfast or an early lunch.

"Uh, OK, Naomi. What is?"

"Well," Naomi replied, "you know that rocker arm die deal? The one where we're upgrading to a 10-stage die? The rough replacement study you did seems to have worked out OK. We're going to do something, sometime, but now we have to be a little more precise."

"OK. What do we do?" Terry was interested now. Terry liked things precise and detailed.

"The main thing is to make sure we are working with the best numbers we can get. I'm getting good cost figures from Tan Wang at Hamilton Tools for this die and future possibilities, and we'll also work out our own costs for the changeover. Your cost calculations are going to have to be more accurate, too."

"You mean to more significant digits?" Naomi couldn't tell whether Terry was making a joke or the idea of more significant digits really did thrill him. She decided it was the former.

"Ha, ha. No, I mean that we had better explicitly look at the tax implications of the purchase. Our rough calculations ignored taxes, and taxes can have a significant effect on the choice of best alternative. And when lots of money is at stake, the details matter."

8.1 | Introduction

Taxes are fees paid by individuals or businesses to support their government. Taxes are generally the primary source of government revenue, and are used to provide public goods and services such as highways and dams, police services, the military, water treatment, and health care and other social programs.

Taxes can have a significant impact on the economic viability of a project because they change the actual cash flows experienced by a company. A vital component of a thorough economic analysis will therefore include the tax implications of an investment decision. This chapter provides an introduction to corporate taxes and shows how they can affect engineering economics decisions.

When a firm makes an investment, the income from the project will affect the company's cash flows. If the investment yields a profit, the profits will be taxed. Since the taxes result as a direct consequence of the investment, they reduce the net profits associated with that investment. In this sense, taxes associated with a project are a disbursement. If the investment yields a loss, the company may be able to offset the loss from this project against the profits from another, and end up paying less tax overall. As a result, when evaluating a loss-generating project, the net savings in tax can be viewed as a negative disbursement. Income taxes thus reduce the benefits of a successful project, while at the same time reducing the costs of an unsuccessful project.

Calculating the effects of income taxes on the viability of a project can be complicated. Prior to exploring the details of such calculations, it is helpful to look at the context of corporate income taxes. In the next two sections we will compare corporate income taxes to the more familiar personal income taxes, and also describe how corporate tax rates are determined.

8.2 | Personal Income Taxes and Corporate Income Taxes Compared

There are substantial differences between personal income taxes and corporate income taxes. Most adults are familiar with the routine of filing income tax returns. For most people, the procedure is rather simple; an employer provides a statement reporting the person's income, income tax already paid, and other amounts for the previous year. These amounts are assembled on the tax form and are used to calculate the total tax owed or the amount of a refund. The tax return is then submitted to the appropriate government tax agency for processing.

Corporate taxes are similar in some ways, but there are several substantial differences. One main difference has to do with the tax rate. Personal income taxes usually exhibit a **progressive tax rate**, meaning that people who earn more are charged a larger percentage of their income. For example, if a person's taxable income is very low, they might not pay any income taxes. At a moderate income, an individual might be expected to pay, say, 25% of his or her income as income tax. A very-high-income individual in some countries might pay 50% or more of his or her income in income tax. Although the exact rate of taxation changes from year to year, and varies from country to country, individual income taxes in developed countries are almost always progressive.

In contrast, corporate taxes are typically levied according to a proportional or "flat" tax. This means that corporations pay the same tax rate, regardless of their income level. Again, rates can change over time and between countries, but the rate will be the same whether a company makes a small profit or a very large profit. The main exception is that most countries give smaller companies favourable tax treatment to encourage new businesses. However, once the business is of a certain size, the tax rate is constant.

Another difference between personal income tax and corporate income tax has to do with how the tax is calculated. The income used to determine the taxes an individual pays is reduced by deductions or **tax credits**, which are real or nominal costs that are not taxed or are taxed at a reduced rate. For example, excess medical expenses, tuition fees, and pension plan contributions generally are all considered eligible deductions, along with a substantial "basic personal amount." Thus an individual's "taxable income" can be quite a bit less than his or her actual income.

A corporation's taxes are calculated in quite a different manner. Net income for tax purposes is calculated by subtracting **expenses**, which are either real costs associated with performing the corporation's business or a portion of the capital expense for an asset, from gross income. Consequently, a company that makes no profit (income less expenses) may pay no income taxes.

NET VALUE 8.1

Government Tax Websites

Most countries maintain websites that provide useful general information as well as details about current regulations governing both personal and corporate taxes. Here are some of the key sites in English:

Australia: **www.ato.gov.au**

Canada: **www.cra-arc.gc.ca**

India: **www.incometaxindia.gov.in**

South Africa: **www.sars.gov.za**

United Kingdom: **www.hmrc.gov.uk**

United States: **www.irs.gov**

Finally, personal and corporate taxes differ in complexity. In particular, a company's taxes will usually be complicated by issues concerning **capital expenses**, which are purchases of assets of significant value. Such assets have a strong effect on the income taxes paid by a company, but how they are treated for tax purposes is complex. This is usually not an issue of concern to individuals.

8.3 | Corporate Tax Rates

In this chapter, we are concerned only with corporate taxes, and in particular the impact of corporate income taxes on the viability of an engineering project. Table 8.1 compares corporate taxes among several countries, but the actual tax rate applied in any circumstance can be fairly complicated and can depend on the size of the firm, whether it is a manufacturer, its location, and a variety of other factors. Comparing the tax rates in different countries is difficult. For example, some countries provide a full range of healthcare services from their tax revenues, while others do not. Certain healthcare costs may be, instead, a significant expense for a company. In some countries the total corporate tax is a combination of federal tax and state or provincial tax. Tax rates change over time, and there are sometimes short-term tax reductions or incentives (see Close-Up 8.1) for

Table 8.1   Corporate Tax Rates Around the World

Argentina	35%	Germany	25%
Australia	30%	India	30–40%
Austria	25%	Ireland	12.5%
Brazil	34%	Israel	31%
Canada	36.1%	South Africa	29%
China	25%	United Kingdom	28%
France	33.33%	United States	15–35%

CLOSE-UP 8.1   Incentives

Governments sometimes try to influence corporate behaviour through the use of *incentives*. These incentives include grants to certain types of projects; for example, projects undertaken in particular geographic areas, or projects providing employment to certain categories of people.

Other incentives take the form of tax relief. For example, several countries allow pollution equipment to be fully expensed in the year of purchase. The ability to depreciate pollution equipment quickly makes it a more desirable investment for a company, and a beneficial investment for society as a whole.

The exact form of incentives changes from year to year as governments change and as the political interests of society change. In most companies there is an individual or department that keeps track of possible programs affecting company projects.

Incentives must be considered when assessing the viability of a project. Grant incentives provide additional cash flow to the project that can be taken into account like any other cash flow element. Tax incentives may be more difficult to assess since sometimes, for example, they use other forms of depreciation, or may result in different tax rates for different parts of the project.

CLOSE-UP 8.2 Small Company Tax Rules

Special tax advantages are given to smaller companies, such as a technically oriented start-up. This may be of particular interest to engineers interested in starting their own firm rather than working for a larger company. The purpose of these incentives is to give a strong motivation to incorporate a new company and provide more after-tax income for reinvestment and expansion.

In Canada, the small business deduction applies to small Canadian-controlled private corporations, and reduces the effective tax rate for a small Canadian company to less than 20%.

In Australia, the entrepreneurs' tax offset (ETO) provides a tax reduction of up to 25% of the income tax for a small business.

In the United Kingdom, small businesses can accelerate depreciation in their first year, which makes capital investment in a small business much more attractive to an investor or entrepreneur.

In the United States, a graduated federal tax rate allows small firms to be taxed at rates as low as 15% based on their income level.

---

companies in specific industries. Rather than focusing on the implications of such differences, our concern here is with the basic approach used in determining the impact of taxes on a project. For special tax rules, it is best to check with the appropriate national tax agency or a tax specialist.

It is also worth noting that tax rules can change suddenly. For example, in the last 30 years there have been fundamental changes to the corporate tax rules in Australia, Canada, the United Kingdom, and the United States. These changes can and have had a significant impact on investments made by companies. Each of these countries also implements less-significant new rules every year due to changes in technology, current economic conditions, or political factors. In particular, there are usually opportunities for small technological companies to take advantage of beneficial tax rules, as detailed in Close-Up 8.2.

8.4 | Before- and After-Tax MARR

Taxes have a significant effect on engineering decision making, so much so that they cannot be ignored. In this text, so far, it seems as though no specific tax calculations have been done. In fact, they have been implicitly incorporated into the computations through the use of a before-tax MARR, though we have not called it such.

The basic logic is as follows. Since taxes have the effect of reducing profits associated with a project, we need to make sure that we set an appropriate MARR for project acceptability. If we do not explicitly account for the impact of taxes in the project cash flows, then we need to set a MARR high enough to recognize that taxes will need to be paid. This is the *before-tax* MARR. If, on the other hand, the impact of taxes is explicitly accounted for in the cash flows of a project (i.e., reduce the cash flows by the tax rate), then the MARR used for the project should be lower, since the cash flows already take into account the payment of taxes. This is the *after-tax* MARR.

In fact, we can express an approximate relationship

$$\text{MARR}_{\text{after-tax}} \cong \text{MARR}_{\text{before-tax}} \times (1 - t) \tag{8.1}$$

where t is the corporate tax rate. The *before-tax* MARR means that the MARR has been chosen high enough to provide an acceptable rate of return without explicitly considering taxes. In other words, since all profits are taxed at the rate t, the *before-tax* MARR has to include enough returns to meet the *after-tax* MARR and, in addition, provide the amount to be paid in taxes. As we can see from the above equation, the after-tax MARR will generally be lower than the before-tax MARR. We will see later in this chapter how the relationship given in this equation is a simplification but a reasonable approximation of the effect of taxes. In practice, the before- and after-tax MARRs are often chosen independently and are not directly related by this equation. Generally speaking, if a MARR is given without specifying whether it is on a before- or after-tax basis, it can be assumed to be a before-tax MARR.

EXAMPLE 8.1

Swanland Gold Mines (SGM) has been selecting projects for investment on the basis of a before-tax MARR of 12%. Sherri feels that some good projects have been missed because the effects of taxation on the projects have not been examined in enough detail, so she proposed reviewing the projects on an after-tax basis. What would be a good choice of after-tax MARR for her review? SGM pays 45% corporate taxes.

Although the issue of selecting an after-tax MARR is likely to be more complicated, a reasonable choice for Sherri would be to use Equation (8.1) as a way of calculating an after-tax MARR for her review. This gives

$$\text{MARR}_{\text{after-tax}} = 0.12 \times (1 - 0.45) = 0.066 = 6.6\%$$

A reasonable choice for after-tax MARR would be 6.6%.∎

It is important to know when to use the after-tax MARR and when to use a before-tax MARR. In general, if you are doing an approximate calculation without taking taxes into account explicitly, then you should use a before-tax MARR. Thus all of the examples in this book up to this point were appropriately done using a before-tax MARR, boosted to account for the fact that profits would be taxed. However, when considering taxes explicitly in calculations, the after-tax MARR should be used. There is no need to increase the decision threshold when the profits that are lost to taxes are already taken into account.

8.5 | The Effect of Taxation on Cash Flows

There is a set of cash flows that arises whenever an investment is made. Consider the following example.

EXAMPLE 8.2(A)

Ebcon Corp. is considering purchasing a small device used to test printed circuit boards that has a first cost of $45 000. The tester is expected to reduce labour costs and improve the defect detection rate so as to bring about a saving of $23 000 per year. Additional operating costs are expected to be $7300 per year. The salvage value of the tester will be $5000 in five years. The corporate tax rate is $t = 42\%$ and the after-tax MARR is 12%.∎

There are three key cash flow elements in this example.

- first cost: the $45 000 investment. This is a negative cash flow made at time 0.
- net annual savings: the $23 000 per year savings less the $7300 additional operating costs for a net $15 700 positive cash flow at the end of each of years 1–5.
- salvage value: the $5000 residual value of the tester at the end of its service life in five years.

Each of these cash flow elements is affected by taxes, and in different ways.

8.5.1 The Effect of Taxes on First Cost

Companies are taxed on net profits, which are revenues less expenses. Consequently, when an expense is incurred, less tax is paid. From this perspective, although the tester in Example 8.2(A) had a first cost of $45 000, if this first cost could be recognized as an immediate expense, there would be a tax savings of $45 000 × 0.42 = $18 900. However, the tax benefits are usually claimed at the end of a tax year. Assuming that the tester was purchased at the beginning of the year, and the tax savings accrue at the end of that same year, then the present worth of the first cost of the tester would only be –$45 000 + $18 900(P/F,12%,1) = –$45 000 + $18 900 (0.89286) ≅ –$28 000.

However, by tax rules, capital purchases such as production equipment cannot usually be fully claimed as an expense in the year in which the purchase occurred. The logic is based on the idea of depreciation. As we saw in Chapter 6, assets retain value over time. Their loss in value, as opposed to their cost, is what tax rules generally recognize as an eligible expense.

Given a choice, a firm would want to write off (i.e., depreciate) an investment as quickly as possible. Since depreciation is considered an expense and offsets revenue, the earlier the expense is recognized, the earlier the tax savings will accrue. Since money earlier in time is worth more than the same amount of money later in time, there is good reason to depreciate capital assets as quickly as allowable. This effect of reducing taxes can be considerable.

Although companies would, of course, like to reduce taxes by writing off their assets as quickly as possible, there are also undesirable effects of doing this. For one thing, it may lead to accounting documents that are not representative of a firm's actual financial situation. For example, if valuable capital assets are assigned—for tax reasons—to have an excessively low book value, this may underrepresent the true value of the company as a whole. It may also create an opportunity for inefficient business decisions. From the perspective of the government imposing the taxes, it can lead to companies improperly exploiting the tax rules.

To counter these undesirable effects, tax authorities have carefully defined the depreciation methods they permit for use in computing taxes. The permitted methods for claiming depreciation expenses may not reflect the true depreciation of an asset, as they are rule-oriented, rather than truly attempting to recognize the diminishment in value of an asset. Consequently, the depreciation charge is often referred to by a different name, such as *capital allowance* or *capital cost allowance* to recognize that the book value implied by the taxation-related depreciation charges does not necessarily represent a good estimate of the asset's market value. However, there is usually a fair correspondence between capital allowance and depreciation, and many firms do not distinguish between them in calculations for their financial statements. It is therefore reasonable to think of them as being the same thing for most purposes. In this book we will typically refer to capital allowances as *depreciation expenses*, and the corresponding undepreciated value as *book value* to avoid confusion. However, we will also introduce specific terms that are used by particular tax regimes when appropriate.

EXAMPLE 8.2(B)

Tax rules in Ebcon Corp.'s country (see Example 8.2(A)) specify that the capital allowance for the tester be calculated as straight-line depreciation over the life of the tester. What is the present worth of the tester's first cost?

This seems like an odd question, but it results from the fact that in a taxed company, the first cost immediately gives rise to future tax savings as defined by the host country's tax rules. This particular example is consistent with one of the depreciation methodologies permitted in Australia.

As illustrated in Figure 8.1, the first cost of $45 000 gives rise to benefits in the form of an annuity of 45 000 × 0.42/5 = $3780. In a sense, the presence of taxes transforms the $45 000 purchase into a more complex set of cash flows consisting of not only the first cost but also a substantial annuity.

Figure 8.1   The Beneficial Effects of a First Cost (for Example 8.2)

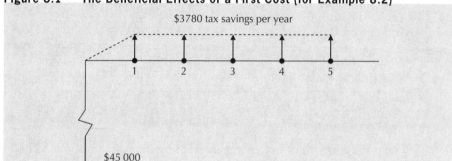

The present worth of the first cost is then:

$$PW_{first\ cost} = -45\ 000 + 3780(P/A,12\%,5) = -45\ 000 + 3780(3.6048) \cong -31\ 400$$

The present worth of the tester's first cost is about –$31 400. ■

Note that this result is more costly for Ebcon by about $31 400 – $28 000 = $3400 than if Ebcon could expense the entire tester in the year of purchase. This difference is the present worth of the increased taxes that the government receives by not allowing Ebcon to count the tester as an expense in the year that it was purchased.

8.5.2 The Effect of Taxes on Savings

Although in a taxed business environment the first cost is reduced, this is balanced by the fact that the savings from the investment are also reduced. An underlying assumption is that any investment decision is made in a profitable company for which this decision is a relatively small part of the overall business. Consequently, if any money is saved, those savings increase the profits of the firm. And since profits are taxed at the tax rate, the net savings are reduced proportionally.

EXAMPLE 8.2(C)

What is the present worth of the annual net savings created by the tester?

The approach used is to reduce the net annual savings of the tester by the tax rate, as illustrated in Figure 8.2, and then bring the amount to a present worth:

$$PW_{savings} = (23\ 000 - 7300) \times (1 - 0.42) \times (P/A, 12\%, 5)$$
$$= 15\ 700 \times 0.58 \times 3.6048$$
$$\cong 32\ 800$$

Figure 8.2   Loss of Net Savings Due to Taxes (for Example 8.2)

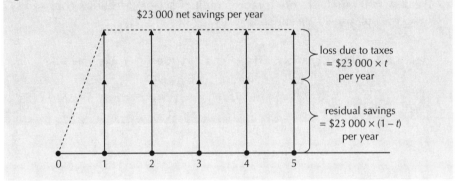

The present worth of net savings created by the tester is about $32 800. ∎

8.5.3 The Effect of Taxes on Salvage or Scrap Value

When an asset is salvaged or scrapped, unless the value received is zero, money comes into the firm as income. Any money received that is in excess of the asset's remaining book value is new revenue, and is taxable. Under the depreciation scheme described for Example 8.2, at the end of the tester's service life, its book value for tax purposes is zero. In this case, all money obtained by salvaging or selling it counts as revenue and is taxed at the tax rate. This is illustrated in Figure 8.3.

Figure 8.3   Loss of Salvage Revenue Due to Taxes (for Example 8.2)

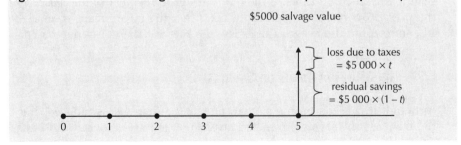

In other circumstances, at the time an asset is sold, it may have remaining book value for taxation purposes. In this case, different things happen depending on the jurisdiction. The specific treatment for several different countries will be discussed later.

However, for Example 8.2(A), the situation is straightforward.

EXAMPLE 8.2(D)

What is the present worth of the salvage value of the tester?

This amount is fully taxable since the book value of the tester will be zero. The approach used is to reduce the amount by the tax rate and then bring the result to a present worth:

$$PW_{salvage} = 5000 \times (1 - 0.42) \times (P/F,12\%,5)$$
$$= 5000 \times 0.58 \times 0.56743$$
$$= 1646$$

The present worth of the salvage value of the tester is $1646. ∎

8.6 | Present Worth and Annual Worth Tax Calculations

As seen in the previous three sections, a complete tax calculation consists of recognizing how the existence of taxes affects each of the components of an investment. Table 8.2 summarizes the procedure for a present worth comparison.

Table 8.2   Components of a Complete Present-Worth Tax Calculation

Component	Treatment
First cost	Add the present worth of tax savings due to depreciation expenses to the first cost. Savings are positive, first cost is negative.
Savings or expenses	Multiply by $(1 - t)$. Convert to present worth.
Salvage value	Multiply by $(1 - t)$ and convert to present worth if book value is zero; otherwise follow specific national regulations.

EXAMPLE 8.2(E)

What is the present worth of the tester investment decision?

$$PW_{first\ cost} = -45\ 000 + 3780(P/A,12\%,5)$$
$$= -45\ 000 + 3780(3.6048)$$
$$= -31\ 374$$
$$PW_{savings} = (23\ 000 - 7300) \times (1 - 0.42) \times (P/A,12\%,5)$$
$$= 15\ 700 \times 0.58 \times 3.6048$$
$$= 32\ 825$$
$$PW_{salvage} = 5000 \times (1 - 0.42) \times (P/F,12\%,5)$$
$$= 5000 \times 0.58 \times 0.56743$$
$$= 1646$$

$$PW_{total} = -31\ 374 + 32\ 825 + 1646$$
$$= 3097$$

The present worth of the tester is $3097. This is an acceptable investment at the given after-tax MARR. ■

Very similar calculations can be done if an annual worth comparison is desired. Table 8.3 summarizes the procedure for annual worth comparisons.

Table 8.3   Components of a Complete Annual Worth Tax Calculation

Component	Treatment
First cost	Convert the first cost (a negative amount) to an annuity and add it to the annual tax savings (positive) due to depreciation expenses.
Savings or expenses	Multiply by $(1 - t)$.
Salvage value	Multiply by $(1 - t)$ and convert to an annuity if book value is zero, otherwise follow specific national regulations.

EXAMPLE 8.2(F)

What is the annual worth of the tester?

$$AW_{first\ cost} = -45\ 000(A/P,12\%,5) + 3780$$
$$= -45\ 000 \times (0.27741) + 3780$$
$$= -8703$$

$$AW_{savings} = (23\ 000 - 7300) \times (1 - 0.42)$$
$$= 15\ 700 \times 0.58$$
$$= 9106$$

$$AW_{salvage} = 5000 \times (1 - 0.42) \times (A/F,12\%,5)$$
$$= 5000 \times 0.58 \times 0.15741$$
$$= 456$$

$$AW_{total} = -8703 + 9106 + 456$$
$$= 859$$

The annual worth of the tester is $859 per year. This is an acceptable investment at the given after-tax MARR. ■

8.7 | IRR Tax Calculations

8.7.1 Accurate IRR Tax Calculations

As has been demonstrated, the effect of taxation is to modify the pre-tax cash flows. The after-tax IRR is simply the IRR calculated on the after-tax cash flows.

EXAMPLE 8.2(G)

What is the accurate after-tax IRR of the tester investment decision?

We answer this question by setting the sum of present worth of receipts and the present worth of disbursements to zero and solving for the unknown interest rate, i^*:

$$45\ 000 = (45\ 000/5) \times 0.42 \times (P/A, i^*, 5) + 15\ 700$$
$$\times (1 - 0.42) \times (P/A, i^*, 5) + 5000(1 - 0.42)(P/F, i^*, 5)$$
$$45\ 000 = (3780 + 15\ 700 \times 0.58)$$
$$\times (P/A, i^*, 5) + 5000 \times 0.58 \times (P/F, i^*, 5)$$

Solving through trial and error results in: $i^* = \text{IRR}_{\text{after-tax}} = 14.7\%$

This exceeds the acceptance threshold of after-tax MARR of 12% and therefore is an acceptable investment.

Alternatively, setting the sum of the annual worth of receipts and the annual worth of the disbursements to zero and solve for i^*:

$$45\ 000(A/P, i^*, 5) = 45\ 000/5 \times 0.42 + 15\ 700$$
$$\times (1 - 0.42) + 5000(1 - 0.42) \times (A/F, i^*, 5)$$
$$45\ 000 = [3780 + 15\ 700 \times 0.58 + 5000 \times 0.58 \times (A/F, i^*, 5)]/(A/P, i^*, 5)$$
$$= [12\ 886 + 5000 \times 0.58 \times (A/F, i^*, 5)]/(A/P, i^*, 5)$$

Solving through trial and error results in: $i^* = \text{IRR}_{\text{after-tax}} = 14.7\%$

This exceeds the acceptance threshold of after-tax MARR of 12% and therefore is an acceptable investment. ■

8.7.2 Approximate After-Tax Rate-of-Return Calculations

The IRR is probably one of the most popular means of assessing the desirability of an investment. Unfortunately, a detailed analysis can be somewhat involved. However, an approximate IRR analysis when taxes are explicitly considered can be very easy. The formula to use is:

$$\text{IRR}_{\text{after-tax}} \cong \text{IRR}_{\text{before-tax}} \times (1 - t) \tag{8.2}$$

The reasons for this are exactly the same as described in Section 8.4 for the before- and after-tax MARR. It is an approximation that works because the IRR represents the percentage of the total investment that is net income. Since the tax rate is applied to net income, it correspondingly reduces the IRR by the same proportion. It is not exactly correct because it assumes that expenses offset receipts in the year that they occur. Consequently, if the after-tax IRR is close to the after-tax MARR, a more precise calculation is advisable so that the correct decision is made.

EXAMPLE 8.2(H)

What is the approximate after-tax IRR of the tester investment?

We first find the before-tax IRR by setting the present worth of the disbursements equal to the receipts and solving for the interest rate:

$$45\ 000 = 15\ 700(P/A,i,5) + 5000(P/F,i,5)$$

Through trial and error, we find that the before-tax IRR is 23.8%. The after-tax IRR is then calculated as:

$$\begin{aligned}
IRR_{after\text{-}tax} &= IRR_{before\text{-}tax} \times (1 - t) \\
&= 0.238(1 - 0.42) \\
&= 0.13804
\end{aligned}$$

The approximate IRR of 13.8% is somewhat lower than the accurate IRR of 14.7%. If the after-tax MARR were 14% rather than 12%, it would have given rise to an incorrect investment decision. ■

When doing an after-tax IRR computation in practice, the approximate after-tax IRR can be used as a first pass on the IRR computation. If the approximate after-tax IRR turns out to be close to the after-tax MARR, a precise after-tax IRR computation may be required to make a fully informed decision about the project.

EXAMPLE 8.3

Essen Industries pays 40% corporate income taxes. Its after-tax MARR is 18%. A project has a before-tax IRR of 24%. Should the project be approved? What would your decision be if the after-tax MARR were 14%?

$$\begin{aligned}
IRR_{after\text{-}tax} &= IRR_{before\text{-}tax} \times (1 - t) \\
&= 0.24(1 - 0.40) \\
&= 0.144
\end{aligned}$$

The after-tax IRR is approximately 14.4%. For an after-tax MARR of 18%, the project should not be approved. However, for an after-tax MARR of 14%, since the after-tax IRR is an approximation, a more detailed examination would be advisable. ■

In summary, we can simplify after-tax IRR computations by using an easy approximation. The approximate after-tax IRR may be adequate for decision making in many cases, but in others a detailed after-tax analysis may be necessary.

8.8 | Specific Tax Rules in Various Countries

The principles of project evaluation in the presence of taxation given in the previous sections are valid for any set of tax regulations. Unfortunately, every country has particular rules that make the calculations more complex than seen in our examples so far. The remaining sections of this chapter give an overview of the current tax rules in Australia, the UK, the USA and Canada. Other countries will have their own particular regulations, but the appropriate method of calculating an economic comparison can be relatively easily synthesized once the fundamental principles are understood.

8.8.1 Australia Tax Calculations

Of the four countries for which we give details in this chapter, Australia probably has the most straightforward regulations. For the most part, they follow the general principles already given in and illustrated through Example 8.1. The key features of Australian tax rules are listed in Table 8.4. The main effects of these rules are on the first cost calculations.

Table 8.4   Key Features of Australian Tax Rules

Specified depreciation method: Taxpayer's choice of straight line (called *prime cost method*) or declining-balance (called *diminishing value rate*) at 150% divided by life of the asset

Asset life: Determined by taxpayer

Treatment of low-cost acquisitions: $300 or less – fully expensed; $1000 or less – pooled together and subject to declining balance depreciation at 37.5%

Special first-year rules: First-year depreciation expense is pro-rated on number of days asset is in use – but this need not be taken into account for project evaluation

Salvage value: If greater than book value, difference is taxable; if less than book value, difference is a deductible expense

The Australian corporate taxpayer can choose either straight-line or declining balance depreciation, and can also choose the life of the asset. With straight-line depreciation, the book value of the asset will be zero at the end of its specified life. If declining balance is used, in the last year of its service life a "balancing charge", that is, an extra depreciation expense, is made to bring the book value to zero.

Assets in the value range of $300 – $1000 are pooled and subject to a uniform depreciation rate of 37.5%. The method for dealing with pooled assets will be covered in the next section on UK rules, and problem 8.11 at the end of this chapter addresses this asset pool issue explicitly.

Australia uses a taxation year that runs from July 1 of one year to June 30 of the next. The amount of depreciation expense that can be claimed in the first year is pro-rated to the number of days of that taxation year for which the asset is owned. For example, if a $10 000 asset with a 5-year life is owned for 120 days of its initial taxation year, the first-year depreciation expense is calculated as $10\,000 \times 120/365 \times (1.5/5) = \$986$, with a similar part-year calculation being made at the end of its specified life.

For project evaluation purposes, this first-year tax treatment actually makes things very easy. Note that the other elements of a project's cash flows are sensibly treated at annual anniversaries of the asset's purchase. For example, if an asset gives rise to annual savings in labour and/or materials, it makes sense to recognize those savings as occurring one year, two years, etc. after the asset's purchase. Similarly, if an asset has a 5-year life, the end of its life would be 5 years from the purchase of the asset. Consequently, the special treatment in the first and last year of purchase essentially aligns the evaluation of the depreciation expenses to the asset's date of purchase. The actual depreciation effects will generally occur at times associated with the tax year, but with only a small error, they can be treated as if they are taken fully at the calendar anniversaries of the purchase dates.

If a company chooses to use straight-line depreciation, the calculations are quite easy and follow the pattern given in Example 8.2. However, if declining-balance depreciation is used, things are a bit more complicated. This is illustrated in the following example.

EXAMPLE 8.4

Broome Diamonds is considering investing in a new high-speed conveyor for one of its mines in Western Australia. The total first cost of the conveyor is $4 500 000. It is expected to provide net savings of $900 000 per year over its 12-year service life. It has an estimated scrap value at that time of $300 000. Broome Diamonds would elect to use the diminishing value rate (declining balance) depreciation. It pays a 44% corporate tax rate and uses an 11% after-tax MARR, and would make the purchase on April 7 of next year. What is the present worth of the conveyor? Should Broome Diamonds make this investment?

Although the actual depreciation Broome Diamonds would pay would be proportional to the number of days remaining in the tax year after April 7, this information is redundant for our tax calculations. We can proceed as if the purchase is made at the beginning of the tax year, with the depreciation benefits accruing at the end of each tax year.

First cost:

If F = first cost, t = tax rate, and d = depreciation rate, then the first year's depreciation savings will be Ftd, the second year's savings will be $Ftd(1 - d)$, the third's $Ftd(1 - d)^2$, etc. This can be recognized as a geometric gradient.

In this case, F = 4 500 000, t = 0.44, and d = 150%/12 = 12.5%. Recall from Chapter 3 the form of the geometric gradient conversion factor is $(P/A,g,i,N)$.

Therefore:

$$\text{PW}_{\text{first cost}} = -4\ 500\ 000 + 4\ 500\ 000 \times 0.44 \times 0.125$$
$$\times (P/A, -12.5\%, 11\%, 12)$$

Recall from Chapter 3 that:

$$(P/A, g, i, N) = (P/A, i^o, N)/(1 + g)$$

where the growth adjusted interest rate i^o in this case is calculated as

$$i^o = (1 + i)/(1 + g) - 1$$
$$= (1 + 0.11)/(1 - 0.125) - 1$$
$$= 1.11/0.875 - 1$$
$$= 0.269$$

and

$$(P/A, -12.5\%, 11\%, 12) = (P/A, 26.9\%, 12) = 3.509$$

Consequently,

$$\begin{aligned} PW_{\text{first cost}} &= -4\,500\,000 + 4\,500\,000 \times 0.44 \times 0.125 \times 3.509 \\ &\cong -3\,630\,000 \end{aligned}$$

Savings:

$$\begin{aligned} PW_{\text{savings}} &= 900\,000 \times (1 - 0.44) \times (P/A, 11\%, 12) \\ &= 900\,000 \times 0.56 \times 6.4924 \\ &\cong 3\,270\,000 \end{aligned}$$

Salvage value:

Under Australian tax rules, a balancing adjustment is made at the end of the service life to bring the book value of the asset to zero. Recall from Chapter 7 that in general,

$$BV_{db}(n) = P(1 - d)^n$$

Consequently, in this case, the remaining depreciation is calculated as $4\,500\,000$ $(1 - 0.125)^{12} \cong 906\,000$

Since this is greater than the salvage value of $300\,000, the difference: $906\,000 - 300\,000 = 606\,000$ is a deductible expense, and the present worth of the salvage value is then

$$\begin{aligned} PW_{\text{salvage value}} &= 606\,000 \times 0.44 \times (P/F, 11\%, 12) \\ &= 606\,000 \times 0.44 \times 0.28584 \\ &\cong 76\,300 \end{aligned}$$

Note that if the salvage value had been greater than the remaining depreciation, the difference would have been considered taxable income instead. With a net expense, we multiply by the tax rate t to recognize taxes saved, while with net revenue, we multiply by $(1-t)$ to recognize taxes lost.

Finally, the total present worth for the conveyor is

$$\begin{aligned} PW_{\text{total}} &= PW_{\text{first cost}} + PW_{\text{savings}} + PW_{\text{salvage value}} \\ &= -3\,630\,000 + 3\,270\,000 + 76\,300 \\ &= -283\,700 \end{aligned}$$

The total present worth of the new conveyor is about $-\$283\,700$. Since this is less than zero, this does not meet the required MARR and would not be a supportable investment. ∎

8.8.2 UK Tax Calculations

Similar to Australia, UK tax rules are fairly straightforward. However, in the UK each asset purchase is not treated in an individual manner. Instead, assets are "pooled", meaning that all assets are treated together for depreciation expense calculations. When a new asset is purchased, the pool is increased by the cost of the asset. At the end of each tax year, 25% of the total value of the pool is recognized as an expense, and the remaining value in the pool—called the **written down value (WDV)**—is correspondingly decreased. If an asset is sold or salvaged, a corresponding adjustment is made to the WDV. Although the assets are pooled for calculation purposes, each asset within that pool is effectively treated in exactly the same manner as the whole pool. The key features of UK tax rules are summarized in Table 8.5.

Table 8.5   Key Features of UK Tax Rules

Specified depreciation method: Declining balance (diminishing value rate) at 25% (in most cases), except buildings which are straight line at 4%
Asset life: Assets are pooled
Treatment of low-cost acquisitions: No special treatment
Special first-year rules: First year depreciation expense 50% for small businesses, 40% for medium-sized businesses
Salvage value: Balancing adjustment to pool is made

EXAMPLE 8.5

Exeter Dodads has a WDV account totalling £120 443 at the beginning of the 2009 taxation year. During that year it purchases depreciable assets totalling £14 987, and disposes of assets of value £3509. In the 2010 taxation year, its purchases total £32 992 and disposals £8987. What is its 2010 depreciation expense claim? What is the value of its WDV account at the end of the 2010 tax year?

The depreciation expense is the value of the WDV account adjusted by the purchases and dispositions that year multiplied by the 25% rate.

The value of its WDV at the end of 2009 [WDV(09)] is

WDV(09) = WDV(08) + purchases + disposals − depreciation expense

WDV(08) is the WDV calculated at the end of 2008, and is the value given as the WDV at the beginning of 2009.

Depreciation expense(09) = (120 433 + 14 987 − 3509) × 0.25 = 32 978

WDV(09) = 120 433 + 14 987 − 3509 − 32 978 = 98 933

Depreciation expense(10) = (98 933 + 32 992 − 8987) × 0.25 = 30 735

WDV(10) = 98 933 + 32 992 − 8987 − 30 735 = 92 203

Exeter Dodads will have a £30 735 expense claim for the 2010 tax year, and at the end of that year its WDV account will total £92 203. ∎

In order to evaluate the merits of a specific investment, it is necessary to recognize the effect of depreciation benefits that continue forever. Unlike the Australian system, where an asset life is recognized and the book value is reduced to zero at the end of that specified life, under the UK system, the book value never gets to zero. Consequently, for any purchase, except buildings, there is an annual benefit from a depreciation expense that goes on forever.

In order to deal with an endless diminishing sequence of depreciation benefits, we use a formula called the **Tax Benefit Factor** (*TBF*) that simplifies the necessary calculations. This formula, which is derived in Appendix 8A, is a value that summarizes the effect of the benefit of all future tax savings due to depreciation; it allows analysts to take these benefits into account when calculating the present worth equivalent of an asset. The *TBF* is constant for a given declining balance rate, interest rate, and tax rate, and allows the determination of the present worth independently of the actual first cost of the asset.

The *TBF* is:

$$TBF = td/(i + d)$$

where t = tax rate

d = declining balance depreciation rate

i = after-tax interest rate

The present worth of the first cost of an asset under simple declining-balance depreciation is easily calculated:

$$\text{Present worth} = \text{First cost} - (\text{First cost} \times TBF)$$
$$= \text{First cost} (1 - TBF)$$

Similarly, the salvage value under UK tax rules cannot be treated as cash flow at the time the asset is scrapped or salvaged because the WDV is adjusted according to the value received for the asset. Rather than increasing the WDV as an asset purchase does, asset disposition reduces the WDV, correspondingly proportionately reducing future depreciation expense claims. The calculation of this effect is identical to that done for the first cost, which is to modify the salvage value by the loss in future depreciation claims as recognized in the *TBF*. In general:

$$\text{Present worth of salvage value at time of disposal} = \text{Salvage value}(1 - TBF)$$

EXAMPLE 8.6

What is the present worth of the first cost of an asset that costs £350 000 and has a 30% tax rate while after-tax interest is at 5%?

The *TBF* is calculated as:

$$TBF = (0.3 \times 0.25)/(0.05 + 0.25) = 0.25$$

$$PW = 350\,000 \times (1 - 0.25) = 262\,500$$

The first cost of the asset has a present worth of £262 500. ∎

For small- and medium-sized businesses, there is an accelerated depreciation rate for the first year. For project evaluation purposes, the sensible way to treat this situation is to consider the first year's depreciation expense as a separate cash flow element, and then use the *TBF* for the remaining depreciation expenses.

EXAMPLE 8.7

The owner of a spring water bottling company in Reading has just purchased an automated bottle capper. What is the after-tax present worth of the new automated bottle capper if it costs £10 000 and saves £3000 per year over its six-year life? Assume a £2000 salvage value and a 50% tax rate. The after-tax MARR is 10%. The company qualifies as a small business under UK tax rules, and is thus eligible for a 50% depreciation expense in the first year.

First cost:

The present worth of the first cost in this case will be the sum of three elements: 1) the (negative) first cost; 2) the (positive) first year depreciation expense (at 50%) times the tax rate; and 3) the residual value after the first year's depreciation times the *TBF*. The *TBF* is calculated as:

$$TBF = (0.5 \times 0.25)/(0.12 + 0.25) = 0.3378$$

So that the present worth of the first cost is:

$$
\begin{aligned}
PW_{\text{first cost}} &= -10\,000 + 10\,000 \times 0.5 \times 0.5 \times (P/F,10\%,1) \\
&\quad + 10\,000(1 - 0.5) \times TBF \times (P/F,10\%,1) \\
&= -10\,000 + 10\,000 \times (P/F,10\%,1) \times [(0.5 \times 0.5) \\
&\quad + (1 - 0.5) \times TBF] \\
&= -10\,000 + 10\,000 \times 0.90909 \times [0.\,25 + (0.5 \times 0.3378)] \\
&= -6192
\end{aligned}
$$

The present worth of the first cost of the automated capper is $-£6192$.

Annual savings:

The £3000 annual savings are taxed at the tax rate and taken to present worth:

$$
\begin{aligned}
PW_{\text{savings}} &= 3000 \times 0.5 \times (P/A,10\%,6) \\
&= 3000 \times 0.5 \times (4.3553) \\
&= 6533
\end{aligned}
$$

The present worth of the annual savings created by the capper is £6533.

Salvage value:

The £2000 salvage value must be modified by the *TBF* to take into account the effect of the adjustment to the WDV, and then taken to present worth:

$$
\begin{aligned}
PW_{\text{salvage}} &= 2000 \times (1 - TBF) \times (P/F,10\%,6) \\
&= 2000 \times (1 - 0.3378) \times (0.56447) \\
&= 748
\end{aligned}
$$

The present worth of the salvage value of the capper is £748.

The present worth of the capper, taking into account all elements of its cash flows, is then:

$$
\begin{aligned}
PW_{\text{total}} &= PW_{\text{first cost}} + PW_{\text{savings}} + PW_{\text{salvage}} \\
&= -6192 + 6533 + 748 = 1089
\end{aligned}
$$

The automated capper has a net present worth of £1089. Since this is positive at the after-tax MARR of 10%, this is an acceptable project. ∎

8.8.3 US Tax Calculations

US tax rules are a bit more complicated than either the Australian or UK models. However, with the aid of a spreadsheet, they can be relatively simple to use in practice. The key features of US tax rules are summarized in Table 8.6.

Table 8.6 Key Features of US Tax Rules

> Specified depreciation method: A particular mixture of declining balance and straight line called MACRS (Modified Accelerated Cost Recovery System).
>
> Asset life: Specified by regulation. For manufacturing equipment, 7 years
>
> Treatment of low-cost acquisitions: Assets with life less than one year are exempt.
>
> Special first-year rules: As defined by the MACRS schedule; assumes an asset is purchased in the middle of the first year.
>
> Salvage value: If greater than book value, difference is taxable; if less than book value, difference is a deductible expense.

The core of the US approach is MACRS, which stands for Modified Accelerated Cost Recovery System. Under MACRS, all assets come under one of six "property classifications", as illustrated in Table 8.7. Each property classification has a nominal asset life associated with it, which is called the **recovery period.**

Table 8.7 MACRS Property Classes

Property Class	Personal Property (all property except real estate)
3-year property	Special handling devices for food and beverage manufacture Special tools for the manufacture of finished plastic products, fabricated metal products, and motor vehicles
5-year property	Information Systems; Computers / Peripherals Aircraft (of non-air-transport companies) Computers Petroleum drilling equipment
7-year property	All other property not assigned to another class Office furniture, fixtures, and equipment
10-year property	Assets used in petroleum refining and certain food products Vessels and water transportation equipment
15-year property	Telephone distribution plants Municipal sewage treatment plants
20-year property	Municipal sewers

Source: U.S. Department of Treasury, Internal Revenue Service, Publication 946, *How to Depreciate Property.* Washington, DC: U.S. Government Printing Office. Available online at http://www.irs.gov/publications/p946/index.html

The general idea is that during the recovery period, an asset depreciates at a specific declining balance rate until a particular point in time, and thereafter by straight-line depreciation, such that the book value reaches zero at the end of the recovery period. Complicating this further is the built-in assumption that an asset is purchased half-way through the first year, reducing the allowable depreciation expense for that year, and adding an extra year of partial expenses at the end.

The consequence of all of this is a rather complex set of depreciation expense cash flows, as illustrated in Table 8.8. However, because of the limited number of classifications and the fixed recovery periods within each classification, it is easy to create a spreadsheet for each property classification to deal with the cash flow on a year-by-year basis. The spreadsheet will have the first cost as a variable. It simply calculates the depreciation expense at the end of each year using the information in Table 8.8, multiplies this by the tax rate, and adds them all together, discounting each cash flow element by the appropriate number of years. A pre-built version of such a spreadsheet is available on the companion website for this book, but it is also very easy to build one for yourself.

Table 8.8   Permitted Depreciation Expenses (in percent) under MACRS

Recovery Year	3-Year Property	5-Year Property	7-Year Property	10-Year Property	15-Year Property	20-Year Property
1	33.33	20.00	14.29	10.00	5.00	3.750
2	44.45	32.00	24.49	18.00	9.50	7.219
3	14.81	19.20	17.49	14.40	8.55	6.677
4	7.41	11.52	12.49	11.52	7.70	6.177
5		11.52	8.93	9.22	6.93	5.713
6		5.76	8.92	7.37	6.23	5.285
7			8.93	6.55	5.90	4.888
8			4.46	6.55	5.90	4.522
9				6.56	5.91	4.462
10				6.55	5.90	4.461
11				3.28	5.91	4.462
12					5.90	4.461
13					5.91	4.462
14					5.90	4.461
15					5.91	4.462
16					2.95	4.461
17						4.462
18						4.461
19						4.462
20						4.461
21						2.231

Source: U.S. Department of Treasury, Internal Revenue Service; Publication 946, *How to Depreciate Property*. Washington, DC: U.S. Government Printing Office. Available online at http://www.irs.gov/publications/p946/index.html

To illustrate the procedure, the following example for an asset with a three-year recovery period will be done manually.

EXAMPLE 8.8

Karen's company is considering investing in some special tools for the manufacture of plastic products. These tools qualify for 3-year property treatment under MACRS. The tools have a first cost of $280 000, a service life of 6 years with $40 000 salvage value, and provide net savings of $80 000 per year. Given a tax rate of 31% and an after-tax MARR of 4%, what is the present worth of this investment decision?

First cost:

The present worth of the first cost will be the sum of the (negative) first cost plus the (positive) savings from reduced taxes due to the depreciation expenses. The savings are the depreciation amounts calculated from Table 8.8 times the tax rate, each taken to present worth. This gives

$$
\begin{aligned}
PW_{\text{first cost}} = {} & -280\,000 + (280\,000 \times 0.3333) \times 0.31 \times (P/F,4\%,1) \\
& + (280\,000 \times 0.4445) \times 0.31 \times (P/F,4\%,2) \\
& + (280\,000 \times 0.1481) \times 0.31 \times (P/F,4\%,3) \\
& + (280\,000 \times 0.0741) \times 0.31 \times (P/F,4\%,4) \\
= {} & -280\,000 + 280\,000 \times 0.31 \times [0.3333 \times (P/F,\,4\%,1) \\
& + 0.4445 \times (P/F,4\%,2) + 0.1481 \times (P/F,4\%,3) \\
& + 0.0741 \times (P/F,4\%,4)] \\
= {} & -280\,000 + 280\,000 \times 0.31 \times [0.3333 \times 0.96154 \\
& + 0.4445 \times 0.92456 + 0.1481 \times 0.889 \\
& + 0.0741 \times 0.8548] \\
= {} & -280\,000 + 80\,416 \\
= {} & -199\,584
\end{aligned}
$$

The present worth of the first cost of the special tools is −$199 584.

The calculation of the total present worth of the investment decision is then:

$$
\begin{aligned}
PW_{\text{total}} = {} & PW_{\text{first cost}} + 80\,000 \times (1 - 0.31) \times (P/A,4\%,6) \\
& + 40\,000 \times (1 - 0.31) \times (P/F,\,4\%,6) \\
= {} & -199\,584 + 80\,000 \times 0.69 \times 5.4221 \\
& + 40\,000 \times 0.69 \times 0.79031 \\
= {} & 121\,528
\end{aligned}
$$

The total present worth of the special tools investment is $121 528. This is a very worthwhile investment. ■

8.8.4 Canadian Tax Calculations

Canada's tax rules are probably the most complicated of the ones reviewed in detail in this chapter. The key features of the Canadian system are shown in Table 8.9.

Table 8.9   Key Features of Canadian Tax Rules

Specified depreciation method: Declining balance at various specified rates. Manufacturing equipment is generally at 20%
Asset life: Assets are pooled by class
Treatment of low-cost acquisitions: $200 or less – fully expensed
Special first-year rules: First-year depreciation expense assumes an asset is purchased in the middle of the first year
Salvage value: A balancing charge is made to the asset pool

Like some other things in Canada, the tax rules have similarities to both the UK and US systems. As in the UK, the specified depreciation method is declining balance, and assets are pooled. However, like the US system, there are many classes of assets. Each of the classes depreciates at a different rate, requiring a separate pool for each. In the Canadian system the remaining value in each pooled account (the WDV in the UK system) is called the **undepreciated capital cost (UCC)**. Finally, like the US, only half of the value of the asset can be depreciated in the first year.

The depreciation rate is called the *Capital Cost Allowance* (CCA) rate. Table 8.10 shows some of the important classes of assets under the Canadian system, and the associated CCA rates. Typical engineering projects fall into class 8, with a 20% declining balance depreciation rate.

Table 8.10   Sample CCA Rates and Classes

CCA Rate (%)	Class	Description
4 to 10	1, 3, 6	Buildings and additions
20	8	Machinery, office furniture, and equipment
25	9	Aircraft, aircraft furniture, and equipment
30	10	Passenger vehicles, vans, trucks, computers, and systems software
40	16	Taxis, rental cars, and freight trucks; coin-operated video games
100	12	Dies, tools, and instruments that cost less than $200; computer software other than systems software

Another feature of the Canadian system is the half-year rule, which assumes that any asset purchase occurs in the middle of the year of purchase. Consequently, rather than being allowed to depreciate the asset by the full declining balance amount, the allowed depreciation expense is only half of the full year amount. So, for example, if the asset was in CCA class 10 (see Table 8.10) the first year depreciation is 15% of the asset's first cost rather than 30%.

Consequently, the procedure to use for calculating the present worth of the first cost of an investment is simply to use the *TBF* directly on half of the first cost, and add to this the tax benefits of the other half of the first cost delayed by one year, also calculated using the *TBF.*

The following example can illustrate this procedure.

EXAMPLE 8.9

An automobile purchased this year by Lestev Corporation for $25 000 has a CCA rate of 30%. Lestev is subject to 43% corporate taxes and the corporate (after-tax) MARR is 12%. What is the present worth of the first cost of this automobile, taking into account the future tax savings resulting from depreciation?

The present worth of the first cost of the car is then calculated as the first cost (negative) plus the present worth of the depreciation benefit from half of the first cost calculated using the *TBF* plus the present worth of the depreciation benefit of the other half of the first cost delayed by one year.

$$PW = -25\ 000 + [25\ 000/2 \times TBF + 25\ 000/2 \times TBF \times (P/F,12\%,1)]$$
$$= -25\ 000 + 25\ 000/2 \times TBF\,[1 + (P/F,12\%,1)]$$

The *TBF* is calculated as

$$TBF = (0.43 \times 0.3)/(0.12 + 0.3)$$
$$= 0.30714$$
$$PW = -25\ 000 + 12\ 500\,(0.30714)\,[1 + (0.89286)]$$
$$= -17\ 733$$

The present worth of the first cost of the car, taking into account all future tax savings due to depreciation, is about −$17 733. The tax benefit due to claiming CCA has effectively reduced the cost of the car from $25 000 to $17 733 in terms of present worth. ■

The *TBF* is also required under the Canadian system to recognize the effects of diminishing the UCC account for an asset that has been scrapped or salvaged. However, the half-year rule does not apply in that case and so the *TBF* can be applied directly.

EXAMPLE 8.10

A chemical recovery system, required under environmental legislation, costs $30 000 and saves $5280 each year of its seven-year life. The salvage value is estimated at $7500. The after-tax MARR is 9% and taxes are at 45%. What is the net after-tax annual benefit or cost of purchasing the chemical recovery system?

First Cost:

This purchase is clearly in CCA class 8, which has a 20% declining balance depreciation rate.

$$PW_{first\ cost} = -30\ 000 + [30\ 000/2 \times TBF \times [1 + (P/F,9\%,1)]$$

Then the *TBF* can be calculated as:

$$TBF = (0.45 \times 0.2)/(0.09 + 0.2) = 0.31$$

$$\text{PW}_{\text{first cost}} = -30\,000 + 15\,000 \times 0.31 \times (1 + 0.9174)$$
$$= -21\,084$$

$$\text{AW}_{\text{first cost}} = \text{PW}_{\text{first cost}} \times (A/P,9\%,7)$$
$$= -21\,084 \times 0.19869$$
$$= -4189$$

The annual worth of the first cost is $\$-4189$.

Savings:

$$\text{AW}_{\text{saving}} = 5280 \times (1 - t)$$
$$= 5280 \times 0.55$$
$$= 2904$$

The annual worth of the saving is $2904.

Salvage value:

The salvage value reduces the UCC and consequently an infinite series of tax benefits are lost. The annual worth of the salvage value is then the amount received reduced by this loss of benefits (calculated using the *TBF*) and then converted to an annuity over the system's seven-year life:

$$\text{AW}_{\text{salvage}} = 7500 \times (1 - TBF) \times (A/F,9\%,7)$$
$$= 7500 \times 0.69 \times 0.10869$$
$$= 562$$

The annual worth of the salvage value is $562.

Total annual cost of the chemical recovery system is then:

$$\text{AW}_{\text{total}} = \text{AW}_{\text{first cost}} + \text{AW}_{\text{savings}} + \text{AW}_{\text{salvage}}$$
$$= -4189 + 2904 + 562$$
$$= -723$$

The chemical recovery system is expected to have an after-tax *cost* of about $723 over its seven-year life. ■

REVIEW PROBLEMS

REVIEW PROBLEM 8.1

International Computing Corporation (ICC) is considering updating its automated inventory control system in its plants in Australia, the UK, the US and Canada. The costs, savings, and salvage values of the capital investments in computers and systems software are the same in all four countries. The UK-based supplier has estimated the first cost as £2 300 000 and the annual savings as £880 000; after its 10-year life it will have a salvage value of £200 000. The tax rate for International Computing in this case is 45% in all four countries, and ICC uses an after-tax MARR of 10%. In which country will the present worth of the new system be most favourable for ICC?

First, note that no information is given to convert from pounds to dollars, whether Australian, Canadian, or US. That information is not necessary to answer the question. The calculations and comparisons can all be made in pounds. Only if we were asked for the present worth in local currency would we need a conversion factor.

Australia:

Australia allows flexibility on choice of depreciation methods and asset life. It would be improper to choose an asset life different from the one specified, but both straight-line and declining-balance depreciation should be tested.

Straight line:

$$PW = -2\,300\,000 + 2\,300\,000/10 \times 0.45 \times (P/A,10\%,10)$$
$$+ 880\,000 \times (1 - 0.45) \times (P/A,10\%,10) + 200\,000 \times (1 - 0.45)$$
$$\times (P/F,10\%,10)$$
$$= -2\,300\,000 + 230\,000 \times 0.45 \times 6.1446 + 880\,000 \times 0.55 \times 6.1446$$
$$+ 200\,000 \times 0.55 \times 0.3855$$
$$\cong 1\,350\,000$$

Straight-line depreciation under Australian tax rules gives a present worth of £1 350 000.

Declining balance:

Noting that $2\,300\,000 \times (1 - 0.125)^{10} = 605\,074 > 200\,000$ so that salvaging the system after 10 years results in net revenue, the present worth calculation is:

$$PW = -2\,300\,000 + 2\,300\,000 \times 0.45 \times 0.125 \times (P/A,-12.5\%,10\%,10)$$
$$+ 880\,000 \times (1 - 0.45) \times (P/A,10\%,10)$$
$$+ [(2\,300\,000 \times (1 - 0.125)^{10}) - 200\,000] \times 0.45 \times (P/A,10\%,10)$$

Noting that

$$i^{o} = (1 + i)/(1 + g) - 1 = (1 + 0.10)/(1 - 0.125) - 1$$
$$= 1.10/0.875 - 1 = 0.257$$

and

$$(P/A,-12.5\%,10\%,10) = (P/A,25.7\%,10) = 3.4959$$

$$PW = -2\,300\,000 + 2\,300\,000 \times 0.45 \times 0.125 \times 3.4959$$
$$+ 880\,000 \times 0.55 \times 6.1446$$
$$+ [(2\,300\,000 \times 0.875^{10}) - 200\,000] \times 0.45 \times 0.38554$$
$$= -2\,300\,000 + 452\,282 + 2\,974\,180 + 70\,277$$
$$\cong 1\,200\,000$$

Declining balance depreciation under Australian tax rules gives a present worth of about £1 200 000.

UK:

There is no reason to believe that ICC will benefit from any special first-year deduction for this investment. The *TBF* is calculated as:

$$TBF = (0.45 \times 0.25)/(0.10 + 0.25) = 0.3214$$

And the present worth of the capital investments in the inventory system upgrade is

$$PW = -2\,300\,000 \times (1 - 0.3214)$$
$$+\, 880\,000 \times (1 - 0.45) \times (P/A,10\%,10)$$
$$+\, 200\,000 \times (1 - 0.3214) \times (P/F,10\%,10)$$
$$= -2\,300\,000 \times 0.6786 + 880\,000 \times 0.55 \times 6.1446$$
$$+\, 200\,000 \times 0.6786 \times 0.38554$$
$$= -1\,560\,780 + 2\,973\,986 + 52\,325$$
$$\cong 1\,470\,000$$

UK tax rules give a present worth of about £1 470 000.

US:

From Table 8.7 it is clear that ICC's capital investments would fall under the 5-year MACRS property class. Using Table 8.8, we can calculate the present worth of the first cost of the inventory system upgrade as:

$$PW_{\text{first cost}} = -2\,300\,000 + (2\,300\,000 \times 0.2) \times 0.45 \times (P/F,10\%,1)$$
$$+\, (2\,300\,000 \times 0.32) \times 0.45 \times (P/F,10\%,2)$$
$$\vdots$$
$$+\, (2\,300\,000 \times 0.0576) \times 0.45 \times (P/F,10\%,6)$$

Using the spreadsheet from the companion website to perform this calculation, we get
$$PW_{\text{first cost}} = 1\,499\,675$$
The overall present worth is then calculated as:

$$PW = -1\,499\,675 + 880\,000 \times (1 - 0.45) \times (P/A,10\%,10)$$
$$+\, 200\,000 \times (1 - 0.45) \times (P/F,10\%,10)$$
$$= -1\,499\,675 + 880\,000 \times 0.55 \times 6.1446 + 200\,000 \times 0.55 \times 0.38554$$
$$= -1\,499\,675 + 2\,973\,986 + 42\,409$$
$$\cong 1\,520\,000$$

US tax rules give a present worth of about £1 520 000.

Canada:

As we can see from Table 8.10, ICC's investments are clearly in CCA class 10, with a depreciation rate (CCA rate) of 30%.

In this case the *TBF* is calculated as:

$$TBF = (0.45 \times 0.3)/(0.10 + 0.3) = 0.3375$$

This is different from the *TBF* determined for the UK calculations because the depreciation rate is 30% in this class in Canada, as compared to the universal 25% rate in the UK.

The present worth of the inventory system investments is then:

$$PW = -2\,300\,000 + 2\,300\,000/2 \times TBF \times [1 + (P/F,10\%,1)]$$
$$+\, 880\,000 \times (1 - 0.45) \times (P/A,10\%,10)$$
$$+\, 200\,000 \times (1 - TBF) \times (P/F,10\%,10)$$

$$= -2\,300\,000 + 1\,150\,000 \times 0.3375 \times [1 + 0.90909]$$
$$+ 880\,000 \times 0.55 \times 6.1446 + 200\,000 \times 0.6625 \times 0.38554$$
$$\cong 1\,470\,000$$

Canadian tax rules give a present worth of £1 470 000.

The calculated present worths for the inventory system upgrades are summarized as follows:

Australia (straight line): = 1 350 000

Australia (declining balance): = 1 200 000

UK: £1 470 000

US: £1 520 000

Canada: £1 470 000

Consequently, in this particular case, the project has the greatest present worth under US tax rules. Of course, this comparison may not reflect reality because actual tax rates are difficult to compare between countries and are unlikely to be the same, as they were assumed to be in this case. ∎

REVIEW PROBLEM 8.2

Putco is a Canadian company that assembles printed circuit boards. Business has been good lately, and Putco is thinking of purchasing a new IC chip-placement machine. It has a first cost of $450 000 and is expected to save the company $125 000 per year in labour and operating costs compared with the manual system now in place. A similar system that also automates the circuit board loading and unloading process costs $550 000 and will save about $155 000 per year. The life of either system is expected to be four years. The salvage value of the $450 000 machine will be $180 000, and that of the $550 000 machine will be $200 000. Putco uses an after-tax MARR of 9% to make decisions about such projects. On the basis of an IRR comparison, which alternative (if either) should the company choose? Putco pays taxes at a rate of 40%, and the CCA rate for the equipment is 20%.

ANSWER

Putco has three mutually exclusive alternatives:

1. Do nothing
2. Buy the chip-placement machine
3. Buy a similar chip-placement machine with an automated loading and unloading process

Following the procedure from Chapter 5, the projects are already ordered on the basis of first cost. We therefore begin with the first alternative: the before-tax (and thus the after-tax) IRR of the "do nothing" alternative is 0%. Next, the before-tax IRR on the incremental investment to the second alternative can be found by solving for i in

$$-450\,000 + 125\,000 \times (P/A,i,4) + 180\,000 \times (P/F,i,4) = 0$$

By trial and error, we obtain an $IRR_{before-tax}$ of 15.92%. This gives an approximate $IRR_{after-tax}$ of $0.1592(1 - 0.40) = 0.0944$ or 9.44%. With an after-tax MARR of 9%, it would appear that this alternative is acceptable, though a detailed after-tax computation may be in order. We need to solve for i in

$$(-450\,000) + [450\,000 \times 0.4 \times 0.2/2 + 450\,000 \times (1 - 0.2/2) \times TBF] \times (P/F,i,1)$$
$$+ 125\,000 \times (P/A,i,4) \times (1 - 0.4) + 180\,000(P/F,i,4)\,(1 - TBF)$$
$$= 0$$

Doing so gives an $IRR_{after-tax}$ of 9.36%. Since this exceeds the required after-tax MARR of 9%, this alternative becomes the current best. We next find the $IRR_{after-tax}$ on the incremental investment required for the third alternative. The $IRR_{before-tax}$ is first found by solving for i in

$$-(550\ 000 - 450\ 000) + (155\ 000 - 125\ 000) \times (P/A,i,4)$$
$$+ (200\ 000 - 180\ 000) \times (P/F,i,4)$$
$$= 0$$

This gives an $IRR_{before-tax}$ of 7.13%, or an approximate $IRR_{after-tax}$ of 4.28%. This is sufficiently below the required after-tax MARR of 9% to warrant rejection of the third alternative without a detailed incremental $IRR_{after-tax}$ computation. Putco should therefore select the second alternative. ■

SUMMARY

Income taxes can have a significant effect on engineering economics decisions. In particular, taxes reduce the effective cost of an asset, the savings generated, and the value of the sale of an asset.

The first cost of an asset is reduced because it generates a series of expense claims against taxes in future years. These expenses offset revenues, and consequently reduce taxes. Thus, for any asset purchase, there are offsetting positive cash flows. When the present worth of these future tax benefits is subtracted from the first cost, there can be a substantial effect.

However, any beneficial savings brought about by the investment are also taxed, and consequently the benefit is reduced. Similarly, if there is a salvage or scrap value at the end of the asset life, the cash flow generated is also reduced in value by taxes.

Generally speaking, the effect of taxes is to alter the cash flows of a project in predictable ways, and the principles of present worth, annual worth, and IRR analysis are simply applied to the modified cash flows. Once this is understood, it is straightforward to apply the principles to any tax regime.

Engineering Economics in Action, Part 8B:
The Work Report

"So what is this, anyhow?" Clem was looking at the report that Naomi had handed him. "A consulting report?"

"Sorry, chief, it is a bit thick." Naomi looked a little embarrassed. "You see, Terry has to do a work report for his university. It's part of the co-op program that they have. He got interested in the 10-stage die problem and asked me if he could make that study his work report. I said OK, subject to its perhaps being confidential. I didn't expect it to be so thick, either. But he's got a good executive summary at the front."

"Hmm . . ." The room was quiet for a few minutes while Clem read the summary. He then leafed through the remaining parts of the report. "Have you read this through? It looks really quite good."

"I have. He has done a very professional job—er, at least what seems to me to be very professional." Naomi suddenly remembered that she hadn't yet gained her professional engineer's designation. She also hadn't been working at Global Widgets much longer than Terry. "I gathered most of the data for him, but he did an excellent job of analyzing it. As you can see from the summary, he set up the replacement problem as a set of mutually exclusive alternatives, involving upgrading the die now or later and even more than once. He did a nice job on the taxes, too."

"Tell me more about how he handled the taxes."

"He was really thorough. He broke down all of the cash flows into capital costs, annual savings, and salvage values, and recognized the appropriate tax effect in each case. It is pretty complicated but I think he got it all right. He even compared what we do here with tax regimes in some of the countries where we have sister plants."

"Yeah, I noticed that. What's that all about?"

"Well, it was his idea. Every country has its own rules, and he thought that a comparison in this particular case might be helpful in future decisions about upgrades in any of our plants. We have knowledge of each country's tax rules, but nobody has ever really compared them internationally for the same engineering decision. And there are some real surprises, to me at least, in the difference."

"Not bad. Not only did he do a great job on a hard project, but he went beyond that to contribute something new and important. I think we've got a winner here, Naomi. Let's make sure we get him back for his next work term."

Naomi nodded. "What about his work report, Clem? Should we ask him to keep it confidential?"

Clem laughed, "Well, I think we should, and not just because there are trade secrets in the report. I don't want anyone else knowing what a gem we have in Terry!"

PROBLEMS

For additional practice, please see the problems (with selected solutions) provided on the Student CD-ROM that accompanies this book.

8.1 The MARR generally used by Collingwood Caskets is a before-tax MARR of 14%. Vincent wants to do a detailed calculation of the cash flows associated with a new planer for the assembly line. What would be an appropriate after-tax MARR for him to use if Collingwood Caskets pays

(a) 40% corporate taxes?

(b) 50% corporate taxes?

(c) 60% corporate taxes?

8.2 Last year the Cape Town Furniture Company bought a new band saw for R360 000. Aside from depreciation expenses, its yearly expenses totalled R1 300 000 versus R1 600 000 in income. How much tax (at 50%) would the company have paid if it had been permitted to use each of the following depreciation schemes?

(a) Straight-line, with a life of 10 years and a 0 salvage value

(b) Straight-line, with a life of five years and a 0 salvage value

(c) Declining-balance, at 20%

(d) Declining-balance, at 40%

(e) Fully expensed that year

8.3 Calculate the *TBF* for each of the following:

(a) Tax rate of 50%, declining-balance depreciation rate of 20%, and an after-tax MARR of 9%

(b) Tax rate of 35%, declining-balance depreciation rate of 30%, and an after-tax MARR of 12%

(c) Tax rate of 55%, declining-balance depreciation rate of 5%, and an after-tax MARR of 6%

8.4 Enrique has just completed a detailed analysis of the IRR of a waste-water treatment plant for Monterrey Meat Products. The 8.7% after-tax IRR he calculated compared favourably with a 5.2% after-tax MARR. For reporting to upper management, he wants to present this information as a before-tax IRR. If Monterrey Meat Products pays 53% corporate taxes, what figures will Enrique report to upper management?

8.5 Warsaw Widgets is looking at a €400 000 digital midget rigid widget gadget. It is expected to save €85 000 per year over its 10-year life, with no scrap value. The company's tax rate is 45%, and its after-tax MARR is 15%. On the basis of an approximate IRR, should it invest in this gadget?

8.6 What is the approximate after-tax IRR on a two-year project for which the first cost is $12 000, savings are $5000 in the first year and $10 000 in the second year, and taxes are at 40%?

8.7 A new binder will cost Revelstoke Printing $17 000, generate net savings of $3000 per year over a nine-year life, and be salvaged for $1000. Revelstoke's *before-tax* MARR is 10%, and it is taxed at 40%. What is its approximate after-tax IRR on this investment? Should the investment be made?

 8.8 Use a spreadsheet program to create a chart showing how the values of the *TBF* change for after-tax MARRs of 0% to 30%. Assume a fixed tax rate of 50% and a declining-balance depreciation rate of 20%.

8.9 Refer to Problem 5.9. Grazemont Dairy has a corporate tax rate of 40% and an after-tax MARR of 10%. Using an approximate IRR approach, determine which alternative Grazemont Dairy should choose.

8.10 A machine has a first cost of $45 000 and operating and maintenance costs of $0.22 per unit produced. It will be sold for $4500 at the end of five years. Production is 750 units per day, 250 days per year. The after-tax MARR is 20%, and the corporate income tax rate is 40%. What is the total after-tax annual cost of the machine:

(a) in Australia using straight-line depreciation?

(b) in Australia using declining-balance depreciation?

(c) in the UK?

(d) in the US, assuming a 5-year MACRS property class?

(e) in Canada using a 30% CCA rate?

Questions 8.11–8.25 are specific to Australian tax rules

8.11 Last year, Melbourne Microscopes made seven purchases valued between $300 and $1000 in the following amounts: $980.30, $344.44, $799.05, $454.55, $334.67, $898.44, and $774.35. The book value of this asset pool at the beginning of the year was $4689.90. What was the depreciation expense claim made by Melbourne Microscopes for tax purposes last year, and what was the book value of the asset pool at the beginning of this year?

8.12 Adelaide Construction (AC) has just bought a crane for $380 000. There are several defensible choices for a service life: 5 years with a $120 000 salvage value; 7 years with a $80 000 salvage value, and 10 years with a $30 000 salvage value. Assuming that AC will actually keep the crane for the chosen service life, which service life is the best choice for minimizing the present cost of the crane? Remember that AC also has a choice of using the prime cost method or diminishing value rate. AC has a tax rate of 35% and an after-tax MARR of 6%.

8.13 Cairns Cams acquired an automated lathe on January 5th of last year (not a leap year), with a total first cost of $165 500. The lathe has a 15-year service life, and Cairns Cams has elected to use declining-balance depreciation. What was the actual depreciation expense Cairns Cams claimed for the tax year in which the purchase was made?

8.14 Kalgoorlie Konstruction is considering the purchase of a truck. Over its five-year life, it will provide net revenues of $15 000 per year, at an initial cost of $65 000 and a salvage value of $20 000. KK pays 35% in taxes and its after-tax MARR is 12%. What is the annual cost or worth of this purchase, given that KK uses a diminishing value rate to calculate depreciation?

8.15 Australian Widgets is looking at a $400 000 digital midget rigid widget gadget. It is expected to save $85 000 per year over its 10-year life, with no scrap value. AW's tax rate is 45%, and its after-tax MARR is 15%. On the basis of an exact IRR, and based on depreciation calculated on a straight-line basis, should AW invest in this gadget?

8.16 What is the after-tax present worth of a chip placer if it costs $55 000 and saves $17 000 per year? After-tax interest is at 10%. Assume the device will be sold for a $1000 salvage value at the end of its six-year life. Declining-balance depreciation is used, and the corporate income tax rate is 54%.

8.17 What is the exact after-tax IRR on a project for which the first cost is $12 000, savings are $5000 in the first year and $10 000 in the second year, taxes are at 40%, and depreciation is calculated using the prime cost method? The project has a two-year life.

8.18 Tasmanian Tools is considering the purchase of production equipment with a first cost of $450 000. After its five-year life, due to increasing demand for this kind of equipment, it will be sold for its original purchase price. The equipment will provide a net value of $450 000 per year during the five-year period. Given that depreciation is declining balance, the after-tax MARR is 20%, and the corporate income tax rate is 40%, what is the total after-tax annual cost of the equipment?

8.19 A new binder will cost Revelstoke Printing $17 000, generate net savings of $4000 per year over a nine-year life, and be salvaged for $1000. Revelstoke's *before-tax* MARR is 10%, it is taxed at 40%. What is the company's exact after-tax IRR on this investment using a diminishing value rate for depreciation? Should the investment be made?

The following Australian Custom Metal Products (ACMP) case is used for Problems 8.20 to 8.25. ACMP must purchase a new tube bender. There are three models:

Model	First Cost	Economic Life	Yearly Net Savings	Salvage Value
T	$100 000	5 years	$50 000	$20 000
A	150 000	5 years	60 000	30 000
X	200 000	3 years	75 000	100 000

ACMP's after-tax MARR is 11% and the corporate tax rate is 52%. ACMP uses straight-line depreciation.

8.20 Using the present worth method and the least-cost multiple of the service lives, which tube bender should they buy?

8.21 ACMP realizes that it can forecast demand for its products only three years in advance. The salvage value for model T after three years is $40 000 and for model A, $80 000. Using the present worth method and a three-year study period, which of the three alternatives is now best?

8.22 Using the annual worth method, which tube bender should ACMP buy?

8.23 What is the approximate after-tax IRR for *each* of the tube benders?

8.24 What is the exact after-tax IRR for *each* of the tube benders?

8.25 Using the approximate after-tax IRR comparison method, which of the tube benders should ACMP buy? (*Reminder:* You must look at the incremental investments.)

Questions 8.26–8.40 are specific to UK tax rules

8.26 A chemical recovery system costs £30 000 and saves £5280 each year of its seven-year life. The salvage value is estimated at £7500. The after-tax MARR is 9% and taxes are at 45%. What is the net after-tax annual benefit or cost of purchasing the chemical recovery system?

8.27 Hull Hulls is considering the purchase of a 30-ton hoist. The first cost is expected to be £230 000. Net savings will be £35 000 per year over a 12-year life. It will be salvaged for £30 000. If the company's after-tax MARR is 8% and it is taxed at 45%, what is the present worth of this project? Hull Hulls qualifies for a 50% FYA.

8.28 Salim is considering the purchase of a backhoe for his pipeline contracting firm. The machine will cost £110 000, last six years with a salvage value of £20 000, and reduce annual maintenance, insurance, and labour costs by £30 000 per year. The after-tax MARR is 9%, and Salim's corporate tax rate is 55%. What is the exact after-tax IRR for this investment? What is the approximate after-tax IRR for this investment? Should Salim buy the backhoe?

8.29 CB Electronix needs to expand its capacity. It has two feasible alternatives under consideration. Both alternatives will have essentially infinite lives.

Alternative 1: Construct a new building of 20 000 square metres now. The first cost will be £2 000 000. Annual maintenance costs will be £10 000. In addition, the building will need to be painted every 15 years (starting in 15 years) at a cost of £15 000.

Alternative 2: Construct a new building of 12 500 square metres now and an addition of 7500 square metres in 10 years. The first cost of the 12 500-square-metre building will be £1 250 000. The annual maintenance costs will be £5000 for the first 10 years (i.e., until the addition is built). The 7500-square-metre addition will have a first cost of £1 000 000. Annual maintenance costs of the renovated building (the original building and the addition) will be £11 000. The renovated building will cost £15 000 to repaint every 15 years (starting 15 years after the addition is done).

Given a corporate tax rate of 45%, and an after-tax MARR of 15%, carry out an annual worth comparison of the two alternatives. Which is preferred?

8.30 British Widgets is looking at a £400 000 digital midget rigid widget gadget. It is expected to save £85 000 per year over its 10-year life, with no scrap value. The tax rate is 45%, and their after-tax MARR is 15%. British Widgets is a medium-sized company. On the basis of an exact IRR, should they invest in this gadget?

8.31 What is the after-tax present worth of a chip placer if it costs £55 000 and saves £17 000 per year? After-tax interest is at 10%. Assume the device will be sold for a £1000 salvage value at the end of its six-year life. The corporate income tax rate is 54%.

8.32 What is the exact after-tax IRR on a two-year project for which the first cost is £12 000, savings are £5000 in the first year and £10 000 in the second year, and taxes are at 40%?

8.33 Devon Devices is considering the purchase of production equipment with a first cost of £450 000. After its five-year life, due to increasing demand for this kind of equipment, it will be sold for its original purchase price. The equipment will provide a net value of £450 000 per year during the five-year period. Given that depreciation is declining balance, the after-tax MARR is 20%, and the corporate income tax rate is 40%, what is the total after-tax annual cost of the equipment?

8.34 A new binder will cost Revelstoke Printing £17 000, generate net savings of £4000 per year over a nine-year life, and be salvaged for £1000. Revelstoke's *before-tax* MARR is 10%, they are taxed at 40%. What is their exact after-tax IRR on this investment? Should the investment be made?

The following British Custom Metal Products (BCMP) case is used for Problems 8.35 to 8.40. BCMP must purchase a new tube bender. There are three models:

Model	First Cost	Economic Life	Yearly Net Savings	Value
T	£100 000	5 years	£50 000	£ 20 000
A	150 000	5 years	60 000	30 000
X	200 000	3 years	75 000	100 000

BCMP's after-tax MARR is 11% and the corporate tax rate is 52%.

8.35 Using the present worth method and the least-cost multiple of the service lives, which tube bender should they buy?

8.36 BCMP realizes that it can forecast demand for its products for only three years in advance. The salvage value for model T after three years is $40 000 and for model A, $80 000. Using the present worth method and a three-year study period, which of the three alternatives is now best?

8.37 Using the annual worth method, which tube bender should BCMP buy?

8.38 What is the approximate after-tax IRR for *each* of the tube benders?

8.39 What is the exact after-tax IRR for *each* of the tube benders?

8.40 Using the approximate after-tax IRR comparison method, which of the tube benders should BCMP buy? (*Reminder:* You must look at the incremental investments.)

Questions 8.41–8.54 are specific to US tax rules

8.41 What is the MACRS property class for each of the following?

(a) A new computer

(b) A 100-ton punch press

(c) A crop dusting attachment for a small airplane

(d) A fishing boat

(e) Shelving for the storage of old paper files

8.42 Bob recently purchased some special tools for the manufacture of finished plastic products. The first cost was $289 500, and the net yearly savings were recognized as

$54 000 over the tools' 10-year life, with zero salvage value. When Bob did his cash flow analysis to determine the merits of acquiring the tools, he mistakenly assumed that they were classified as MACRS 7-year property. He has since discovered that they are actually in the 3-year property class. What effect does this error have on the after-tax annual worth of the tools? Bob's company pays 37% tax and has an after-tax MARR of 12%.

8.43 A slitter for sheet sandpaper owned by Abbotsford Abrasives (AA) requires regular maintenance costing $7500 per year. Every five years it is overhauled at a cost of $25 000. The original capital cost was $200 000, with an additional $25 000 in non-capital expenses that occurred at the time of installation. The machine has an expected life of 20 years and a $15 000 salvage value. The machine will not be overhauled at the end of its life. AA pays taxes at a rate of 45% and expects an after-tax rate of return of 10% on investments. Given that the slitter is in the MACRS 7-year property class, what is the after-tax annual cost of the slitter?

8.44 American Widgets is looking at a $400 000 digital midget rigid widget gadget (3-year MACRS property class). It is expected to save $85 000 per year over its 10-year life, with no scrap value. Their tax rate is 45%, and its after-tax MARR is 15%. On the basis of an exact IRR, should they invest in this gadget?

8.45 What is the after-tax present worth of a chip placer if it costs $55 000 and saves $17 000 per year? After-tax interest is at 10%. Assume the device will be sold for a $1000 salvage value at the end of its six-year life. The chip placer is in the 5-year MACRS property class, and the corporate income tax rate is 54%.

8.46 What is the exact after-tax IRR on a two-year project for which the first cost is $12 000, savings are $5000 in the first year and $10 000 in the second year, and taxes are at 40%? It is in the 5-year MACRS property class.

8.47 Mississippi Machines is considering the purchase of production equipment with a first cost of $450 000. After its five-year life, due to increasing demand for this kind of equipment, it will be sold for its original purchase price. The equipment will provide a net value of $450 000 per year during the five-year period. Given that depreciation is declining balance, the after-tax MARR is 20%, and the corporate income tax rate is 40%, what is the total after-tax annual cost of the equipment?

8.48 A new binder will cost Revelstoke Printing $17 000, generate net savings of $4000 per year over a nine-year life, and be salvaged for $1000. Revelstoke's *before-tax* MARR is 10%, it is taxed at 40%, and the binder is in the 7-year MACRS property class. What is the company's exact after-tax IRR on this investment? Should the investment be made?

The following Columbia Custom Metal Products (CCMP) case is used for Questions 8.49–8.54. CCMP must purchase a new tube bender. There are three models:

Model	First Cost	Economic Life	Yearly Net Savings	Salvage Value
T	$100 000	12 years	$50 000	$20 000
A	150 000	12 years	60 000	30 000
X	200 000	8 years	75 000	100 000

CCMP's after-tax MARR is 11% and the corporate tax rate is 52%. A tube bender is a MACRS 7-year property class asset.

8.49 Using the present worth method and the least-cost multiple of the service lives, which tube bender should they buy?

8.50 CCMP realizes that it can forecast demand for its products only eight years in advance. The salvage value for model T after eight years is $40 000 and for model A, $80 000. Using the present worth method and an eight-year study period, which of the three alternatives is now best?

8.51 Using the annual worth method, which tube bender should CCMP buy?

8.52 What is the approximate after-tax IRR for *each* of the tube benders?

8.53 What is the exact after-tax IRR for *each* of the tube benders?

8.54 Using the approximate after-tax IRR comparison method, which of the tube benders should CCMP buy? (*Reminder:* You must look at the incremental investments.)

Questions 8.55–8.70 are specific to Canadian tax rules

8.55 Identify each of the following according to their CCA class(es) and CCA rate(s):

    (a) A soldering gun costing $75

    (b) A garage used to store spare parts

    (c) A new computer

    (d) A 100-ton punch press

    (e) A crop dusting attachment for a small airplane

    (f) An oscilloscope worth exactly $200

8.56 A company's first year's operations can be summarized as follows:

Revenues: $110 000

Expenses (except CCA): $65 000

Its capital asset purchases in the first year totalled $100 000. With a CCA rate of 20% and a tax rate of 55%, how much income tax did the company pay?

8.57 Chrétien Brothers Salvage made several equipment purchases in the 1980s. Their first asset was a tow truck bought in 1982 for $25 000. In 1984, a van was purchased for $14 000. A second tow truck was bought in 1987 for $28 000, and the first one was sold the following year for $5000. What was the balance of CBS's 30% CCA rate (automobiles, trucks, and vans) UCC account at the end of 1989?

8.58 Rodney has discovered that for the last three years his company has been classifying as Class 8, items costing between $100 and $200 that should be in CCA Class 12. If an estimated $10 000 of assets per year were misclassified, what is the present worth today of the cost of this mistake? Assume that the mistake can only be corrected for assets bought in the future. Rodney's company pays taxes at 50% and its after-tax MARR is 9%.

8.59 Refer to Problem 5.9. Grazemont Dairy has a corporate tax rate of 40% and the filling machine for the dairy line has a CCA rate of 30%. The firm has an after-tax MARR of 10%. On the basis of the exact IRR method, determine which alternative Grazemont Dairy should choose.

8.60 Canada Widgets is looking at a $400 000 digital midget rigid widget gadget (CCA Class 8). It is expected to save $85 000 per year over its 10-year life, with no scrap value. CW's tax rate is 45%, and its after-tax MARR is 15%. On the basis of an exact IRR, should the company invest?

8.61 What is the after-tax present worth of a chip placer if it costs $55 000 and saves $17 000 per year? After-tax interest is at 10%. Assume the device will be sold for a $1000 salvage value at the end of its six-year life. The CCA rate is 20%, and the corporate income tax rate is 54%.

8.62 What is the exact after-tax IRR on a project for which the first cost is $12 000, savings are $5000 in the first year and $10 000 in the second year, taxes are at 40%, and the CCA rate is 30%?

8.63 Saskatchewan Systems is considering the purchase of production equipment with a first cost of $450 000. After its five-year life, due to increasing demand for this kind of equipment, it will be sold for its original purchase price. The equipment will provide a net value of $450 000 per year during the five-year period. Given that depreciation is declining balance, the after-tax MARR is 20%, and the corporate income tax rate is 40%, what is the total after-tax annual cost of the equipment?

8.64 A new binder will cost Revelstoke Printing $17 000, generate net savings of $4000 per year over a nine-year life, and be salvaged for $1000. Revelstoke's *before-tax* MARR is 10%, it is taxed at 40%, and the binder has a 20% CCA rate. What is Revelstoke's exact after-tax IRR on this investment? Should the investment be made?

The following Dominion Custom Metal Products (DCMP) case is used for Questions 8.65–8.70. DCMP must purchase a new tube bender. There are three models:

Model	First Cost	Economic Life	Yearly Net Savings	Salvage Value
T	$100 000	5 years	$50 000	$20 000
A	150 000	5 years	60 000	30 000
X	200 000	3 years	75 000	100 000

DCMP's after-tax MARR is 11% and the corporate tax rate is 52%. A tube bender is a CCA Class 8 asset.

8.65 Using the present worth method and the least-cost multiple of the service lives, which tube bender should they buy?

8.66 DCMP realizes that it can forecast demand for its products only three years in advance. The salvage value for model T after three years is $40 000 and for model A, $80 000. Using the present worth method and a three-year study period, which of the three alternatives is now best?

8.67 Using the annual worth method, which tube bender should DCMP buy?

8.68 What is the approximate after-tax IRR for *each* of the tube benders?

8.69 What is the exact after-tax IRR for *each* of the tube benders?

8.70 Using the approximate after-tax IRR comparison method, which of the tube benders should DCMP buy? (*Reminder:* You must look at the incremental investments.)

More Challenging Problem

8.71 Global Widgets will be investing in a new integrated production line for its Mark II widgets. It has a choice of placing the line in its plant in Liverpool, England, or in Mississauga, Ontario, Canada. The costs and savings associated with each installation are shown in the table below:

	Liverpool	Mississauga
First cost	£4 500 000	$6 500 000
Annual savings	1 200 000	2 600 000
Service life	15 years	15 years
Salvage value	500 000	280 000

Global's after-tax MARR is 11%. The UK tax rate is 42% while the Canadian rate is 48%. The production line is a CCA Class 8 investment.

At what exchange rate between the British pound and Canadian dollar does the best location for the plant shift from one country to the other?

MINI-CASE 8.1

Flat Taxes

As discussed in this chapter, personal income taxes in most developed countries are **progressive,** meaning that the rate of taxation increases at higher income levels. However, in several countries a **flat (or, proportional) tax** is used. In its basic form, a flat tax requires that a fixed percentage of income (for individuals) or profit (for corporations) be paid as tax. It is simple, and said to be fairer for everyone, as well as more lucrative for governments.

Following the fall of communism, Russia and most of the former communist Eastern European countries instituted flat tax regimes. These countries have since had strong economic growth, and many economists credit the flat tax as an important factor in this growth. Other countries that have flat tax include Hong Kong, Saudi Arabia, Nigeria, Uruguay, and the Bahamas.

For individuals, a flat tax is very easy to manage. For most workers, no tax return is required since deductions made by employers on their behalf complete their tax compliance. Also there is usually a minimum income before any tax is payable, so that low income earners are protected.

There are several important consequences of a flat tax on businesses. One is that in most implementations profit is taxed only once, as opposed to some countries in which it is taxed at the company level, and then again at the taxpayer level when profits are distributed as dividends or capital gains. Another important consequence for most implementations of a flat tax is that all capital purchases for a company are expensed in the year of purchase. Flat taxes also reduce the administration costs for companies, individuals and governments.

Discussion:

The concept of flat taxes has generated heated debate in some developed countries. The simplicity and efficiency of flat taxes make them attractive alternatives to bloated tax bureaucracies, massively complex tax rules, and frustrating reporting requirements that are the norm. Also, there is evidence that flat taxes promote compliance and generate more tax revenue than progressive taxes. However, there are those who would be adversely affected by a change to a flat tax, and there is considerable resistance generated whenever a serious effort to institute flat tax is made. This is why the only major countries to move to a flat tax in recent years are ones like Russia and the Eastern European countries who are newly instituting income taxes, and consequently are not threatening entrenched interests.

Questions:

1. Which of the following groups would likely welcome a flat tax? Which would resist it?

    (a) Accountants

    (b) Small business owners

    (c) Individuals with investments in the stock market

    (d) Government tax department workers

    (e) Low-wage earners

    (f) Rich people

    (g) Capital equipment manufacturers

    (h) Leasing companies

    (i) Welfare recipients

    (j) You

2. If a country changed from a tax regime in which corporate profits are taxed twice to a regime in which they are taxed only once, what would likely be the effect on the stock market?

3. In the long run, would the ability to expense capital equipment fully in the year of purchase dramatically affect a company's investment decisions? If so, do you think society is better off as a consequence of these effects, or worse off?

Appendix 8A Deriving the Tax Benefit Factor

Under some tax regimes, depreciable assets give rise to depreciation expenses that continue indefinitely. The tax benefit factor permits the present value of this infinite series of benefits to be calculated easily.

The tax benefit that can be obtained for a depreciable asset with a declining balance rate d and a first cost P, when the company is paying tax at rate t is:

Ptd for the first year

$Ptd(1 - d)$ for the second year

$Ptd(1 - d)^{N-1}$ for the Nth year

Taking the present worth of each of these benefits and summing gives

$$\text{PW(benefits)} = Ptd\left(\frac{1}{(1+i)} + \frac{(1-d)}{(1+i)^2} + \ldots + \frac{(1-d)^{N-1}}{(1+i)^N} + \ldots\right)$$

$$= \frac{Ptd}{(1+i)}\left(1 + \frac{(1-d)}{(1+i)} + \frac{(1-d)^2}{(1+i)^2} + \ldots + \frac{(1-d)^N}{(1+i)^N} + \ldots\right)$$

Noting that for $q < 1$

$$\lim_{n\to\infty}(1 + q + q^2 + \ldots + q^n) = \frac{1}{1-q}$$

and

$$\frac{1-d}{1+i} < 1$$

then

$$\text{PW (benefits)} = \frac{Ptd}{1+i}\left[\frac{1}{1 - \dfrac{(1-d)}{(1+i)}}\right]$$

$$= \frac{Ptd}{1+i}\left[\frac{1}{\dfrac{(1+i)}{(1+i)} - \dfrac{(1-d)}{(1+i)}}\right]$$

$$= \frac{Ptd}{1+i}\left(\frac{(1+i)}{(i+d)}\right)$$

$$= P\frac{td}{(i+d)}$$

The factor $\dfrac{td}{(i+d)}$ is the *TBF*.

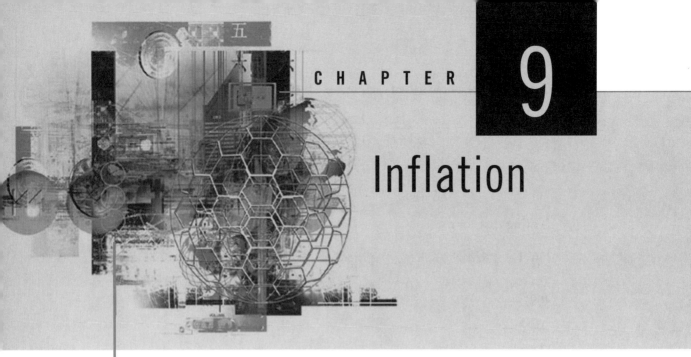

CHAPTER 9

Inflation

Engineering Economics in Action, Part 9A: The Inflated Expert

Review Problems
Summary

Engineering Economics in Action, Part 9B: Exploiting Volatility

Problems

Mini-Case 9.1

Appendix 9A: Computing a Price Index

Engineering Economics in Action, Part 9A:
The Inflated Expert

Terry left Global Widgets to go back for his final term of school. Naomi and Terry had worked through several projects over Terry's last few work terms, and Naomi had been increasingly taking part in projects involving sister companies of Global Widgets; all were owned by Global Conglomerate Inc., often referred to as "head office."

"There's a guy from head office to see you, Naomi." It was Carole announcing the expected visitor, Bill Astad. Bill was one of the company's troubleshooters. His current interest concerned a sister company, Mexifab, a maquiladora on the Mexican border with Texas. (A maquiladora is an assembly plant that manufactures finished goods in northern Mexico under special tariff and tax rules.) After a few minutes of socializing, Bill explained the concern.

"It's the variability in the Mexican inflation rate that causes the problems. Mexico gets a new president every six years, and usually, about the time the president changes, the economy goes out of whack. And we can't price everything in dollars or euros. We do some of that, but we are located in Mexico and so we have to use Mexican pesos for a lot of our transactions.

"I understand from Anna Kulkowski that you know something about how to treat problems like that," Bill continued.

Naomi smiled to herself. She had written a memo a few weeks earlier pointing out how Global Widgets had been missing some good projects by failing to take advantage of the current very low inflation rates, and suddenly she was the expert!

"Well," she said, "I might be able to help. What you can do is this."

9.1 | Introduction

Prices of goods and services bought and sold by individuals and firms change over time. Some prices, like those of agricultural commodities, may change several times a day. Other prices, like those for sugar and paper, change infrequently. While prices for some consumer goods and services occasionally decrease (as with high-tech products), on average it is more typical for prices to increase over time.

Inflation is the increase, over time, in average prices of goods and services. It can also be described as a decrease in the purchasing power of money over time. While most developed countries have experienced inflation in most years since World War II, there have been short periods when average prices in some countries have fallen. A decrease, over time, in average prices is called **deflation**. It can also be viewed as an increase in the purchasing power of money over time.

Because of inflation or deflation, prices are likely to change over the lives of most engineering projects. These changes will affect cash flows associated with the projects. Engineers may have to take predicted price changes into account during project evaluation to prevent the changes from distorting decisions.

In this chapter, we shall discuss how to incorporate an expectation of inflation into project evaluation. We focus on inflation because it has been the dominant pattern of price changes since the beginning of the twentieth century. The chapter begins with a discussion of how inflation is measured. We then show how to convert cash flows that occur at different points in time into dollars with the same purchasing power. We then consider how inflation affects the MARR, the internal rate of return, and the present worth of a project.

9.2 | Measuring the Inflation Rate

The **inflation rate** is the rate of increase in average prices of goods and services over a specified time period, usually a year. If prices of all goods and services moved up and down together, determining the inflation rate would be trivial. If all prices increased by 2% over a year, it would be clear that the average inflation rate would also be 2%. But prices do not move in perfect synchronization. In any period, some prices will increase, others will fall, and some will remain about the same. For example, candy bars are about ten times as costly now as they were in the 1960s, but some electronics are about the same price or cheaper.

Because prices do not move in perfect synchronization, a variety of methods have been developed to measure the inflation rate. Most countries' governments track movement of average prices for a number of different collections of goods and services and calculate inflation rates from the changes in prices in these collections over time.

One set of prices typically tracked consists of goods and services bought by consumers. This set forms the basis of the **consumer price index (CPI)**. The CPI for a given period relates the average price of a fixed "basket" of these goods in the given period to the average price of the same basket in a *base period*. For example, Figure 9.1 shows the CPI for the United States, the United Kingdom, Japan, and Australia for the period 1986 to 2006. The CPI for this chart has the year 2000 as the base year. The base year index is set at 100.

N E T   V A L U E   9 . 1

Government Collection of Statistics

A prime source of statistics is data collected by national organizations. In addition to CPI and inflation rates, different countries report various additional statistics that may be useful for engineering economics studies. These might include gross domestic product (GDP), personal spending on consumer goods and services, manufacturing and construction figures, and statistics on population, labour, and employment.

Below are some governmental statistics collection websites.

Australia: **www.abs.gov.au/ausstats**

Canada: **www.statcan.ca/english**

China: **www.stats.gov.cn/english**

Europe: **www.epp.eurostat.ec.europa.eu**

India: **www.labourbureau.nic.in**

Japan: **www.stat.go.jp/english**

United Kingdom: **www.statistics.gov.uk**

United States: **www.stats.bls.gov**

The index for any other year indicates the number of dollars (or other currency) needed in that year to buy the fixed basket of goods that cost $100 in 2000. For example, Figure 9.1 shows that a basket of goods that cost $100 in the United States in 2000 (the base year) would have cost about $65 in 1986. In the United Kingdom, a basket costing £100 in 2000 would have similarly cost about £65 in 1986. $100 in value in Australia in 2000 would cost about $120 in 2006, while a ¥100 basket of goods in the year 2000 in Japan could be purchased for only about ¥98 in 2006.

A national inflation rate can be estimated directly from the CPI by expressing the changes in the CPI as a year-by-year percentage change. This is probably the most commonly used estimate of a national inflation rate. Figure 9.2 shows the national inflation rate for the period from 1986 to 2006 as derived from the CPI quantities in Figure 9.1.

It is important to note that, although the CPI is a commonly accepted inflation index, many different indexes are used to measure inflation. The value of an index depends on

Figure 9.1   CPI for Four Selected Countries 1986–2006

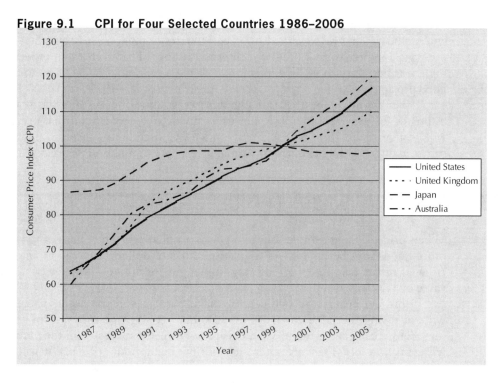

the method used to compute the index and the set of goods and services for which the index measures price changes. To judge whether an index is appropriate for a particular purpose, the analyst should know how the goods and services for which he or she is estimating inflation compare with the set of goods and services used to compute the index. For this reason, we provide Appendix 9A, in which we illustrate the computation of one popularly used index.

Figure 9.2   Inflation Rates for Four Selected Countries 1986–2006

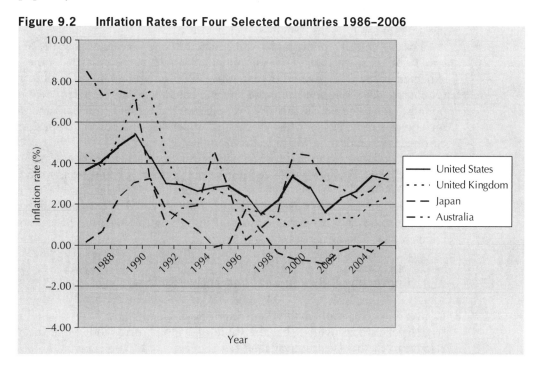

CPI values and inflation rates vary considerably over time, as seen in Figures 9.1 and 9.2, and can be even more extreme. Low expected rates of inflation may be safely ignored, given the typical imprecision of predicted future cash flows. However, when expected inflation is high, it is necessary to include inflation in detailed economic calculations to avoid rejecting good projects.

Throughout the rest of this chapter, we assume that an analyst is able to obtain estimates for expected inflation rates over the life of a project and that project cash flows will change at the same rate as average prices. Consequently, the cash flows for a project can be assumed to increase at approximately the rate of inflation per year.

9.3 | Economic Evaluation With Inflation

When prices change, the amount of goods a dollar will buy changes too. If prices fall, more goods may be bought for a given number of dollars, and the value of a dollar has risen. If prices rise, fewer goods may be bought for a given number of dollars, and the value of a dollar has fallen.

In project evaluation, we cannot make comparisons of dollar values across time without taking the price changes into account. We want dollars, not for themselves, but for what we can get for them. Workers are not directly interested in the money wages they will earn in a job. They are interested in how many hours of work it will take to cover expenses for their families, or how long it will take them to accumulate enough to make down payments on houses. Similarly, investors want to know if making an investment now will enable them to buy more real goods in the future, and by how much the amount they can buy in the future will increase. To know if an investment will lead to an increase in the amount they can buy in the future, they must take into account expected price changes.

We can take price changes into account in an approximate way by measuring the cash flows associated with a project in monetary units of constant purchasing power called **real dollars** (sometimes called **constant dollars**). This is in contrast to the **actual dollars** (sometimes called **current dollars** or **nominal dollars**), which are expressed in the monetary units at the time the cash flows occur.

For example, if a photocopier will cost $2200 one year from now, the $2200 represents *actual* dollars since that is the amount that would be paid at that time. If inflation is expected to be 10% over the year, the $2200 is equivalent to $2000 in real dollars. Of course, depending on the currency, one could also use the term "real euros" or "real pounds," for example. In this text we will use the terms *real dollars* and *actual dollars* to include other currencies when discussing principles, and revert to speaking about other currencies when dealing with specific examples.

Real dollars always need to be associated with a particular date (usually a year), called the **base year**. The base year need not be the present; it could be any time. People speak of "2000 dollars" or "1985 dollars" to indicate that real dollars are being used as well as indicating the base year associated with them. Provided that cash flows occurring at different times are converted into real dollars with the same base year, they can be compared fairly in terms of buying power.

9.3.1 Converting Between Real and Actual Dollars

Converting actual dollars in year N into real dollars in year N relative to a base year 0 is straightforward, provided that the value of a global price index like the CPI at year N relative to the base year is available. Let

A_N = actual dollars in year N

$R_{0,N}$ = real dollars equivalent to A_N relative to year 0, the base year

$I_{0,N}$ = the value of a global price index (like the CPI) at year N, relative to year 0

Then the conversion from actual to real dollars is

$$R_{0,N} = \frac{A_N}{I_{0,N}/100} \tag{9.1}$$

Note that in Equation (9.1), $I_{0,N}$ is divided by 100 to convert it into a fraction because of the convention that a price index is set at 100 for the base year.

Transforming actual dollar values into real dollars gives only an approximate offset to the effect of inflation. The reason is that there may be no readily available price index that accurately matches the "basket" of goods and services being evaluated. Despite the fact that available price indexes are approximate, they do provide a reasonable means of converting actual cash flows to real cash flows relative to a base year.

An alternative means of converting actual dollars to real dollars is available if we have an estimate for the average yearly inflation between now (year 0) and year N. Let

A_N = actual dollars in year N

$R_{0,N}$ = real dollars equivalent to A_N relative to year 0, the base year

f = the inflation rate per year, assumed to be constant from year 0 to year N

Then the conversion from actual dollars in year N to real dollars in year N relative to the base year 0 is

$$R_{0,N} = \frac{A_N}{(1 + f)^N}$$

When the base year is omitted from the notation for real dollars, it is understood that the current year (year 0) is the base year, as in

$$R_N = \frac{A_N}{(1 + f)^N} \tag{9.2}$$

Equation (9.2) can also conveniently be written in terms of the present worth compound interest factor

$$R_N = A_N(P/F,f,N) \tag{9.3}$$

Note that here A_N is the actual dollar amount in year N, that is, a future value. It should not be confused with an annuity A.

EXAMPLE 9.1

Elliot Weisgerber's income rose from $40 000 per year in 2004 to $42 000 in 2007. At the same time the CPI (base year 1992) in his country rose from 113.5 in 2004 to 122.3 in 2007. Was Elliot worse off or better off in 2007 compared with 2004?

We can convert Elliot's actual dollar income in 2004 and 2007 into real dollars. This will tell us if his total purchasing power increased or decreased over the period from 2004 to 2007. Since the base year for the CPI is 1992, we shall compare his 2004 and 2007 incomes in terms of real 1992 dollars.

His real incomes in 2004 and 2007 in terms of 1992 dollars, using Equation (9.1), were

$R_{92,04} = 40\ 000/1.135 = 35\ 242$

$R_{92,07} = 42\ 000/1.223 = 34\ 342$

Even though Elliot's actual dollar income rose between 2004 and 2007, his purchasing power fell, since the real dollar value of his income, according to the CPI, fell about 3%.■

EXAMPLE 9.2

The cost of replacing a storage tank one year from now is expected to be €2 000 000. If inflation is assumed to be 5% per year, what is the cost of replacing the storage tank in real (today's) euros?

First, note that the €2 000 000 is expressed in actual euros one year from today. The cost of replacing the tank in real (today's) euros can be found by letting

$A_1 = 2\ 000\ 000$ = the actual cost one year from the base year (today)

R_1 = the real euro cost of the storage tank in one year

f = the inflation rate per year

Then, with Equation (9.2)

$$R_1 = \frac{A_1}{1 + f} = \frac{2\ 000\ 000}{1.05} = €1\ 904\ 762$$

Alternatively, Equation (9.3) gives

$R_1 = A_1(P/F,5\%,1) = 2\ 000\ 000\ (0.9524) = 1\ 904\ 762$

The €2 000 000 actual cost is equivalent to €1 904 762 real (today's) euros at the end of one year.■

EXAMPLE 9.3

The cost of replacing a storage tank in 15 years is expected to be €2 000 000. If inflation is assumed to be 5% per year, what is the cost of replacing the storage tank 15 years from now in real (today's) euros?

The cost of the tank 15 years from now in real dollars can be found by letting

$A_{15} = 2\ 000\ 000$ = the actual cost 15 years from the base year (today)

R_{15} = the real euro cost of the storage tank in 15 years

f = the inflation rate per year

Then, with the use of Equation (9.2), we have

$$R_{15} = \frac{A_{15}}{(1 + f)^{15}} = \frac{2\ 000\ 000}{(1.05)^{15}} = 962\ 040$$

Alternatively, Equation (9.3) gives

$R_{15} = A_{15}(P/F,5\%,15) = 2\ 000\ 000\ (0.48102) = 962\ 040$

In 15 years, the storage tank will cost €962 040 in real (today's) euros. Note that this €962 040 is money to be paid 15 years from now. What this means is that the new storage tank can be replaced at a cost that would have the same purchasing power as about €962 040 today.∎

Now that we have the ability to convert from actual to real dollars using an index or an inflation rate, we turn to the question of how inflation affects project evaluation.

9.4 | The Effect of Correctly Anticipated Inflation

The main observation made in this section is that engineers must be aware of potential changes in price levels over the life of a project. We shall see that when future inflation is expected over the life of a project, the MARR needs to be increased. Engineers need to recognize this effect of inflation on the MARR to avoid rejecting good projects.

9.4.1 The Effect of Inflation on the MARR

If we expect inflation, the number of actual dollars that will be returned in the future does not tell us the value, in terms of purchasing power, of the future cash flow. The purchasing power of the earnings from an investment depends on the *real* dollar value of those earnings.

The **actual interest rate** is the stated or observed interest rate based on actual dollars. If we wish to earn interest at the actual interest rate, i, on a one-year investment, and we invest $\$M$, the investment will yield $\$M(1 + i)$ at the end of the year. If the inflation rate over the next year is f, the real value of our cash flow is $\$M(1 + i)/(1 + f)$. We can use this to define the *real* interest rate. The **real interest rate**, i', is the interest rate that would yield the same number of real dollars in the absence of inflation as the actual interest rate yields in the presence of inflation.

$$M(1 + i') = M\left(\frac{1 + i}{1 + f}\right)$$
$$i' = \frac{1 + i}{1 + f} - 1 \tag{9.4}$$

We may see terms like *real rate of return* or *real discount rate*. These are just special cases of the real interest rate.

The definition of the real interest rate can be turned around by asking the following question: If an investor wants a real rate of return, i', over the next year, and the inflation rate is expected to be f, what actual interest rate, i, must be realized to get a real rate of return of i'?

The answer can be obtained with some manipulation of the definition of the real interest rate in Equation (9.4):

$$i = (1 + i')(1 + f) - 1 \text{ or, equivalently, } i = i' + f + i'f \tag{9.5}$$

Therefore, an investor who desires a real rate of return i' and who expects inflation at a rate of f will require an actual interest rate $i = i' + f + i'f$. This has implications for the actual MARR used in economic analyses. The **actual MARR** is the minimum acceptable rate of return when cash flows are expressed in actual dollars. If investors expect inflation, they require higher actual rates of return on their investments than if inflation were not expected. The actual MARR then will be the real MARR plus an upward adjustment that reflects the effect of inflation. The **real MARR** is the minimum acceptable rate of return when cash flows are expressed in real, or constant, dollars.

If we denote the actual MARR by $MARR_A$ and the real MARR by $MARR_R$, we have from Equation (9.5)

$$MARR_A = MARR_R + f + MARR_R \times f \qquad (9.6)$$

Note that if $MARR_R$ and f are small, the term $MARR_R \times f$ may be ignored and $MARR_A = MARR_R + f$ can be used as a "back of the envelope" approximation.

The real MARR can also be expressed as a function of the actual MARR and the expected inflation rate:

$$MARR_R = \frac{1 + MARR_A}{1 + f} - 1 \qquad (9.7)$$

Figure 9.3 shows the Canadian experience with inflation, the actual prime interest rate, and the real interest rate for the 1961–2003 period. From 1961 to 1971, when inflation was moderate and stable, the real interest rate was also stable, except for one blip in 1967. In the 1970s, conditions were very different when inflation exploded. This was due partly to large jumps in energy prices. Real interest rates were negative for the period 1972 to 1975 and 1977 to 1978. In the 1980s and early 1990s, real interest rates were quite high. The rest of the 1990s and early 2000s experienced lower inflation rates and real interest rates.

The high inflation rates of the 1970s were very unusual. Inflation in the range of 2% to 4% per year is more typical of developed countries like Canada. Averages of real interest rates and actual inflation rates over subperiods are shown in Table 9.1.

Figure 9.3   Canadian Inflation Rate and Actual and Real Interest Rates 1961–2003

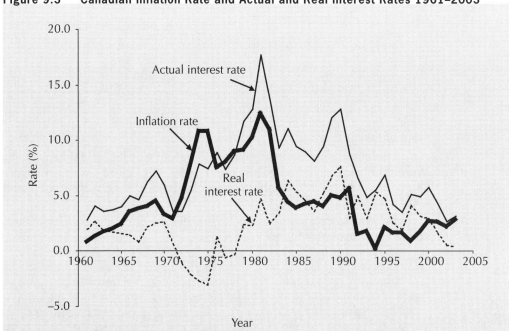

Table 9.1   Average Canadian Real and Actual Interest Rates

Period	Average Real Interest Rate (%)	Average Actual Interest Rate (%)
1961–1971	1.82	2.75
1972–1981	0.08	8.47
1982–1991	4.84	5.29
1992–2003	2.95	1.81

EXAMPLE 9.4

Security Trust is paying 12% on one-year guaranteed investment certificates (GICs). The inflation rate is expected to be 5% over the next year. What is the real rate of interest? For a $5000 GIC, what will be the real dollar value of the amount received at the end of the year?

The real interest rate is

$$i' = \frac{1 + i}{1 + f} - 1 = \frac{1.12}{1.05} - 1 = 0.067, \text{ or } 6.7\%$$

A $5000 GIC will return $5600 at the end of the year. The real value of the $5600 in *today's* dollars is $5600/1.05 = $5333. This is the same as if there were no inflation and the investment earned 6.7% interest.■

EXAMPLE 9.5

Susan got a $1000 present from her aunt on her 16th birthday. She has noticed that Security Trust offers 6.5% on one-year guaranteed investment certificates (GICs). Her mother's business newspaper indicates that analysts are predicting an inflation rate of about 3.5% for the coming year. Susan's real MARR for such investments is 4%. If the analysts are correct, what is Susan's actual MARR? Should she invest?

If the analysts are correct, Susan's actual MARR is

$$\begin{aligned}
\text{MARR}_A &= \text{MARR}_R + f + \text{MARR}_R \times f \\
&= 0.04 + 0.035 + (0.04)(0.035) \\
&= 0.0764
\end{aligned}$$

Susan's actual MARR is about 7.64%. Since the actual interest rate on the GIC is only 6.5%, she should not invest in the GIC.■

9.4.2 The Effect of Inflation on the IRR

The effect of expected inflation on the actual internal rate of return of a project is similar to the effect of inflation on the actual MARR. Suppose that we are considering a T-year investment. The **actual internal rate of return** on the project, IRR_A, is the rate of return of the project on the basis of actual dollar cash flows. It can be found by solving for i^* in

$$\sum_{t=0}^{T} \frac{A_t}{(1 + i^*)^t} = 0$$

where

A_t = the actual cash flow in period t (receipts – disbursements)
T = the number of time periods
i^* = the actual internal rate of return

Suppose further that a yearly inflation rate of f is expected over the T-year life of the project. In terms of real dollars (with a base year of the time of the first cost), the actual cash flow in period t can be written as $A_t = R_t(1 + f)^t$ where R_t refers to the *real* dollar amount equivalent to the cash flow A_t. The expression that gives the actual internal rate of return can be rewritten as

$$\sum_{t=0}^{T} \frac{R_t(1 + f)^t}{(1 + i^*)^t} = 0 \tag{9.8}$$

In contrast, the **real internal rate of return** for the project, IRR_R, is the rate of return obtained on the real dollar cash flows associated with the project. It is the solution for i' in

$$\sum_{t=0}^{T} \frac{R_t}{(1 + i')^t} = 0 \tag{9.9}$$

What is the relationship between IRR_R and IRR_A? We have from Equations (9.4) and (9.5)

$$\frac{1}{1 + i'} = \frac{1 + f}{1 + i} \quad \text{or} \quad i = i' + f + i'f$$

and thus, analogous to Equation (9.5),

$$IRR_A = IRR_R + f + IRR_R \times f \tag{9.10}$$

Or, analogous to Equation (9.4), the real IRR can be expressed in terms of the actual IRR and the inflation rate:

$$IRR_R = \frac{1 + IRR_A}{1 + f} - 1 \tag{9.11}$$

In summary, the effect of inflation on the IRR is that the actual IRR will be the real IRR plus an upward adjustment that reflects the effect of inflation.

EXAMPLE 9.6

Consider a two-year project that has a 10 000 000 rupee first cost and that is expected to bring about a saving of 15 000 000 rupees at the end of the two years. If inflation is expected to be 5% per year and the real MARR is 13%, should the project be undertaken? Base your answer on an IRR analysis.

From the information given, $A_0 = -10\,000\,000$, $A_2 = 15\,000\,000$, and $f = 0.05$. The actual IRR can be found by solving for i in

$$A_0 + \frac{A_2}{(1 + i)^2} = 0$$

$$10\,000\,000 = 15\,000\,000/(1 + i)^2$$

which leads to an actual IRR of 22.475%.

The real IRR is then

$$\begin{aligned}
IRR_R &= \frac{1 + IRR_A}{1 + f} - 1 \\
&= \frac{1 + 0.22475}{1 + 0.05} - 1 \\
&= 0.1664 \text{ or } 16.64\%
\end{aligned}$$

Since the real IRR exceeds the real MARR, the project should be undertaken.∎

In conclusion, the impact of inflation on the actual MARR and the actual IRR is that both have an adjustment for expected inflation implicitly included in them. The main implication of this observation is that, since both the actual MARR and the actual IRR increase in the same fashion, any project that was acceptable without inflation remains acceptable when inflation is expected. Any project that was unacceptable remains unacceptable.

9.5 | Project Evaluation Methods With Inflation

The engineer typically starts a project evaluation with an observed (actual) MARR and projections of cash flows. As we have seen, the actual MARR has two parts: the real rate of return on investment that investors require to put money into the company, plus an adjustment for the expected rate of inflation. The engineer usually observes only the sum and not the individual parts.

As for the projected cash flows, these are typically based on current prices. Because the projected cash flows are based on prices of the period in which evaluations are being carried out, they are in *real* dollars. They do not incorporate the effect of inflation. In this case, the challenge for the engineer is to correctly analyze the project when he or she has an *actual* MARR (which incorporates inflation implicitly) and *real* cash flows (which do not take inflation into account).

Though it is common to do so, the engineer or analyst does not always start out with an *actual* MARR and *real* cash flows. The cash flows may already have inflation implicitly factored in (in which case the cash flows are said to be actual amounts). To carry out a project evaluation properly, the analyst must know whether inflation has been accounted for already in the MARR and the cash flows or whether it needs to be dealt with explicitly.

As a brief example, consider a one-year project that requires an investment of $1000 today and that yields $1200 in one year. The actual MARR is 25%. Whether the project is considered acceptable will depend on whether the $1200 in one year is understood to be the actual cash flow in one year or if it is the real value of the cash flow received in one year.

If the $1200 is taken to be the actual cash flow, the actual internal rate of return, IRR_A, is found by solving for i^* in

$$-1000 + \frac{1200}{(1 + i^*)} = 0$$

$$i^* = IRR_A = 20\%$$

Hence, the project would not be considered economical. However, if the $1200 is taken to be the real value of the cash flow in one year, and inflation is expected to be 5% over the year, then the actual internal rate of return is found by solving for i^* in:

$$-1000 + \frac{1200\,(1 + 0.05)}{(1 + i^*)} = 0$$

$$i^* = \text{IRR}_A = 26\%$$

Hence, the project *would* be considered acceptable. As seen in this example, the economic viability of the project may depend on whether the $1200 in one year has inflation implicitly factored in (i.e., is taken to be the actual amount). This is why it is important to know what type of cash flows you are dealing with.

If the engineer has an estimate of inflation, there are two equivalent ways to carry out a project evaluation properly. The first is to work with actual values for cash flows and actual interest rates. The second is to work with real values for cash flows and real interest rates. *The two methods should not be mixed.*

These two methods of dealing with expected inflation, as well as two incorrect methods, are shown in Table 9.2.

The engineer must have a forecast of the inflation rate over the life of the project in order to adjust the MARR or cash flows for inflation. The best source of such forecasts may be the estimates of experts. Financial publications regularly report such predictions for relatively short periods of up to one year. Because there is evidence that even the short-term estimates are not totally reliable, and estimates for longer periods will necessarily be imprecise, it is good practice to determine a range of possible inflation values for both long- and short-term projects. The engineer should test for sensitivity of the decision to values in the range. The subject of sensitivity analysis is addressed more fully in Chapter 11. Close-Up 9.1 discusses atypical patterns of price changes that may be specific to certain industries.

Table 9.2   Methods of Incorporating Inflation Into Project Evaluation

1. Real MARR and Real Cash Flows
The real MARR does not include the effect of expected inflation.
Cash flows are determined by today's prices.
Correct

2. Actual MARR and Actual Cash Flows
The actual MARR includes the effect of anticipated inflation.
Cash flows include increases due to inflation.
Correct

3. Actual MARR and Real Cash Flows
The actual MARR includes the effect of anticipated inflation.
Cash flows are determined by today's prices.
Incorrect: Biased against investments

4. Real MARR and Actual Cash Flows
The real MARR does not include the effect of expected inflation.
Cash flows include increases due to inflation.
Incorrect: Biased in favour of investments

| CLOSE-UP 9.1 | Relative Price Changes |

Engineers usually expect prices associated with a project to move with the general inflation rate. However, there are situations in which it makes sense to expect prices associated with a project to move differently from the average. This can happen when there are atypical forces affecting either the supply of or the demand for the goods.

For example, reductions in the availability of logs in North America caused a decrease in the supply of wood for construction, furniture, and pulp and paper in the 1980s. Average wood prices more than doubled between 1986 and 1995. This was about twice the increase in the CPI over that period. Since then, average wood prices have been decreasing while the CPI continues to increase. Another example is the price of computers. Product development and increases in productivity have led to increases in the supply of computers. This, in turn, has led to reductions in the relative price of computing power.

Changes in the relative prices of the goods sold by a specific industry will generally not have a noticeable effect on a MARR because investors are concerned with the overall purchasing power of the dollars they receive from an investment. Changes in the relative prices of the goods of one industry will not have much effect on investors' abilities to buy what they want.

Because the relative price changes will not affect the MARR, the analyst must incorporate expected relative price changes directly into the expected cash flows associated with a project. If the rate of relative price change is expected to be constant over the life of the project, this can be done using a geometric gradient to present worth conversion factor.

EXAMPLE 9.7

Jagdeep can put his money into an investment that will pay him £1000 a year for the next four years and £10 000 at the end of the fifth year. Inflation is expected to be 5% over the next five years. Jagdeep's real MARR is 8%. What is the present worth of this investment?

The present worth may be obtained with real pound cash flows and a real MARR or with actual pound cash flows and an actual MARR.

The first solution approach will be to use real pounds and MARR_R. The real pound cash flows in terms of *today's* pounds are

$$R_1, R_2, R_3, R_4, R_5 = \frac{A_1}{(1+f)}, \frac{A_2}{(1+f)^2}, \frac{A_3}{(1+f)^3}, \frac{A_4}{(1+f)^4}, \frac{A_5}{(1+f)^5}$$

$$= \frac{1000}{(1.05)}, \frac{1000}{(1.05)^2}, \frac{1000}{(1.05)^3}, \frac{1000}{(1.05)^4}, \frac{10\ 000}{(1.05)^5}$$

The present worth of the real cash flows, discounted by $\text{MARR}_R = 8\%$, is

$$PW = \frac{1000}{(1.05)(1.08)} + \frac{1000}{(1.05)^2(1.08)^2} + \frac{1000}{(1.05)^3(1.08)^3}$$

$$+ \frac{1000}{(1.05)^4(1.08)^4} + \frac{10\ 000}{(1.05)^5(1.08)^5}$$

$$\cong 8282$$

The present worth of Jagdeep's investment is about £8282.

Alternatively, the present worth can be found in terms of actual pounds and $MARR_A$:

$$PW = \frac{1000}{(1 + MARR_A)} + \frac{1000}{(1 + MARR_A)^2} + \frac{1000}{(1 + MARR_A)^3}$$

$$+ \frac{1000}{(1 + MARR_A)^4} + \frac{10\,000}{(1 + MARR_A)^5}$$

where

$$MARR_A = MARR_R + f + MARR_R \times f$$

Note that this is the sum of a four-period annuity with equal payments of £1000 for four years and a single payment of £10 000 in period 5. With this observation, the present worth computation can be simplified by the use of compound interest formulas:

$$PW = 1000(P/A, MARR_A, 4) + 10\,000(P/F, MARR_A, 5)$$

With a real MARR of 8% and an inflation rate of 5%, the actual MARR is then

$$MARR_A = MARR_R + f + MARR_R \times f$$

$$= 0.08 + 0.05 + (0.08)(0.05)$$

$$= 0.134$$

and the present worth of Jagdeep's investment is

$$PW = 1000(P/A, 13.4\%, 4) + 10\,000(P/F, 13.4\%, 5)$$

$$\cong 8282$$

The present worth of Jagdeep's investment is about £8282, as was obtained through the use of the real MARR and a conversion from actual to real pounds. ∎

Though there are two distinct means of correctly adjusting for inflation in project analysis, the norm for engineering analysis is to make comparisons with the actual MARR. One reason this is done has to do with how a MARR is chosen. As discussed in Chapter 3, the MARR is based on, among other things, the cost of capital. Since lenders and investors recognize the need to have a return on their investments higher than the expected inflation rate, they will lend to or invest in companies only at a rate that exceeds the inflation rate. In other words, the cost of capital already has inflation included. A MARR based on this cost of capital already includes, to some extent, inflation.

Consequently, if inflation is fairly static (even if it is high), an *actual* dollar MARR is sensible and will arise naturally. On the other hand, if changes in inflation are foreseen, or if sensitivity analysis specifically for inflation is desired, it may be wise to set a *real* dollar MARR and recognize an inflation rate explicitly in the analysis.

EXAMPLE 9.8

Lethbridge Communications is considering an investment in plastic moulding equipment for its product casings. The project involves $150 000 in first costs and is expected to generate net savings (in actual dollars) of $65 000 per year over its three-year life. They forecast an inflation rate of 15% over the next year, and then inflation of 10% thereafter. Their real dollar MARR is 5%. Should this project be accepted on the basis of an IRR analysis?

In this problem, the inflation rate is not constant over the life of the project, so it is easiest to consider the cash flows for each year separately and to work in real dollars. First, as shown in Table 9.3, the actual cash flows are converted into real cash flows.

Table 9.3   Converting From Actual to Real Dollars for Lethbridge Communications

Year	Actual Dollars	Real Dollars	
0	−$150 000	−$150 000	
1	65 000	56 522	= 65 000(P/F,15%,1) = 65 000(0.86957)
2	65 000	51 384	= 65 000(P/F,15%,1)(P/F,10%,1) = 65 000(0.86957)(0.90909)
3	65 000	46 713	= 65 000(P/F,15%,1)(P/F,10%,2) = 65 000(0.86957)(0.82645)

Then, the real IRR can be found by solving for i' in

$$56\ 522(P/F,i',1) + 51\ 384(P/F,i',2) + 46\ 713(P/F,i',3) = 150\ 000$$

At $i' = 1\%$, LHS (left-hand side) = 151 673

At $i' = 2\%$, LHS = 148 821

The real IRR is between 1% and 2%. This is less than the real dollar MARR of 5%, so the project should not be undertaken. ∎

EXAMPLE 9.9

Glasgow Resources has been offered a contract to sell land to the government at the end of 20 years. The contract states that Glasgow will get £500 000 after 20 years from today, with no costs or benefits in the intervening years. A financial analyst for the firm believes that the inflation rate will be 4% for the next two years, rise to 15% for the succeeding 10 years, and then go down to 10%, where it will stay forever. Glasgow's real pound MARR is 10%. What is the present worth of the contract?

In this case, it is easiest to proceed by calculating the actual pound MARR for each of the different inflation periods:

$$\text{MARR}_A, \text{ years 13 to 20} = 0.10 + 0.10 + (0.10)(0.10)$$
$$= 0.21 \text{ or } 21\%$$

$$\text{MARR}_A, \text{ years 3 to 12} = 0.10 + 0.15 + (0.10)(0.15)$$
$$= 0.265 \text{ or } 26.5\%$$

$$\text{MARR}_A, \text{ years 0 to 2} = 0.10 + 0.04 + (0.10)(0.04)$$
$$= 0.144 \text{ or } 14.4\%$$

With the individual MARRs, the present worth of the £500 000 for each of years 12, 2, and 0 can be found:

$$\text{PW(year 12)} = 500\ 000(P/F,21\%,8) = 500\ 000 \times 1/(1.21)^8 = 108\ 815$$

$$PW(\text{year 2}) = 108\ 815\ (P/F,26.5\%,10) = 108\ 815 \times 1/(1.265)^{10} = 10\ 370$$

$$PW(\text{year 0}) = 10\ 370\ (P/F,14.4\%,2) = 10\ 370 \times 1/(1.144)^2 = 7924$$

The present worth of the contract is approximately £7924. ∎

EXAMPLE 9.10

Bildmet is an extruder of aluminum shapes used in construction. It is experiencing a high scrap rate of 5%. The manager, Greta Kehl, estimates that reprocessing scrap costs about $0.30 per kilogram. The high scrap rate is due partly to operator error. Ms. Kehl believes that a short training course for the operator would reduce the scrap rate to about 4%. The course would cost about $1100. Bildmet is now working with a before-tax actual MARR of 22%. Past experience suggests that operators quit their jobs after about five years; the correct time horizon for the retraining project is therefore five years. The data pertaining to the training course are summarized in Table 9.4. Should Bildmet retrain its operator?

Table 9.4 Training Course Data

Output (kilograms/year)	125 000
Scrap (kilograms/year)	6250
Reprocessing cost ($/kilogram)	0.30
Scrap cost ($/year)	1875
Savings due to training ($/year)	375
First cost of training ($)	1100
Inflation rate (%/year)	5
Actual MARR (%/year)	22

First, note that the actual MARR $i = 22\%$ incorporates an estimate by investors of inflation of $f = 5\%$ per year over the next five years. If this estimate of future inflation is correct, Ms. Kehl needs to make an adjustment to take inflation into account. Either the projected annual saving from reduced scrap needs to be increased by the 5% rate of inflation, or she needs to reduce the MARR to its real value. We shall illustrate the first approach with actual cash flows and the actual MARR.

Increasing savings to take inflation into account leads to projected (actual) savings as shown in Table 9.5.

For example, using Equation (9.2), the expected saving in year 3 is $375(1 + f)^3 = 434.11. The present worth of the savings in year 3 is $434.11/(1 + i)^3 = 239.07.

The present worth of the savings over the five-year time frame, when discounted at the actual MARR of 22%, is $1222. This makes the project viable since its cost is $1100.

We note that the same result could have been reached by working with the real MARR and the constant cost savings of $375 per year. MARR_R is given by

$$\text{MARR}_R = 1 + \frac{\text{MARR}_A}{1 + f} - 1 = \frac{1.22}{1.05} - 1 = 0.1619$$

Table 9.5   Savings Due to the Training Course

Year	Actual Savings	Present Worth
1	$393.75	$322.75
2	413.44	277.77
3	434.11	239.07
4	455.81	205.75
5	478.61	177.08

The present worth of the real stream of returns, when these are discounted by the real MARR, is given by

$$PW = 375(P/A,0.1619,5) \cong 1222$$

which is the same result as the one obtained with the actual MARR and actual cash flows.∎

REVIEW PROBLEMS

REVIEW PROBLEM 9.1

Athabaska Engineering was paid $100 000 to manage a construction project in 1985. How much would the same job have cost in 2005 if the average annual inflation rate between 1985 and 2005 was 5%?

ANSWER

The compound amount factor can be used to calculate the value of 100 000 1985 dollars in 2005 dollars, using the inflation rate as an interest rate:

$$2005 \text{ dollars} = 100\ 000(F/P,5\%,20)$$
$$= 100\ 000(2.6533)$$
$$= 265\ 330$$

The same job would have cost about $265 330 in 2005 dollars.∎

REVIEW PROBLEM 9.2

A computerized course drop-and-add program is being developed for a local community college. It will cost $300 000 to develop and is expected to save $50 000 per year in administrative costs over its 10-year life. If inflation is expected to be 4% per year for the next 10 years and a real MARR of 5% is required, should the project be adopted?

ANSWER

First, assuming that $50 000 in administrative costs are actual dollars, we can calculate the actual IRR for the project. The actual IRR is the solution for i in

$$300\ 000 = 50\ 000(P/A,i,10)$$
$$(P/A,i,10) = 6$$

For $i = 11\%$, $(P/A,i,10) = 5.8892$

For $i = 10\%$, $(P/A,i,10) = 6.1445$

The actual IRR of 10.55% is found by interpolating between these two points. We then convert the actual IRR into a real IRR to determine if the project is viable:

$$\text{IRR}_R = \frac{1 + \text{IRR}_A}{1 + f} - 1 = \frac{1.1055}{1.04} - 1 = 0.06298 \text{ or } 6.3\%$$

Since the real IRR of 6.3% exceeds the MARR of 5%, the project should be undertaken.■

REVIEW PROBLEM 9.3

Robert is considering purchasing a bond with a face value of €5000 and a coupon rate of 8%, due in 10 years. Inflation is expected to be 5% over the next 10 years. Robert's real MARR is 10%, compounded semiannually. What is the present worth of this bond to Robert?

ANSWER

This problem can be done with either real interest and real cash flows or actual interest and actual cash flows. It is somewhat easier to work with actual cash flows, so we must first convert the real interest rate given to an actual interest rate.

Robert's annual real MARR is $(1 + 0.10/2)^2 - 1 = 0.1025$. (Recall that the 10% is a nominal rate, compounded semiannually.)

If annual inflation is 5%, Robert's actual *annual* MARR is

$$\text{MARR}_A = \text{MARR}_R + f + \text{MARR}_R \times f$$

$$= 0.1025 + 0.05 + (0.1025)(0.05)$$

$$= 0.15763 \text{ or } 15.763\%$$

The present worth of the €5000 Robert will get in 10 years is then

$$\text{PW} = 5000(P/F,\text{MARR}_A,10)$$

$$= 5000(0.23138) \cong 1157$$

Next, the bond pays an annuity of €5000 × 0.08/2 = €200 every six months. To convert the annuity payments to their present worth, we need an actual six-month MARR. This can be obtained with a six-month inflation rate and Robert's six-month real MARR of 10%/2 = 5%. With $f = 5\%$ per annum, the inflation rate per six-month period can be calculated with

$$f_{12} = (1 + f_6)^2 - 1$$

$$f_6 = (1 + f_{12})^{1/2} - 1$$

$$= (1 + 0.05)^{1/2} - 1 = 0.0247 \text{ or } 2.47\%$$

The actual MARR per six-month period is then given by

$$\text{MARR}_A = \text{MARR}_R + f + \text{MARR}_R \times f = 0.05 + 0.0247 + (0.05)(0.0247)$$

$$= 0.07593 \text{ or } 7.593\%$$

The present worth of the dividend payments is

$$\text{PW(dividends)} = 200(P/A,7.59\%,20)$$
$$= 200(10.125)$$
$$= 2025$$

Finally,

$$\text{PW(bond)} = 1157 + 2025$$
$$= 3182$$

The present worth of the bond is €3182.∎

REVIEW PROBLEM 9.4

Trimfit, a manufacturer of automobile interior trim, is considering the addition of a new product to its line. Data concerning the project are given below. Should Trimfit accept the project?

New Product Line Information	
First cost ($)	11 500 000
Planned output (units/year)	275 000
Actual MARR	20%
Range of possible inflation rates	0% to 4%
Study period	10 years

Current-Year Prices ($/unit)	
Raw materials	16.00
Labour	6.25
Product sales price	32.00

ANSWER

First, we note that the expected net revenue per unit (not counting amortization of first costs) is $9.75 = $32 – $16 – $6.25. The project is potentially viable.

In doing the project evaluation, we can proceed with either actual dollars or real dollars. Since we do not know what the inflation rate will be, the easiest way to account for inflation is to keep all prices in real dollars and adjust the actual MARR to a real MARR by using values for the inflation rate within the potential range given. The project can then be evaluated with one of the standard methods. Since many of the figures are given in terms of annual amounts, an annual worth analysis will be carried out. Inflation rates of 0%, 1%, and 4% will be used. The results are shown in Table 9.6.

In Table 9.6, the annual worth of the project depends on the inflation rate assumed. Since the actual MARR of 20% implicitly includes anticipated inflation, different trial

Table 9.6   Annual Worth Computations for Trimfit

Annual Worth Comparisons for Various Inflation Rates					
Inflation Rate Per Year	Real MARR	Fixed Cost Per Year ($)	Variable Cost Per Year ($)	Revenue Per Year ($)	Annual Worth (Profit) Per Year ($)
0%	20.00%	2 743 012	6 118 750	8 800 000	− 61 762
1%	18.81%	2 633 122	6 118 750	8 800 000	48 128
4%	15.38%	2 325 083	6 118 750	8 800 000	356 167

inflation rates imply different values for the real MARR. For example, at 1% inflation, the real MARR implied is

$$\text{MARR}_R = \frac{1 + \text{MARR}_A}{1 + f} - 1 = \frac{1.20}{1.01} - 1 = 0.1881 \text{ or } 18.81\%$$

The fixed cost per year is obtained by finding the annual amount over 10 years equivalent to the first cost when the appropriate real MARR is used. For example, with 1% inflation, the fixed cost per year is

$$A = P(A/P, \text{MARR}_R, 10) = 11\ 500\ 000 \left(\frac{0.1881\ (1.1881)^{10}}{(1.1881)^{10} - 1} \right)$$

$$= 2\ 633\ 122$$

Next, the variable cost per year is the sum of the raw material cost and the labour cost per unit multiplied by the total expected output per year, that is, $22.25 × 275 000 = $6 118 750. Revenue per year is the sales price multiplied by the expected output: $32 × 275 000 = $8 800 000. Notice that the variable cost and the revenue are the same for all three values of the inflation rate. This is because they are given in real dollars.

Finally, the annual worth of the project is determined by the revenue per year less the fixed and variable costs per year. The annual worth is negative for zero inflation, but is positive for both 1% and 4% inflation rates. Since periods of at least 10 years in which there has been zero inflation have been rare in the twentieth and twenty-first centuries, it is probably safe to assume that there will be some inflation over the life of the project. Therefore, the project appears to be acceptable, since its annual worth will be positive if inflation is at least 1%.■

SUMMARY

In this chapter, the concept of inflation was introduced, and we considered the impact that inflation has on project evaluation. We began by discussing methods of measuring inflation. The main result here was that there are many possible measures, all of which are only approximate.

The concept of actual cash flows and interest rates and real cash flows and interest rates was introduced. Actual dollars are in currency at the time of payment or receipt, while real dollars are constant over time and are expressed with respect to a base year. Compound amount factors can be used to convert single payments between real and actual dollars.

Most of the chapter was concerned with the effect of correctly anticipated inflation on project evaluation and on how to incorporate inflation into project evaluation correctly. We showed that, where engineers have no reason to believe project prices will behave differently from average prices, project decisions are the same with or without correctly anticipated inflation. Finally, we pointed out that predicting inflation is very difficult. This implies that engineers should work with ranges of values for possible future inflation rates. The engineer should test for sensitivity of decisions to possible inflation rates.

Engineering Economics in Action, Part 9B:
Exploiting Volatility

Bill Astad of head office had been asking Naomi about how to deal with the variable inflation rates experienced by a sister company in Mexico. "OK, Naomi, let's see if I have this straight. For long-term projects of, say, six years or more, it makes sense to use a single inflation figure—the average rate. I can just add that to the real MARR to get an actual MARR. Boy, it's easy to get confused between the real and the actual. But I do understand the principle."

"And the short-term projects?" Naomi prompted.

"For the short-term ones, it makes more sense to break them up into time periods. For each period, select a 'best guess' inflation rate, and do a stepwise calculation from period to period. So the inflation rate in the middle of the presidential cycle would be relatively low, while near the changeover time it would be a higher estimate. Of course, the actual values used would depend on the political and economic situation at the time the decision is made. I understand that one, too, but it is complicated."

"I agree," said Naomi. "I guess we're lucky things are more predictable here."

"We are," Bill replied. "On the other hand, if we can make good decisions in spite of a volatile economy in Mexico, Mexifab may have an advantage over its competitors. Thanks for your help, Naomi."

PROBLEMS

For additional practice, please see the problems (with selected solutions) provided on the Student CD-ROM that accompanies this book.

9.1 Which of the following are real and which are actual?

(a) Allyson has been promised a €10 000 inheritance when her Uncle Bill dies.

(b) Bette's auto insurance will pay the cost of a new windshield if her current one breaks.

(c) Cory's meal allowance while he is in university is $2000 per term.

(d) Dieter's company promises that its prices will always be the same as they were in 1975.

(e) Engworth will construct a house for Zolda, and Zolda will pay Engworth $150 000 when the house is finished.

(f) Fran's current salary is $3000 per month.

(g) Greta's retirement plan will pay her $1500 per month, adjusted for the cost of living.

9.2 Find the real amounts (with today as the base year) corresponding to the actual amounts shown below, for a 4% inflation rate.

   (a) $400 three years from now

   (b) €400 three years ago

   (c) $10 next year

   (d) $350 983 ten years from now

   (e) £1 one thousand years ago

   (f) $1 000 000 000 three hundred years from now

9.3 Find the present worth today in real value corresponding to the actual values shown below, for a 4% inflation rate and a 4% interest rate.

   (a) $400 three years from now

   (b) €400 three years ago

   (c) $10 next year

   (d) $350 983 ten years from now

   (e) £1 one thousand years ago

   (f) $1 000 000 000 three hundred years from now

9.4 An investment pays $10 000 in five years.

   (a) If inflation is 10% per year, what is the real value of the $10 000 in today's dollars?

   (b) If inflation is 10% and the real MARR is 10%, what is the present worth?

   (c) What actual dollar MARR is equivalent to a 10% real MARR when inflation is 10%?

   (d) Compute the present worth using the actual dollar MARR from part (c).

9.5 An annuity pays $1000 per year for 10 years. Inflation is 6% per year.

   (a) If the real MARR is 8%, what is the actual dollar MARR?

   (b) Using the actual dollar MARR from part (a), calculate the present worth of the annuity.

9.6 An annuity pays $1000 per year for 12 years. Inflation is 6% per year. The annuity costs $7500 now.

   (a) What is the actual dollar internal rate of return?

   (b) What is the real internal rate of return?

9.7 A bond pays $10 000 per year for the next ten years. The bond costs $90 000 now. Inflation is expected to be 5% over the next 10 years.

   (a) What is the actual dollar internal rate of return?

   (b) What is the real internal rate of return?

9.8 Inflation is expected to average about 4% over the next 50 years. How much would we expect to pay 50 years from now for each of the following?

   (a) $1.59 hamburger

(b) €15 000 automobile

(c) $180 000 house

9.9 The average person now has assets totalling $38 000. If the average real wealth per person remains the same, and if inflation averages 5% in the future, when will the average person become a millionaire?

9.10 How much is the present worth of $10 000 ten years from now under each of the following patterns of inflation, if interest is at 5%? On the basis of your answers, is it generally reasonable to use an average inflation rate in economic calculations?

(a) Inflation is 4%.

(b) Inflation is 0% for five years, and then 8% for five years.

(c) Inflation is 8% for five years, and then 0% for five years.

(d) Inflation is 6% for five years and then 2% for five years.

(e) Inflation is 0% for nine years and then 40% for one year.

9.11 The actual dollar MARR for Jungle Products Ltd. of Parador is 300%. The inflation rate in Parador is 250%. What is the company's real MARR?

9.12 Krystyna has a long-term consulting contract with an insurance company that guarantees her £25 000 per year for five years. She believes inflation will be 3% this year and 5% next year, and then will stay at 10% indefinitely. Krystyna's real MARR is 12%. What is the present worth of this contract?

9.13 I have a bond that will pay me $2000 every year for the next 30 years. My first payment will be a year from today. I expect inflation to average 3% over the next 30 years. My real MARR is 10%. What is the present worth of this bond?

9.14 Ken will receive a $15 000 annual payment from a family trust. This will continue until Ken is 30; he is now 20. Inflation averages 4%, and Ken's real MARR is 8%. If the first payment is a year from now and a total of 10 payments are to be made, what is the present worth of his remaining income from the trust?

9.15 Inflation in Russistan currently averages 40% per month. It is expected to diminish to 20% per month following the presidential elections 12 months from now. The Russistan Oil Company (ROC) has just signed an agreement with the Global Petroleum Group for the sale of future shipments. The ROC will receive 500 million rubles per month over the next two years, and also 500 million rubles per month indexed to inflation (i.e., real rubles). If the ROC has a real MARR of 1.5% per month, what is the total present worth of this contract?

9.16 The widget industry maintains a price index for a standard collection of widgets. The base year was 1997 until 2007, when the index was recomputed with 2007 as the base year. The following data concerning prices for the years 2005 to 2008 are available:

Year	Price Index 1997 Base	Price Index 2007 Base
2005	125	N/A
2006	127	N/A
2007	130	100
2008	N/A	110

What was the percentage increase in prices of widgets between 2005 and 2008?

9.17 A group of farmers in Inverness is considering building an irrigation system from a water supply in some nearby mountains. They want to build a concrete reservoir with a steel pipe system. The first cost would be £200 000 with (actual) annual maintenance costs of £2000. They expect the irrigation system will bring them £22 000 per year in additional (actual) revenues due to better crop production. Their real pound MARR is 4% and they anticipate inflation to be 3% per year. Assume that the reservoir will have a 20-year life.

   (a) Using the actual cash flows, find the actual IRR on this project.

   (b) What is the actual MARR?

   (c) Should they invest?

9.18 Refer to Problem 9.17.

   (a) Convert the actual cash flows into real cash flows.

   (b) Find the present worth of the project using the real MARR.

   (c) Should they invest?

9.19 Go to your country's government's statistical collection webpage (see the Net Value box for this chapter).

   (a) What is the current base year for the CPI? What is the trend over the last two years?

   (b) Summarize, if the information is given, what the commodities in the CPI basket are, the relative importance of the commodities in the CPI basket, how the CPI basket is updated, and how prices are collected for the CPI.

   (c) Which of the subcategories (e.g., food, transportation, etc.) appears to be contributing the most to the overall CPI index?

   (d) Are there regions of the country with a different CPI than other regions? Why?

9.20 Go to your country's government's statistical collection webpage (see the Net Value box for this chapter).

   (a) Find or calculate the most recent inflation rate. What has been the trend in the inflation rate over the last two years?

   (b) How does the current inflation rate compare with the interest rates offered at banks and other financial institutions for savings accounts? Can you estimate the real interest rate available for your savings?

   (c) How does inflation in your country compare with inflation elsewhere? Find a country with particularly low inflation (or high deflation) and a country with particularly high inflation.

9.21 Bosco Consulting is considering a potential contract with the Upper Sobonian government to advise them on exploration for oil in Upper Sobonia. Bosco would make an investment of 1 500 000 Sobonian zerts to set up a Sobonian office in 2010. The Upper Sobonian government would pay Bosco 300 000 zerts in 2011. In the years 2012 to 2017, the actual zerts value of the payments would increase at the rate of inflation in Upper Sobonia. The following data are available concerning the project:

Investment in Upper Sobonia	
Expected Sobonian inflation rate (2010–2017)	15%/year
Expected inflation rate (2010–2017)	3%/year
Value of Sobonian zerts in 2010	$0.25
Expected decline in value of zerts (2010–2017)	10%/year
First cost in 2010 (zerts)	1 500 000
Cash flows in 2011–2017 (real 2011 zerts)	300 000
Bosco's actual dollar MARR	22%

(a) Construct a table with the following items:

Real (2010) zert cash flows

Actual zert cash flows

Actual dollar cash flows

Real dollar cash flows

(b) What is the present worth in 2010 dollars of this project?

9.22 Bildkit, a building products company, is considering an agreement with a distributor in the foreign country of Maloria to supply kits for constructing houses in Maloria. Sales would start next year. The expected receipts from the sale of the kits next year is 30 000 000 Malorian yen. The number of units sold is expected to grow by 10% per year over the life of the contract. The actual yen price is expected to grow at the rate of Malorian inflation.

There will be a first cost for Bildkit. As well, there will be operating costs over the life of the contract. Operating cost per unit will be constant in real dollars over the life of the contract. Since the number of units sold will rise by 10% per year, real operating costs will rise by 10% per year. Actual operating costs per unit will rise at the rate of inflation in Bildkit's country.

The value of the Malorian yen is expected to increase over the life of the contract. Data concerning the proposed contract are shown in the table below.

Bildkit in Maloria			
Receipts in first year (actual yen)	30 000 000	First cost now (actual $)	200 000
Growth of receipts (real yen)	10%/year	Operating cost in first year of operation (actual $)	350 000
Malorian inflation rate	1%/year	Inflation rate in Bildkit's country	3%
Value of yen year 0 ($)	0.015	Actual dollar MARR	22%
Rate of increase in value of yen	2%	Study period	8 years

(a) What is the present worth of receipts in dollars?

(b) What is the present worth of the cash outflows in dollars?

9.23 Leftway Information Systems is considering a contract with the Ibernian government to supply consulting services over a five-year period. The following real Ibernian pound cash flows are expected:

Cash Flows in Year 2010 Ibernian Pounds	
First cost	1 800 000
Net revenue 2011 to 2015	550 000

Further information is in the table below:

Expected Ibernian inflation rate	10%
Value of Ibernian pound in 2010 ($)	1.25
Expected annual rate of decline in the value of the Ibernian pound	5%
Expected inflation rate in Leftway's country	2.50%
Leftway's real MARR	15%

(a) What is the real Ibernian pound internal rate of return on this project? (*Hint:* Leftway's country can be ignored in answering this question.)

(b) What is the actual pound internal rate of return? (*Hint:* Leftway's country can be ignored in answering this question.)

(c) Use the internal rate of return in dollars to decide if Leftway should accept the proposed contract.

9.24 Sonar warning devices are being purchased by the St. James department store chain to help trucks back up at store loading docks. The total cost of purchase and installation is £220 000. There are two types of saving from the system. Faster turnaround time at the congested loading docks will save £50 000 per year in today's pounds. Reduced damage to the loading docks will save £30 000 per year in today's pounds. St. James has an observed actual pound MARR of 18%. The sonar system has a life of four years. Its scrap value in today's pounds is £20 000. The inflation rate is expected to be 6% per year over the next four years.

(a) What is St. James's real MARR?

(b) What is the real internal rate of return? (This is most easily done with a spreadsheet.)

(c) Compute the actual internal rate of return using Equation (9.10).

(d) Compute the actual internal rate of return from the actual pound cash flows. (This is most easily done with a spreadsheet.)

(e) What is the present worth of the system?

9.25 Johnson Products now buys a certain part for its chain saws. The managers are considering the production of the part in-house. They can install a production system that would have a life of five years with no salvage value. They believe that over the next five years the real price of purchased parts will remain fixed. They expect the real price of labour and other inputs to production to rise over the next five years. Further information about the situation is in the table below.

Annual cost of purchase ($/year)	750 000
Expected real change in cost of purchase	0%
Expected real change in labour cost	4%
Expected real change in other operating costs	2%
Labour cost/unit (first year of operation) ($)	10.5
Other operating cost/unit (first year of operation) ($)	9
In-house first cost ($)	200 000
Use rate (units/year)	25 000
Actual dollar MARR	20%
Study period (years)	5

(a) Assume inflation is 2% per year in the first year of operation. What will be the actual dollar cost of labour for in-house production in the second year?

(b) Assume inflation is 2% per year in the first two years of operation. What will be the actual dollar cost of other operating inputs for in-house production in the third year?

(c) Assume that inflation averages 2% per year over the five-year life of the project. What is the present worth of costs for purchase and for in-house production?

9.26 Lifewear, a manufacturer of women's sports clothes, is considering adding a line of skirts and jackets. The production would take place in a part of its factory that is now not being used. The first output would be available in time for the 2011 fall season. The following information is available:

New Product Line Information	
First cost in 2010 ($)	15 500 000
Planned output (units/year)	325 000
Observed, actual dollar MARR before tax	0.25
Study period	6 years
Year 2010 Prices ($/unit)	
Materials	12
Labour	7.75
Output	35

(a) What is the real internal rate of return?

(b) What inflation rate will make the real MARR equal to the real internal rate of return?

(c) Calculate the present worth of the project under three possible future inflation rates. Assume that the inflation rate will be 1%, 2%, or 3% per year.

(d) Decide if Lifewear should add this new line of skirts and jackets. Explain your answer.

9.27 Century Foods, a producer of frozen meat products, is considering a new plant near Essen, Germany, for its sausage rolls and frozen meat pies. The company has estimates of production cost and selling prices in the first year. It expects the real value of operating costs per unit to fall because of improved operating methods. It also expects competitive pressures to cause the real value of product prices to fall. The following data are available:

Century Food Plant Data	
Output price in 2011 (€/box)	22
Operating cost in 2011 (€/box)	15.5
Planned output rate (boxes/year)	275 000
Fall in real output price	1.5% per year
Fall in real operating cost per box	1.0% per year
First cost in 2010 (€)	7 500 000
Study period	10 years
Actual euro MARR before tax	20%

(a) Assume that there is zero inflation. What is the present worth, in 2010, of the project?

(b) Assume that there is zero inflation. What is the internal rate of return? (This is most easily done with a spreadsheet.)

(c) At what inflation rate would the actual euro internal rate of return equal 20%?

(d) Should Century Foods build the new plant? Explain your answer.

9.28 Metcan Ltd.'s smelter produces its own electric power. The plant's power capacity exceeds its current requirements. Metcan has been offered a contract to sell excess power to a nearby utility company. Metcan would supply the utility company with 17 500 megawatt-hours per year (MWh/a) for 10 years. The contract would specify a price of $22.75 per megawatt-hour for the first year of supply. The price would rise by 1% per year after this. This is independent of the actual rate of inflation over the 10 years.

Metcan would incur a first cost to connect its plant to the utility system. There would also be operating costs attributable to the contract. Metcan believes these costs would track the actual inflation rate. The terms of the contract and Metcan's costs are shown in the tables below.

Metcan Sale of Power	
Output price in 2011 ($/MWh)	22.75
Price adjustment (2012–2020)	1% per year
Power to be supplied (MWh/a)	17 500
Contract length	10 years
Metcan's Costs	
First cost in 2010 ($)	175 000
Operating cost in 2005($)	332 500
Actual dollar MARR before tax	20%

(a) Find the present worth of the contract under the assumption that there is no inflation over the life of the contract.

(b) Find the present worth of the contract under four assumptions: inflation is (i) 1% per year, (ii) 2% per year, (iii) 3% per year, and (iv) 4% per year.

(c) Should Metcan accept the contract?

 9.29 Clarkwood is a wood products manufacturer. Its managers are considering a modification to the production line that would enable an increase in output. One of Clarkwood's concerns is that the price of wood is rising more rapidly than inflation. The managers expect that because of this the operating cost per unit will rise at a rate 4% higher than the rate of inflation. That is, if the rate of inflation is f, Clarkwood's operating cost will rise at the rate $f_c = 1.04(1 + f) - 1$. However, competitive pressures from plastics will prevent the prices of Clarkwood's products from rising more than 1% above the inflation rate. The particulars of the project are shown in the table below.

Clarkwood's Project	
Output price in 2011 ($/unit)	30
Price increase	2% above inflation
Operating cost in 2011 ($/unit)	24
Operating cost increase	4% above inflation
Expected output due to project (units/year)	50 000
First cost in 2010 ($)	900 000
Observed actual dollar MARR	0.25
Time horizon (years)	10

(a) Find the present worth of the project under the assumption of zero inflation.

(b) Find the present worth of the project under these assumptions: the expected inflation is (i) 1% per year, and (ii) 2% per year.

(c) Should Clarkwood accept the project?

 9.30 Smooth-Top is a manufacturer of desktops. It is considering an increase of capacity. Consulting engineers have submitted two routes to accomplish this: (1) install a new production line that would produce wooden desktops finished with hardwood veneer and (2) install a new production line that would produce wooden desktops finished with simulated wood made from hard plastic.

Smooth-Top is concerned about the price of hardwood veneer. They believe the price of veneer will rise over the next 10 years. However, they believe the price of veneer-finished desktops will rise by less than the rate at which the price of veneer rises. Information about the two potential projects is in the following table.

(a) Compute the present worth of each option under the assumption that the real price of hardwood-finished desktops and real cost of hardwood veneer do not change (rather than as stated in the table). Assume zero inflation.

Smooth-Top Desktop Project	
Plastic-finish real price and real cost change	0%
Veneer-finish expected real price change	1%
Veneer-finish expected real cost change	5%
Wood cost/unit ($)	12.5
Plastic cost/unit ($)	9
Wood price/unit ($)	32
Plastic price/unit ($)	26
Wood first cost ($)	2 050 000
Plastic first cost ($)	2 700 000
Wood output rate (units/year)	30 000
Plastic output rate (units/year)	45 000
Study period	10 years
Actual dollar MARR	25%

(b) Compute the present worth of each option under the assumption that the real price of hardwood-finished desktops and the real value of hardwood veneer desktop operating costs increase as indicated in the table. Assume that inflation is expected to be 2% over the study period.

 9.31 Belmont Grocers has a distribution centre in Brisbane. The manual materials-handling system at the centre has deteriorated to the point at which it must be either replaced or substantially refurbished. Replacement with an automated system would cost about $240 000. Refurbishing the manual system would cost about $50 000. In either case, capital expenditures would take place this year. Operating either the new system or the refurbished system would begin next year. It is expected that either the new system or the refurbished system will operate for 10 years with no further capital expenditures. Belmont is concerned that labour costs in Brisbane may rise in real terms over the next 10 years. The range of increases in real terms that appears possible is from 4% to 7% per year. Inflation rates between 2% and 4% are expected over the next 10 years. Complete data on the two alternatives are given in the table that follows.

(a) Find the total costs per unit for each of the two alternatives under the assumption of zero inflation and no increase in costs for the manual system.

(b) Make a recommendation about which alternative to adopt. Base the recommendation on the present worth of costs for the two systems under various assumptions concerning inflation and the rate of change in the real operating cost of the manual system. Explain your recommendation.

Materials Handling Data	
Automated expected real operating cost change	0%
Manual expected real operating cost change	4% to 7%
Manual operating cost/unit (first year of operation) ($)	10.5
Automated operating cost/unit (first year of operation) ($)	9
Manual first cost ($)	50 000
Automated first cost ($)	240 000
Output rate (units/year)	15 000
Actual dollar MARR	20%
Study period	10 years
Possible inflation rates	2% to 4%

9.32 The United Gum Workers have a cost-of-living clause in their contract with Mont-Gum-Ery Foods. The contract is for two years. The contract states that, if the inflation rate in the first year exceeds 1%, wages in the second year will increase by the inflation rate of the first year. Does this clause increase or decrease risk? Explain.

 9.33 Free Wheels has a plant that assembles bicycles. The plant now has a small cafeteria for the workers. The kitchen equipment is in need of substantial overhaul. Free Wheels has been offered a contract by Besteats to supply food to the workers. The particulars of the situation are shown in the table. Should Free Wheels continue with the in-house food service or contract the service to Besteats?

Food Service: In-House Versus Contract	
Food service labour (hours/year)	6000
Wage rate (real, time 1, $/hour)	7.5
Overhead cost (real, time 1, $/year)	18 000
Kitchen equipment first cost (actual, time 0, $)	25 000
Contract cost, years 1 to 3 (actual $)	55 000
Contract cost, years 4 to 6 (actual $)	63 700
Actual dollar MARR	22%
Expected annual inflation rate	5%
Study period (years)	6

More Challenging Problem

9.34 In 10 years, Mid-Atlantic Corp. will be investing $200 000 either in Columbo or in Avalon. The exchange rate between the Columboan dollar and the Avalonian pound is fixed at $2 = £1. Dollars and pounds can be exchanged at no cost at any time.

If the $200 000 is invested in Columbo, each dollar invested will return $0.30 per year for each of the following five years. If the $200 000 is converted to pounds and invested in Avalon, each pound invested will return £0.24 per year for the following seven years.

Columbo is subject to ongoing average inflation of 4%, while inflation in Avalon averages 2%. If the real MARR for Mid-Atlantic is 10%, which investment is preferred? How much money should Mid-Atlantic set aside now (invested at the MARR) to ensure that it has enough money to make the investment in 10 years?

MINI-CASE 9.1

Economic Comparison of High Pressure and Conventional Pipelines: Associated Engineering

Associated Engineering conducted an evaluation of sources of water supply for a municipality. One of the considerations was the choice of high-pressure or conventional pipelines for transmitting treated water to the municipality from a distant water source.

Conventional pipelines, most often made of concrete, have a limited maximum tensile strength, which for analysis purposes was taken to be 200 pounds per square inch (psi). High-pressure pipe, made of steel, can withstand up to 60 000 psi, although the pipe examined by Associated Engineering had a strength of 42 000 psi.

The major advantage of the steel pipe is that fewer pumping stations are needed than with the concrete pipe. The distance to be pumped is 85 kilometres; this requires either one pumping station for high-pressure pipe or six pumping stations for concrete pipe.

Each pipeline type was analyzed over a range of pipeline diameters ranging from 24" to 72". Construction costs included the pipe, pumping stations, and a reception reservoir, with the time of the cost taken to be the commissioning date of 2025. Operating and maintenance costs starting in 2026 were included, and administration, engineering fees, contingencies, and taxes were also accounted for.

The best alternative was chosen on the basis of a present worth comparison with a 4% discount rate. In the analysis, real 1993 dollars were used and an inflation rate of 2% was assumed for the period of study. The result was that a 360-diameter high-pressure pipeline was economically best, at a present cost $7.5 million lower than for the best conventional pipeline.

Discussion

Estimating future inflation is difficult. The average inflation in many developed countries over the last 50 years has been about 4%, but there have been periods of several years when it has been 2% to 2.5%. For some other periods, inflation has averaged over 10%. Historically, many countries have experienced hyperinflation (extreme inflation) or even deflation. How can we estimate future inflation?

One way is simply to assume that inflation will remain at the current value. This is probably wrong; as has been seen, inflation typically changes over time. However, there are factors that are controlling the inflation rate. Lacking knowledge of any reason why these controlling factors might change, the current rate seems to be a reasonable choice.

A second approach is to use the long-term average. Knowing that inflation will change over time suggests that the long-term average is a good choice even if inflation is lower or higher than the average right now. After all, those controlling factors have changed in the past and are likely to change again.

A third way is to take into account the controlling factors for inflation. These include government policy: a government committed to social welfare is likely to induce more inflation than one committed to fiscal responsibility. Trends in business and consumer behaviour affect inflation: large labour contract increases presage inflation, as does high consumer borrowing. Social trends like the aging of the baby boomers also have an effect on inflation.

Understanding the effect of the controlling factors for inflation in detail is very difficult. So usually decision makers make a broad judgment based on both the current inflation rate and the historical average, and perhaps informed by a general understanding of the contributing factors.

Questions

1. How significant would the difference have been to the savings of the high-pressure pipeline if an inflation rate of 4% had been used instead? Assume the only difference between the concrete- and steel-pipe systems was the capital cost, expended in 2025. Would the decision be any different? Could it be different for any assumed inflation rate?

2. Design two cash flow structures for projects that start in 2025, such that the present worth in the current year at a discount rate of 4% is higher for one project than the other at an inflation rate of 2%, but lower at an inflation rate of 4%. Is there a significant opportunity to control the best choice in a decision situation by selecting the appropriate inflation rate?

3. Why would the analysts have chosen to separate the inflation rate from the discount rate for this problem, rather than combining them into an actual dollar discount rate? Do you think the analysts estimated the actual dollar cost of the alternatives in 2025, or would they have used the real costs?

Appendix 9A Computing a Price Index

We can represent changes in average prices over time with a **price index**. A price index relates the average price of a given set of goods in some time period to the average price of the same set of goods in another period. Commonly used price indexes work with weighted averages because simple averages do not reflect the differences in importance of the various goods and services in which we are interested.

Many different ways of weighting changes in prices may be used, and each method leads to a different price index. We shall discuss only the most commonly used index, the **Laspeyres price index**. It can be explained as follows.

Suppose there are n goods in which we are interested. We want to represent their prices at a time, t_1, relative to a **base period**, t_0, the period from which the expenditure shares are calculated.

The prices of the n goods at times t_0 and t_1 are denoted by $p_{01}, p_{02}, \ldots, p_{0n}$ and $p_{11}, p_{12}, \ldots, p_{1n}$. The quantities of the n goods purchased at t_0 are denoted by $q_{01}, q_{02}, \ldots, q_{0n}$. The share, s_{0j}, of good j in the total expenditure for the period, t_0, is defined as

$$s_{0j} = \frac{p_{0j}q_{0j}}{p_{01}q_{01} + p_{02}q_{02} + \ldots + p_{0n}q_{0n}}$$

Note that

$$\sum_{j=1}^{n} s_{0j} = 1$$

A Laspeyres price index, π_{01}, is defined as a weighted average of relative prices.

$$\pi_{01} = \left(\frac{p_{11}}{p_{01}} s_{01} + \frac{p_{12}}{p_{02}} s_{02} + \ldots + \frac{p_{1n}}{p_{0n}} s_{0n} \right) \times 100$$

The term in the brackets is a weighted average because the weights (the expenditure shares in the base period) sum to one. The relative prices are the prices of the individual goods in period t_1 relative to the base period, t_0. The weighted average is multiplied by 100 to put the index in percentage terms.

EXAMPLE 9A.1

A student uses four foods for hamburgers: (1) ground beef, (2) hamburger buns, (3) onions, and (4) breath mints. Suppose that, in one year, the price of ground beef fell by 10%, the price of buns fell by 1%, the price of onions rose by 5%, and the price of breath mints rose by 50%.

The price and quantity data for the student's hamburger are shown in Table 9A.1.

Table 9A.1   Price and Quantity Data for Hamburger

	Quantity at t_0	Price at t_0 ($)	Price at t_1 ($)
Ground beef (kg)	0.25	3.5/kg	3.15/kg
Buns	1	0.40	0.396
Onions	1	0.20	0.21
Breath mints	1	0.10	0.15

The Laspeyres price index is calculated in four steps:

1. Compute the base period expenditure for each ingredient.
2. Compute the share of each ingredient in the total base period expenditure.
3. Compute the relative price for each ingredient.
4. Use the shares to form a weighted average of the relative prices.

These computations are shown in Table 9A.2.

Table 9A.2   The Laspeyres Price Index Calculation

	Price at t_0 ($)	Share at t_0	Relative Price	Weighted Relative Price
Ground beef (kg)	0.875	0.556	0.900	0.500
Buns	0.400	0.254	0.990	0.251
Onions	0.200	0.127	1.050	0.133
Breath mints	0.100	0.063	1.500	0.095
Sums	1.575	1.000		0.980

As an example of the computations, the price of the ground beef per hamburger at t_0 is found by multiplying the price per kilogram by the weight of the hamburger used: $3.50/kg × 0.25 kg = $0.875. Similar computations for each of the other ingredients lead to a total cost of $1.575 per hamburger. The ground beef then represents a share of 0.875/1.575 = 0.556 of the total cost. The relative price for the hamburger is 3.15/3.5 = 0.9 and thus the weighted relative price is 0.556 × 0.9 = 0.50. Similar computations for the other ingredients lead to a total weighted average of 0.98. After multiplying by 100, the Laspeyres price index is 98 (it is understood that this is a percentage). Therefore, the cost of the hamburger ingredients at t_1 was 2% lower than in the base period.■

Governments compile many Laspeyres price indexes. The consumer price index (CPI) is a Laspeyres price index in which the weights are the shares of urban consumers' budgets in the base year. Another well-known Laspeyres price index is the gross national product (GNP) deflator. For the GNP deflator, the weights are the shares of total output in the base year.

The CPI and the GNP deflator are global indexes in that they represent an economy-wide set of prices. As well, Laspeyres price indexes can be calculated by sector. For example, there are price indexes for durable consumer goods, for exports, and for investment by businesses. It is up to the analyst to know the composition of the different indexes and to decide which is best for his or her purposes.

EXAMPLE 9A.2

We can classify consumer goods and services into four classes: durable goods, semi-durable goods, non-durable goods, and services. Assume the classes had the following prices in 2001 and 2008:

Category	Price ($) 2001	2008
Durable	2.421	2.818
Semi-durable	2.849	3.715
Non-durable	4.926	6.404
Services	4.608	6.263

Quantities in 2001 were

Quantity in 2001 (Units)	
Durable	21.304
Semi-durable	11.315
Non-durable	19.159
Services	31.422

Find the Laspeyres price index for 2008 with 2001 as a base. We first calculate the relative prices:

	Price ($)		
Category	**2001**	**2008**	**Relative Price**
Durable	2.421	2.818	1.164
Semi-durable	2.849	3.715	1.304
Non-durable	4.926	6.404	1.300
Services	4.608	6.263	1.359

We next determine expenditure shares in 2001:

Category	**Expenditure**	**Share**
Durable	51.583	0.1597
Semi-durable	32.235	0.0998
Non-durable	94.381	0.2922
Services	144.801	0.4483
Total	323.000	

We then multiply the relative prices by the shares and sum. We get the index by multiplying the sum by 100. For example, the term for durable goods is given by $1.164(0.1597) = 0.186$.

	Index
Durable	0.186
Semi-durable	0.130
Non-durable	0.380
Services	0.609
Total	1.305

This gives a Laspeyres price index of 130.5.■

CHAPTER

10

Public Sector Decision Making

"Hi, Naomi. How's it going down under?" Naomi could easily imagine Bill with his feet up on his desk, leaning back in his chair, the telephone wedged against his ear. Naomi was in Melbourne, Australia, checking into plans for how Melbourne Manufacturing, a division of Global Widgets, was dealing with a change to the fuel tax credit system recently introduced by the Australian government. The fuel tax credit system provides rebates to Melbourne Manufacturing provided that its vehicles satisfy specific emissions-control-related environmental criteria. Melbourne Manufacturing had an extensive distribution system in Eastern Australia that used a large fleet of trucks, so the fuel tax rebates were a significant issue. The fuel tax credit system had been in place for some time, but what was new was that in order to claim more than $3 million in credits, Melbourne Manufacturing had to be a member of the Greenhouse Challenge Plus program.

Naomi answered, "Melbourne is great. I can look out my window and see the rowing sculls in the Yarra River, and I got a chance to go to St. Kilda Beach on the weekend. Not bad for late February!" She continued, "Things look pretty good at Melbourne Manufacturing but we're up against some tough decisions. Melbourne Manufacturing has an ongoing program of truck replacement—they've spent over $5 million in the past few years on updating the fleet. Some of this was because of government emissions regulations, but mostly it was just part of their regular replacement program." She paused, "But now it looks like we might have to make some pretty big investments."

Carbon dioxide, unburned hydrocarbons, and nitrous oxides are the main gaseous pollutants produced by the combustion process in gasoline and diesel-powered vehicles. Carbon dioxide is a "greenhouse gas" that traps the Earth's heat and contributes to global warming. Nitrous oxides contribute to the formation of ozone, a lung and eye irritant, and are a major source of acid rain. In 2000, Australia agreed to use the United Nations Economic Commission for Europe targets to reduce emissions, and had introduced measures to either require or encourage emissions reductions. "So far, Melbourne Manufacturing has been able to comply with government restrictions just by purchasing fuel efficient trucks."

"So, what's the problem, then?" Bill interjected.

"Well," Naomi began, "New legislation requires Melbourne Manufacturing to be a member of the Greenhouse Challenge Plus program if they want to claim more than $3 million in fuel tax credits in a financial year. They're easily beyond that amount. The problem is, to become a member of Greenhouse Challenge Plus, they have to measure and monitor greenhouse gas emissions and report annually to the government. That's not a minor proposition—it means revamping all kinds of procedures and processes. It's going to be expensive, and it's not clear that the extra tax credits are going to be enough to make it worth doing."

"Well maybe there is more to it," Bill responded. "Being required to monitor greenhouse gas emissions might uncover opportunities for reducing waste and saving energy. Let's explore things further. I'll try to get some thoughts from the people here. See if you can get more information from the folks there. I'll set up a meeting for us with Anna Kulkowski when you are back. I suspect she may have some helpful ideas, too."

10.1 | Introduction

All organizations—public or private—and the engineers who work for them need to take into account the effects of what they do on society as a whole. Consider these two examples.

1. Lead has been known since ancient times. It is a malleable, corrosion-resistant element with a low melting point. The ancient Romans used lead to make water pipes, some of which are still in use today. Today, lead is used as a liner for tanks

that hold corrosive liquids, and as a coating on wires to prevent corrosion. Its density makes it useful as a shield against X-ray and gamma-ray radiation.

Lead alloys are also widely used. Solder, an alloy of lead and tin, has a relatively low melting point and is used, for example, to join electrical components to circuit boards. Other lead alloys are used to reduce friction in bearings. Lead compounds are used in glass, rubber, and paint. The lead makes the paint more vibrant in colour, more weather resistant, and faster drying.

Unfortunately, this useful element is also toxic. Lead causes a variety of adverse effects, such as vomiting, learning and behavioural problems, seizures, and death. Children six years old and under are most at risk of permanent brain damage. Lead-contaminated household dust and exposure to deteriorating lead-based paints are the major sources of exposure in children in developed countries, while air pollution in countries still using leaded gas has been a leading cause of poisoning in children. Most adult exposure to lead comes from occupational hazards in lead-related industries such as smelting and battery production, or is due to work with lead-based paints.

The toxicity of lead was recognized as early as 200 B.C., but it was not until the late 1800s and early 1900s that medical studies of children exposed to lead-based paints prompted some countries to ban them. In 1909, France, Belgium, and Austria were among the earliest to ban lead in paint. Many other European countries followed suit quickly. Canada adopted a voluntary phase-out in 1972 and a complete ban in 2006. The United States banned lead paint in 1978.

Governments in countries all over the world have introduced regulations on the use of lead in manufactured products. Since the early 1900s, regulations that limit workers' exposure to hazardous workplace substances—such as lead—have also been introduced. Unfortunately, there are no global standards for these limits or global regulations regarding the use of lead in consumer products. This poses a growing issue to companies that subcontract work to other countries. Mattel, one of the world's largest toy producers, had three major product recalls in 2007 and removed almost 21 million toys from shelves worldwide due to lead paint having been used on the toys. A class action lawsuit has also been filed demanding that Mattel pay for blood lead-level tests for children whose toys were recalled.

2. The extraction of coal and the production of steel once was the mainstay of the economy in Nova Scotia, Canada. The centre of the steel prosperity, Sydney, is now the site of one of North America's largest environmental disasters—the Sydney Tar Ponds. The contaminants in the tar ponds are mainly the result of steel-making operations—the core of which is the coke oven. As coal is heated in a coke oven, tar and gases are separated off from the desired coke. Over more than 80 years, these toxic wastes were discharged into an estuary of Sydney Harbour in an area known as Muggah Creek. The water in the area has been seriously contaminated with arsenic, lead, benzene, and other toxins such as PCBs (polychlorinated biphenyls), many of which are believed to cause cancer. Though concerns about air and ground pollution had been evident for decades, little money was spent to reduce the large volumes of water and air pollution. The steel mills, owned privately until the 1960s, were finally closed in 2000 after 40 years of operation by Sydney Steel (SYSCO), a provincial Crown corporation.

Residents of the area say that when it rains, puddles in the area turn fluorescent green. Others who live near the ponds report orange ooze in their basements, massive headaches, nosebleeds, and serious breathing problems. Sydney has one of the highest rates of cancer, birth defects, and miscarriages in Canada. In March 2004, residents filed a $1 billion lawsuit, the largest class-action suit to date in Canada, for damages allegedly caused by years of pollution from the SYSCO operations. In May 2004, the federal and provincial governments agreed to commit $400 million to assist with the cleanup. It remains to be seen what the outcome of the lawsuit will be and how the cleanup operation will remediate the long-polluted site.

The lead-poisoning and Sydney Tar Ponds examples illustrate a phenomenon that has important implications for engineers. It is not enough to produce goods and services at a cost that customers are willing to pay. Engineers must also pay attention to broader social values. This is because the market prices that guide most production decisions may not reflect all the social benefits and costs of engineering decisions adequately. Where markets fail to reflect all social benefits and costs, society uses other means of attaining social values.

Up to this point in our coverage of engineering economics, we have been concerned with project evaluation methods for private sector organizations. The private sector's frame of reference is profitability of the firm itself—a well-understood objective. To be profitable, firms sell goods and services to customers with the goal of realizing a return on investment. The focus in project evaluation is usually on the impact it has on the financial well-being of the firm, even if the impact of the project may have effects on individuals or groups external to the firm. For example, building a new pulp mill will provide additional production capacity, but it also may have an impact on local water quality. The firm may limit its economic analysis to the direct impact of the plant on its own profitability. This would include, for example, the construction and operating costs of the mill and the additional revenues associated with expanded production capacity. It may not consider the effect their emissions may have on the local fisheries.

When we consider the broader social context, the concept of profitability is extremely difficult to define because the frame of reference is society at large. This is particularly true for projects where there may not be an open market in which customers have free choice over what products they buy. Costs and benefits for these projects can be difficult to identify and value in a quantifiable manner. For example, if the UK Department for Transport were to consider construction of a new highway to improve traffic flow in a congested area, the analysis should take into account, in addition to the costs of construction, the benefits and costs to society at large. This includes assigning a value to social benefits such as increased safety and reduced travel time for business and recreational travellers. It also includes assigning values to social costs such as the inconvenience to owners due to loss of land and houses that lie on the highway route and increased noise to nearby houses.

In this chapter, we shall look at the social aspects of engineering decision making. First, we shall consider the reasons markets may fail in such areas as the environment and health. We shall also consider different methods that society may use to correct these failures. Second, we shall consider decision making in the public sector. Here we shall be concerned mainly with government projects or government-supported projects.

10.2 | Market Failure

A **market** is a group of buyers and sellers linked by trade in a particular product or service. The prices in a market that guide most production decisions usually reflect all the social benefits and costs of engineering decisions adequately. This is not always true, however. When prices do not reflect all social benefits and costs of a decision, we say that there has been *market failure*. When this occurs, society will seek a means of correcting the failure. In this section, we define market failure and give examples of its effects. Then we discuss a number of ways in which society seeks remedies for market failure.

10.2.1 Market Failure Defined

Most decisions in the private sector lead to market behaviour that has desirable outcomes. This is because these decisions affect mainly those people who are party to those decisions. Since people can generally freely choose to participate in markets, it is reasonable to assume that the individuals who participate must somehow benefit by their actions. In most cases, this is the end of the story. In an **efficient market**, decisions are made so that it is impossible to find a way for at least one person to be better off and no person to be worse off.

In some cases, however, decisions have important effects on people who are not party to the decisions. In these cases, it is possible that the gains to the decision makers, and any others who might benefit from the decisions, are less than the losses imposed on those who are affected by the decisions. Such situations are clearly undesirable. These decisions are instances of market failure. **Market failure** occurs when a market, left on its own, fails to make decisions in which resources are allocated efficiently. When decisions are made in which aggregate benefits to all persons who benefit from the decision are less than aggregate costs imposed on persons who bear the costs that result from the decision, the decision is inefficient and hence market failure has occurred. Market failure can occur if the decision maker does not correctly take into account the gains and losses imposed on others by the consequences of a decision.

There are several reasons for market failure. First, there may be no market through which those affected by the decision can induce the decision maker to take their situations into account. However, losses may exceed gains even when there is a market, if the market has insufficient information about the gains and losses resulting from decisions. Market failure can also occur whenever a single buyer or seller (a monopolist) can influence prices or output. Market failure can even occur when someone decides *not* to do something that would create benefits to others. The market would fail if the cost of creating the benefits was less than the value of the benefits.

Acid rain is an example of the effects of market failure. The burning of high-sulphur coal by thermal-electric power plants is believed to be one of the causes of acid rain. These plants could burn low-sulphur fuels, but they do not, partly because low-sulphur fuels are more expensive than high-sulphur fuels. If a market existed through which those affected by acid rain could buy a reduction in power plant sulphur emissions, they could try to make a deal with the power plants. They would be able to offer the power plants enough to offset the higher costs of low-sulphur fuel and still come out ahead. But there is no such market. The reason for this is that there is no single private individual or group whose loss from acid rain is large enough to make it worthwhile to offer the power plants payment to reduce sulphur emissions. It would require a large number of those affected by acid rain to form a coalition to make the offer. There are markets for electric power and for coal. However, they do not lead to socially desirable decisions about the use of power or coal. This is because the market prices for power and coal do not reflect the costs related to acid rain. If the prices for power and high-sulphur coal reflected these costs, less power would be used, and less of it would be made with high-sulphur coal. Both would reduce acid rain.

The damage to the residents of Sydney, Nova Scotia, due to air pollution and toxic waste is another example of the effects of market failure. The market failed to take into account the health and environmental costs in the price of steel. In the early years of steel production, there was little information about the deleterious health effects of air pollution and of the toxic coke oven by-products. However, as early as the 1960s the government started to exert pressure on the Dominion Steel and Coal Company (DOSCO), then operating the steel mill, to reduce its emissions. DOSCO balked, however, indicating that the estimated cost would put it out of business. Not long after,

DOSCO announced its intention to close the steel mills due to its inability to compete in international markets. This led the local residents to lobby DOSCO and the provincial government to keep the mills open. They claimed that closing the mill would have a devastating effect on the local economy. SYSCO then took over the operation of the steel mill. SYSCO continued to contribute to the pollution problem, and it was only in 1982, when contamination in the local lobster catch forced the fisheries to close, that the effects of market failure really started to become evident. The loss to the fishermen certainly had not been factored into the costs of running the steel mills. Over the 1980s other costs associated with the steel mills, unaccounted for in the price of steel, began to become more evident. Health problems such as birth defects, cancer, severe headaches, and lung disease in local residents were the subject of numerous scientific studies. The government spent $250 million between 1980 and 2000 in several unsuccessful attempts to clean up the toxic wastes. Finally, a decision was made in 2000 to close the steel mill as it became evident that the true costs of the mill far exceeded the benefits to the steel mill and to the local economy. Estimates range from $400 million to almost $1 billion.

We can see how market failure has caused socially undesirable outcomes such as acid rain and contaminated industrial sites. When markets fail, as in cases such as these, society will seek to remedy these problems through a variety of mechanisms. These remedies are the subject of the next section.

10.2.2 Remedies for Market Failure

There are three main formal methods of eliminating or reducing the impact of market failure:

1. Policy instruments used by the government
2. The ability of persons or companies adversely affected by the actions of others to seek compensation in the courts
3. Government provision of goods and services

We shall discuss the first two methods in this section. Government provision of goods and services is discussed in the next section under decision making in the public sector.

The first and most common means of trying to deter or reduce the effects of market failure is the use of what are commonly called **policy instruments**—government-imposed rules intended to modify behaviour.

One class of policy instruments includes *regulations* such as standards, bans, permits, or quotas. The regulations are backed by penalties for non-compliance. Most countries have regulations concerning such widely differing areas as product labelling, automobile emissions, and the use of bodies of water to dispose of waste. A challenge associated with developing regulations is that they may be inefficient. For example, suppose that we wish to improve the quality of a lake by reducing the amount of effluents dumped into it. These effluents may contain material with excessive biological oxygen demand (BOD). A typical regulation to control dumping would require all those who dump to meet the same BOD standards. But the costs of meeting these standards are likely to differ among the producers of the effluent. To attain a regulated reduction in BOD in their effluent, some producers will have to make expensive changes to their procedures, while others can respond with a lower cost. The most efficient way to obtain the reduction in BOD in the lake is to have those with low effluent-cleaning costs make the greatest reduction in BOD. Uniform regulation is not likely to do this. A common means of implementing regulations so that they are reasonably efficient is the use of a tradeable permit system. These permits allow producers emissions (as an example) up to a certain level, and if the producer is able to find ways to reduce pollution, it can sell its "spare" emissions capacity to others that find achieving such pollution reduction to be more difficult. Mini-Case 10.1 at the end of this chapter provides some additional information about such systems.

A second class of policy instruments meant to overcome market failure is *monetary incentives* or *deterrents* to induce desired behaviour. Monetary incentives or deterrents, often more efficient than regulations, may be subsidies or special tax treatments. For example, referring back to effluent dumping, there could be subsidies for the installation of equipment that would reduce the amount of effluent produced. In this way, producers for whom the cost of reducing BOD is low will do so since this would be cheaper than paying fees. By setting an appropriate subsidy, the desired reduction can be attained. The Australian fuel excise tax rebate noted in the opening vignette is another example of a monetary incentive.

Policy instruments are constantly being evaluated and modified. The reason is that market failure is a complex issue and policy makers may not fully understand its causes, or the implications of implementing a given policy. Nonetheless, their use is widespread and they can be highly effective in mitigating the effects of market failure.

A second formal means of reducing the effects of market failure is *litigation*. In the last several decades the use of courts as a means of reducing the health and safety effects of market failure has grown all over the world, particularly in the United States. Most countries have established legal regulations by which regular sellers of a product implicitly guarantee that the product is fit for reasonable use. Where the cost of reducing a risk in the use of their product is less than the objectively estimated expected loss, sellers are supposed to reduce the risk. Moreover, these sellers are held legally responsible for having expertise in the production and use of the products. It is not enough for sellers to say they did not know that use of the product was risky. Sellers are supposed to make reasonable efforts to determine potential risks in the use of the products they sell.

While the name by which these regulations are known varies by country—*Uniform Commercial Code* (United States), *Trade Practices Act* (Australia), *Consumer Code* (Britain, Canada)—they all fall into the general category of consumer law. The development of consumer law has been made more complex in recent years due to the prevalence of global markets where buyers and sellers are in different countries, each with different regulations.

The third formal method of reducing the effects of market failure is *government provision of goods and services*. This provision may be direct, as in the case of police services, or indirect, as in the case of health care given by physicians. Market failure is remedied when public sector analysts take into account all parties affected by a decision through a comprehensive assessment of total costs and benefits of a decision. Health care provision, transportation, municipal services, and electric and gas utilities are some examples of goods and services provided by the public sector. Each service requires numerous economic decisions to be made in the best interests of the public. This is of sufficient importance for us to devote a separate section of this chapter to the topic.

In addition to formal methods for dealing with market failure, *informal methods* can be used. Groups of individuals, whether formally organized or not, can exert pressure on companies or government to change their policies with respect to a specific issue. Direct pressure tactics can include lobbying companies, politicians, or the legal system itself. Demonstrations and boycotts are a peaceful means of social protest, but pressure tactics such as civil disobedience or criminal acts have also been used in attempts to change policy or practice. Another informal method is information dissemination to industry and to the general public, encouraging or discouraging certain behaviours.

10.3 | Decision Making in the Public Sector

This section is devoted to the decision-making process for public provision of goods and services. Public (government) production generally occurs mainly in two classes of goods and services. The first class includes those services for which there is no market because it is not practical to require people to pay for the service. Police and fire protection, defence, and the maintenance of city streets are examples of government services that it is not practical for users to pay for.

The second class includes those services for which scale economies make it inefficient to have more than one provider. Where there is only a single provider of a service, there is no market competition to enforce efficiency and low prices. There is a danger that the single provider, called a *monopolist*, will charge excessive prices and/or be inefficient. To ameliorate this potential problem, governments are often the provider.

For example, local deliveries of natural gas and electric power are situations where economies of scale are important enough that having more than one provider is not efficient. These services may be provided by publicly owned monopolies. For example, Royal Mail is the state-owned national postal service for the United Kingdom. An alternative to government provision of services where there is one provider is for the government to monitor and regulate the performance of a private monopolist. The British Gas Corporation, a public monopoly, was privatized in 1986. The Office of Gas and Electricity Markets (now OFGEM) was created to regulate the industry. British Gas was restructured in 1997 and then again in 2000 to form three separate companies with gas exploration, generation, and distribution functions.

Privatization of traditionally government-run functions has been a growing trend worldwide in the past several decades. Proponents of privatization argue that the private sector is more efficient than the government in its operations. Those against privatization argue that private organizations will not take into account all the social costs and benefits of their actions and thus contribute to the potential market failure. Close-Up 10.1 describes **public–private partnership** (P3) projects, in which private sector companies invest in public sector projects and recoup their investments through user-based tolls, fees, or tariffs.

In this section, we shall consider **benefit–cost analysis (BCA)**, a method of project evaluation widely used in the public sector. BCA provides a general framework for assessing the gains and losses associated with alternative projects when a broad societal view is necessary. This method is commonly used for public projects because the government is responsible for applying public resources in a manner that balances the costs and benefits of a project in such a way as to produce the greatest overall social benefit. BCA can also be used for project evaluation in the private sector where the broader impact on society must be taken into account.

The starting point for a BCA is a clear understanding of what alternative courses of action, or projects, are being considered. Once these are established, there are two main sets of issues to resolve. The first is to identify and measure the benefits associated with each project, and to clearly understand to whom these benefits accrue. The second is to identify and measure the costs associated with each project, and to establish who pays the costs. Once these issues are resolved, then the projects can be fairly evaluated by a variety of comparison methods. Benefit–cost ratios are a commonly used comparison method for public sector projects, though the methods of Chapters 4 and 5 are also applicable.

| CLOSE-UP 10.1 | Public–Private Partnerships |

Traditionally, large-scale public infrastructure projects—such as roads, bridges, power plants, and public utilities—have been financed, built, and maintained by government. In recent years, many countries have moved increasingly to public–private partnerships.

A public–private partnership, also known as a PPP or P3, is a contractual agreement between the government and a private organization to provide a service or product to society. The private partner may be a single company or consortium of several companies. The public sector typically maintains an oversight and quality-assessment role, while the private sector is responsible for provision of the service or project.

Numerous categories of P3s exist. Each is categorized on the basis of the extent of public and private sector involvement and the allocation of financial responsibility between the two. The most popular forms are the BOOT (build–own–operate–transfer) and BOT (build–operate–transfer) models, which are used most frequently to build government facilities such as schools, hospitals, bridges, and roads. In another category of PPP, the government sells an asset and then leases it back from the private partner.

In India, PPPs have been used to build/upgrade airports in Bangalore, Hyderabad, and Delhi. Some high-profile examples of PPPs in Canada are the Confederation Bridge construction between Nova Scotia and Prince Edward Island, and Ontario's 407 ETR toll highway. Some 190 PPPs were launched in Germany between 2000 and 2005 for projects such as hospitals, schools, and transport. A recently announced German PPP is an €89 million project to construct a 23 km six-lane A4 highway to be part of a Trans European Transport Network.

PPPs are not without problems, however. Critics worry that privatization may not bring about more efficient operations, and that the traditionally profit-oriented private sector may not take social good into account in its decision making. In addition, the private sector is not willing to use interest rates as low as the government typically uses, thus making the projects appear more expensive than when they are undertaken by the government.

10.3.1 The Point of View Used for Project Evaluation

In carrying out a benefit–cost analysis, it is important to clearly establish what point of view is to be taken, and to use this point of view consistently. The basic questions to address are: "Who will benefit from the project?" and "Who will pay for the project?" By identifying these two points of view clearly at the outset of the evaluation, confusion about what to include and what not to include can be reduced. The point of view defines who is "in" and who is "out."

Generally, it is members of society who are the users and beneficiaries of project services, and the government, the sponsor, who pays for the project. The evaluation will thus take into account the impact of the project on both the users and the sponsors. For example, if a government is considering a highway improvement project to reduce traffic congestion and to improve safety, the point of view taken would focus on the impact of the project on the government's finances, and on the benefits to the users most affected by the highway improvement. Therefore the analysis would include government costs such as construction and maintenance expenses. Savings (reduction in costs) to the government could include new property tax income if land values were to go up due to increased economic activity in the area. Social benefits could include factors such as reduced travel

NET VALUE 10.1

International Benefit–Cost Analysis Guides

Various branches of the government make benefit–cost analysis guides available to assist analysts and managers with the task of evaluating public projects. Several are listed below.

Handbook of Cost–Benefit Analysis (Department of Finance and Administration, Australian Government): This guide provides a framework for evaluating public projects, programs, or policies that have major social, economic, or environmental impacts. The guide indicates the steps to be taken in a thorough benefit–cost analysis and sets out best practices at each stage. It also provides insights into identifying and measuring costs and benefits associated with public projects. Three case studies illustrate the process of project evaluation. The guide can be found at www.finance.gov.au.

Cost Benefit Analysis Guidance Unit (Department for Transport, United Kingdom): This guide provides a framework for evaluating alternative projects when they have a transportation focus. It sets out the factors to be included in a cost–benefit analysis, such as business and consumer travel time, impacts of noise, safety, and security. The 2006 guide notes that they will be making the monetary valuation of air quality and greenhouse gas levels available in future. It also gives guidance on the discount rate that should be applied for projects of different periods. The guide can be found at www.webtag.org.uk.

Economic Analysis of Federal Regulations (US Office of Management and Budget): This guide provides a description of best practices for preparation of an economic analysis of a regulatory action. The guide helps analysts to prepare a thorough package to inform decision makers of the consequences of regulatory alternatives. It provides an overview of alternative ways in which regulations can be implemented (e.g., setting dates of compliance, methods of ensuring compliance, and various stringency levels). Principles for valuing benefits and costs are also detailed. This guide can be found at www.whitehouse.gov/omb. A similar guide for Canada, *Benefit Cost Analysis Guide for Regulatory Programs* (Privy Council Office, Canada), can be found at www.pco-bcp.gc.ca.

time, reduced vehicle operating costs, and increased safety for users of the goods and services provided by the project. Social costs (also called disbenefits) include the costs of travel delays and disruption during the construction period, increased traffic noise in nearby neighbourhoods, and possibly negative environmental impacts, all of which can affect users. Careful definition of the point of view taken ensures that all the effects of the project are taken into account.

One of the challenges associated with the analysis of public projects is that there may be several reasonable points of view, each with a different set of users or sponsors to be included in the analysis. Consider the decision to construct a dam in an agricultural area in order to provide hydroelectric power and irrigation. An analyst providing advice on such a project would need to consider the impact of relocating families from the flooded areas as well as the increased productivity and income to farmers due to improved irrigation. Aside from the costs and benefits to the local community, the analyst would also need to consider the potential impact on the wider environment due to the potential of moving from coal-fired electricity production to hydroelectricity. The point of view chosen delineates whose costs and benefits are to be taken into account in a benefit–cost analysis. The fact that several points of view may be taken can lead to ambiguity and controversy over the results. Nonetheless, BCA is a common framework for analysis.

Another important factor to consider in project evaluation is the frame of reference for measuring the impacts of the project. The analysis should consider costs and benefits associated with the project based on the difference between what would occur *with* the project versus what would occur *without* the project. In other words, we concern ourselves with the marginal benefits and costs associated with the project so that its impact is fairly measured.

10.3.2 Identifying and Measuring the Costs of Public Projects

Once we have a clear view of who will benefit from a project (the users) and who will pay for a project (the sponsor), the next step in a BCA is to identify and measure the various costs and benefits associated with the project. The fundamental questions are "In what ways will users benefit from the project, and by how much will they benefit?" and "What costs will be incurred by the sponsor, and how much will these costs be?"

This section deals with identifying and measuring the costs of public projects to the project sponsor. For the most part, these costs are straightforward to identify and measure. This is not the case with benefits, which will be covered in the next section.

The sponsor costs broadly include all of the resources, goods, and services required to develop, implement, and maintain a project or program. These can be generally classified into the initial capital costs and ongoing operating and administration costs. If the project creates savings, these are deducted from costs to produce a net cost to the sponsor.

For the highway-widening project mentioned in previous sections, the following is a classification of some of the sponsor's costs and savings:

Costs to the Sponsor
• Construction costs
• Operating and maintenance costs
• Administrative costs

Savings to the Sponsor
• Increased tax revenues due to higher land values

Some costs can be *directly* attributed to a project, while others may be stimulated indirectly by the project. These are referred to as *direct* and *indirect* costs, respectively. For example, in the above project, construction and operating costs are direct costs, and increased tax revenues are indirect savings.

While identifying and valuing the costs of a public project can be relatively straightforward, the same cannot generally be said about identifying and valuing social benefits.

10.3.3 Identifying and Measuring the Benefits of Public Projects

This section deals with identifying and measuring the benefits associated with public projects. We begin by providing an overview of the process, and then indicate some of the challenges associated with this portion of a benefit–cost analysis.

The benefits of a project to society include the value of all goods and services that result from the project or program. Generally, the benefits of a public project will be positive. However, some of the effects from a project may be negative. These are referred to as social costs, and are subtracted from the benefits to obtain a net measure of social benefits.

For the example of a highway improvement project, the following is a classification of some of the social benefits and social costs:

Social Benefits
• Reduced travel time for business and recreational users
• Increased safety
• Reduced vehicle operating costs for business and recreational users

Social Costs
• Increased noise and air pollution
• Disruption to the local environment
• Disruption of traffic flows or business transactions during construction
• Loss of business elsewhere due to traffic rerouting onto the new highway

As with sponsor costs, some of the social benefits or social costs can be *directly* attributed to a project, while others may be stimulated indirectly by the project. These are referred to as *direct* and *indirect* benefits, respectively. For example, direct benefits of an improved highway may be decreased travel time and increased safety. Indirect benefits could include lost business to other areas if traffic is rerouted from these areas onto the new highway. A thorough benefit–cost analysis should always include the direct benefits, and will include indirect benefits if they have an important effect on the overall project.

The task of identifying and measuring the social benefits and costs of public projects can be challenging. The challenge arises because the benefits may not be reflected in the monetary flows of the project. We are concerned with the real effects of a project, but the cash flows may or may not reflect all these real effects. For example, in the highway improvement project, there are cash flows for the wages of the workers who construct the road. This is a real cost to the government, and the wages reflect these costs well. In contrast, consider the intangible social cost of disruption of traffic during road construction. These costs are not reflected in the cash flows of a project, but are nonetheless an important cost of putting the road in place.

Beyond the challenge of identifying certain costs and benefits is that of assigning a value or measure to each. For goods and services that are distributed through markets, we have prices to measure values. Valuation is relatively straightforward for these items. However, many public projects create or use goods and services for which there is no market in which prices reflect values. These intangible, non-market goods and services are challenging to value.

Consider the highway improvement project mentioned earlier. Obvious sponsor costs are the labour, materials, and equipment used for the project. These costs can be relatively easy to estimate, as there are markets for these goods and services through which appropriate prices have been established. However, there are several intangible social costs and benefits that are somewhat more difficult to measure. First, there are the social costs associated with temporary travel-time disruptions due to construction. Measuring the cost of traffic disruption during the work requires valuing the time delays incurred by car passengers and trucks. There are approximations to these delay costs based on earnings per hour of passengers and the hourly cost of running trucks. The approximation for the value of travellers' time is based on the idea that a person who can earn, for example, $35 an hour working should be willing to pay $35 an hour for time saved travelling to work. The disruption costs may be large enough that they make it more efficient to have the work done at night, despite the fact that this will raise the explicit construction cost. The intangible social benefits of reduced travel time once the project is completed can also be factored into the benefit–cost analysis in a similar way.

An intangible social benefit associated with many public projects is improved health and safety. A common method for assessing health and safety benefits is to estimate the reduction in the number and type of injuries expected with a project or program and then to put a value on these injuries. A variety of methods exist, but one approach is to use lost wages, treatment costs, or insurance claims as the basis for estimating the value of certain injuries. Estimates of the value of a human life using this method range from several hundred thousand to several million dollars.

Other examples of intangible items commonly valued in a benefit–cost analysis are the value of noise abatement and the value of the environment. The United Kingdom's *Cost-Benefit Analysis Guidance Unit* contains useful information on these and other non-market costs and benefits associated with typical public projects, and helpful examples of how to value impacts where there are no market prices.

Two basic methods have been developed to value intangible social benefits and costs. The first is *contingent valuation*. This method uses surveys to ask members of the public what they would be willing to pay for the good or service in question. For example, members of the public may be asked how much they would be willing to pay for each minute of reduced travel time due to a highway improvement. Contingent valuation has the benefit of being a direct approach to valuation, but has the drawback that individuals may not provide accurate or truthful responses.

The second commonly used method for assessing the value of non-market goods and services is called the *hedonic price* method. This method involves deducing indirectly individuals' valuations based on their behaviour in other markets for goods and services. For example, the value of noise pollution due to a busy road may be estimated by examining the selling prices of houses near other busy roads and comparing these prices to those of similar houses away from busy roads. The value of the noise pollution cannot be higher than the difference in prices; otherwise, the owners would move. The advantage of this method is that members of the public reveal their true preferences through their behaviour. The disadvantage of the hedonic price method is that it requires reasonably sophisticated knowledge.

The following example provides further illustration of how the measurement of costs and benefits without market prices can be challenging, and also how these challenges can be dealt with.

EXAMPLE 10.1

Consider the construction of a bridge across a narrow part of a lake that gives access to a park. The major benefit of the bridge will be reduced travel time to get to the park from a nearby urban centre. This will lower the cost of camping trips at the park. As well, more people are expected to use the park because of the lower cost per visit. How can these benefits be measured?

Data concerning the number of week-long visits and their costs are shown in Table 10.1.

Table 10.1   Average Cost Per Visit and Number of Visits Per Year

	Without Bridge	With Bridge
Travel cost per visit (£)	140.00	87.50
Use of equipment (£)	50.00	50.00
Food cost per visit (£)	100.00	100.00
Total cost per visit (£)	290.00	237.50
Number of visits/year	8000	11 000

First, the reduction in cost for the 8000 visits that would have been made even without the bridge creates a straightforward benefit:

Travel cost saving on 8000 visits = $(140 - 87.50) \times 8000 = 420\ 000$

There is a benefit of £420 000 per year from reduced travel cost on the 8000 visits that would have been taken even without the bridge.

Next, we see that the number of visits to the park is expected to rise from 8000 per year to 11 000 per year. But how much is this worth? We do not have prices for park visits, but we do have data that enable estimates of actual costs to campers. These costs may be used to infer the value of visits to campers.

We see that before the bridge, the cost of a week-long park visit, including travel and other costs, averaged £290.

It is reasonable to assume that a week spent camping was worth at least £290 to anyone who incurred that cost. The average cost of a week-long visit to a park would fall from £290 per visit to £237.50 per visit if the bridge were built. We are concerned with the value of the incremental 3000 visits per year. Clearly, none of these visits would be made if the cost were £290 per trip. And each of them is worth at least £237.50 or else the trip would not have been taken. The standard approximation in cases like this is halfway between the highest and lowest possible values. This gives an aggregate benefit of the increased use of the park of

$$\frac{(290.00 + 237.50)}{2} \times 3000 = 791\ 250$$

Therefore, the value of the incremental 3000 visits per year is estimated as approximately £791 000 per year. However, there is also a cost of £237.50 per visit. The net benefit of the incremental 3000 visits is therefore

$$791\ 250 - 237.50(3000) = 78\ 750$$

The total value of benefits of the bridge is the sum of the reduced travel cost plus increased use:

$$420\ 000 + 78\ 750 = 498\ 750$$

The total value of the benefits yielded by the bridge is almost £500 000 per year. These benefits are the shaded area shown in Figure 10.1. These benefits must then be weighed against the costs of the bridge.■

Figure 10.1   Benefits From Bridge

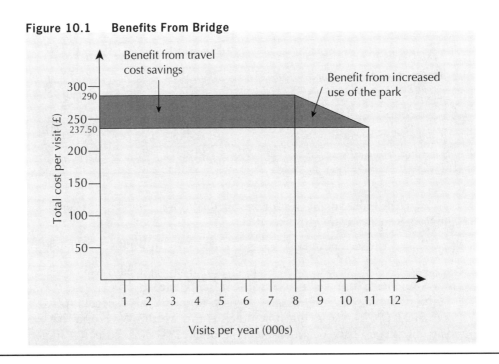

10.3.4 Benefit–Cost Ratios

Once we have identified and obtained measures for the costs and benefits associated with a public project, we can evaluate its economic viability. The same comparison methods that are used for private sector projects are appropriate for government sector projects. That is, we can use the present worth, annual worth, and internal-rate-of-return performance measures in the same ways in both the private and government sectors. It is important to emphasize this, because other methods based on ratios of benefits to costs have been used frequently in government project evaluations, almost to the exclusion of present worth, annual worth, and internal-rate-of-return methods. Because of the prevalent use of these ratios, this section is devoted to a discussion of several benefit–cost ratios that are commonly used in public sector decision making. We then point out several problems associated with the use of benefit–cost ratios so that the reader is aware of them and understands the correct way of using them.

Benefit–cost ratios can be based on either the present worths or the annual worths of benefits and costs of projects. The resulting ratios are equivalent in that they lead to the same decisions. We shall discuss the ratios in terms of present worths, but the reader should be aware that everything we say about ratios based on present worths applies to ratios based on annual worths.

The conventional **benefit–cost ratio (BCR)** is given by the ratio of the present worth of net users' benefits (social benefits less social costs) to the present worth of the net sponsor's costs for a project. That is,

$$BCR = \frac{PW(\text{users' benefits})}{PW(\text{sponsors' costs})}$$

A **modified benefit-cost ratio**, also in common use and denoted BCRM, is given by the ratio of the present worth of net benefits minus the present worth of operating costs to the present worth of capital costs, that is,

$$BCRM = \frac{PW(\text{users' benefits}) - PW(\text{sponsors' operating costs})}{PW(\text{sponsors' capital costs})}$$

Though the modified benefit–cost ratio is less used than the conventional benefit–cost ratio, it has the advantage that it provides a measure of the net gain per dollar invested by the project sponsor.

In general, a project is considered desirable if its benefit–cost ratio exceeds one, which is to say, its benefits exceed its costs.

EXAMPLE 10.2

The town of Mandurah, West Australia, has limited parking for cars near its main shopping street. The town plans to pave a lot for parking near Main Street. The main beneficiaries will be the merchants and their customers. The present worth of expected benefits is $3 000 000. The cost to the town of buying the lot, clearing, paving, and painting is expected to be $500 000. The present worth of expected maintenance costs over the lifetime of the project is $50 000. In the period in which the lot is being cleared and paved, there will be some disbenefits to the local merchants and customers due to disruption of traffic in the Main Street area. The present worth of the disruption is expected to be about $75 000. What is the benefit–cost ratio?

$$BCR = \frac{PW(\text{users' benefits})}{PW(\text{sponsors' costs})} = \frac{3\ 000\ 000 - 75\ 000}{500\ 000 + 50\ 000} = 5.3$$

We can see that the benefit–cost ratio exceeds 1, and thus the proposal is economically justified. ∎

EXAMPLE 10.3

A fire department in a medium-sized British city is considering a new dispatch system for its firefighting equipment responding to calls. The system would select routes, taking into account recently updated traffic conditions. This would reduce response times. An indirect effect would be a reduction in required fire-fighting equipment. The present worths of benefits and costs are:

Users' Benefits

PW(benefits) = £37 500 000

Sponsors' Costs

PW(operating costs) = £3 750 000

PW(capital costs for the city) = £13 500 000

PW(reduced equipment requirements) = £2 250 000

What is the benefit–cost ratio for the dispatch system? What is the modified benefit–cost ratio for the dispatch system? Is the project economically justifiable?

$$BCR = \frac{37\ 500\ 000}{(13\ 500\ 000 + 3\ 750\ 000 - 2\ 250\ 000)} = 2.50$$

$$BCRM = \frac{(37\ 500\ 000 - 3\ 750\ 000)}{(13\ 500\ 000 - 2\ 250\ 000)} = 3.00$$

We see for this example that both the conventional and modified cost-benefit ratio are greater than one, indicating that the project, under either criterion, is economically justified. ∎

Examples 10.2 and 10.3 demonstrate the basic use of benefit–cost ratios. It is worth noting that if a conventional benefit–cost ratio is greater (less) than one, it follows that the modified benefit–cost ratio will be greater (less) than one. This means that the two types of benefit–cost ratios will lead to the same decision. We present both because the reader may encounter either ratio.

For *independent* projects then, the following decision rule may be used: *Accept all projects with a benefit–cost ratio greater than one*. In other words, accept a project if

$$BCR = \frac{PW(\text{users' benefits})}{PW(\text{sponsors' costs})} > 1, \text{ or}$$

$$BCRM = \frac{PW(\text{users' benefits}) - PW(\text{sponsors' operating costs})}{PW(\text{sponsors' capital costs})} > 1$$

This rule, using either the benefit–cost ratio or the modified benefit–cost ratio, is equivalent to the rule that all projects with a present worth of benefits greater than the present worth of costs should be accepted. This is, then, the same rule that was presented in Chapter 4, which accepts a project if its present worth is positive. This is easily shown. If

$$\frac{\text{PW(users' benefits)}}{\text{PW(sponsors' costs)}} > 1$$

it follows that

PW(users' benefits) > PW(sponsors' costs)

which is equivalent to

PW(users' benefits) – PW(sponsors' costs) > 0

Also recall that, as shown in Chapter 5, the present/annual worth method and the internal rate of return method give the same conclusion if an independent project has a unique IRR.

To use benefit–cost ratios to choose among *mutually exclusive* projects, we must evaluate the increment between projects, just as we did in Chapter 5 with the internal rate of return method. Suppose we have two mutually exclusive projects, X and Y, with present worths of benefits, B_X and B_Y, and present worths of costs, C_X and C_Y. We first check to see if the individual benefit–cost ratios are greater than one. We discard a project with a benefit–cost ratio of less than one. If both projects have benefit–cost ratios greater than one, we rank the projects in ascending order by the present worths of costs. Suppose $C_X \geq C_Y$. We then form the ratio of the differences in benefits and costs,

$$\text{BCR(X} - \text{Y)} = \frac{B_X - B_Y}{C_X - C_Y}$$

If this ratio is greater than one, we choose project X. If it is less than one, we choose project Y. If $C_X = C_Y$ (in which case the ratio is undefined), we choose the project with the greater present worth of benefits. If $B_X = B_Y$, we choose the project with the lower present worth of costs.

This rule is the same as comparing two mutually exclusive projects using the internal rate of return method, which was presented in Chapter 5. In order to choose a project, we saw in Chapter 5 that not only the IRR of the individual project, but also the IRR of the incremental investment, must exceed the MARR. This rule is also the same as choosing the project with the largest present worth, as presented in Chapter 4, provided the present worth is positive.

The following example, concerning two mutually exclusive projects, summarizes our discussion of benefit–cost ratios and illustrates their use.

EXAMPLE 10.4

A medium-sized city is considering increasing its airport capacity. At the current airport, flights are frequently delayed and congestion at the terminal has limited the number of flights. Two mutually exclusive alternatives are being considered. Alternative A is to construct a new airport 65 kilometres from the city. Alternative B is to enlarge and otherwise upgrade the current airport that is only 15 kilometres from the city. The advantage of a new airport is that there are essentially no limits on its size. The disadvantage is that it will require travellers to spend additional travel time to and from the airport.

There are two disadvantages for upgrading and enlarging the current airport. One disadvantage is that the increase in size is limited by existing development. The second disadvantage is that the noise of airplanes in a new flight path to the current airport will reduce the value of homes near that flight path. A thousand homes will be affected, with the average loss in value about $25 000. Such losses are large enough that it is not reasonable to expect that the owners' losses will be offset by gains elsewhere. If the city wishes to ensure that their losses are offset, it must compensate these owners. Note that such compensation would not be an additional social cost. It would merely be a transfer from taxpayers (through the government) to the affected owners. Benefit and cost data are shown in Table 10.2.

Table 10.2   Airport Benefits and Costs

Effect	Alternative A New Airport (millions of $)	Alternative B Current Airport (millions of $)
Improved service/year	55	28.5
Increased travel cost/year	15	0
Cost of highway improvements	50	10
Construction costs	150	115
Reduced value of houses	0	25

The city will use a MARR of 10% and a 10-year time horizon for this project. What are the benefit–cost ratios for the two alternatives? Which alternative should be accepted?

Before we start the computation of benefit–cost ratios, we need to convert all values to present worths. Two effects are shown on an annual basis: improved service, which occurs under both alternatives, and increased travel cost, which appears only with the new airport. We get the present worths (in millions of $) by multiplying the relevant terms by the series present worth factor:

$$PW(\text{improved service of A}) = 55(P/A,10\%,10)$$
$$= 55(6.1446)$$
$$= 337.95$$

$$PW(\text{improved service of B}) = 28.5(6.1446) = 175.10$$

$$PW(\text{increased travel cost of A}) = 15(6.1446) = 92.17$$

The remainder of the costs are already in terms of present worth. The benefit–cost ratio for alternative A is

$$BCR(\text{alternative A}) = \frac{337.95}{(150 + 50 + 92.17)} \cong 1.16$$

and for alternative B is

$$BCR(\text{alternative B}) = \frac{175.10}{(10 + 115 + 25)} = 1.1673$$

First, we note that both benefit–cost ratios exceed one. Both alternatives are viable. Since they are mutually exclusive, we must choose one of the two. To decide which alternative is better, we must compute the benefit–cost ratio of the incremental investment between the alternatives. We use the alternative with the smallest present worth of costs as the starting point (alternative B), and compute the benefit–cost ratio associated with the marginal investment:

$$\text{BCR(A – B)} = (B_A – B_B)/(C_A – C_B)$$

$$= \frac{(337.95 – 92.17 – 175.10)}{(50 + 150) – (10 + 115 + 25)} = 1.4136$$

The ratio of the benefits of the new airport minus the benefits of the current airport modification over the difference in their costs is greater than one. We interpret this to mean that the benefit–cost ratio ranks the new airport ahead of the current airport.

We also note that the same ranking would result if we were to compare the present worths of the two alternatives. The present worth of the new airport (in millions of $) is

$$\text{PW(alternative A)} = 337.95 – 92.17 – 150 – 50 = 45.78$$

The present worth of modifying the current airport (in millions of $) is

$$\text{PW(alternative B)} = 175.10 – 10 – 115 – 25 = 25.10$$

In addition to the quantifiable aspects of each of the two airports, recall that the new airport is preferred in terms of the unmeasured value of space for future growth. As well, the new airport does not entail the difficulties associated with compensating home owners for loss in value of their homes. Together with the benefit–cost ratios (or equivalently, the present worths), this means that the city should build the new airport.■

A word of caution is needed before we conclude this section. There may be some ambiguity in how to correctly calculate benefit–cost ratios, either conventional or modified. The reason is that for some projects, it may not be clear whether certain positive effects of projects are benefits to the public or reductions in cost to the government, or whether certain negative effects are disbenefits to the public or increases in costs to the government. As a result, benefit–cost ratios for a given project may not be unique. This lack of uniqueness means that a *comparison of the benefit–cost ratios of two projects is meaningless.* It is for this reason that we need to exercise caution when comparing projects with the benefit–cost ratio. To avoid this ambiguity, many experts recommend the use of present worth or annual worth for comparisons rather than benefit–cost ratios. Despite this recommendation, benefit–cost ratios are still used extensively for project comparisons, and since they are still in use we present readers with enough material to properly construct and interpret them.

To illustrate the ambiguity in constructing benefit–cost ratios, consider Example 10.4. The reduction in the value of residential properties could have been treated as a disbenefit to the public instead of a cost to the government for compensating homeowners. As a result, there are two reasonable benefit–cost ratios for alternative B. The second is

$$\text{BCR}_2\text{(alternative B)} = \frac{175.10 – 25}{(10 + 115)} = 1.2008$$

The lack of uniqueness of benefit–cost ratios does not mean that the ratios cannot be used, but it does mean that some care needs to be given to their correct application. Comparison methods based on benefit–cost ratios remain valid because the comparison methods depend only on whether the benefit–cost ratio is less than or greater than one. Whether it is greater or less than one does not depend upon how positive and negative effects are classified. This is clearly illustrated in the following example.

EXAMPLE 10.5

A certain project has present worth of benefits, B, and present worth of costs, C. As well, there is a positive effect with a present worth of d; the analyst is unsure of whether this positive effect is a benefit or a reduction in cost. There are two possible benefit–cost ratios,

$$\text{BCR}_1 = \frac{B + d}{C}$$

and

$$\text{BCR}_2 = \frac{B}{C - d}$$

For a ratio to exceed one, the numerator must be greater than the denominator. This means that, for BCR_1,

$$\text{BCR}_1 = \frac{B + d}{C} > 1 \quad \Leftrightarrow \quad B + d > C \quad \Leftrightarrow \quad B > C - d$$

But this is the same as

$$\frac{B}{C - d} = \text{BCR}_2 > 1$$

Consequently, any BCR that is greater than one or less than one will be so regardless of whether any positive effects are treated as positive benefits or as negative costs. A similar analysis would show that the choice of classification of a negative effect as a cost or as a reduction in benefits does not affect whether the benefit–cost ratio is greater or less than one. ■

To conclude this section, we can note several things about the evaluation of public sector projects. First, the comparison methods developed in Chapters 4 and 5 are fully applicable to public sector projects. Second, the reader may encounter the use of benefit–cost ratios for public projects, despite the fact that benefit–cost ratios can be ambiguous. Next, the reader should be wary of decisions based on the absolute magnitude of benefit–cost ratios, as these ratios may be changed by reclassification of some of the positive effects of projects as benefits or as reductions in cost, or by reclassification of some of the negative effects as reductions in benefits or increases in costs. However, since the classifications do not affect whether a ratio is greater or less than one, it is possible to use benefit–cost ratios to reach the same conclusions as those reached using the methods of Chapters 4 and 5.

10.3.5 The MARR in the Public Sector

There are significant differences between the private sector and public (government) sector of society with respect to investment. As observed in the previous sections, public sector organizations provide a mechanism for resources to be allocated to projects believed beneficial to society in general. These include projects for which scale economies make it inefficient to have more than one supplier, and projects in markets that would otherwise suffer market failure. The government also regulates markets and collects and redistributes taxes toward the goal of maximizing social benefits. Typical projects include health, safety, education programs, cultural development, and infrastructure development. Profits generated by public projects are not taxed.

Private institutions and individuals, in contrast to public institutions, are more concerned with generating wealth (profits) and are taxed by the government on this wealth. We would therefore expect the MARR for a public institution to be lower than that of a private institution, because the latter has a substantial extra expense acting to reduce its profits.

In evaluating public sector projects, the MARR is used in the same way as in evaluating private sector projects—it captures the time value of money. In the private sector, the MARR expresses the minimum rate of return required on projects, taking into account that those profits will ultimately be taxed. The MARR used for public projects, often called the *social discount rate*, reflects the more general investment goal of maximizing social benefits. Of course, as with private projects, public projects will sometimes be chosen on the basis of lowest present cost (or the equivalent) rather than highest present worth.

There is substantial debate as to what an appropriate social discount rate should be. Several viewpoints can shed some light on this issue.

The first is simply that the MARR should be the interest rate on capital borrowed to fund a project. In the public sector, funds are typically raised by issuing government bonds. Hence, the current bond rate might be an appropriate MARR to use. The second is that the MARR used to evaluate public projects should take into account that government spending on public projects consumes capital that might otherwise be used by taxpayers for private purposes. This viewpoint says that funds taken away from individuals and private organizations in the form of taxes might otherwise be used to fund private projects. This line of thinking might lead to a social discount rate that is the same as the MARR for the taxpayers. The two viewpoints are lower and upper bounds, respectively, to what can be seen as reasonable MARRs to apply when evaluating public projects.

In practice, a range of MARRs is used for evaluating public projects so that the impact of these differing points of view can be understood. (See Close-Up 10.2.)

CLOSE-UP 10.2 MARR Used in the Public Sector

Public projects in different countries are evaluated using various MARRs. Some examples of the MARR applied in the public sector are shown in the table below. Due to uncertainty about what the *correct* MARR is, sensitivity analysis is recommended for public projects.

Location	Project Type	Recommended MARR
United Kingdom (Treasury)	Projects < 30 years	3.5%
	Projects 31–75 years	3.0%
	Projects 76–125 years	2.5%
European Union	Depending on nation	3.5%–6%
Canada (Treasury)	Projects > 30 years	3%–7%
	Projects < 30 years	8%–12%
United States	Projects 3–30+ years	2.5%–3%
New Zealand	All projects	10%

REVIEW PROBLEMS

REVIEW PROBLEM 10.1

This review problem is adapted from an example in the Canadian Treasury Board's *Benefit–Cost Analysis Guide* (1976).

There are periodic floods in the spring and drought conditions in the summer that cause losses in a 15 000-square-kilometre Prairie river basin that has a population of 50 000 people. The area is mostly farmland, but there are several towns. Several flood control and irrigation alternatives are being considered:

1. Dam the river to provide flood control, irrigation, and recreation.
2. Dam the river to provide flood control and irrigation without recreation.
3. Control flooding with a joint Canada–United States water control project on the river.
4. Develop alternative land uses that would not be affected by flooding.

The constraints faced by the government are the following:

1. The project must not reduce arable land.
2. Joint Canada–United States projects are subject to delays caused by legal and political obstacles.
3. Damming of the river in the United States will cause damage to wildlife refuges.
4. The target date for completion is three years.

Taking into account the constraints, alternatives 3 and 4 above can be eliminated, leaving two:

1. Construct a dam for flood control, irrigation, and recreation.
2. Construct a dam for flood control and irrigation only.

A number of assumptions were made with respect to the dam and the recreational facilities:

1. An earthen dam will have a 50-year useful life.
2. Population and demand for recreational facilities will grow by 3.25% per year.
3. A three-year planning and construction period is reasonable for the dam.
4. Operating and maintenance costs for the dam will be constant in real dollars.
5. Recreational facilities will be constructed in year 2.
6. It will be necessary to replace the recreational facilities every 10 years. This will occur in years 12, 22, 32, and 42. Replacement costs will be constant in real dollars.
7. Operating and maintenance costs for the recreational facilities will be constant in real dollars.
8. The real dollar opportunity cost of funds used for this project is estimated to be in the range of 5%–15%.

The benefits and costs of the two projects are shown in Tables 10.3 and 10.4.

Notice that the benefits and costs are estimated averages. For example, the value of reduced flood damages will vary from year to year, depending on such factors as rainfall and snowmelt. It is not possible to predict actual values for a 50-year period.

Table 10.3   Estimated Average Benefits of the Two Projects

Year	Flood Damage Reduction	Irrigation Benefits	Recreation Benefits
0	$ 0	$ 0	$ 0
1	0	0	0
2	0	0	0
3	182 510	200 000	27 600
4	182 510	200 000	28 497
⋮	⋮	⋮	⋮
52	182 510	200 000	132 288

Table 10.4   Estimated Average Costs of the Two Projects

Year	Dam Construction	Operating and Maintenance Dam	Recreation Construction	Operating and Maintenance Recreation
0	$ 300 000	$ 0	$ 0	$ 0
1	750 000	0	0	0
2	1 500 000	0	50 000	0
3	0	30 000	0	15 000
4	0	30 000	0	15 000
⋮	⋮	⋮	⋮	⋮
11	0	30 000	0	15 000
12	0	30 000	20 000	15 000
13	0	30 000	0	15 000
⋮	⋮	⋮	⋮	⋮
21	0	30 000	0	15 000
22	0	30 000	20 000	15 000
23	0	30 000	0	15 000
⋮	⋮	⋮	⋮	⋮
31	0	30 000	0	15 000
32	0	30 000	20 000	15 000
33	0	30 000	0	15 000
⋮	⋮	⋮	⋮	⋮
41	0	30 000	0	15 000
42	0	30 000	20 000	15 000
43	0	30 000	0	15 000
⋮	⋮	⋮	⋮	⋮
52	0	30 000	0	15 000

(a) What is the present worth of building the dam only? What is the benefit–cost ratio? What is the modified benefit–cost ratio? Use 10% as the MARR.

(b) What is the present worth of building the dam plus the recreational facilities? Use 10% as the MARR.

(c) What is the benefit–cost ratio for building the dam and recreation facilities together? What is the modified benefit–cost ratio?

(d) Which project, 1 or 2, is preferred, on the basis of your benefit–cost analysis? Use 10% as the MARR.

ANSWER

(a) We need to determine the present worth of benefits and costs of the dam alone. There are two benefits from the dam alone. They are those resulting from reduced flood damage and those associated with the benefits of irrigation. Both are approximated as annuities that start in year 3. We get the present worths of these benefits by multiplying the annual benefits by the series present worth factor and the present worth factor. The present worth of benefits resulting from reduced flood damage is

$$PW(\text{flood control}) = 182\ 510(P/A,10\%,50)(P/F,10\%,2)$$
$$= \frac{182\ 510(9.99148)}{(1.1)^2}$$
$$= 1\ 495\ 498$$

Similar computations give the present worth of irrigation as

$$PW(\text{irrigation}) = 1\ 638\ 812$$

There are two costs for the dam: capital costs that are incurred at time 0 and over years 1 and 2, and operating and maintenance costs that are approximated as an annuity that begins in year 3. Capital costs are given by

$$PW(\text{dam, capital cost}) = 300\ 000 + 750\ 000(P/F,10\%,1)$$
$$+ 1\ 500\ 000(P/F,10\%,2)$$
$$= 2\ 221\ 487$$

The present worth of operating and maintenance costs is obtained in the same way as the present worths of flood control and irrigation benefits. The result is

$$PW(\text{operating and maintenance}) = 245\ 822$$

The present worth of the dam alone is

$$PW(\text{dam}) = 1\ 495\ 498 + 1\ 638\ 812 - (2\ 221\ 488 + 245\ 822)$$
$$= 667\ 001$$

The benefit–cost ratio for the dam is

$$BCR(\text{dam}) = \frac{1\ 495\ 498 + 1\ 638\ 812}{2\ 221\ 488 + 245\ 822} = 1.27$$

The modified benefit–cost ratio is given by

$$BCRM(\text{dam}) = \frac{1\ 495\ 498 + 1\ 638\ 812 - 245\ 822}{2\ 221\ 488} = 1.30$$

The present worth of the dam alone is positive, and both benefit–cost ratios are greater than one. The dam alone appears to be economically viable.

(b) We already have the present worths of benefits and costs for the dam alone. Therefore, we need only compute the present worths of benefits and costs for the recreation facilities. The capital costs for the recreation facilities consist of five outlays, in years 2, 12, 22, 32, and 42. The present worth of capital costs for the recreational facilities is given by

PW(recreation facilities, capital cost)

$$= 50\ 000(P/F,10\%,2)$$
$$+ 20\ 000[(P/F,10\%,12) + (P/F,10\%,22)$$
$$+ (P/F,10\%,32) + (P/F,10\%,42)]$$
$$= 51\ 464$$

Operating and maintenance costs are estimated as an annuity that starts in year 3. The computation is the same as that of similar annuities that were used for the benefits and operating and maintenance costs of the dam. Thus, the present worth of recreation operating and maintenance costs is

PW(recreation facilities, operating and maintenance) = 122 911

To obtain the present worth of the benefits from recreation, we need to define a growth-adjusted interest rate with $i = 10\%$ and $g = 3.25\%$ per year.

$$i^\circ = \frac{1 + i}{1 + g} - 1 = \frac{1 + 0.10}{1 + 0.0325} - 1 = 0.0653$$

We then use this to get the present worth geometric gradient series factor,

$$(P/A,g,i,N) = \left(\frac{(1 + i^\circ)^N - 1}{i^\circ(1 + i^\circ)^N} \right) \left(\frac{1}{1 + g} \right)$$

$$= \left(\frac{(1.0653)^{50} - 1}{0.0653(1.0653)^{50}} \right) \left(\frac{1}{1.0325} \right) = 14.19$$

To bring this to the end of year 0, we multiply by $(P/F,10\%,2)$. We then multiply by the initial value to get the present worth of recreation benefits.

PW(recreation benefits) $= [(P/A,3.25\%,10\%,N)(27\ 600)\ (P/F,10\%,2)]$
$$= 14.19(27\ 600)(1.1)^{-2}$$
$$= 323\ 679$$

Another way to get this result is to use a spreadsheet. First, a column that contains the benefits in each year is created. The benefits start at $27 600 in the third year and grow at 3.25% each year. Each benefit is then multiplied by the appropriate present worth factor in another column to obtain its present worth. The individual present worths are then summed to obtain the overall total of $323 679. Some of the spreadsheet computations are shown in Table 10.5.

The total present worth of the recreation facilities is

PW(total recreation facilities) $= 323\ 679 - 51\ 464 - 122\ 911$
$$= 149\ 304$$

The present worth of the dam plus recreation facilities is obtained by adding the present worth of the dam alone to the present worth of the recreation facility. The final result is given by

PW(dam and recreation facility) $= 667\ 001 + 149\ 304$
$$= 816\ 305$$

Table 10.5   Spreadsheet Computations

Year	Recreation Benefits	PW of Recreation Benefits
0	$ 0.00	$ 0.00
1	0.00	0.00
2	0.00	0.00
3	28 600.00	20 736.29
4	28 497.00	19 463.83
5	29 423.15	18 269.46
6	30 379.40	17 148.38
⋮	⋮	⋮
52	132 288.48	931.33
Total PW		323 678 .84

In present worth terms, the present worth of the dam *and* recreation facility exceeds that of the dam alone, so the dam and the recreation facility should be chosen.

(c) The benefit–cost ratio for the dam and recreation facilities together is

BCR(dam and recreation)

$$= \frac{(1\ 495\ 498 + 1\ 638\ 812 + 323\ 679)}{(2\ 221\ 488 + 245\ 822 + 51\ 464 + 122\ 911)} = 1.31$$

The modified benefit–cost ratio is

BCRM(dam and recreation)

$$= \frac{(1\ 495\ 498 + 1\ 638\ 812 + 323\ 679 - 245\ 822 - 122\ 911)}{(2\ 221\ 488 + 51\ 464)}$$

$$= 1.36$$

The dam and recreation facilities project appears to be viable, since the benefit–cost ratios are greater than one.

(d) On the basis of benefit–cost ratios, which of the two projects, project 1 or project 2, should be chosen? The dam and recreation facility is more costly, so the correct benefit–cost ratio for comparing the two is

$$BCR = \frac{\text{Benefits(dam and recreation facility)} - \text{Benefits(dam)}}{\text{Costs(dam and recreation facility)} - \text{Costs(dam)}}$$

$$= \frac{323\ 679}{(51\ 464 + 122\ 911)} = 1.86$$

The ratio exceeds one, and hence the dam and recreation facilities project should be chosen. This is consistent with the original present worth computations. ■

SUMMARY

Chapter 10 concerns decision making in the public sector. We started by considering why markets may fail to lead to efficient decisions. We presented three formal methods by which society seeks to remedy market failure, one of which is to have production by government. Next, we laid out three issues in decision making about government production. First, we saw that the identification and measurement of costs and benefits in the public sector are more difficult than in the private sector. Identification may be difficult because there may not be cash flows that reflect the costs or benefits. Measurement may be difficult because there are no prices to indicate values. Second, we discussed the use of benefit–cost ratios in public sector project evaluation. While it is possible to use benefit–cost ratios so as to give the same conclusions as those obtained using the comparison methods discussed in Chapters 4 and 5, care must be taken because there may be ambiguity in the way in which costs and benefits are classified. Third and last, we considered the MARR for government sector investments. The result of this discussion is that it is wise to use a range of MARRs corresponding to the opportunity cost of funds used in the public sector.

Engineering Economics in Action, Part 10B:
Look at It Broadly

"How was your trip, Naomi?" Anna Kulkowski asked as she walked into the conference room. Bill and Naomi were already there waiting.

"Well, I'm here," responded Naomi, "but my jets are still lagging."

"You'll get used to it." Anna looked at Bill. "So, what's it going to cost to implement the Greenhouse Challenge Plus program for the folks down under?"

"Naomi got some additional information from the people at Melbourne Manufacturing," Bill answered. "They did some checking around with a couple of freight companies based in the same region. Apparently it does not cost anything, directly, to join the Greenhouse Challenge Plus program, but there are lots of costs. As it turns out, there are lots of benefits as well. The big positive is that the program's focus is on reducing greenhouse gas emissions overall, which is much broader than just reducing fuel emissions from our truck fleet."

Naomi continued, "It does require Melbourne Manufacturing to do an energy audit and progress report each year, but there is quite a bit of information available on the internet about this, and energy reducing ideas, too." She paused, "So things are not as bad as I first told Bill at all. Once we found out how broad the program actually is, lots of possibilities opened up. Melbourne Manufacturing has not focused much on making the warehouse and offices more energy efficient yet, so they can benefit quite a bit from simple things like more efficient light bulbs and computer monitors, or setting the air conditioning at a higher temperature. The office staff had an impromptu focus group discussion the day before I left that was really productive. A lot of no-cost ideas came up. One of the newest engineers even suggested that she might be able to increase the amount shipped per truckload by some clever delivery route redesign. That could reduce fuel consumption without changing our service levels or delivery volumes."

"That all makes sense," Anna said. "It's a combination of zero-cost initiatives and some changes that will need Melbourne Manufacturing to make some relatively small investments. Once they are able to project the savings due to energy reduction, they should be off to a good start."

"Yes," Naomi said. "By the time I left, they had started to look into some of the government incentive plans for investing in energy efficient office equipment and to estimate the potential savings. Based on reports from other companies that have implemented some of these changes, they should be able to be a lot greener and save money doing it."

"And to think," Naomi mused, "that this all started with trying to make sure that Melbourne Manufacturing would be able to claim its fuel tax credits. Who would have thought it could have such a broad impact on operations?"

"That's true, Naomi," Anna commented. "Good work!"

PROBLEMS

For additional practice, please see the problems (with selected solutions) provided on the Student CD-ROM that accompanies this book.

10.1 The following data are available for a project:

Present worth of benefits	€17 000 000
Present worth of operating and maintenance costs	€5 000 000
Present worth of capital costs	€6 000 000

   (a) Find the benefit–cost ratio.

   (b) Find the modified benefit–cost ratio.

10.2 The following data are available for two mutually exclusive projects:

	Project A	Project B
PW(benefits)	$19 000 000	$15 000 000
PW(operating and maintenance costs)	5 000 000	8 000 000
PW(capital cost)	5 000 000	1 000 000

   (a) Compute the benefit–cost ratios for both projects.

   (b) Compute the modified benefit–cost ratios for both projects.

   (c) Compute the benefit–cost ratio for the increment between the projects.

   (d) Compute the present worths of the two projects.

   (e) Which is the preferred project? Explain.

10.3 The following data are available for two mutually exclusive projects:

	Project A	Project B
PW(benefits)	£17 000 000	£17 000 000
PW(operating and maintenance costs)	5 000 000	11 000 000
PW(capital cost)	6 000 000	1 000 000

   (a) Compute the benefit–cost ratios for both projects.

   (b) Compute the modified benefit–cost ratios for both projects.

   (c) Compute the benefit–cost ratio for the increment between the projects.

   (d) Compute the present worths of the two projects.

   (e) Which is the preferred project? Explain.

10.4 The following data are available for two mutually exclusive projects:

   (a) Compute the benefit–cost ratios for both projects.

   (b) Compute the modified benefit–cost ratios for both projects.

	Project A	Project B
PW(benefits)	$17 000 000	$15 000 000
PW(operating and maintenance costs)	5 000 000	8 000 000
PW(capital cost)	6 000 000	3 000 000

(c) Compute the benefit–cost ratio for the increment between the projects.

(d) Compute the present worths of the two projects.

(e) Which is the preferred project? Explain.

10.5 There are two beef packing plants, A and B, in the city of Rybinsk, Russia. Both plants put partially treated liquid waste into the Volga river. The two plants together dump over 33 000 kilograms of BOD5 per day. (BOD5 is the amount of oxygen used by microorganisms over five days to decompose the waste.) This is more than half the total BOD5 dumped into the river at this location. The city wants to reduce the BOD5 dumped by the two plants by 10 000 kilograms per day.

The following data are available concerning the two plants:

	Outputs of the Two Plants	
	Steers/Day	**BOD5/Steer (kg)**
Plant A	20 000	1.0
Plant B	9 000	1.5

The costs of making reductions in BOD5 per steer in Russian rubles (RUB) are shown below:

	Incremental Cost of Reducing BOD (rubles/kg/steer)						
Reduction (kg/steer)	**0.1**	**0.2**	**0.3**	**0.4**	**0.5**	**0.6**	**0.7**
Plant A	1.5	2.4	3.6	7.5	13.5	19.5	28.5
Plant B	4.5	4.5	4.5	4.5	4.5	10.5	13.5

For example, to reduce the BOD5 of plant A by 0.25 kilograms per steer, the cost is calculated as

$$(0.1 \times 1.5) + (0.1 \times 2.4) + (0.05 \times 3.6) = \text{RUB } 0.57/\text{steer}$$

The council is considering three methods of inducing the plants to reduce their BOD5 dumping: (1) a regulation that limits BOD5 dumping to 0.81 kilograms/steer, (2) a tax of RUB 4.8/kilogram of BOD5 dumped, and (3) a subsidy paid by the town to the plants of RUB 4.8/kilogram reduction from their current levels in BOD5 dumped.

(a) Verify that, if both plants reduce their BOD5 dumping to 0.81 kilograms/steer, there will be a 10 000 kilograms/day reduction in BOD5 dumped. What will this cost?

(b) Under a tax of RUB 4.8/kilogram, how much BOD5 will plant A dump? How much will plant B dump? (Assume that outputs of steers would not be affected by the tax.)

Verify that this will lead to more than a 10 000 kilograms/day reduction in BOD5. What will this cost?

(c) Under a subsidy of RUB 4.8/kilogram reduction in BOD5, how much will plant A dump? How much will plant B dump? Verify that this will lead to more than a 10 000 kilograms/day reduction in BOD5. What will this cost?

(d) Explain why the tax and subsidy schemes lead to the same behaviour by the meat packing plants.

(e) Explain why the tax and subsidy schemes have lower costs for the company than the regulation.

10.6 There are three petrochemical plants, A, B, and C, in Houston, Texas. The three plants produce Good Stuff. Unfortunately, they also dump Bad Stuff into Houston air. Data concerning their outputs of Stuff are shown below:

	Outputs	
	Good (kg/day)	Bad/Good (cL/kg)
Plant A	17 000	10
Plant B	11 000	15
Plant C	8000	18

The city wants to reduce the dumping of Bad Stuff by 150 000 centilitres per day. Costs for reducing the concentration of Bad Stuff in output are shown below.

The city is considering two methods: (1) Require all plants to meet the performance level of the best-practice plant, plant A, which is 10 centilitres of Bad Stuff per kilogram of Good Stuff. (2) Impose a tax of $0.20/centilitre of Bad Stuff dumped.

Incremental Cost of Reducing Bad Stuff/Good Stuff ($/cL/kg)								
Reduction (cL/kg)	1	2	3	4	5	6	7	8
Plant A	0.02	0.032	0.048	0.1	0.18	0.26	0.38	0.57
Plant B	0.06	0.06	0.063	0.068	0.075	0.193	0.27	0.405
Plant C	0.25	0.25	0.25	0.25	0.25	0.25	0.25	0.375

(a) What will be the reduction in dumping of Bad Stuff under the best-practice regulation? What will be the cost of this reduction?

(b) Under the tax, how much Bad Stuff will be dumped from the three plants combined? What will be each plant's reduction in dumping per kilogram of Good Stuff?

(c) How much will the reduction of dumping cost for each company under the tax?

10.7 In the summer of 2004, the city of Kitchener, Ontario, often experienced the worst air pollution in Canada, affecting the health of hundreds of thousands of people. Explain how air pollution is an example of market failure. Give an example of how each of the four remedies for market failure listed in Subsection 10.2.2 might be applied to the case of air pollution.

10.8 The North Sea fishing industry has been devastated in recent years because of overfishing. Tens of thousands of jobs have been lost in Scotland as quotas have been cut drastically in an attempt to save fish such as cod, haddock, and whiting from extinction. How is over-fishing an example of market failure? Give an example of how each of the four remedies for market failure listed in Subsection 10.2.2 might be applied to the case of overfishing.

10.9 Consider these situations in which cutting of trees is relevant:

1. The Brown family owns a view house in Perth, Australia. Their neighbours across the street, the Smith family, have trees on their lot that are obstructing the Browns' view of Matilda Bay. The Browns are the only ones affected by the Smiths' trees. The Browns have asked the Smiths to top their trees, but the Smiths refuse to do so. The Browns have also offered to pay the Smiths for the topping and an additional $500 to cover any loss they might feel because their tall trees were topped. The Smiths still refuse.

2. The Brown family, the Green family, the White family, and the Blue family own view houses in Perth, Australia. Their neighbours across the street, the Smith family, have trees on their lot that are obstructing everyone's view of Matilda Bay. The Browns, Greens, Whites, and Blues have asked the Smiths to top their trees, but the Smiths refuse to do so.

3. Timber companies in southwestern Australia sometimes use clear-cutting on old-growth forests. Environmentalists have asked the companies to change this practice because it leads to reduced biodiversity.

   Why is there no market failure in the first situation involving the Smiths and the Browns? Why may there be market failure in the situation with several families? Why is there market failure in the third situation involving the timber companies?

10.10 Technical changes in electricity supply and information transmission have made it efficient for consumers of both services to be served by suppliers using different technologies and operating in different locations. Does this increase or decrease the need for government regulation in these industries?

10.11 An electric utility company is considering a re-engineering of a major hydroelectric facility. The project would yield greater capacity and lower cost per kilowatt-hour of power. As a result of the project, the price of power would be reduced. This is expected to increase the quantity of power demanded. The following data are available:

Effect of Reduced Price of Power	
Current price ($/kWh)	0.07
Current consumption (kWh/year)	9 000 000
New price ($/kWh)	0.05
Expected consumption (kWh/year)	12 250 000

What is the annual benefit to consumers of power from this project?

10.12 Brisbane and Johnsonburg are two towns separated by the Wind River. Traffic between them crosses the river by a ferry run by the Johnsonburg Ferry Company, which charges a toll. The two towns are considering building a bridge somewhat upstream from the ferry crossing; there would be no toll on the bridge. Travel time between the towns would be about the same with the bridge as with the ferry because of the bridge's upstream location. The following information is available concerning the crossing.

Ferry/Bridge Information	
Ferry crossings (number/year)	60 000
Average cost of ferry trip ($/crossing)	1
Ferry fare ($/crossing)	1.5
Bridge toll ($/crossing)	0
Expected bridge crossings (number/year)	90 000
EAC of bridge ($/year)	85 000

Note that all data are on an annual basis. The cost of the bridge is given as the equivalent annual cost of capital and operating costs. We assume that all bridge costs are independent of use, that is, there are no costs that are due to use of the bridge. The average cost per crossing of the ferry includes capital cost and operating cost.

(a) If the bridge were built, what would be the annual benefits to travellers?

(b) How much would the owners of the Johnsonburg Ferry Company lose if the bridge were built?

(c) What would be the effect on taxpayers if the bridge were built? (Assume that Johnsonburg Ferry pays no taxes.)

(d) What would be the net social gains or losses if the bridge were built? Take into account the effects on travellers, Johnsonburg Ferry owners, and taxpayers.

(e) Would the net social gains or losses be improved if there were a toll for crossing the bridge?

10.13 It is common for cities in cold countries to provide snowplow service for public roads. The major benefit of such services is to allow the convenient movement of vehicles over public roads following snow accumulation, at a cost of the snowplow (capital and operating) and driver. Are there other costs and benefits? List all you can think of, along with how they could be measured.

10.14 Most countries provide travel information kiosks alongside major highways just across the border from neighbouring countries. These kiosks provide maps and brochures on attractions, and some will make hotel and campground reservations. There are obvious costs, such as staffing costs and building capital and maintenance costs. What other costs and benefits are associated with this government service? How can these costs and benefits be measured?

10.15 In order to encourage the use of carpooling, the city of Beijing is considering putting carpool parking lots near major transportation arteries into downtown Beijing. Each parking lot allows commuters to meet in separate cars, park all but one, and proceed in one car to a joint destination. Studies estimate that an average of 200 cars will be parked at each lot on weekdays, with one-quarter of that number on weekends. The average commuting distance from the parking lots is 75 kilometres, and the marginal cost of driving an average car is 2 yuan per kilometre.

(a) If it is assumed that all the cars that are parked in the lot would otherwise have been driven to work, how much will be saved by this commuter parking lot per year? Assume an interest rate of 0.

(b) How could you find out how many would have been parked somewhere else?

(c) How could you calculate, per parking lot, the benefit to all drivers of having fewer cars on the road?

10.16 A medium-sized city (population 250 000, 45 000 families) is considering introducing a recycling program. The program would require them to separate newspaper, cardboard, and cans from their regular waste so that it could be collected weekly, sorted, and sold, rather than being put into the local landfill site. The program would also require households to separate "wet," or compostable, waste, which they would then be responsible for composting. The city would provide free composting units to the households. What kinds of potential costs and benefits can you identify for this project? How might this information be gathered?

10.17 Consider Problem 10.5. We saw that an effluent tax enabled the same reduction in BOD5 as a regulation, but with a lower cost. We did not consider the distribution of this cost.

(a) How much tax does plant A pay under the tax of RUB 20 per kilogram dumped? How much does plant B pay?

(b) What are the two total effluent costs (tax plus cleaning cost) for plant A? For plant B?

(c) Compare these costs with the costs under regulation.

(d) Suppose that the city used the proceeds of the tax to provide benefits that had equal value to each plant. How great would these benefits have to be to ensure that both plants would be better off with the combination of tax and benefits?

10.18 A four-day school week has been advocated as a means of reducing education costs. School days would be longer so as to maintain the same number of hours per week as under the current five-day-a-week system. The main expected cost savings are in school cleaning and maintenance and in school bus operation. The main effects that are anticipated are

(a) Reduced school cleaning and maintenance

(b) Reduced use of school buses on the off-day

(c) Reduced driving to school by parents, students, and staff on the off-day

(d) Reduced public transportation use on the off-day

(e) Increased performance of part-time work by some high school students and school staff for the off-day

(f) Reduced absences of students and staff; this is mainly because some required personal activities could be scheduled for the off-day

(g) Reduced subsidized school lunch requirements

(h) Increased need for day care on the off-day for working parents; about a third of elementary schools could be opened for day care, and the costs would be covered by fees

(i) Reduced learning by elementary students is possible because of their limited attention spans

(j) Reduced school taxes

Which of these effects are benefits? Who receives the benefits? Which are costs? Who bears the costs? Which are neither costs nor benefits?

10.19 A new suburban development project is being planned to the west of Monterrey, Mexico. There is now a two-lane road from the site of the development, along the base of the mountain *Cerro de la Silla* (Saddle Hill). The new development will require additional road capacity. Two alternatives are being considered. The first is to upgrade the existing road to four wide modern lanes. The second is to build a new four-lane highway east through Saddle Hill tunnel. The following data are available concerning the two routes:

Route	Distance	First Cost (pesos)	Operating and Maintenance Costs Per Year (pesos)
Base of mountain	20 km	210 million	900 000
Tunnel	10 km	450 million	1 300 000

The planning period is 40 years for both routes. The MARR is 10%. Cars will travel at 100 kilometres per hour along either route. Operating cost for the cars is expected to be about 2.50 pesos per kilometre along either route. About 400 000 trips per year are expected on either route.

(a) Which route should be built? Use only the data given. Use annual worth to make your decision.

(b) What important benefit of the Saddle Hill tunnel route has been left out of the analysis?

(c) Do you need more information about travellers to determine if the benefit that has been left out of the analysis from part (b) would change the recommendation? Explain.

(d) How would the possibility of collecting a toll on the tunnel route affect your recommendation?

10.20 The Principality of Upper Pigovia has just one export, pig crackling. The crackling is produced in two plants, Old Gloria and New Gloria. Both plants give off a delightful, mouth-watering odour while in operation. This odour has created a health problem. The citizens' appetites are huge, and, consequently, so are the citizens. Princess Piglet has decreed that the daily emission of odour from the two plants must be reduced by 7000 Odour Units (OU) per day. The Princess prides herself on her even-handedness. The decree specifies that each plant is to reduce its emission by 3500 OU per day. The following table shows the incremental cost in Upper Pigovian dollars (U$) per 1000 OU of attaining various levels of odour reduction in the two plants. To help in interpreting the table, note that the cost for Old Gloria to remove 3000 OU per day would be 2(U$25) + U$30 = U$80.

Quantity of Odour Removed (1000 OU/Day)	Incremental Cost of Removing Odour in $U/1000 OU			
	0 to 2	Over 2	Over 4	Over 5
Old	25	30	40	50
New	20	20	25	30

(a) What is the cost per day of implementing Princess Piglet's decree?

(b) Can you suggest a tax scheme that will yield the same reduction in odour emission as the decree at a lower cost?

(c) Can you suggest a subsidy scheme that will yield the same reduction in odour emission as the decree, but at a lower cost?

10.21 The transportation department in an Israeli town is considering the construction of a bridge over a narrow point in a lake. Traffic now goes around the lake. The bridge will save 30 kilometres in travel distance. Three alternatives are being considered: (1) do nothing, (2) build the bridge, and (3) build the bridge and charge a toll. If the bridge is built, the present road will be maintained, but its use will decline. One effect of the reduced use of the present road is a loss of revenue by businesses along that road. Available data, in thousands of shekels (ILS, or ₪) is given in the following table.

Costs and Benefits	Do Nothing (000s ₪)	Bridge With Toll (000s ₪)	Bridge Without Toll (000s ₪)
First cost	0	184 000	184 000
Annual road and/or bridge operating and maintenance costs	640	320	40
Annual vehicle operating cost	13 200	400	400
Annual driver and passenger time cost	10 000	2000	2000
Annual accident cost	2000	40	40
Annual revenue lost by roadside businesses	0	4000	4000
Annual toll revenues	N/A	4800	N/A

(a) Identify social benefits and costs of the bridge.

(b) Which of the costs and revenues in the table are neither social benefits nor social costs?

(c) The table makes an implicit assumption about the effect of the toll on bridge traffic. This assumption is probably incorrect. What is the assumption?

(d) How would you expect the toll to affect benefits and costs? Explain.

10.22 An example concerning the effect of a flood control project is found in the benefit–cost analysis chapter of an imaginary engineering economics text. The benefits of the project are stated as:

Benefit	Cost
Prevented losses due to floods in the Amahata River Basin	¥48 000 000/year
Annual worth of increased land value in the Amahata River Basin	¥4 800 000

Comment on these two items.

10.23 A province in Canada is considering the construction of a bridge. The bridge would cross a narrow part of a lake near a provincial park. The major benefit of the bridge would be reduced travel time to a campsite from a nearby urban centre. This lowers the cost of camping trips at the park. As well, an increase in the number of visits resulting from the lower cost per visit is expected.

Data concerning the number of week-long visits and their costs are shown below:

Inputs	Number of Visits and Average Cost Per Visit to Park	
	Without Bridge	With Bridge
Travel cost ($)	140	87.5
Use of equipment ($)	50	50
Food ($)	100	100
Total ($)	290	237.5
Number of visits/year	8000	11 000

The following data are available as well:

1. The bridge will take one year to build.
2. The bridge will have a 25-year life once it is completed. This means that the time horizon for computations is 26 years.
3. Construction cost for the bridge is $3 750 000. Assume that this cost is incurred at the beginning of year 1.
4. Annual operating and maintenance costs for the bridge are given by

   $$\$7500 + 0.25q$$

   where $7500 is the fixed operating and maintenance cost per year and q is the number of crossings.
5. Operating and maintenance costs are incurred at the end of each year over which the bridge is in operation. This is at the ends of years 2, 3, ..., 26.
6. The MARR is 10%.

(Notice that the annual benefits for this project were computed as part of the discussion of Example 10.1.)

(a) Compute the present worth of the project.

(b) Compute the benefit–cost ratio.

(c) Compute the modified benefit–cost ratio.

10.24 Consider the bridge project of Problem 10.23. We assumed there would be no toll for crossing. Now suppose the province is considering a toll of $7 per round trip over the bridge. They estimate that, if the toll is charged, the number of park visits will rise to only 10 600 per year instead of 11 000.

(a) Compute the present worth of the project if the toll is charged.

(b) Why is the present worth of the project reduced by the toll?

10.25 The town of Magdeburg has a new subdivision, Paradise Mountain, at its outskirts. The town wants to encourage the growth of Paradise Mountain by improving transportation between Paradise Mountain and the centre of Magdeburg. Two alternatives are being considered: (1) new buses on the route between Paradise Mountain and Magdeburg and (2) improvement of the road between Paradise Mountain and Magdeburg.

Both projects will have as their main benefit improved transportation between Paradise Mountain and Magdeburg. Rather than measuring the value of this benefit directly to the city, engineers have estimated the benefit in terms of an increase in the value of land in Paradise Mountain. That is, potential residents are expected to show their evaluations of the present worth of improved access to the town centre by their willingness to pay more for homes in Paradise Mountain.

The road improvement will entail construction cost and increased operating and maintenance costs. As well, the improved road will require construction of a parking garage in the centre of Magdeburg. The new buses will have a first cost as well as operating and maintenance costs. Information about the two alternatives is shown below.

	Road Improvement	New Buses
First cost	€15 000 000	€ 4 500 000
PW (operating and maintenance cost)	5 000 000	12 000 000
Parking garage cost	4 000 000	
Estimated increased land value	26 000 000	18 000 000

(a) Compute the benefit–cost ratio of both alternatives. Is each individually viable?

(b) Using an incremental benefit–cost ratio approach, which of the two alternatives should be chosen?

(c) Compute the present worths of the two alternatives. Compare the decision based on present worths with the decisions based on benefit–cost ratios.

10.26 A local government is considering a new two-lane road through a mountainous area. The new road would improve access to a city from farms on the other side of the mountains. The improved access would permit farmers to switch from grains to perishable soft fruits that would be either frozen at an existing plant near the city or sold in the city. Two routes are being considered. Route A is more roundabout. Even though the speed on route B would be less than that on route A, the trip on route B would take less time. Almost all vehicles using either road would go over the full length of the road. A department of transport engineer has produced information shown in the table below. The government uses a 10% MARR for road projects. The road will take one year to build. The government uses a 21-year time horizon for this project, since it is not known what the market for perishable crops will be in the distant future. Comment on the engineer's list of benefits; there may be a couple of errors.

Costs and Benefits of the New Road		
	Route A	Route B
Properties		
Distance (km)	24	16
Construction cost ($)	53 400	75 000
Operating and maintenance cost per year ($)	60	45
Resurfacing after 10 years of use ($)	3 100	2 350
Road Use		
Number of vehicles per year	1 000 000	1 200 000
Vehicle cost per kilometre ($)	0.3	0.3
Speed (km/h)	100	80
Value of time per vehicle hour ($)	15	15
Benefits		
Increased crop value per year ($)	13 500	18 000
Increased land value ($)	104 484.6	139 312.8
Increased tax collections per year ($)	811.21	1 081.61

10.27 Consider the road project in Problem 10.26. (*Note:* Correct for the errors mentioned.)

    (a) Compute a benefit–cost ratio for route A with road use costs counted as a cost.

    (b) Compute a benefit–cost ratio for route A with road use costs counted as a reduction in benefits.

    (c) In what way are the two benefit–cost ratios consistent, even though the numerical values differ?

    (d) Make a recommendation as to what the province should do regarding these two roads. Explain your answer briefly.

10.28 Find the net present worths for the dam and the dam plus recreation facilities considered in Review Problem 10.1. Use a MARR of 15%. Make a recommendation as to which option should be adopted.

10.29 The recreation department for Southampton is trying to decide how to develop a piece of land. They have narrowed the choices down to tennis courts or a swimming pool. The swimming pool will cost £2.5 million to construct, and will cost £300 000 per year to operate, but will bring benefits of £475 000 per year over its 25-year expected life. Tennis courts would cost £200 000 to build, cost £20 000 per year to operate, and bring £60 000 per year in benefits over their eight-year life. Both projects are assumed to have a salvage value of zero. The appropriate MARR is 5%.

    (a) Which project is preferable? Use a BCR and an annual worth approach.

    (b) Which project is preferable? Use a BCRM and an annual worth approach.

10.30 The environmental protection agency of a county would like to preserve a piece of land as a wilderness area. The owner of the land will lease the land to the county for a 20-year period for the sum of $1 750 000, payable immediately. The protection agency estimates that the land will generate benefits of $150 000 per year, but they will forgo $20 000 per year in taxes. Assume that the MARR for the county is 5%.

    (a) Calculate a BCR using annual worth and classify the forgone taxes as a cost to the government.

    (b) Repeat part (a), but consider the forgone taxes a reduction in benefits.

    (c) Using your results from part (b), determine the most the county would be willing to pay for the land (within $10 000) if they accept projects with a BCM of one or more.

10.31 The data processing centre at a Swiss tax centre has been troubled recently by the increasing incidence of repetitive strain injuries in the workplace. Health and safety consultants have recommended to management that they invest in upgrading computer desks and chairs at a cost of 500 000 Swiss francs (CHF). They advise that this would reduce the number and severity of medical costs by CHF 70 000 per year and that productivity losses and sick leaves could be reduced by a further CHF 80 000 per year. The furniture has a life of eight years with zero scrap value. The city uses a MARR of 9%. Should the centre purchase the furniture? Use a benefit–cost analysis.

10.32 Several new big-box stores have created additional congestion at an intersection in north Buffalo, New York. City engineers have recommended the addition of a turn lane, a computer-controlled signal, and sidewalks, at an estimated cost of $1.5 million. The annual maintenance costs at the new intersection will be $8000, but users will save $50 000 per year due to reduced waiting time. In addition, accidents are expected to decline, representing a property and medical savings of $175 000 per year. The renovation is expected to handle traffic adequately over a 10-year period. The city uses a MARR of 5%.

(a) What is the BCR of this project?

(b) What is the BCRM of the project?

(c) Comment on whether the project should be done.

10.33 What determines the MARR used on government-funded projects?

10.34 How will a decrease in the tax rate on investment income affect the MARR used for evaluating government-funded projects?

10.35 How does an expectation of inflation affect the MARR for public sector projects?

10.36 The department of road transport and highways in a certain district in India has 660 000 000 rupees (INR) that it can commit to highway safety projects. The goal is to maximize the total life-years saved per rupee. The potential projects are: (1) flashing lights at 10 railroad crossings; (2) flashing lights and gates at the same 10 railroad crossings; (3) widening the roadway on an existing rural bridge from 3m to 3.5m; (4) widening the roadway on a second and third rural bridge from 3m to 3.5m; (5) reducing the density of utility poles on rural roads from 30 to 15 poles per kilometre; and (6) building runaway lanes for trucks on steep downhill grades.

Highway Safety Projects		Total Cost (000s INR)	Life-Years Saved Per Year
1	Flashing lights	18 000	14
2	Flashing lights and gates	30 000	20
3	Widening bridge #1	48 000	14
4	Widening bridge #2	28 000	10
5	Widening bridge #3	44 000	18
6	Pole density reduction	120 000	96
7	Runaway lane #1	240 000	206
8	Runaway lane #2	240 000	156

The data for the flashing-lights and flashing-lights-with-gates projects reflect the costs and benefits for the entire set of 10 crossings. Portions of the projects may also be completed for individual crossings at proportional reductions in costs and savings. At any single site, the lights and lights-with-gates projects are mutually exclusive. Any fraction of the reduction of utility pole density project can be carried out. Data concerning costs and safety effects of the projects are shown above. (A life-year saved is one year of additional life for one person.)

Advise the department of transportation how best to commit the 660 000 000 rupees. Assume that the money must be used to increase highway safety.

10.37 A department of transportation is considering widening lanes on major highways from 6m to 7.5m. The objective is to reduce the accident rate. Accidents have both material and human costs. Data for highway section XYZ are found on the next page.

(a) Compute the present worth of costs of lane-widening.

(b) Compute the present worth of savings of non-human accident costs.

(c) What minimum value for a serious personal injury would justify the project?

Lane Widening on Section XYZ	
Accidents per 100 000 000 vehicle-kilometres in 6 m lanes	150
Accidents per 100 000 000 vehicle-kilometres in 7.5 m lanes	90
Serious personal injuries per accident	10%
Average non-human cost per accident ($)	2500
Annual road use (vehicles)	7 500 000
First cost per kilometre ($)	175 000
Operating and maintenance costs per kilometre/year ($)	7500
Project life (years)	25
MARR	10%

10.38 The Chinese government is considering three new flood control projects on the Huaihe River. Projects A and B consist of permanent dikes. Project C is a small dam. The dam will have recreation and irrigation benefits as well as the flood control benefits. Facts about the three projects are shown in the following table. Each project has a life of 25 years. The MARR is 10%.

	Project A	Project B	Project C
First cost (millions of yuan)	200	256	416
Annual benefits (millions of yuan)	26.4	33.6	56.8
Annual operating and maintenance costs (millions of yuan)	2.4	2.4	4.0

(a) Use present worth to choose the best project.

(b) Compute the benefit–cost ratios for the three projects.

(c) Use incremental benefit–cost to choose the best project.

(d) It is possible that the dam (project C) will cause some wildlife damage. This damage might be prevented by an additional expenditure of 24 million yuan when the dam is constructed. Does this change the choice of best project? Compute the benefit–cost ratio of project C assuming that the additional 24 million yuan is an addition to the first cost.

10.39 A highway department needs to upgrade a 20-kilometre rural road to accommodate increased traffic. There will be 5000 cars per day in the first year after upgrading. The number of cars will increase each year by 200 cars per day for the next 20 years. A modern two-lane road will be adequate for current traffic levels, but it will be inadequate after about 10 years. At that time, a four-lane road will be required. A four-lane road will permit greater speed even at the current traffic level.

Two alternatives are being considered. One proposal is for a modern two-lane road now, followed by addition of lanes to make a four-lane road after 10 years. The second alternative is for a four-lane road now. The planning horizon for both alternatives is 20 years. Costs for the alternatives are shown in the table on the next page. All values are in real dollars.

	Two Lanes	Four Lanes
First cost (millions of $)	20	32
Adding lanes after 10 years (millions of $)	18	0
Annual operating and maintenance costs (years 1–10) (in $000s)	30	40
Annual operating and maintenance costs (years 11–20) (in $000s)	40	40

The average value of travel time for each car is estimated at $40 an hour. Speeds will be as shown below.

(a) Which alternative has a greater present worth if the MARR is 5%?

(b) Which alternative has a greater present worth if the MARR is 20%?

(c) Suppose the real after-tax rate of return on government savings bonds is 5% and the real before-tax rate of return on investment in the private sector is 20%. Use your answers to parts (a) and (b) to decide which alternative is better. Explain your answer.

Years	Two Lanes	Four Lanes
1 to 5	80 km/h	100 km/h
6 to 10	70 km/h	100 km/h
11 to 20	N/A	100 km/h

10.40 A small town needs to upgrade the town dump to meet environment standards. Two alternatives are being considered. Alternative A has a first cost of €420 000 and annual operating and maintenance costs of €52 500. Alternative B has a first cost of €315 000 and annual operating and maintenance costs of €74 000. Both alternatives have 15-year lives with no salvage. An increase in dumping fees for households and businesses in the town is expected to yield €50 000 per year.

(a) Which alternative has the lower cost if the MARR is 5%? Use annual worth.

(b) Which alternative has the lower cost if the MARR is 20%? Use annual worth.

(c) The after-tax return on government savings bonds is 5%. The average rate of return before taxes in private sector investment is 20%. Which alternative should be chosen?

More Challenging Problem

10.41 Consider the following situation. The production and use of a certain pesticide poses both health and environmental risks to the workers who make the pesticide, those who use it, and the environment. The government regulates certain aspects of its production and distribution and limits its use and release into the environment. These control measures require the manufacturer to pay considerable monitoring and compliance costs, and require users and transport companies to take expensive precautions around its use and transport.

The government is considering a total ban on the production and use of the pesticide. In evaluating the social costs and benefits of the ban, you recognize that the manufacturer, distributors, and users will save substantial amounts because they will no longer need to monitor and control the production and use of the pesticide.

(a) In general terms, what do you consider to be the social benefits of the ban? What are the social costs?

(b) In computing the overall social costs of the ban, is it reasonable to deduct from other social costs the savings arising from not having to monitor and control the pesticide's production and use? Why, or why not?

MINI-CASE 10.1

Emissions Trading

Fossil fuel-burning electric power plants in the United States and Canada produce air pollution as a by-product of their operation. The pollutants produced by these plants are linked to high levels of acid rain in the northeastern United States, and the Canadian provinces of Ontario and Quebec, as well as a growing number of "smog" days in Ontario. The primary pollutants associated with acid rain are nitrogen oxide and sulphur dioxide (NO_X, SO_2). Acid rain is linked to increased fish morbidity rates, destruction of forests, and building decay. Overall increases in smog levels have led to a host of respiratory illnesses such as asthma and lung cancer in the northeastern United States and Ontario.

Air quality is a public good. Though there are markets for fossil fuels and electricity, they fail to take into account the public costs of smog, acid rain, and climate change. Government regulation in the form of an emissions trading system has been used as a mechanism to produce more socially desirable outcomes.

Emissions trading refers to a market-based approach to controlling pollution in which companies can buy and sell allowances for emitting specified pollutants. Alternatively, companies can get credits for reducing emissions. In such trading systems, a central agency—typically a government—sets goals or limits on the allowable volume of pollutants that can be emitted. Companies must hold sufficient allowances for that volume. Those that reduce their emissions can sell their allowances and those that have emissions above their allowable limit must purchase credits from those that pollute less.

Since 1995, the US Environmental Protection Agency (EPA) has used an emissions trading system to permit plants to adopt the most cost-efficient approach to reduce their sulphur dioxide emissions. The EPA sets caps, or limits, on allowable total emissions levels, and then issues a limited number of permits that are auctioned off annually. These permits can be traded or saved. In 2003, a permit for the atmospheric release of one tonne of sulphur dioxide cost approximately US $170. By 2004, the cost was $225 per tonne, and by 2006, it was more than $1500 per tonne. In 2001, Ontario introduced a "Cap, Credit, and Trade" emission trading system for controlling sulphur dioxide.

Emissions trading systems apply to controlling other pollutants. The recent rise in temperature of the earth's atmosphere, global warming, is understood to be the result of higher concentrations of certain "greenhouse gases" in our atmosphere. The Kyoto Protocol is a 1997 international agreement that went into effect in 2005. It commits developed countries to reduce their emissions of certain greenhouse gases to below 1990 levels by 2012. The Kyoto Protocol uses an emissions trading scheme to coordinate reductions in emissions. In conjunction with the Kyoto accord, the European Union (EU) has introduced a mandatory carbon trading program to control emissions of carbon dioxide.

In the United States, which is the only major country that has not ratified the Kyoto accord, a number of trading schemes for carbon dioxide and other pollutants are underway or planned. These systems are other examples of a growing trend for countries around the world to implement emissions trading systems so that the overall levels of greenhouse gases can be reduced in coming years. These trading systems are not without their implementation challenges, however, as the discussion questions to follow indicate.

Sources: "EPA's Clean Air Markets—Acid Rain Program," Environmental Protection Agency site, www.epa.gov/airmarkets/progregs/arp/index.html; "Ontario Emissions Trading Registry Introduction," Ontario Ministry of the Environment site, www.ene.gov.on.ca/envision/air/etr/index.htm; United Nations Framework Convention on Climate Change, www.unfccc.int/2860.php; "European Union Emission Trading Scheme," European Commission site, ec.europa.eu/environment/climat/emission.htm; "Emissions Trading Analysis," Australian Greenhouse Office site, www.greenhouse.gov.au/emissionstrading/index.html. All accessed November 10, 2007.

Discussion

Despite the early successes of these systems, some issues remain. First, these regulations may be successful in reducing overall emission levels, but they may not address local "hot spots." Second, the global targets themselves may not be sufficient to truly reflect the social costs of acid rain or greenhouse gases. Finally, there may be more efficient ways of distributing the allowances so that the overall cost of operating the system is lowered.

Government regulatory programs such as the emissions trading systems are established to remediate the impact of market failure. These programs are intended to act as an incentive to allow plants to adopt the most cost-efficient means of emissions reductions. But despite the best intentions of the government, individuals or companies can exploit poorly designed programs for their own gain. At the best of times, such as with SO_2 or carbon emission controls, the inherent process of establishing the system incurs inefficiencies. On the other hand, the benefits to society of correcting situations of market failure can nevertheless be enormous.

Questions

1. What are the implications of having too high a level of carbon allowances in an emissions trading system?
2. What problem could arise if an emissions trading system allowed companies to sell carbon credits to other companies at a mutually acceptable price? Why would an open-market auction be preferable?
3. Emissions trading schemes require some form of enforcement. One kind of enforcement uses regulators that can fine companies that do not have sufficient allowances for their emissions. What are potential shortcomings of this form of enforcement?
4. An alternative to an emissions trading system is an emissions tax. Some economists argue that a tax would be simpler to implement, would have more predictable costs, and would be less prone to corruption. Can you see any disadvantages to an emissions tax?

CHAPTER 11

Dealing With Uncertainty: Sensitivity Analysis

Engineering Economics in Action, Part 11A: Filling a Vacuum

Review Problems
Summary

Engineering Economics in Action, Part 11B: Where the Risks Lie

Problems

Mini-Case 11.1

Engineering Economics in Action, Part 11A:
Filling a Vacuum

"I have something new for you, Naomi. It's going to require some imagination and disciplined thinking at the same time." Anna's tone indicated to Naomi that something interesting was coming.

"As you may know, Global Widgets has been working toward getting into consumer products for some time." Anna continued, "An opportunity has come up to cooperate on a new vacuum cleaner with Powerluxe. They have some potential designs for a vacuum with electronic sensing capabilities. It will sense carpet height and density. It will then adjust the power and the angle of the head to optimize cleaning on a continuous basis. Our role in this would be to design the manufacturing system and to do the actual manufacturing for North America. Sound interesting so far?"

"Yes, Ms. Kulkowski," Naomi answered. Naomi couldn't help being respectful in front of Anna, who was Global Widgets' president, among other things. "What would my role be?"

"For one thing, you will be working with Bill Astad from head office. I think you know him." Naomi did; she had given Bill some advice on handling variable rates of inflation a few weeks earlier. "First, we want to establish some idea of the demand for the product. We have to see how this might affect our manufacturing capacity and capital costs to determine if the whole idea is even feasible. Later on, we will have to make some design decisions, but the general feasibility comes first."

"Do we have any market studies from Powerluxe?" Naomi asked.

"Yes, but they seem to be guesswork and magic, not hard figures. After all, no one has sold a product like this before."

Naomi looked pensive. "Sounds as though we'll need to do some sensitivity analysis on this one." She muttered more to herself than Anna. Then to Anna, "Right. We'll do what we can. Thanks for the nice opportunity, Anna. This is interesting!"

11.1 | Introduction

To this point in our coverage of engineering economics, we have assumed that project parameters such as prices, interest rates, and the magnitude and timing of cash flows have values that are known with certainty. In fact, many of these values are estimates and are subject to some uncertainty. Since the results of an evaluation can be influenced by variations in uncertain parameters, it is important to know how *sensitive* the outcome is to variations in these parameters.

There are several reasons why there may be uncertainty in estimating project parameter values. Technological change can unexpectedly shorten the life of a product or piece of equipment. A change in the number of competing firms may affect sales volume or market share or the life of a product. In addition, the general economic environment may affect inflation and interest rates and overall activity levels within an industry. All of these factors may result in cash flows different from what was expected in both timing and size, or in other changes to the parameters of an evaluation.

Making decisions under uncertainty is challenging because the overall impact of uncertainty on project evaluation may initially not be well understood. **Sensitivity analysis** is an approach to project evaluation that can be used to gain a better understanding of how uncertainty affects the outcome of the evaluation by examining how sensitive the outcome is to changes in the uncertain parameters. It can help a manager decide whether it is worthwhile to get more accurate data as part of a more detailed evaluation, or whether it may be necessary to control or limit project uncertainties. The problem of decision making under *risk*, where there is information about the probability distribution of parameter values, is discussed in the next chapter.

Economic analyses are not complete unless we try to assess the potential effects of project uncertainties on the outcomes of the evaluations. Because parameter estimates can be so hard to determine, analysts usually consider a range of possible values for uncertain components of a project. There is then naturally a range of values for present worth, annual worth, or whatever the relevant performance measure is. In this way, the analyst can get a better understanding for the range of possible outcomes and can make better decisions.

In this chapter, we will consider three basic sensitivity analysis methods commonly used by analysts. These methods are used to better understand the effect that uncertainties or errors in parameter values have on economic decisions. The first method is the use of *sensitivity graphs*. Sensitivity graphs illustrate the sensitivity of a particular measure (e.g., present worth or annual worth) to one-at-a-time changes in the uncertain parameters of a project. Sensitivity graphs can reveal key parameters that have a significant impact on the performance measures of interest and hence we should be particularly careful to get good estimates for these key parameters.

The second method for sensitivity analysis is the use of *break-even analysis*. Break-even analysis can answer such questions as "What production level is necessary in order for the present worth of the project to be greater than zero?" or "Below what interest rate will the project have a positive annual worth?" Break-even analysis can also give insights into comparisons between projects. With break-even analysis, we can answer questions like "What scrap value for the proposed forklift will cause us to be indifferent between replacing the old forklift and not replacing it?"

The third method we will introduce is called *scenario analysis*. Both sensitivity graphs and break-even analyses have the drawback that we can look at parameter changes only one at a time. Scenario analysis allows us to look at the overall impact of different sets of parameter values, referred to as "scenarios," on project evaluation. In other words, they allow us to look at the impact of varying several parameters at a time. In this way, the analyst comes to understand the range of possible economic outcomes.

Each of these three sensitivity analysis methods—sensitivity graphs, break-even analysis, and scenario analysis—tries to assess the sensitivity of an economic measure to uncertainties in estimates in the various parameters of the problem. A thorough economic evaluation should include aspects of all three types of analysis.

11.2 | Sensitivity Graphs

The first sensitivity analysis tool we will look at is the sensitivity graph. Sensitivity graphs are used to assess the effect of one-at-a-time changes in key parameter values of a project on an economic performance measure. We usually begin with a "base case" where all the estimated parameter values are used to evaluate the present worth, annual worth, or IRR of a project, whatever the appropriate measure is. We then vary parameters above and below the base case one at a time, *holding all other parameters fixed*. A graph of the changes in a performance measure brought about by these one-at-a-time parameter changes is called a **sensitivity graph**. From the graph, the analyst can see which parameters have a significant impact on the performance measure and which do not.

EXAMPLE 11.1

Cogenesis Corporation is replacing its current steam plant with a six-megawatt cogeneration plant that will produce both steam and electric power for operations. The new plant will use wood as a source of fuel, which will eliminate the need for Cogenesis to purchase a large amount of electric power from a public utility. To move to the new system, Cogenesis will have to integrate a new turbogenerator and cooling tower with its current system. The estimated first cost of the equipment and installation is $3 000 000, though there is some uncertainty surrounding this estimate. The plant is expected to have a 20-year life and no scrap value at the end of this life. In addition to the first cost, the turbogenerator will require an overhaul with an estimated cost of $35 000 at the end of years 4, 8, 12, and 16. The cooling tower will need an overhaul at the end of 10 years. This is expected to cost $17 000.

The cogeneration system is expected to have higher annual operating and maintenance costs than the current system, and will require the use of chemicals to treat the water used in the new plant. These incremental costs are estimated to be $65 000 per year. The incremental annual costs of wood fuel are estimated to be $375 000. The cogeneration plant will save Cogenesis from having to purchase 40 000 000 kilowatt-hours of electricity per year at $0.025 per kilowatt-hour, an annual savings of $1 000 000. Cogenesis uses a MARR of 12%. What is the present worth of the incremental investment in the cogeneration plant? What is the impact of a 5% and 10% increase and decrease in each of the parameters of the problem?

PW(cogeneration plant)

$= -3\ 000\ 000 - (65\ 000 + 375\ 000 - 1\ 000\ 000)\ (P/A,12\%,20)$

$\quad - 17\ 000(P/F,12\%,10)$

$\quad - 35\ 000[(P/F,12\%,4) + (P/F,12\%,8) + (P/F,12\%,12)$

$\quad + (P/F,12\%,16)]$

$= 1\ 126\ 343$

The present worth of the incremental investment is $1 126 343. On the basis of this assessment, the project appears to be economically viable.

In order to better understand the situation, analysts for Cogenesis have also completed some sensitivity graphs that indicate how sensitive the present worth is to changes in some of the parameters. In particular, they feel that some of the cash flows may turn out to be different from their estimates, and they would like to get a feel for what impact these errors may have on the evaluation of the cogeneration plant. To investigate, they have labelled their current estimates the "base case" and have generated other cash flow estimates that are 5% and 10% above and below the base case for each major cash flow category. These are summarized in Table 11.1.

For example, the initial investment may be more than the estimate of $3 000 000 if they run into unforeseen difficulties in the installation. Or the savings in electricity costs may be overestimated if the cost per kilowatt-hour drops in the future. The analysts would like to get a better understanding of which of these changes would have the greatest impact on the evaluation of the plant.

To keep the illustration simple, we will consider changes to the initial investment; annual chemical, operations, and maintenance costs; the MARR; and the savings in electrical costs. Each of these is varied one at a time, leaving all other cash flow estimates at the base case values. For example, if the initial investment is 10% below the initial estimate of $3 000 000, and all other estimates are as in the base case, the present worth of the project will be $1 426 343 (see the first row of Table 11.2, under –10%). Similarly, if the first cost is 10% more than the original estimate, the present worth drops to $826 343.

Table 11.1   Summary Data for Example 11.1

Cost Category	−10%	−5%	Base Case	+5%	+10%
Initial investment	$2 700 000	$2 850 000	$3 000 000	$3 150 000	$3 300 000
Annual chemical, operations, and maintenance costs	58 500	61 750	65 000	68 250	71 500
Cooling tower overhaul (after 10 years)	15 300	16 150	17 000	17 850	18 700
Turbogenerator overhauls (after 4, 8, 12, and 16 years)	31 500	33 250	35 000	36 750	38 500
Annual wood costs	337 500	356 250	375 000	393 750	412 500
Annual savings in electricity costs	900 000	950 000	1 000 000	1 050 000	1 100 000
MARR	0.108	0.114	0.12	0.126	0.132

Interest rate uncertainty will almost always be present in an economic analysis. If Cogenesis's MARR increases by 10% (with all other parameters at their base case values), the present worth of the project drops to $835 115, about the same impact as if the first cost ended up being 10% more than expected. Other variations are shown in Table 11.2. A sensitivity graph, shown in Figure 11.1, illustrates the impact of one-at-a-time parameter variations on the present worth.

Small changes in the annual chemical, operations, and maintenance costs do not have much of an impact on the present worth of the project, as can be seen from Table 11.2 and Figure 11.1. What appears to have the greatest impact on the viability of the project is the savings in electricity costs. A 10% drop in the savings causes the

Table 11.2   Present Worth of Variations from Base Case in Example 11.1

Cost Category	−10%	−5%	Base Case	+5%	+10%
Initial investment	$1 426 343	$1 276 343	$1 126 343	$ 976 343	$ 826 343
Annual chemical, operations, and maintenance costs	1 174 894	1 150 619	1 126 343	1 102 067	1 077 792
Cooling tower overhaul (after 10 years)	1 126 890	1 126 617	1 126 343	1 126 069	1 125 796
Turbogenerator overhauls (after 4, 8, 12, and 16 years)	1 131 450	1 128 897	1 126 343	1 123 789	1 121 236
Annual wood costs	1 406 447	1 266 395	1 126 343	986 291	846 239
Savings in electricity costs	379 399	752 871	1 126 343	1 499 815	1 873 287
MARR	1 456 693	1 286 224	1 126 343	976 224	835 115

Figure 11.1   Sensitivity Graph for Example 11.1

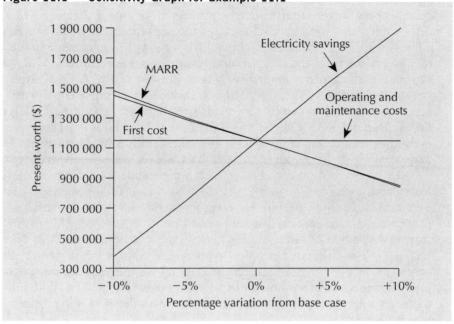

present worth of the project to drop to about one-third of the base case estimate. This change could occur because of a drop in electricity rates or a drop in demand. Alternatively, the present worth of the project increases to almost $1 900 000 if the savings are higher than anticipated. This could, once again, occur because of a change in either rates or demand for power. Clearly, if Cogenesis is to expend effort in getting better forecasts, it should be for energy consumption and power rates.

One final point about this example should be noted. If management feels that, individually, the cash flow estimates will fall within the ±10% range, the investment looks economically viable (i.e., yields a positive present worth) and they should go ahead with it.∎

As we can see from Example 11.1, the benefit of a sensitivity graph is that it can be used to select key parameters in an economic analysis. It is easy to understand and communicates a lot of information in a single diagram. There are, however, several shortcomings of sensitivity graphs. First, they are valid only over the range of parameter values in the graph. The impact of parameter variations outside the range considered may not be simply a linear extrapolation of the lines in the graph. If you need to assess the impact of greater variations, the computations should be redone. Second, and probably the greatest drawback of sensitivity graphs, is that they do not consider the possible interaction between two or more parameters. You cannot simply "add up" the impacts of individual changes when several parameters are varied, producing an interaction effect. We will come back to this issue in the section on scenario analysis, where we do consider entire "packages" of changes from the base case.

11.3 | Break-Even Analysis

In this section, we cover a second type of sensitivity analysis called break-even analysis. Once again, we are trying to answer the question of what impact changes (or errors) in parameter estimates will have on the economic performance measures we use in our analyses, or on a decision made on the basis of an economic performance measure. In general, **break-even analysis** is the process of varying a parameter of a problem and determining what parameter value causes the performance measure to reach some threshold or "break-even" value. In Example 11.1, we saw that an increase in the MARR caused the present worth of the cogeneration plant to decrease. If the MARR were to increase sufficiently, the project might have a zero present worth. A break-even analysis could answer the question "What MARR will result in a zero present worth?" This analysis would be particularly useful if Cogenesis were uncertain about the MARR and wanted to find a threshold MARR above which the project would not be viable. Other such break-even questions could be posed for the cogeneration problem, to try to get a better understanding of the impact of changes in parameter values on the economic analysis.

Break-even analysis can also be used in the comparison of two or more projects. We have already seen in Chapter 4 that the best choice among mutually exclusive alternatives may depend on the interest rate, production level, or a variety of other problem parameters. Break-even analysis applied to multiple projects can answer questions like "Over what range of interest rates is project A the best choice?" or "For what output level are we indifferent between two projects?" Notice that we are varying a single parameter in two or more projects and asking when the performance measure for the projects meets some threshold or break-even point. The point of doing this analysis is to try to get a better understanding of how sensitive a decision is to changes in the parameters of the problem.

11.3.1 Break-Even Analysis for a Single Project

In this section, we show how break-even analysis can be applied to a single project to illustrate how sensitive a project evaluation is to changes in project parameters. We will continue with Example 11.1 to expand upon the information provided by the sensitivity graphs.

EXAMPLE 11.2

Having completed the sensitivity graph in Example 11.1, management recognizes that the present worth of the cogeneration plant is quite sensitive to the savings in electricity costs, the MARR, and the initial costs. Since there is some uncertainty about these estimates, they want to explore further the impact of changes in these parameters on the viability of the project. You are to carry out a break-even analysis for each of these parameters to find out what range of values results in a viable project (i.e., PW > 0) and to determine the "break-even" parameter values that make the present worth of the project zero. You are also to construct a graph to illustrate the present worth of the project as a function of each parameter.

First, Figure 11.2 shows the present worth of the project as a function of the MARR. It shows that the break-even MARR is 17.73%. In other words, the project has a positive present worth for any MARR less than 17.73% (all other parameters fixed) and a negative present worth for a MARR more than 17.73%. Notice that the break-even interest rate is, in fact, the IRR for the project.

A similar break-even chart for the first cost, Figure 11.3, shows that the first cost can be as high as $4 126 350 before the present worth declines to zero. Assuming that all other cost estimates are accurate, the project will be viable as long as the first cost is below this break-even amount. One issue management should assess is the likelihood that the first cost will exceed $4 126 350.

Figure 11.2   Break-Even Chart for the MARR

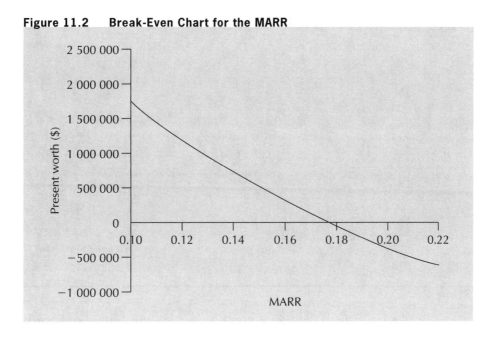

Finally, a break-even chart for the savings in electrical power costs is shown in Figure 11.4. We have already seen from the sensitivity graph that the viability of the project is very sensitive to the savings in electricity produced by the cogeneration plant. Provided that the annual savings are above $849 207, the project is viable. Below this break-even level, the present worth of the project is negative. If the actual saving in electrical power costs is likely to be much below the estimate, this will put the project's viability at risk. Given the particular sensitivity of the present worth to the savings, it may be worthwhile to spend additional time looking into the two factors that make up these savings: the cost per kilowatt-hour and the total kilowatt-hours of demand provided for by the new plant.■

Figure 11.3   Break-Even Chart for First Cost

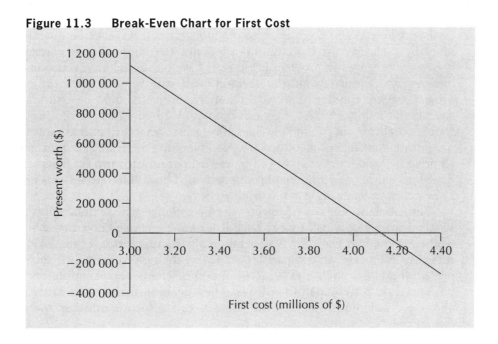

Figure 11.4   Break-Even Chart for Electricity Savings

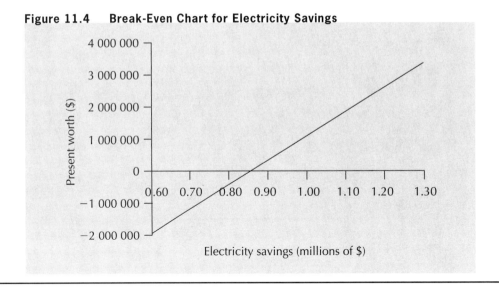

Break-even analysis done for a single project expands upon the information sensitivity graphs provide. It has the advantage that it is easy to apply and allows us to determine the range of values for a parameter within which the project is viable or some other criteria are met. It can provide us with break-even parameter values that give an indication of how much a parameter can change from its original estimate before the project's viability becomes a concern. Graphical presentation of the break-even analysis, as in Figures 11.2, 11.3, and 11.4, summarizes the information in an easily understood way.

11.3.2 Break-Even Analysis for Multiple Projects

In the previous section, we saw how break-even analysis can be applied to a single project in order to understand more clearly the impact of changes in parameter values on the evaluation of the project. This analysis may influence a decision on whether the project should be undertaken. When there is a choice among several projects, be they independent or mutually exclusive, the basic question remains the same. We are concerned with the impact that changes in problem parameters have on the relevant economic performance measure, and, ultimately, on the decision made with respect to the projects. With one project, we are concerned with whether the project should be undertaken and how changes in parameter values affect this decision. With multiple projects, we are concerned about how changes in parameter values affect which project or projects are chosen.

For multiple independent projects, assuming that there are sufficient funds to finance all projects, break-even analysis can be carried out on each project independently, as was done for a single project in the previous section. This will lead to insights into how robust a decision is under changes in the parameters.

For mutually exclusive projects, the best choice will seldom stand out as clearly superior from all points of view. Even if we have narrowed down the choices, it is still likely that the best choice may depend on a particular interest rate, level of output, or first cost. A break-even comparison can reveal the range over which each alternative is preferred and can show the break-even points where we are indifferent between two projects. Break-even analysis will provide a decision maker with further information about each of the projects and how they relate to one another when parameters change.

EXAMPLE 11.3

Westmount Waxworks (see Problem 4.18) is considering buying a new wax melter for its line of replicas of statues of government leaders. Westmount has two choices of suppliers: Finedetail and Simplicity. The proposals are as follows:

	Finedetail Wax Melter	Simplicity Wax Melter
Expected life	7 years	10 years
First cost	$200 000	$350 000
Maintenance	$10 000/year + $0.05/unit	$20 000/year + $0.01/unit
Labour	$1.25/unit	$0.50/unit
Other costs	$6 500/year + $0.95/unit	$15 500/year + $0.55/unit
Salvage value	$5 000	$20 000

The marketing manager has indicated that sales have averaged 50 000 units per year over the past five years. In addition to this information, management thinks that they will sell about 30 000 replicas per year if there is stability in world governments. If the world becomes very unsettled so that there are frequent overturns of governments, sales may be as high as 200 000 units per year. There is also some uncertainty about the "other costs" of the Simplicity wax melter. These include energy costs and an allowance for scrap. Though the costs are estimated to be $0.55 per unit, the Simplicity model is a new technology, and the costs may be as low as $0.45 per unit or as high as $0.75 per unit. Westmount Waxworks would like to carry out a break-even analysis on the sales volume and on the "other costs" of the Simplicity wax melter. They want to know which the preferred supplier would be as sales vary from 30 000 per year to 200 000 per year. They also wish to know which is the preferred supplier if the "other costs" per unit for the Simplicity model are as low as $0.45 per unit or as high as $0.75 per unit. Westmount Waxworks uses an after-tax MARR of 15% for equipment projects. They pay taxes in Canada with a tax rate of 40%. The CCA rate for such equipment is 30%.

Assuming that the "other costs" of the Simplicity wax melter are $0.55 per unit, a break-even chart that shows the present worth of the projects as a function of sales levels can give much insight into the supplier selection. Table 11.3 gives the annual cost of each of the two alternatives, and Figure 11.5 shows the break-even chart for sales level.

Table 11.3   Annual Cost as a Function of Sales

Sales (Units)	Annual Costs ($) Finedetail	Annual Costs ($) Simplicity
20 000	72 658	85 651
60 000	126 658	111 091
100 000	180 658	136 531
140 000	234 658	161 971
180 000	288 658	187 411
220 000	342 658	212 851

Figure 11.5   Break-Even Chart for Sales Level

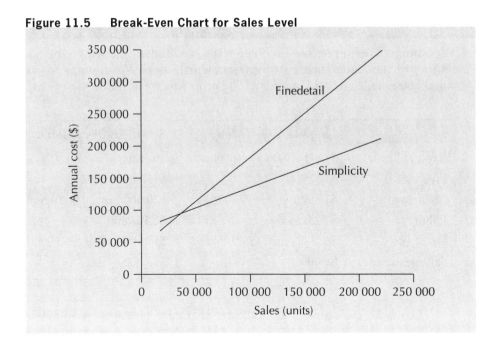

A sample computation for the Finedetail wax melter at the 60 000 sales level is

$$\begin{aligned}
\text{AW(Finedetail)} &= [200\,000 - [(200\,000/2 \times TBF)(1 + (P/F,15\%,1))]](A/P,15\%,7) \\
&\quad - 5000(1 - TBF)(A/F,15\%,7) + (1 - t)\,[10\,000 + 6500 \\
&\quad + (0.05 + 1.25 + 0.95)(\text{sales level})] \\
&= [200\,000 - [(10\,000 \times 0.2667)(1 + 0.86957)]](0.24036) \\
&\quad + 5000(1 - 0.2667)(0.09036) \\
&\quad + (1 - 0.4)[16\,500 + 2.25(60\,000)] \\
&\cong 126\,658
\end{aligned}$$

where

$$TBF = (0.4 \times 0.3)/(0.15 + 0.3) \cong 0.2667$$

If sales are 30 000 units per year, the Finedetail wax melter is slightly preferred to the Simplicity melter. At a sales level of 200 000 units per year, the preference is for the Simplicity wax melter. Interpolation of the amounts in Table 11.3 indicates that the break-even sales level is 38 199 units. That is to say, for sales below 38 199 per year, Finedetail is preferred, and Simplicity is preferred for sales levels of 38 199 units and above.

Since 30 000 units per year is the lowest sales will likely be, and sales have averaged 50 000 units per year over the past five years, it appears that the Simplicity wax melter would be the preferred choice, assuming that its "other costs" per unit are $0.55. The robustness of this decision may be affected by the other types of costs, such as maintenance and labour, of the Simplicity melter.

Table 11.4   Annual Cost as a Function of Simplicity's Other Costs Per Unit

Other Costs Per Unit ($)	Sales = 30 000 Units/Year Annual Costs ($)		Sales = 50 000 Units/Year Annual Costs ($)		Sales = 200 000 Units/Year Annual Costs ($)	
	Finedetail	Simplicity	Finedetail	Simplicity	Finedetail	Simplicity
0.45	86 158	90 211	113 158	101 731	315 658	188 131
0.55	86 158	92 011	113 158	104 731	315 658	200 131
0.65	86 158	93 811	113 158	107 731	315 658	212 131
0.75	86 158	95 611	113 158	110 731	315 658	224 131

To assess the sensitivity of the choice of wax melter to the variable other costs of Simplicity, a break-even analysis similar to that for sales level can be carried out. We can vary the "other costs" from the estimate of $0.45 per unit to $0.75 per unit and observe the effect on the preferred wax melter. Table 11.4 gives the annual costs for the two wax melters as a function of the "other costs" of the Simplicity model for sales levels of 30 000, 50 000, and 200 000 units per year. In each case, we see that the best choice is not sensitive to the "other costs" of the Simplicity wax melter. In fact, for a sales level of 30 000 units per year, the break-even "other costs" are less than $0.25, as shown in Figure 11.6. This means that the other costs per unit would have to be lower than $0.25 for the best choice to change from Finedetail to Simplicity. For a sales level of 200 000 per year, the break-even "other costs" are much higher, at $1.51 per unit, and for a sales level of 50 000 units per year the break-even cost per unit is $0.83. For both of the latter sales levels, the Simplicity model is preferred.

Having done the break-even analysis for both sales level and "other costs" per unit for the Simplicity wax melter, it would appear that the Simplicity model is the better choice if sales are at all likely to exceed the break-even sales level of 38 199. Historically, sales have exceeded this amount. Even if sales in a particular year fall below the

Figure 11.6   Break-Even Chart for Simplicity's Other Costs Per Unit (Sales = 30 000 Units)

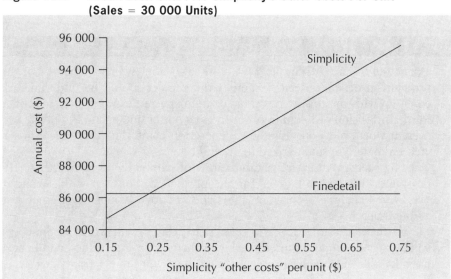

break-even level, the Simplicity wax melter does not have annual costs far in excess of those of the Finedetail model, so the decision would appear to be robust with respect to possible sales levels. Similarly, the decision is not sensitive to the other costs per unit of the Simplicity wax melter.■

We have seen in this section that break-even analysis for either a single project or multiple projects is a simple tool and that it can be used to extract insights from a modest amount of data. It communicates threshold (break-even) parameter values where preference changes from one alternative to another or where a project changes from being economically justified to not justified. Break-even analysis is a popular means of assessing the impact of errors or changes in parameter values on an economic performance measure or a decision.

The main disadvantage of break-even analysis is that it cannot easily capture interdependencies among variables. Although we can vary one or two parameters at a time and graph the results, more complicated analyses are not often feasible. This disadvantage can be overcome to some degree by what is referred to as scenario analysis, the subject of the next section.

11.4 | Scenario Analysis

The third type of sensitivity analysis tool that we will look at is **scenario analysis**, which is the process of examining the consequences of several possible sets of variables associated with a project. Scenario analysis recognizes that many estimates of cash flows or other project parameters may vary from what is projected. It is useful to look at several "what if" scenarios in order to understand the effect of changes in values of whole sets of parameters. Commonly used scenarios are the "optimistic" (or "best case") outcome, the "pessimistic" (or "worst case") outcome, and the "expected" (or "most likely") outcome. The best-case and worst-case outcomes can, in some sense, capture the entire range of possible outcomes for a project or a comparison among projects and provide an enriched view of the decision.

EXAMPLE 11.4

Cogenesis (refer to Example 11.1) wishes to do a scenario analysis of its cogeneration problem in order to decide whether the project should be undertaken. Analysts have come up with optimistic, pessimistic, and expected estimates of each of the parameters for their decision problem in order to get a better understanding of the possible range of present worth outcomes for the cogeneration plant. The three scenarios and the associated estimates are summarized in Table 11.5.

The scenarios capture combinations of parameter estimates that reflect the worst, best, and expected outcomes for the project. In contrast with sensitivity graphs and break-even analysis, scenario analysis allows entire groups of parameters to be changed at one time.

Evaluation of each scenario reveals that the present worth of the cogeneration plant will be negative if all parameters take on their worst-case values, and hence, the project is not advisable. The major problem is that the savings in electricity costs are insufficient to make up for the high first cost of the project. In contrast, both the expected-case and best-case scenarios lead to positive present worths, and hence, the project would be viable. (To put the present worths into context, the expected-case and best-case scenarios

Table 11.5   Present Worth of Cogeneration Plant Scenarios

Cost Category	Pessimistic Scenario	Expected Scenario	Optimistic Scenario
Initial investment ($)	3 300 000	3 000 000	2 700 000
Annual chemical, operations, and maintenance costs ($)	75 000	65 000	60 000
Cooling tower overhaul (after 10 years) ($)	21 000	17 000	13 000
Turbogenerator overhauls (after 4, 8, 12, and 16 years) ($)	40 000	35 000	30 000
Additional annual wood costs ($)	400 000	375 000	350 000
Savings in annual electricity costs ($)	920 000	1 000 000	1 080 000
MARR	0.13	0.12	0.11
Present worth of cogeneration plant ($)	–234 639	1 126 343	2 583 848

have IRRs of 17.73% and 24.29%, respectively.) From an overall point of view, there is some risk that the cogeneration project will have a negative present worth, but this will occur only if the worst-case scenario does occur. Even if the worst-case outcome does occur, the loss is not huge compared with the potential gain in the other two cases. What Cogenesis needs to do to look further into the project's viability is assess the risk (or likelihood) that the worst outcome will occur. Decision making under risk is discussed further in Chapter 12.■

As we can see from Example 11.4, scenario analysis allows us to look at the effect of multiple changes in parameter values on an individual project's viability. It can also be used to evaluate the effect of scenarios in a case where there are several alternatives.

EXAMPLE 11.5

Westmount Waxworks analysts have carried out a scenario analysis for three possible outcomes they feel represent pessimistic, optimistic, and expected outcomes for sales levels and the Simplicity wax melter's other costs per unit. The scenarios and the annual costs of the two wax melters are summarized in Table 11.6. From the scenario analysis, we see that the Simplicity wax melter is the preferred choice for the expected and optimistic scenarios. The Finedetail wax melter is preferred only if the pessimistic scenario occurs. In terms of the opportunity cost of making the wrong choice, it is far larger if the optimistic outcome occurs ($315 658 – $188 301 = $127 357) than if the pessimistic outcome occurs ($95 611 – $86 158 = $9453). ■

Table 11.6   Scenario Analysis for Westmount Waxworks

	Pessimistic Scenario	Expected Scenario	Optimistic Scenario
Sales level (units)	30 000	50 000	200 000
Other costs per unit (Simplicity)	$0.75	$0.55	$0.45
Annual cost: Finedetail	$86 158	$113 158	$315 658
Annual cost: Simplicity	$95 611	$104 731	$188 131

As was seen in Examples 11.4 and 11.5, scenario analysis allows us to take into account the interrelationships among parameters when making a choice by examining likely groupings of parameter values in scenarios. The most commonly used scenarios are the pessimistic, optimistic, and expected outcomes. The use of scenarios allows an analyst to capture the range of possible outcomes for a project or group of projects. Done in combination with sensitivity graphs and break-even analysis, a great deal of information can be obtained regarding the economic viability of a project.

The one drawback common to each of the three sensitivity analysis methods covered in this chapter is that they do not capture the likelihood that a parameter will take on a certain value or the likelihood that a certain scenario will occur. This information can further guide a decision maker and is often crucial to assessing the risk of the worst case outcome. Chapter 12 will describe how these concerns are addressed.

REVIEW PROBLEMS

The following case is the basis of Review Problems 11.1 through 11.3.

Betteryet Insurance Inc. is considering two independent energy efficiency improvement projects. Each has a lifetime of 10 years and will have a scrap value of zero at the end of this time. Betteryet can afford to do both if both are economically justified. The first project involves installing high-efficiency motors in the air conditioning system. High-efficiency units use about 7% less electricity than the current motors, which represents annual savings of 70 000 kilowatt-hours. They cost $28 000 to purchase and install and will require maintenance costs of $700 annually.

The second project involves installing a heat exchange unit in the current ventilation system. During the winter, the heat exchange unit transfers heat from warm room air to the cold ventilation air before the air is sent back into the building. This will save about 2 250 000 cubic feet of natural gas per year. In the summer, the heat exchange unit removes heat from the hot ventilation air before it is added to the cooler room air for recirculation. This saves about 29 000 kilowatt-hours of electricity annually. Each heat exchange unit costs $40 000 to purchase and install and annual maintenance costs are $3200.

Betteryet Insurance would like to evaluate the two projects, but there is some uncertainty surrounding what the electricity and natural gas prices will be over the life of the project. Current prices are $0.07 per kilowatt-hour for electricity and $3.50 per thousand cubic feet of natural gas, but some changes are anticipated. They use a MARR of 10%.

REVIEW PROBLEM 11.1

Construct a sensitivity graph to determine the effect that a 5% and 10% drop or increase in the cost of electricity and the cost of natural gas would have upon the present worth of each project.

ANSWER

Table 11.7 gives the costs of electricity and natural gas with 5% and 10% increases and decreases from the base case of $0.07 per kilowatt-hour for electricity and $3.50 per 1000 cubic feet of natural gas. The table also shows the present worths of the two

**Table 11.7   Costs Used as the Basis of the Sensitivity Graph
for Review Problem 11.1**

	–10%	–5%	0%	+5%	+10%
Cost of electricity ($/kWh)	0.063	0.0665	0.07	0.0735	0.077
Cost of natural gas ($/1000 cubic feet)	3.15	3.325	3.5	3.675	3.85
PW of high-efficiency motor ($) With changes to electricity costs	–5204	–3698	–2193	–687	818
With changes to natural gas cost	–2193	–2193	–2193	–2193	–2193
PW of heat exchanger ($) With changes to electricity costs	–48	576	1199	1823	2447
With changes to natural gas cost	–3640	–1220	1199	3619	6038

energy efficiency projects as the costs vary. A sample calculation for the heat exchange unit with base case costs is

$$PW(\text{Heat exchanger}) = -40\,000 + (P/A, 10\%, 10)$$
$$\times [29\,000(0.07) + 2250(3.50) - 3200]$$
$$\cong 1199$$

Figure 11.7 is a sensitivity graph for the high-efficiency motor. It graphically illustrates the effect of changes in the costs of electricity and natural gas on the present worth of a motor. The high-efficiency motor is not economically viable at the current prices for electricity and gas. Only if there is an increase of almost 10% in electricity costs for the life of the project will the motor produce sufficient savings for the project to have a positive present worth.

Figure 11.7   Sensitivity Graph for the High-Efficiency Motor

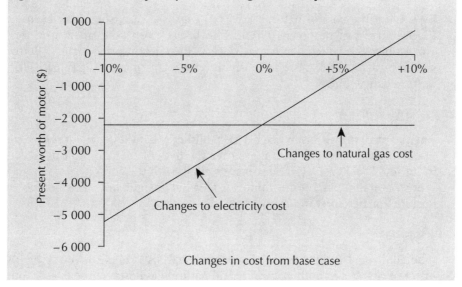

Figure 11.8 is the sensitivity graph for the heat exchange unit. The heat exchange unit has a positive present worth for the current prices, but the present worth is quite sensitive to the price of natural gas. A drop in the price of natural gas in the range of only 2% to 3% (reading from the graph) will cause the project to have a negative present worth.■

Figure 11.8   Sensitivity Graph for the Heat Exchanger

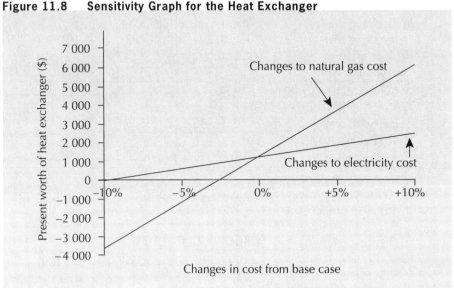

REVIEW PROBLEM 11.2

Refer to Review Problem 11.1. How much of a drop in the cost of natural gas will result in the heat exchange unit's having a present worth of zero? Construct a break-even graph to illustrate this break-even cost.

ANSWER

By varying the cost of natural gas from the base case, the break-even graph shown in Figure 11.9 can be constructed. The break-even cost of natural gas is $3.41 per 1000 cubic feet, which is not much below the current price for gas. Betteryet Insurance should probably look more seriously into forecasts of natural gas prices for the life of the heat exchange unit.■

REVIEW PROBLEM 11.3

Analysts at Betteryet Insurance have established what they think are three scenarios for the prices of electricity and natural gas over the lives of the two projects under consideration in Review Problem 11.1. The scenarios along with the appropriate present worth computations are summarized in Table 11.8. What insight does this add to the investment decision for Betteryet Insurance?

ANSWER

The additional insight that the scenario analysis brings to Betteryet Insurance is the effect of changes in both electricity and natural gas costs on the two proposed projects.

Figure 11.9   Break-Even Chart for the Cost of Natural Gas

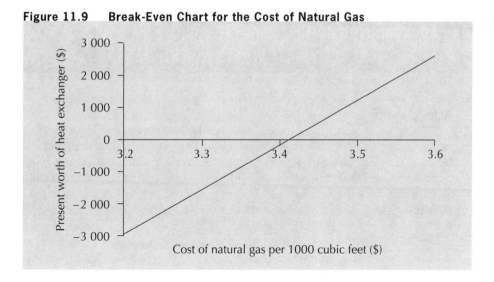

Cost of natural gas per 1000 cubic feet ($)

Table 11.8   Scenario Analysis for Betteryet Insurance

	Pessimistic Scenario	Expected Scenario	Optimistic Scenario
Cost of electricity ($/kWh)	0.063	0.070	0.077
Cost of natural gas ($/1000 cubic feet)	3.35	3.50	3.65
PW of high efficiency motor ($)	–5204	–2193	818
PW of heat exchange unit ($)	–2122	1199	4520

Sensitivity graphs and the break-even analysis can look only at the effect of one-at-a-time parameter changes on the present worth computations. It appears that the high-efficiency motor is a bad investment, as its present worth is not much above zero even if the optimistic scenario occurs. The heat exchange unit appears to be a better investment, but even that has a chance of having a negative present worth if the pessimistic outcome occurs. What Betteryet really needs to know is the likelihood of each of these scenarios occurring, or some other means of assessing the likelihood of what energy prices will be in the future.■

SUMMARY

In this chapter, we considered three basic methods used by analysts in order to better understand the effect that uncertainties in estimated cash flows have on economic decisions. The first was the use of sensitivity graphs. Sensitivity graphs illustrate the sensitivity of a particular measure (e.g., present worth or annual worth) to changes in one or more of the parameters of a project. The second method was the use of break-even analysis for evaluating both individual projects and comparisons among projects. Finally, scenario analysis allowed us to look at the overall impact of a variety of outcomes, usually optimistic, expected, and pessimistic.

Engineering Economics in Action, Part 11B:
Where the Risks Lie

Bill Astad and Naomi were working through the market demand figures provided by Powerluxe for the new self-adjusting vacuum.

"These figures are pretty ambiguous," Bill said. "We have three approaches: a set of opinions taken from focus groups and surveys of customers, the same thing from dealers and distributors, and an analysis of trends in a set of parallel products such as fuzzy-logic appliances. Like Anna said, nothing hard."

"What we really want to know," said Naomi, "is whether we have the capacity to handle the manufacturing for the product. Based on the surveys and the trend information, let's come up with three scenarios: low demand, expected demand, and high demand. If we behave according to expected demand, and the true demand is low, we will lose money because our capital investments won't be recouped as fast, and we may have passed up other opportunities. Similarly, if the demand is high, we will lose by having to pay overtime, paying for contracting out, or losing customers. But if we make money in all three cases, there really isn't much of a problem."

"And if it turns out we don't make money in all three cases?" Bill asked. "What then?"

"I know it's a lot of work, Bill, but let's do it and find out," Naomi replied. "At minimum, we will know where our risks lie."

PROBLEMS

For additional practice, please see the problems (with selected solutions) provided on the Student CD-ROM that accompanies this book.

11.1 Identify possible parameters that are involved in economic analysis for the following situations:

(a) Buying new equipment

(b) Supplying products to a foreign country with a high inflation rate

11.2 For the following examples of parameters, how would you assign a reasonable base case and a range of variation so that you can carry out sensitivity analysis? Assign specific numerical figures wherever you can.

(a) Your country's inflation rate

(b) US dollar exchange rate for your country's (or another country's) currency

(c) Expected annual savings from a new piece of equipment similar to the one you already have

(d) Expected annual revenue from an internet-based business

(e) Salvage value of a personal computer

11.3 Which sensitivity analysis method may be appropriate for analyzing the following uncertain situations?

(a) Corral Cartage leases trucks to service its shipping contracts. Larger trucks have cheaper operating costs if there is sufficient business, but are more expensive if they are not full. Corral Cartage is not certain about future demand.

(b) Joan runs a dog kennel. She is considering installing a heating system for the interior runs that will allow her to operate all year. Joan is not sure how much the annual heating expenses will be.

(c) Pushpa runs a one-person company producing custom paints for hobbyists. She is considering buying printing equipment to produce her own labels. However, she is not sure if she will have enough orders in the future to justify the purchase of the new equipment.

(d) Lemuel is an engineer working for the electric company. He is estimating the total cost for building transmission lines from a distant nuclear plant to new industrial parks north of the city. Lemuel is uncertain about the construction cost (per kilometre) of transmission lines.

(e) Thanh's company is growing very quickly and has a hard time meeting its orders. An opportunity to purchase additional production equipment has arisen. She is not certain if the company will continue to grow at the same rate in the future, and she is not even certain how long the growth may last.

11.4 The Hanover Go-Kart Klub has decided to build a clubhouse and track several years from now. The club needs to accumulate €50 000 by setting aside a uniform amount at the end of every year. They believe it is possible to set aside €7000 every year at 10% interest. They wish to know how many years it is will take to save €50 000 and how sensitive this result is to a 5% and a 10% increase or decrease in the amount saved per year and in the interest rate. Construct a sensitivity graph to illustrate the situation.

11.5 A new software package is expected to improve productivity at Suretown Insurance. However, because of training and implementation costs, savings are not expected to occur until the third year of operation. Annual savings of approximately $10 000 are expected, increasing by about $1000 per year for the following five years. After this time (eight years from implementation), the software will be abandoned with no scrap value. Construct a sensitivity graph showing what would happen to the present worth of the software with 7.5% and 15% increases and decreases in the interest rate, the $10 000 base savings, and the $1000 savings gradient. MARR is 15%.

11.6 A regional municipality is studying a water supply plan for the area to the end of the year 2050. To satisfy the water demand, one suggestion is to construct a pipeline from a distant lake. It is now the end of 2010. Construction would start in the year 2015 (five years from now) and take five years to complete at a cost of $20 million per year. Annual maintenance and repair costs are expected to be $2 million and will start the year following project completion (all costs are based on current estimates). From a predicted inflation rate of 3% per year, and the real MARR, city engineers have determined that a MARR of 7% per year is appropriate. Assume that all cash flows take place at the end of the year and that there is no salvage value at the end of 2055.

(a) Find the present worth of the project.

(b) Construct a sensitivity graph showing the effects of 5% and 10% increases and decreases in the construction costs, maintenance costs, and inflation rate. To which is the present worth most sensitive?

11.7 The city of Brussels is installing a new swimming pool in the downtown recreation centre. One design being considered is a reinforced concrete pool that will cost €6 000 000 to install. Thereafter, the inner surface of the pool will need to be refinished and painted every 10 years at a cost of €40 000 per refinishing. Assuming that the pool will have essentially an infinite life, what is the present worth of the costs associated with the pool design? The city uses a MARR of 5%. If the installation costs, refinishing costs, and MARR are subject to 5% or 10% increases or decreases, how is the present worth affected? To which parameter is the present worth most sensitive?

11.8 You and two friends are thinking about setting up a grocery delivery service for local residents to finance your last two years at university. In order to start up the business, you will need to purchase a car. You have found a used car that costs $6000 and you expect to be able to sell it for $3000 at the end of two years. Insurance costs are $600 for each six months of operation, starting now. Advertising costs (e.g., flyers, newspaper advertisements) are estimated to be $100 per month, but these might vary as much as 20% above or below the $100, depending on the intensity of your advertising. The big questions you have now are how many customers you will have and how much of a service fee to charge per delivery. You estimate that you will have 300 deliveries every month, and are thinking of setting a $2-per-delivery fee, payable at the end of each month. The interest rate over the two-year period is expected to be 8% per year, compounded monthly, but may be 20% above or below this figure.

Using equivalent monthly worth, construct a sensitivity graph showing how sensitive the monthly worth of this project will be to the interest rate, advertising costs, and the number of deliveries you make each month. To which parameter is the equivalent monthly worth most sensitive?

11.9 Timely Testing (TT) does subcontracting work for printed circuit board manufacturers. They perform a variety of specialized functional tests on the assembled circuit boards. TT is considering buying a new probing device that will assist the technicians in diagnosing functional defects in the printed circuit boards. Two vendors have given them quotes on first costs and expected operating costs over the life of their equipment.

	Vendor A	**Vendor B**
Expected life	7 years	10 years
First cost	$200 000	$350 000
Maintenance costs	$10 000/year + $0.05/unit	$20 000/year + $0.01/unit
Labour costs	$1.25/unit	$0.50/unit
Other costs	$6 500/year + $0.95/unit	$15 000/year + $0.55/unit
Salvage value	$5 000	$20 000

Production levels vary for TT. They may be as low as 20 000 boards per year or as high as 200 000 boards per year if a contract currently under negotiation comes through. They expect, however, that production quantities will be about 50 000 boards. Timely Testing uses a MARR of 15% for equipment projects, and will be using an annual worth comparison for the two devices.

Timely Testing is aware that the equipment vendors have given estimates only for costs. In particular, TT would like to know how sensitive the annual worth of each device is to the first cost, annual fixed costs (maintenance + other), variable costs (maintenance + labour + other), and the salvage value.

(a) Construct a sensitivity graph for vendor A's device, showing the effects of 5% and 10% decreases and increases in the first cost, annual fixed costs, variable costs, and the salvage value. Assume an annual production level of 50 000 units.

(b) Construct a sensitivity graph for vendor B's device, showing the effects of 5% and 10% decreases and increases in the first cost, annual fixed costs, variable costs, and the salvage value. Assume an annual production level of 50 000 units.

11.10 Merry Metalworks would like to implement a local area network (LAN) for file transfer, email, and database access throughout its facility. Two feasible network topologies have been identified, which they have labelled alternative A and alternative B. The three main components of costs for the network are (1) initial hardware and installation costs, (2) initial software development costs, and (3) software and hardware maintenance costs. The installation and hardware costs for both systems are somewhat uncertain, as prices for the components are changing and Merry Metalworks is not sure of the installation costs for the LAN hardware. The costs for each alternative are summarized below.

	Alternative A	Alternative B
Initial hardware and installation costs ($):		
Optimistic estimate	70 000	86 000
Average estimate	92 500	105 500
Pessimistic estimate	115 000	125 000
Initial software cost ($)	138 750	158 250
Annual maintenance costs ($)	9250	10 550
Annual benefits ($):		
Optimistic estimate	80 000	94 000
Average estimate	65 000	74 000
Pessimistic estimate	50 000	54 000

Benefits from the LAN are increased productivity because of faster file transfer times, reduced data redundancy, and improved data accuracy because of the database access. The benefits were difficult to quantify and are stated below as only a range of possible values and an average.

Merry Metalworks uses a 15% MARR and has established a 10-year study period for this decision. They wish to compare the projects on the basis of annual worth.

(a) Construct a sensitivity graph for alternative A. For the base case, use the average values for the initial hardware cost and the annual benefits. Each graph should indicate the effect of a 5% and a 10% drop or increase in the initial hardware cost and the annual benefits. Which of the two factors most affects the annual worth of alternative A?

(b) Construct a sensitivity graph for alternative B. For the base case, use the average values for the initial hardware cost and the annual benefits. Each graph should indicate the effect of a 5% and a 10% drop or increase in the initial hardware cost and the annual benefits. Which of the two factors most affects the annual worth of alternative B?

11.11 Refer back to Problem 11.8.

(a) Assuming base case figures for advertising costs and interest rates, what is the break-even number of deliveries per month? Construct a graph showing the break-even number.

(b) Assuming base case figures for advertising costs and number of deliveries per month, what is the break-even interest rate? Construct a graph illustrating the break-even interest rate.

11.12 Refer back to Problem 11.4. Members of the Go-Kart Klub do not wish to wait for more than five years to build their clubhouse. They have decided to start a fundraising campaign to increase their ability to save each year between €7000 and whatever is necessary to have €50 000 saved in five years. Construct a table and a graph that illustrate how the number of years they must wait depends on the amount they save each year. What additional funds per year will allow them to save €50 000 in five years? Use a 10% interest rate.

11.13 Refer back to Problem 11.10 in which Merry Metalworks is considering two LAN alternatives.

(a) For alternative A, by how much will the installation cost have to rise before the annual worth becomes zero? In other words, what is the break-even installation cost? Is the break-even level within or above the range of likely values Merry Metalworks has specified?

(b) What is the break-even annual benefit for alternative A? Use the average installation costs. Is the break-even level within or above the range of likely values Merry Metalworks has specified?

11.14 Repeat Problem 11.13 for alternative B.

11.15 Refer back to Timely Testing, Problem 11.9.

(a) TT charges $3.25 per board tested. Assuming that costs are as in vendor A's estimates, what production level per year would allow TT to break even if they select vendor A's equipment? That is, for what production level would annual revenues equal annual costs? Construct a graph showing total revenues and total costs for various production levels, and indicate on it the break-even production level.

(b) Repeat part (a) for vendor B's equipment.

11.16 The Bountiful Bread Company produces home bread-making machines. Currently, they pay a custom moulder £0.19 per piece (not including material costs) for the clear plastic face on the control panel. Demand for the bread-makers is forecast to be 200 000 machines per year, but there is some uncertainty surrounding this estimate. Bountiful is considering installing a plastic moulding system to produce the parts itself. The moulder costs £20 000 plus £7000 to install, and has an expected life of six years. Operating and maintenance costs are expected to be £30 000 in the first year and to rise at the rate of 5% per year. Bountiful estimates its capital costs using a declining-balance depreciation model with a rate of 40%, and uses a MARR of 15% for such investments.

Determine the total equivalent annual cost of the new moulder. What is the cost per unit, assuming that production is 200 000 units per year? Also, determine the break-even production quantity. That is, what is the production quantity below which it is better to continue to purchase parts and above which it is better to purchase the moulder and make the parts in-house?

11.17 Tenspeed Trucking (TT) is considering the purchase of a new $65 000 truck. The truck is expected to generate revenues between $12 000 and $22 000 each year, and will have a salvage value of $20 000 at the end of its five-year life. TT pays taxes at the rate of 35%. TT is located in Canada and the CCA rate for trucks is 30% [note to instructors: substitute other appropriate tax information as desired]. TT's after-tax MARR is 12%. Find the annual worth of the truck if the annual revenues are $12 000, and for each $1000 revenue increment up to $22 000. What is the break-even annual revenue? Provide a graph to illustrate the break-even annual revenue.

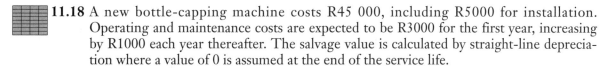

11.18 A new bottle-capping machine costs R45 000, including R5000 for installation. Operating and maintenance costs are expected to be R3000 for the first year, increasing by R1000 each year thereafter. The salvage value is calculated by straight-line depreciation where a value of 0 is assumed at the end of the service life.

(a) Construct a spreadsheet that computes the equivalent annual cost (EAC) for the bottle capper. What is the economic life if the expected service life is 6, 7, 8, 9, or 10 years? Interest is 12%.

(b) How sensitive is the economic life to the different length of service life? Construct a sensitivity graph to illustrate this point.

11.19 A chemical plant is considering installing a new water purification system that costs $21 500. The expected service life of the system is 10 years and the salvage value is computed using the declining-balance method with a depreciation rate of 20%. The operating and maintenance costs are estimated to be $5 per hour of operation. The expected savings are $10 per operating hour.

(a) Find the annual worth of the new water purification system if the current operating hours are 1500 per year on average. MARR is 10%.

(b) What is the break-even level of operating hours? Construct a graph showing the annual worth for various levels of operating hours.

11.20 Fantastic Footwear can invest in one of two different automated clicker cutters. The first, A, has a 100 000 yuan first cost. A similar one, B, with many extra features, has a first cost of 400 000 yuan. A will save 50 000 yuan per year over the cutter now in use. B will save between 120 000 yuan and 150 000 yuan per year. Each clicker cutter will last five years and have a zero scrap value.

(a) If the MARR is 10%, and B will save 150 000 yuan per year, which alternative is better?

(b) B will save between 120 000 yuan and 150 000 yuan per year. Determine the IRR for the incremental investment from A to B for this range, in increments of 5000 yuan. Plot savings of B versus the IRR of the incremental investment. Over what range of savings per year is your answer from part (a) valid? What are the break-even savings for alternative B such that below this amount, A is preferred and above this amount, B is preferred?

11.21 Sam is considering buying a new lawnmower. He has a choice between a "Lawn Guy" model or a Bargain Joe's "Clip Job" model. Sam has a MARR of 5%. The mowers' salvage values at the end of their respective service lives are zero. Sam has collected the following information about the two mowers.

	Lawn Guy	Clip Job
First cost	$350	$120
Life	10 years	4 years
Annual gas	$60	$40
Annual maintenance	$30	$60

Although Sam has estimated the maintenance costs of the Clip Job at $60, he has heard that the machines have had highly variable maintenance costs. One friend claimed that her Clip Job had maintenance costs comparable to those of the Lawn Guy, but another said the maintenance costs could be as high as $80 per year. Construct a table that shows the annual worth of the Clip Job for annual maintenance costs varying from $30 per year to $80 per year. What Clip Job maintenance costs would make Sam indifferent between the two mowers, on the basis of annual worth? Construct a graph showing the break-even maintenance costs. Which mower would you recommend to Sam?

11.22 Ganesh is considering buying a $24 000 car. After five years, he thinks, he will be able to sell the car for $8000, but this is just an estimate that he is not certain about. He is confident that gas will cost $2000 per year, insurance $800 per year, and parking $600 per year, and that maintenance costs for the first year will be $1000, rising by $400 per year thereafter.

The alternative is for Ganesh to take taxis everywhere. This will cost an estimated $7000 per year. If he has no car, Ganesh will rent a car for the family vacation every year at a total (year-end) cost of $1000. Ganesh values money at 11% annual interest. If the salvage value of the car is $8000, should he buy the car? Base your answer on annual worth. Determine the annual worth of the car for a variety of salvage values so that you can help Ganesh decide whether this uncertainty will affect his decision. For what break-even salvage value will he be indifferent between taking taxis and buying a car? Construct a break-even graph showing the annual worth of both alternatives as a function of the salvage value of the car. What advice would you give Ganesh?

11.23 Ridgely Custom Metal Products (RCMP) must purchase a new tube bender. It is considering two alternatives that have the following characteristics:

	Model T	**Model A**
First cost	$100 000	$150 000
Economic life	5 years	5 years
Yearly savings	$50 000	$62 000
Salvage value	$20 000	$30 000

Construct a break-even graph showing the present worth of each alternative as a function of interest rates between 6% and 20%. Which is the preferred choice at 8% interest? Which is the preferred choice at 16% interest? What is the break-even interest rate?

11.24 Julia must choose between two different designs for a safety enclosure. Model A has a life of three years, has a first cost of $8000, and requires maintenance of $1000 per year. She believes that a salvage value can be estimated for model A using a depreciation rate of between 30% and 40% and declining-balance depreciation. Model B will last four years, has a first cost of $10 000, and has maintenance costs of $700 per year. A salvage value for model B can be estimated using straight-line depreciation and the knowledge that after one year the salvage value will be $7500. Interest is at 11%. Which of the two models would you suggest Julia choose? What break-even depreciation rate for model A will make her indifferent between the two models? Construct a sensitivity graph showing the break-even depreciation rate.

11.25 Your neighbour, Kelly Strome, is trying to make a decision about his growing home-based copying business. He needs to acquire colour copiers able to handle maps and other large documents. He is looking at one set of copiers that will cost $15 000 to purchase. If he purchases the equipment, he will need to buy a maintenance contract that will cost $1000 for the first year, rising by $400 per year afterward. He intends to keep the copiers for five years, and expects to salvage them for $2500. Kelly's business is located in Canada and the CCA rate for office equipment is 20% [note to instructors: substitute other appropriate tax information as desired].

Rather than buy the copiers, Kelly could lease them for $5500 per year with no maintenance fee. His business volume has varied over the past few years, and his tax rate has varied from a low of 20% to a high of 40%. Kelly's current cost of capital is 8%. Kelly has asked you for some help in deciding what to do. He wants to know whether he should lease or buy the copiers, and, moreover, he wants to know the impact of his tax rate on the decision. Evaluate both alternatives for him for a variety of tax rates between 20% and 40% so that you can advise him confidently. What do you advise?

11.26 Western Insurance wants to introduce a new accounting software package for its human resources department. A small-scale version is sufficient and economical if the number of employees is less than 50. A large-scale version is effective for managing 80 employees or more. All relevant information on the two packages is shown in the following table. Western Insurance's business is growing, and the number of employees has increased from 10 to 40 in the past three years. Construct a graph showing the annual worth of the two software packages as a function of the number of employees ranging from 40 to 100. On the basis of break-even analysis, which accounting package is a better choice for Western Insurance? MARR is 12%.

Parameter		Small-Scale	Large-Scale
First cost ($)		6000	10 000
Training cost at the time of installation ($)		1500	3500
Service life (years)		5	5
Salvage value		0	0
Expected annual savings ($ per employee) if the average number of employees over the next 5 years is:	Less than 50	200	250
	Between 50 and 80	170	300
	Greater than 80	120	400

11.27 Refer back to Problem 11.10 in which Merry Metalworks is looking at several LAN alternatives. Conduct a scenario analysis for each alternative, using the pessimistic, expected (average), and optimistic outcomes for both installation costs and annual benefits. Which of the two alternatives would you choose? Why?

11.28 Refer back to Problem 11.4 in which the Hanover Go-Kart Klub is trying to save €50 000 in order to build a clubhouse and track. They have established optimistic, expected, and pessimistic estimates for both the interest rate they will earn on their savings and the amount they will be able to save per year. Conduct a scenario analysis that shows the number of years required to save €50 000 for the three scenarios, using the data below.

Parameter	Pessimistic Scenario	Expected Scenario	Optimistic Scenario
Savings per year (€)	6000	7000	8000
Interest rate	8.00%	11.00%	12.00%
Number of years to save €50 000	6.64	5.66	4.94

11.29 Timely Testing (refer back to Problem 11.9) has established pessimistic, expected, and optimistic figures for the first cost and other costs of the two testing devices offered by vendors A and B. They would like you to carry out a scenario analysis to determine which of the two alternatives to choose. They charge $3.25 per board tested. What recommendations would you give Timely Testing?

Parameter	Alternative A		
	Pessimistic Scenario	Expected Scenario	Optimistic Scenario
First cost ($)	220 000	200 000	190 000
Annual fixed costs ($)	18 000	16 500	13 000
Annual variable costs ($ per board)	2.35	2.25	2.20
Salvage value ($)	2000	5000	7000
Annual production volume (boards)	40 000	50 000	80 000

Parameter	Alternative B		
	Pessimistic Scenario	Expected Scenario	Optimistic Scenario
First cost ($)	365 000	350 000	320 000
Annual fixed costs ($)	45 000	35 500	25 000
Annual variable costs ($ per board)	1.100	1.060	1.010
Salvage value ($)	17 000	20 000	23 000
Annual production volume (boards)	40 000	50 000	80 000

11.30 The Bountiful Bread Company (Problem 11.16) currently pays a custom moulder £0.19 per piece (not including material costs) for the clear plastic face on the control panel of bread-maker machines it manufactures. Demand for the bread-makers is estimated at 200 000 machines per year, but there is some uncertainty surrounding this estimate. Bountiful is considering installing a plastic moulding system to produce the parts itself. Installation costs are £7000, and the moulder has an expected life of six years. Operating and maintenance costs are somewhat uncertain, but are expected to be £30 000 in the first year and to rise at the rate of 5% per year. Bountiful estimates its capital costs with a declining-balance depreciation model with a rate of 40%, and uses a MARR of 15% for such investments.

Parameter	Pessimistic Scenario	Expected Scenario	Optimistic Scenario
First cost (£)	25 000	20 000	18 000
Base annual operating and maintenance costs (£)	35 000	30 000	27 000
Production volume (units)	170 000	200 000	240 000

The project engineers have come up with pessimistic, expected, and optimistic figures for the first cost, operating and maintenance costs, and production levels. Determine the total equivalent annual cost of the new moulder for each scenario and then the cost per unit. What advice would you give to Bountiful regarding the purchase of the moulder?

More Challenging Problem

11.31 SoftWaire Inc., a United States-based software development company, is contemplating outsourcing a portion of its operations to Bangalore, India. The primary attractions for SoftWaire are the lower salaries in India and the availability of English-speaking engineers. The average salary of the 20 people they are thinking of replacing with an off-shore operation is US $75 000 per year, while the salary of a comparable engineer in India is 600 000 rupees. With an exchange rate of 40 rupees per dollar, the salary costs in India are about 20% of what they are in the United States. This could bring substantial savings to SoftWaire; however, there appear to be numerous uncertainties associated with outsourcing. You have been asked to conduct an analysis so SoftWaire Inc. can better understand the potential benefits and risks of the project.

Based on your initial data gathering, you have found the following information.

1. The salaries of software developers in India are indeed lower than those in the United States, however, demand from other local and international employers is strong and is causing salaries to increase at a rate of 10%–15% per year, with some forecasts as high as 20% per year. By comparison, American salaries are expected to grow at a rate of roughly 5% per year.

2. Due to the high level of competition for good software developers in India, there is a high turnover rate in the industry. Estimates run from 20% to 30%. The American market is stable by comparison.

3. Most foreign companies use the services of a recruiter to find suitable employees. Recruiters typically charge a fee of about 30% of the salary of the successful employee to solicit applications, screen candidates, and present alternatives for final selection.

4. Training costs for new employees in their first year can be up to 70% of their salary for education in best practices and procedures used by the employer. Their productivity is typically about 50% of a trained employee's due to the learning process.

5. For development outsourcing to be successful, other firms report that they spend considerable time ensuring that software specifications are precise, that quality assurance procedures are clearly spelled out and followed, and that developers adhere to best practices. The cost of this additional project management has been estimated to add 10%–20% to salary costs.

6. Periodic problems with infrastructure and cultural differences have been reported to reduce overall development productivity to about 70% of that in the United States.

7. Setting up an outsourced team can entail considerable start-up costs. Travel, recruiter selection, facilities selection, and equipment installation for a team of 20 can range from $300 000 to $500 000.

8. SoftWaire would lay off 20 software engineers at the start of the project. This will cost one year's salary per person.

9. Office space lease costs will be roughly the same whether SoftWaire outsources or not.

10. SoftWaire uses a MARR of 25%.

11. The American dollar has fluctuated against the Indian rupee in recent years, and there is some concern that the rupee may strengthen against the dollar by as much as 10% in coming years.

You have decided to set up a spreadsheet to help with your analysis. From the information above, you plan to use the following as variables in a sensitivity analysis:

(1) salary growth rates for both American and overseas software developers, (2) turnover rate of overseas employees, (3) overseas recruiting costs, as a percent of salary, (4) training costs, as a percent of salary, (5) productivity rate of an employee in training, (6) productivity rate of a trained employee, (7) additional project management costs, as a percent of salary, (8) SoftWaire's MARR, (9) the exchange rate between the rupee and the American dollar (currently 40 rupees per dollar) and (10) start-up costs.

(a) Construct a spreadsheet with the above variables to project the costs of outsourcing for a study period of 10 years. Start with the basic projected salaries, and adjust for recruiting and training costs, then for training and long-term productivity costs, and finally for project management, start-up, and layoff costs. Assume that the overseas developers are hired at time zero. When a range of values has been given for a variable, use the midpoint as the base case.

(b) Determine the present worth of outsourcing.

(c) Determine the present worth of insourcing, that is, using the United States-based developers.

(d) What is the present worth of the benefits of outsourcing?

(e) Using the spreadsheet, conduct a sensitivity analysis for each of the uncertain factors, varying each factor by 10% to determine which has the greatest impact on the present worth of the benefits of outsourcing.

(f) What factors have the greatest impact on the present worth of the benefits of outsourcing? What would you advise at this point?

(g) At what exchange rate would the present worth of outsourcing be zero?

(h) At what productivity rate would the present worth of outsourcing be zero?

(i) Drawing on specific examples from this problem, describe some of the drawbacks of sensitivity analysis when making the decision to outsource or not.

(j) Formulate what you think would be a worst-case, expected, and best-case scenario for SoftWaire in terms of whether outsourcing is a good decision or not. What are your recommendations based on the information you have? If you could invest in obtaining better parameter estimates, which ones would you pick, and why?

MINI-CASE 11.1

China Steel Australia Limited

Stainless steel is an alloy of iron, chromium, and other metals. It has valuable uses for applications where strength and resistance to corrosion are important. Uses include kitchen cookware, surgical tools, building structures, aircraft, food processing equipment, and automotive components.

There are numerous grades of stainless steel, depending on the needs of the particular application. They vary by the alloying elements and the consequent crystalline structure of the steel. About 70% of stainless steel production, on a global basis, is *austenitic* stainless steel, which is composed of a maximum of 0.15% carbon and a minimum of 18% chromium, along with a significant percentage of nickel. A typical grade used for household flatware has 18% chromium and 10% nickel, for example.

In recent years, the cost and availability of nickel have created a bottleneck for stainless steel producers. There are a limited number of nickel ore deposits, and the majority of deposits consist of *laterite* nickel ore, which is generally not used for producing refined nickel because of its low nickel content.

China Steel has a plant in Linyi, Shandong Province, China, that produces nickel pig iron from laterite nickel ore. This nickel pig iron contains 5%–11% nickel, which can substitute for refined nickel in the production of austenitic stainless steel. The growth of Chinese manufacturing has created very strong demand for stainless steel in general, and for China Steel's nickel pig iron in particular. The Linyi plant is running at full capacity and has offtake commitments in place for the next five years. There is currently a substantial demand for nickel pig iron that cannot be met. Consequently, China Steel wants to expand its plant to increase capacity in order to take advantage of this opportunity.

In February 2008, China Steel released a prospectus seeking investors to fund its expansion plans. A prospectus is a document that lays out an investment opportunity according to specific rules legislated to ensure that potential investors are treated fairly. The prospectus in this case sought AUS $15 000 000 (Australian dollars) to provide a foundation for plant expansion by setting up a structure that would facilitate access to capital markets in Australia and elsewhere.

In order to demonstrate the economic viability of the company, the prospectus included a forecast of earnings for 2008. The forecast was based on expected costs and revenues, barring unusual deviations in the economic and competitive environment. Under these conditions, the profit after income tax expenditures for 2008 was estimated to be AUS $12 304 000.

However, to enrich this estimate, the prospectus also shows the effect of changes in key assumptions on the estimated profit. These changes are as shown in Table 11.9.

Table 11.9   Sensitivity Analysis for China Steel Australia

Parameter	Variation	Increase ($)	Decrease ($)
Revenue	10% change	2 105 000	–2 105 000
Cost of sales	10% change	–1 185 000	1 185 000
Administrative expenses	10% change	–102 000	102 000
Corporate overheads	10% change	–51 000	51 000
Cost of borrowing	1%	442 000	–461 000
Australian dollar/yuan exchange rate	10% change	–745 000	909 000

Discussion

Predicting the future can never be done with complete assurance because everything is uncertain. We can only make our best guess at the future, recognizing that things might turn out differently than we thought.

Engineering design often assumes that the world is much simpler than it really is. When an engineer designs a roof truss, for example, he or she often assumes that the lumber making up the truss will behave in a standard, predictable manner. Similarly, the engineer who designs a circuit will assume that the electrical components will behave according to their nominal values. But lumber is a natural product, and individual pieces will be weaker or stronger than expected. Electrical components, similarly, will have actual values and behaviour different, in general, from their nominal values and mathematical models.

Good engineers understand this and design accordingly. The truss builder specifies a certain grade of lumber, or makes sure that redundant support is built into the design.

The circuit designer similarly specifies the tolerances of significant components, or designs the circuit in a robust way.

The role of sensitivity analysis in economic studies is exactly the same. We don't know the cash flow of a project exactly, just as we don't know the behaviour of a piece of wood or a circuit component exactly. We want to design the project to control the uncertainty of the economic elements as well as the physical ones. In the China Steel Australia prospectus, the expected future was insufficient to adequately explain what a prospective investor might expect in terms of financial performance. This needed to be augmented by explaining the effects of variations on this future view.

Questions

1. Most of the changes made by China Steel were for plus or minus 10%. Does this make sense? If not, what amounts would have been a better choice? Why was the variation on the cost of borrowing only 1%?
2. It is clear that a 10% change in revenue has the biggest effect on profit. Does it necessarily mean that this is also the biggest risk to China Steel's viability as a company?
3. If you were considering investing in China Steel, what would you do to help reduce the uncertainty of the investment?
4. Derek has been assigned the task of designing a parking facility for an insurance company. He must keep in mind a number of different issues, including land acquisition costs, building costs (if a parking building is required), expected usage, fee method (monthly fees, hourly fees, or in-and-out fees), whether the company will subsidize the facility in part or completely, etc. His boss is particularly concerned about reducing the uncertainty of the future cash flows associated with the project. How would you advise Derek?

1. A Morning Meeting

Carole smiled and said a cheerful "Good morning" as Naomi and Dave came up the hallway. "Clem's not here yet, but you might as well have a seat in his office."

"Just like a manager to be late for his own meeting," observed Naomi with a smile, as she sank into her favourite green paisley chair. "Looks like we know who'll be buying the coffee this morning."

No sooner had they sat down when Clem came bounding in with a breathless "Sorry I'm late."

"No problem, Clem. Oh, don't forget the sugar in my coffee this time."

"I deserve that, Naomi," replied Clem unexpectedly. "Tell you what. You guys did such a good job on that cold-former evaluation that I'll even spring for doughnuts today."

Dave raised his eyebrows in mock surprise. "What is this, Clem, an early Christmas?"

"Don't be so sure, Dave," Naomi interjected. "I have a feeling we're being buttered up for something." Looking at Clem, she continued, "Do I feel some onerous task coming on?"

"'Opportunity,' Naomi. We only have 'opportunities' around here," corrected Clem, smiling. "And speaking of opportunities, it looks like we have a good one in this project." He was pointing to Naomi and Dave's cold-former report. "I have a feeling that Anna Kulkowski and Ed Burns will be very happy to get this kind of cost savings."

"So we now do a full engineering economics evaluation for buying a cold-former?" asked Naomi.

"Right. In fact, I'd like you to do evaluations of both the single E1 and single E2 options. No matter what you said in your report, I think that looking at these more carefully is worthwhile."

Dave thought for a minute. "Why both? If we get a better estimate of selling price from Prabha, we'll know whether the E1 or E2 is the better option."

"Yes," replied Clem, "but I am going to recommend performing a market study on selling prices to Ed Burns right after this meeting. With the data you have here, I'm sure he'll approve the $5000 expense right away. The trouble is, it will be at least two weeks before Prabha can give us an answer. I'd like to be able to make a decision as soon as we get it."

"So . . . you'd like us to prepare an evaluation for each machine . . . leaving the selling price as a variable . . . so that when we get the better estimate of what the price will be, we can make an immediate decision about which machine to buy. That sounds like a 'break-even' analysis, Clem," finished Naomi.

"Can't you just see those mental wheels spinning?" said Clem to Dave with a grin. "That's exactly what I'd like. Not only would a break-even analysis show us which machine would be best at whatever selling price we end up getting, but it would also give us an idea of how sensitive the benefits of this project are to the price at which we sell excess production."

Naomi and Dave were quiet for a few seconds while they thought this over.

Dave glanced up. "Would this mean that if we show enough extra benefit from selling our excess capacity, marketing would be asked to put some effort into getting better prices for us?"

"That's right, Dave. A break-even analysis would give us the data to demonstrate to management how much the company would benefit and what the payoff would be as a function of the price marketing gets for those parts."

Naomi broke in. "But for which measure should we do a break-even analysis: present worth, IRR, or payback period?"

"Anna and Ed base their decisions mostly on PW and IRR values. I know they will ask how each behaves as a function of selling price, so we should do a break-even analysis on both."

"Both?"

"Both. We'll still have to calculate payback periods as per the company capital justification procedure, but we can calculate it over a few values of selling price and demand growth if that's more convenient than a break-even analysis."

"'A few values of demand growth'?" echoed Naomi.

"That's right. We still only have a poor estimate of the demand growth rate. So, just like you've already done, we will need to do all the analyses over three possible growth rates: 5, 10, and 15%. It didn't look like the optimal decision was affected by demand growth in your analysis, but we'd better check anyway."

Naomi and Dave glanced at each other, both recognizing a few long work days coming. "Anything else?" asked Dave, with only the slightest hint of sarcasm.

"Glad you asked, Dave. Of course, these will all be after-tax calculations, and you'll have to take salvage values into account. Oh, and write up your results in a full engineering report; the president and everyone between her and us are going to read it. Am I forgetting anything?"

"Inflation, too?" Dave couldn't keep the sour tone out of his voice completely.

"No," Clem laughed. "You're off the hook on that one. Inflation looks like it will continue to stay low for the foreseeable future, so we will ignore it."

"Are we limited to just two options, or can we look for better alternatives?" Naomi asked.

"Good question, Naomi. These two are currently the most likely options to produce the best solution, depending on the selling price we get. We definitely need to do a full study of them. If you want, you can look for a better solution; if you find one it will certainly be a feather in your cap. On the other hand, remember that you still have all your other work to do."

"Don't worry, Clem, we certainly remember that we have other work to do, too," Dave said, dryly but with a smile.

"Well, I've got one last question, Clem," Dave said, a few seconds later. "Who did you say was buying the coffee today?"

"And doughnuts, too!" piped in Naomi.

2. Down to Details

Later that morning, Naomi met Dave in his office to divide up the work. His office was enclosed, and thankfully so, because the walls protected the rest of the offices from Dave's unique organizing style.

"Looks like we're going to be crunching numbers for some time," Dave said, while clearing a chair for Naomi to sit in.

"Maybe not, Dave. We've got most of the spreadsheets for the calculations already set up. We'll have to include salvage values and tax effects. It turns out that the spreadsheet already has PW and IRR financial functions built in."

Dave continued to clear space on the table in front of Naomi as they talked. "That will certainly save a lot of trial-and-error to come up with IRRs, but what about all the cases we have to do to make break-even graphs?"

Naomi gestured at Dave's computer, almost the only thing in the office not hidden by paper. "I've found a feature in Excel called 'Data Tables'—I suppose most spreadsheet programs have something like it. It allows us to vary the contents of several spreadsheets based on data in a table. It looks like that will allow us to calculate all the values we need in no time."

"Sounds like I know who'll be doing the spreadsheets this time," said Dave. "How about if I concentrate on writing the report?" Dave had a clump of papers in his hand that he had not yet repositioned, and was pacing about the room looking for a place to put them.

"OK. Oh, before I forget, we should assume Canadian tax rules? CCA asset class 8 with a 20% CCA rate?" [note to instructors: substitute other appropriate tax information as desired].

"That's right. We can also get a good estimate of salvage value with a declining-balance depreciation rate of 20%. By the way, what was that you said to Clem about better options?"

"Oh, I've got some ideas. I'll let you know if they pan out."

Finally Dave sat down with a clear workspace in front of both of them. "Fair enough. According to my stomach, it must be lunchtime. Want to see if Clem wants to go over to the Grand China?" He jumped out of his chair again, after spending no more than three seconds sitting down.

"Sure," said Naomi. "Do we have to spread these papers out again before we leave?"

QUESTIONS

1. Using data provided in both parts of the Extended Case, update your spreadsheets to include salvage value, tax effects, PW, IRR, and payback calculations. PW, IRR, and payback analyses are required

for each project (single E1 machine or single E2 machine purchased at time 0). Summarize PW, IRR, and payback values in tables for each of 5%, 10%, and 15% demand growth and $0.03, $0.035, and $0.04 selling prices per piece.

2. Perform a break-even analysis for each of PW and IRR for each project over the range of selling prices from $0.03 to $0.04. Repeat for demand growth rates of 5%, 10%, and 15% (i.e., 2 comparison methods × 2 projects × 3 growth rates = 12 break-even calculations). Present the results graphically.

3. On the basis of your current analysis results, can you make a clear recommendation to the company's management which of E1 and E2 should be purchased? If so, why? If not, why not?

4. Write a full engineering economics report to the president, Anna Kulkowski, about this project. The report should follow the format of an engineering report. See the guidelines given below. In the report, you should present the results of your analyses so that company management can make a defensible decision about which machine they should purchase.

OPTIONAL

5. Currently we only consider purchasing one machine, either E1 or E2, at time 0. Find out if a better solution exists by varying the number of machines purchased at a time and/or the timing of the purchase (i.e., time 1, 2, etc.). Include your findings in the report.

Guidelines for an Engineering Report

A typical engineering report may include:

- Cover letter ("Here is the report you ordered" etc.)
- Title page
- Table of contents
- Summary (of the contents of the report)
- Introduction (e.g., description of the problem, background, and purpose)
- Body (e.g., procedure, calculations, results, possible errors, and unanswered questions)
- Conclusions
- Recommendations
- References
- Appendixes (important but too lengthy or disorderly to be included in the body)

Notes for Excel Users (Similar Issues Exist for Other Spreadsheet Programs)

Circular references: When calculating after-tax IRRs, you might run across a problem in Excel with "Circular Reference" warnings. The way to get around this problem is to go to the Excel "Tools" menu, then select "Options." Select the "Calculation" tab and click on the word "Iteration," then "OK." This will set the worksheet to perform iterative calculations that will converge for the value of the IRR. You might try pressing the F9 (recalculate) key a few times to make sure the values have converged.

IRR and NPV (present worth) built-in functions: Excel has built-in functions for calculating IRR and PW (Excel uses the term "NPV: Net Present Value" for present worth). Check Excel's Help file for details. Be careful with the NPV function: it calculates a PW for one period in the future. You have to adjust the value for time 0.

Data tables: You can run analyses for many different cases very quickly by using the Data Table feature, found under the "Data/Table . . . " menu.

CHAPTER 12

Dealing With Risk: Probability Analysis

Engineering Economics in Action, Part 12A: Trees From Another Planet

Review Problems
Summary

Engineering Economics in Action, Part 12B: Chances Are Good

Problems

Mini-Case 12.1

Engineering Economics in Action, Part 12A:
Trees From Another Planet

The coffee cups had been cleaned up, but the numerous brown rings that remained reported the hours of work Bill and Naomi had put in calculating the economic effects of manufacturing the new self-adjusting vacuum cleaner in partnership with the Powerluxe company.

"So, the bottom line is: if the demand is low, we lose money, but otherwise we gain. And more than that, if demand is high, we gain big-time." Bill was a bit plaintive because he knew that this did not solve their problems—he had hoped that they would make money no matter what. "So what it really comes down to is this: What is the chance that demand is low?"

Naomi looked into space for a second, and then thoughtfully started, "No, Bill, it's a little more complicated than that. It's . . . "

Just then, she was interrupted by Clem, who stepped in the lunchroom door. Glancing at the papers spread over the table, he remarked, "Not much of a lunch! Anyhow, I've been talking to Ms. Kulkowski, and she said that you should take into account in your Powerluxe project the potential for a competitive product."

"You mean another vacuum cleaner, just like the—what are they calling it, the 'Adaptamatic'—with the sensing capability?" asked Naomi.

"Yeah," replied Clem. "I guess the big shots are worried that those guys over at Erie Gadgets have some sort of similar thing under study. Well, that's all I had to tell you. Work hard." Clem winked and disappeared as quickly as he arrived.

Bill leaned back in his chair with his arms in the air. "For heaven's sake! We don't have any chance of coming to a clear recommendation now!"

"No, I think we're OK, Bill. In fact, since you mention it, 'chance' is pretty important here. So are trees."

Bill looked at Naomi as if she were from another planet. "Trees! Trees?"

12.1 | Introduction

In our day-to-day life, we often encounter situations where we don't know for sure what events will happen in the future. We talk about the "chance" that the weather will be rainy tomorrow, the "likelihood" that our favourite team will win the championship, or the "probability" of a railcar spill. Of course, once the event occurs, we know the outcome: either it did or did not rain, our team won or didn't, or the railcar spill happened or didn't happen.

In carrying out an economic analysis of a project, it is often the case that the engineer must estimate various parameters of a situation—the life of an asset, the interest rate, the magnitude and timing of cash flows, or factors such as the likely success of a new product. Chapter 11 dealt with sensitivity analysis tools appropriate for decision making under uncertainty—situations in which we can characterize a range of possible outcomes, but do not have a probability assessment associated with each outcome. Sensitivity analysis allowed us to look at the impact of this lack of probability information in a general way. Chapter 12, on the other hand, deals with decision making under *risk*. Decisions made under risk are those where the analyst can characterize a possible range of future outcomes and has available an estimate of the probability of each outcome. The term *risk* is also often used to refer to the probability distribution of outcomes associated with a project, or the probability of an undesirable outcome (this last definition of risk is not utilized in this text, though it is commonly used in the financial literature). Knowledge of the probability distribution of outcomes often permits us to draw more authoritative conclusions than we can draw using sensitivity analysis alone.

This chapter deals with a variety of approaches used by engineers to structure and make decisions when the alternatives involve risk. The first approach, decision trees, is commonly used to decompose a problem clearly into its decision alternatives and uncertain events. After presenting the use of decision trees, the chapter focuses on some frequently used decision criteria as a basis upon which alternatives can be evaluated. Of these criteria, expected value is the most commonly used, but dominance concepts such as mean-variance, outcome, and stochastic dominance are also useful for either screening alternatives or selecting the most preferable alternative. Finally, the chapter closes with a brief coverage of Monte Carlo simulation, a powerful approach for analyzing complex problems that have a large number of decision alternatives or many uncertain components. This chapter is intended only to provide introductory coverage of commonly used methods. More advanced approaches are beyond the scope of this book.

12.2 | Basic Concepts of Probability

In this section, we will define more precisely what we mean by "chance" and "likelihood" so that we are able to make useful predictions in the context of engineering economics. The branch of mathematics that formalizes this common notion of "chance" is called probability theory.

Suppose you are concerned with the market success of a new fuel-cell technology, but you are uncertain about its outcome. It could be that the fuel cell fails to gain market acceptance, it may produce adequate sales, or it could gain market dominance. In the terminology of decision analysis, "market success" is referred to as a *random variable*. A **random variable** is a parameter or variable that can take on a number of possible outcomes. Only one of these outcomes will eventually occur, but which one will occur is unknown at the time a decision is being made. For example, if market success is considered to be a random variable, then the set of possible outcomes are failure to gain market acceptance, adequate sales, or market dominance. To construct a model of this uncertainty, you will need to know the probability associated with each possible outcome. This is accomplished through a probability distribution function.

Consider a random variable X (e.g., the market success of a fuel cell) that can take on m discrete outcomes $x_1, x_2, \ldots, x_m$. If these events are mutually exclusive (if one occurs, another cannot) and collectively exhaustive (one of them must occur), a **probability distribution function** $p(x)$ is a set of numerical measures $p(x_i)$ such that:

$$0 \le p(x_i) \le 1 \quad \text{for } i = 1, \ldots, m$$

and

$$\sum_{i=1}^{m} p(x_i) = 1$$

with the intuitive interpretation that the higher $p(x_i)$, the more likely it is that x_i will occur.

The first statement above says that the **probability** associated with each outcome must be positive, and must be between 0 and 1 (inclusive). Intuitively, this means that any outcome cannot have a chance of occurring less than 0% or more than 100%. The second statement above says that since the outcomes are mutually exclusive and collectively exhaustive, only one of the outcomes will occur, and one *must* occur.

Over the years, various views on probability have emerged. Each is appropriate in different circumstances. Close-Up 12.1 summarizes different views on probability and when they are useful.

CLOSE-UP 12.1	Views on Probability

Classical or symmetric probability: This was the first view of probability and relies on games of chance such as dice, where the outcomes are equally likely. For example, if there are m possible outcomes for an uncertain event, since only one can and must occur, and each are assumed to be equally likely, then the chance of each occurring is $1/m$. For example, the probability that a coin toss will result in "heads" is 1/2 because there are two sides and each is equally likely.

Relative frequency: The outcome of an random event E is observed over a large number of experiments, N. If the number of times the outcome e_i occurs is n_i, then we can estimate the probability of event $p(e_i)$ by n_i/N. More formally, the relative frequency view on probability says that $p(e_i) = \lim_{N \to \infty} n_i / N$. An example of this is flipping a coin 1000 times and discovering that it lands on its edge five times in 1000. An estimate of the probability of the coin landing on its edge is then $5/1000 = 0.005$.

Subjective probability: Subjective or personal probability is an attempt to deal with unique events which cannot be repeated and hence can't be given a frequency interpretation. In rough terms, subjective probability can be interpreted as the odds one would personally give in betting on an event, or it may be a matter of human judgment and intuition as formed by physical relationships and experimental results. An example of this is a person who judges that the chance of winning a coin toss with one of the authors of this text is very low, say 1/1000, because, in that person's experience, the authors usually cheat.

Axiomatic probability: One of the problems associated with defining probabilities is that the definition of probability requires using probability itself. To get around this circular logic, axiomatic probability makes no attempt to define probability, but simply states the rules or axioms it follows. Other properties can then be derived from the basic axioms.

Each of the above methods may be correct, given the circumstances. When the physics of a process suggest a clear judgment of probability, the classical approach makes sense. Where formal experimentation is possible, the relative frequency method may be justified. In many real-world cases, subjective probability supported by historical information and other data is frequently used.

12.3 | Random Variables and Probability Distributions

When the number of outcomes for a random variable X is discrete, $p(x)$ is referred to as a **discrete probability distribution function** (PDF). Examples of discrete random variables are the number of good items in a batch of 100 tested products, the number of car accidents at an intersection each year, the number of days since the last plant shutdown, or the number of bugs found in software testing. Whether $p(x)$ is estimated using the classical, the relative frequency, or the subjective approach, the same terminology for the probability distribution function, $p(x)$, is used.

Various symbols may be used to define a random variable. The normal convention is to capitalize the symbol used for the random variable, and to use subscript lowercase letters to denote its various outcomes. For example, for a discrete random variable X, outcomes are denoted by $x_1, x_2, x_3, \ldots$ and its probability distribution function is $p(x)$. The probability that X takes on the value x_1 is written $\Pr(X = x_1) = p(x_1)$.

EXAMPLE 12.1

Suppose that you are testing solder joints on a printed circuit board and that you are interested in determining the probability distribution function for the random variable X, the number of open joints in three tested joints (to keep it simple). Prior to testing, you don't know how many open joints there will be. Since there are three joints and each will be either open or closed, X is a discrete random variable that can take on four possible values: $x_1 = 0$, $x_2 = 1$, $x_3 = 2$, and $x_4 = 3$.

Note that there are eight distinct test result sequences that can occur. Denoting the result of a single test by O for open and C for closed, the set of possible test results are: (O,O,O), (O,O,C), (O,C,O), (C,O,O), (O,C,C), (C,C,O), (C,O,C), and (C,C,C). We must look through the set of individual test results to see which corresponds to $x_1 = 0$, $x_2 = 1$, $x_3 = 2$, and $x_4 = 3$ open solder joints.

You know from previous data collection efforts that the result of a single test is uncertain. The probability that a single tested joint will be open is 20%. In other words, the outcome of the test is a random variable, say, Y, where the result of a single test can have two outcomes: $y_1 = O =$ open and $y_2 = C =$ closed, and $\Pr(Y = O) = p(O) = 0.2$ and $\Pr(Y = C) = p(C) = 0.8$. Further, suppose it is reasonable to assume that the quality of a solder joint does not change from joint to joint (i.e., the test results are independent of one another). Then the probability that a test sequence results in three open joints is calculated by

$$\Pr(X = 3) = p(x_4) = p(O) \times p(O) \times p(O) = (0.2) \times (0.2) \times (0.2) = 0.008$$

Similar calculations, shown in Table 12.1, yield the probabilities for each of the eight possible test sequences.

Table 12.1   Probability Corresponding to the Outcomes of the Solder Joint Testing With a 20% Chance of an Open Joint

Test Sequence	Number of "Opens"	Probability
(O,O,O)	3	$0.008 = 0.2 \times 0.2 \times 0.2$
(O,O,C)	2	$0.032 = 0.2 \times 0.2 \times 0.8$
(O,C,O)	2	$0.032 = 0.2 \times 0.8 \times 0.2$
(C,O,O)	2	$0.032 = 0.8 \times 0.2 \times 0.2$
(O,C,C)	1	$0.128 = 0.2 \times 0.8 \times 0.8$
(C,C,O)	1	$0.128 = 0.8 \times 0.8 \times 0.2$
(C,O,C)	1	$0.128 = 0.8 \times 0.2 \times 0.8$
(C,C,C)	0	$0.512 = 0.8 \times 0.8 \times 0.8$

Finally, the probability distribution function of X, the number of "open" joints in the three tests, is

$$\Pr(X = 0) = p(x_1) = 0.512$$
$$\Pr(X = 1) = p(x_2) = 0.384$$
$$\Pr(X = 2) = p(x_3) = 0.096$$
$$\Pr(X = 3) = p(x_4) = 0.008$$

Note that the two important properties of probabilities hold: $p(x_i) \geq 0$ for all i, and $\sum p(x_i) = 1$. ∎

It is often useful to display a probability distribution function in graphical format. Such a graph is referred to as a histogram. Figure 12.1 shows the probability distribution function associated with the solder joint testing results.

In contrast to a discrete random variable, a continuous random variable can take on any real value over a defined interval. For example, daily demand for drinking water in a municipality might be anywhere between 10 million litres and 200 million litres. The actual amount consumed—the outcome—is a continuous random variable with a minimum value of 10 million litres and a maximum value of 200 million litres.

In this chapter, we focus on applications of discrete random variables in engineering economics analysis. We do not use continuous random variables because proper treatment requires more advanced mathematical concepts such as differential and integral calculus. Also, continuous random variables can be well approximated as discrete random variables by grouping the possible output values into a number of categories or ranges. For example, rather than treating demand for drinking water as a continuous random variable, demand could be characterized as high, medium, or low. Figure 12.2 shows an example of a probability distribution associated with future demand for water, approximated by a discrete random variable denoted by D.

Another way to characterize a random variable is through its **cumulative distribution function (CDF)**. If X is the random variable of interest, and x is a specific value, then the cumulative distribution function (for a discrete random variable) is defined as follows:

$$P(x) = \text{Pr}(X \leq x) = \sum_{x_i \leq x} p(x_i)$$

For Example 12.1, the CDF for the number of open joints is:

$$\text{Pr}(X \leq 0) = P(x_1) = 0.512$$

$$\text{Pr}(X \leq 1) = P(x_2) = 0.896$$

Figure 12.1   Probability Distribution for the Solder Joint Example

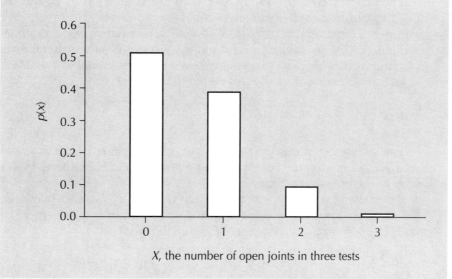

Figure 12.2   **Probability Distribution Function of the Demand for Drinking Water**

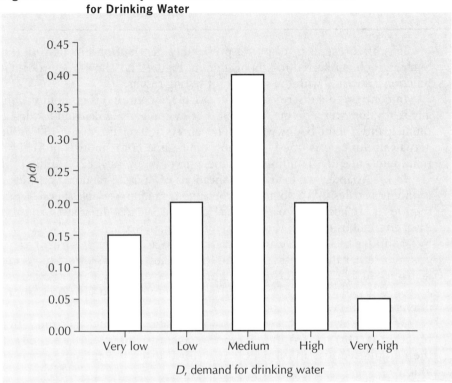

$\Pr(X \leq 2) = P(x_3) = 0.992$

$\Pr(X \leq 3) = P(x_4) = 1.000$

Figure 12.3 shows the cumulative distribution function associated with the solder joint testing example.

The probability distribution of a random variable contains a great deal of information that can be useful for decision-making purposes. However, certain summary statistics are often used to capture an overall picture rather than working with the entire distribution. One particularly useful summary statistic is the **expected value**, or **mean**, of a random variable. The expected value of a discrete random variable X, $E(X)$, which can take on values $x_1, x_2, \ldots, x_m$, is defined as follows:

$$E(X) = \sum_{i=1}^{m} x_i p(x_i)$$

You will no doubt observe that computing the expected value of a random variable is much like computing the centre of mass for an object. The expected value is simply the centre of the probability "mass."

Other useful summary statistics are the variance and standard deviation of a random variable. Both capture the degree of spread or dispersion in a random variable around the mean. The **variance** of a discrete random variable X is

$$Var(X) = \sum_{i=1}^{m} p(x_i)(x_i - E(X))^2$$

Figure 12.3 Cumulative Distribution Function for the Solder Joint Example

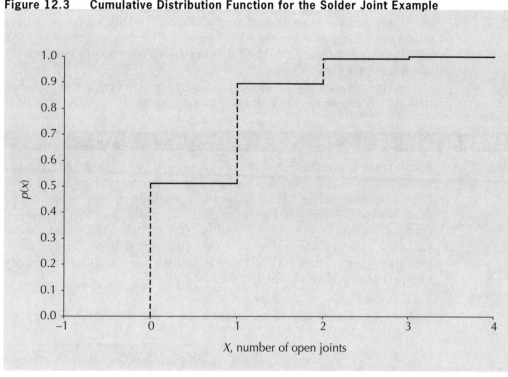

and the square root of the variance is the **standard deviation**. Distributions that have outcomes far above or below the mean will have higher variances than those with outcomes clustered near the mean.

The mean and variance of the number of open solder joints from Example 12.1 can be easily calculated:

$$E(X) = \sum_{i=1}^{m} x_i p(x_i)$$

$$= 0 \times 0.512 + 1 \times 0.384 + 2 \times 0.096 + 3 \times 0.008 = 0.60$$

The mean number of open solder joints is 0.6.

$$Var(X) = \sum_{i=1}^{m} p(x_i)(x_i - E(X))^2$$

$$= 0.512 \times (0 - 0.6)^2 + 0.384 \times (1 - 0.6)^2 + 0.096 \times (2 - 0.6)^2 + 0.008$$
$$\times (3 - 0.6)^2 = 0.48$$

The variance of the number of open solder joints is 0.48.

How are random variables, probability distributions, and their summary measures relevant to engineering economics? In conducting engineering economic studies, costs and benefits can be influenced by several uncertain factors. Often, an engineer does not know the outcome of a project or may not know with certainty the actual value of one or more parameters important to a project. The life of a product may shorten or lengthen unexpectedly due to market forces, the maintenance costs may be difficult to state with certainty, a car may have safety features that may increase or decrease the liability for the vendor, or product demand may be more or less than anticipated. In each case, if the engineer can determine the range of outcomes and their associated probabilities, this information can present a much richer view for the decision-making process. Random variables and their probability distributions are, in fact, the building blocks for many tools that are useful in decision making under risk.

Examples 12.2 and 12.3 illustrate how the expected value and probability distribution information may be used in a decision-making context.

EXAMPLE 12.2

Recall from Example 11.5 that the management of Westmount Waxworks had some uncertainty about the future sales levels of their line of statues of government leaders. Expert opinion helped them assess the probability of the pessimistic, expected, and optimistic sales scenarios. They think that the probability that sales will be 50 000 per year for the next few years is roughly 50% and that the pessimistic and optimistic scenarios have probabilities of 20% and 30%, respectively. Table 12.2 reproduces the annual cost information for the two wax melters, Finedetail and Simplicity. On the basis of expected annual costs, which is the best choice?

Table 12.2   Annual Cost Information for the Finedetail and Simplicity Wax Melters

Scenario	Annual Cost for Finedetail	Annual Cost for Simplicity	Probability
Pessimistic	$ 85 314	$ 94 381	0.2
Expected	112 314	103 501	0.5
Optimistic	314 814	186 901	0.3

The sales level can be represented by a discrete random variable, X. The possible values for X are: x_1 = pessimistic, x_2 = expected, and x_3 = optimistic.

The expected annual cost of the Finedetail wax melter is:

E(Finedetail, annual cost)

$$= (85\ 314)p(x_1) + (112\ 314)p(x_2) + (314\ 814)p(x_3)$$

$$= (85\ 314)(0.2) + (112\ 314)(0.5) + (314\ 814)(0.3)$$

$$= 167\ 663$$

The expected annual cost of the Simplicity wax melter is:

E(Simplicity, annual cost)

$$= (94\ 381)p(x_1) + (103\ 501)p(x_2) + (186\ 901)p(x_3)$$

$$= (94\ 381)(0.2) + (103\ 501)(0.3) + (186\ 901)(0.4)$$

$$= 126\ 697$$

The expected annual cost of the Simplicity wax melter is lower than that of the Finedetail. Hence, the Simplicity melter is preferred.■

EXAMPLE 12.3

Regional Express is a small courier service company. At the central office, all parcels from the surrounding area are collected, sorted, and distributed to the appropriate destinations. Regional Express is considering the purchase of a new computerized sorting device for their central office. The device is so new—in fact, it is still under continuous improvement—that its maximum capacity is somewhat uncertain at the present time. They are told that the possible capacity can be 40 000, 60 000, or 80 000 parcels per month, regardless of the size of the parcels. They have estimated the probabilities corresponding to the three capacity levels. Table 12.3 shows this information. What is the expected capacity level for the new sorting device? Regional Express is growing steadily, so such a computerized sorting device will be a necessity in the future. However, if Regional Express currently deals with an average of 50 000 parcels per month, should they seriously consider purchasing the device now or should they wait?

Table 12.3   Probability Distribution Function for Capacity Levels of the New Sorting Device

i	Capacity Level (Parcels/Month)	$p(x_i)$
1	40 000	0.3
2	60 000	0.6
3	80 000	0.1

If the discrete random variable X denotes the capacity of the device, then the expected capacity level $E(X)$ is

$$E(X) = (40\ 000)p(x_1) + (60\ 000)p(x_2) + (80\ 000)p(x_3)$$
$$= (40\ 000)(0.3) + (60\ 000)(0.6) + (80\ 000)(0.1)$$
$$= 56\ 000 \text{ parcels per month}$$

The expected capacity level exceeds the average monthly demand of 50 000 parcels per month, so according to the expected value analysis alone, Regional Express should consider buying the sorting device now. However, by studying the probability distribution, we see that there is a 30% probability that the capacity level may fall below 50 000 parcels per month. Perhaps Regional Express should include this information in their decision making, and ask themselves whether a 30% chance of not meeting their demand is too risky or costly for them if they decide to purchase the sorting device.■

In summary of this section, engineers can be faced with a variety of uncertain events in project evaluation. When the outcomes of each event can be characterized by a probability distribution, this greatly enhances the analyst's ability to develop a deeper understanding of the risks associated with various decisions. This section provided an introduction to random variables and probability distributions as a starting point for an analysis of project risk.

12.4 | Structuring Decisions With Decision Trees

Many different types of uncertainties exist in decision making. When an economic analysis becomes complex due to these uncertainties, formal analysis methods can help in several ways. First, formal methods can help by providing a means of decomposing a problem and structuring it clearly. Second, formal methods can help by suggesting a variety of decision criteria to help with the process of selecting a preferred course of action. This section provides an introduction to decision trees, which are a graphical means of structuring a decision-making situation where the uncertainties can be characterized by probability distributions. Other means of structuring decisions such as influence diagrams are available, but are beyond the scope of our coverage.

Decision trees help decompose and structure problems characterized by a sequence of one or more decisions and event outcomes. For example, a judgment about the chance of a thunderstorm tomorrow will affect your decision to plan for a picnic tomorrow afternoon. Similarly, the success or failure of a new product may largely depend on future demand for the product. As another example, a decision on the replacement interval for an asset relies on an assessment of its economic life, which can be highly uncertain if the equipment employs an emerging technology.

When a decision is influenced by outcomes of one or more random events, the decision maker must anticipate what those outcomes might be as part of the process of analysis. This section presents a useful tool for structuring such problems, called a decision tree. It is particularly suited to decisions and events that have a natural sequence in time or space.

A **decision tree** is a graphical representation of the logical structure of a decision problem in terms of the sequence of decisions to be made and outcomes of chance events. It provides a mechanism to decompose a large and complex problem into a sequence of small and essential components. In this way, a decision tree clarifies the options a decision maker has and provides a framework with which to deal with the risk involved.

Example 12.4 introduces the overall approach to constructing a decision tree. A detailed explanation of the components and structure of the decision tree is included in the example.

EXAMPLE 12.4

Edwin Electronics (EE) has a factory for assembling TVs. One of the key components is the TV screen. EE does not currently produce TV screens onsite; they are outsourced to a supplier elsewhere. Recently, EE's industrial engineering team asked if they should continue outsourcing the TV screens or produce them in-house. They realized that it was important to consider the uncertainty in demand for the company's TVs. If the future demand is low, outsourcing seems to be the reasonable option in order to save production costs. On the other hand, if the demand is high, then it may be worthwhile to produce the screens onsite, thus getting economies of scale. EE's engineers analyzed the effect of the demand uncertainty in their decision making. They represented their decision problem in a graphical manner with a decision tree. Figure 12.4 represents EE's decision tree.

There are four main components in a decision tree: decision nodes, chance nodes, branches, and leaves. A decision node represents a decision to be made by the decision maker and is denoted by a square in the tree diagram. In Figure 12.4, the single square node represents the decision to produce or outsource TV screens (node 1). A chance node represents an event whose outcome is uncertain, but which has to be considered during decision making. The outcome of a chance node is a discrete random variable, as it has a number of distinct outcomes and each outcome has an associated probability. The circle in the diagram denotes a chance node. The chance node in Figure 12.4 represents the uncertain demand for TV screens (node 2). The branches of a tree are the lines

Figure 12.4   Decision Tree for Edwin Electronics

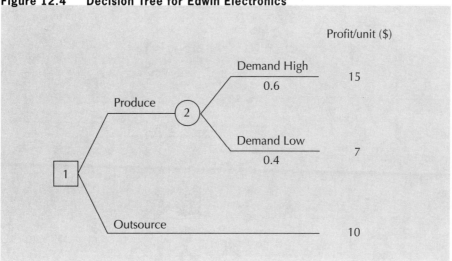

connecting nodes depicting the sequence of possible decisions and chance events. Finally, the leaves indicate the values, or payoffs, associated with each branch of the decision tree.

A decision tree grows from left to right and usually begins with a decision node. The leftmost decision node represents an immediate decision faced by the decision maker. The branches extending from a decision node represent the decision options available for the decision maker at that node, whereas the branches extending from a chance node represent the possible outcomes of the chance event. Each branch extending from a chance node has an associated probability. In Edwin Electronics' case, the two decision options, to produce or to outsource, are represented by the two branches extending from the decision node. The two branches from the chance node indicate that the future demand may be high or low. It is important in decision making that all branches out of a node, whether a decision node or chance node, constitute a set of mutually exclusive and collectively exhaustive consequences. In other words, when a decision is made, exactly one option is taken, or when uncertainty is resolved, exactly one outcome occurs as a result.

Whenever a chance node follows a decision node, as in Figure 12.4, it implies that the decision maker must anticipate the outcome of future uncertain events in decision making. On the other hand, when a decision node follows a chance node, it implies that a decision must be made assuming that a particular outcome of a chance event has occurred. Finally, the rightmost branches lead to the leaves of the decision tree, indicating all possible outcomes of the overall decision situation represented by the tree. Each leaf has an associated valuation, referred to as a "payoff"; quite typically, the payoff is a monetary value. Edwin Electronics uses profit per TV unit as its performance measure.■

The decision tree for Edwin Electronics from Figure 12.4 can be modified to show more complex decision situations.

EXAMPLE 12.5

The EE engineering team from Example 12.4 has realized that the cost per TV screen may vary in the future, especially since EE is subject to purchasing conditions set by the supplier. How does this affect the decision tree?

Figure 12.5 includes the additional uncertainty in the TV screen cost charged by the supplier. The cost may increase, remain the same, or decrease in the future, as shown at node 3.■

Figure 12.5   Edwin Electronics' Modified Decision Tree 1

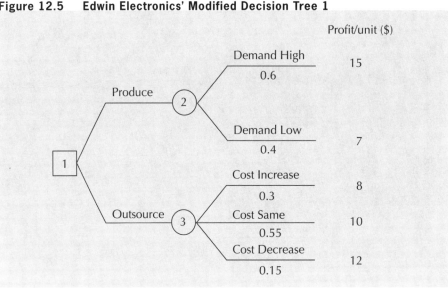

EXAMPLE 12.6

The EE engineering team then considered increasing in-house production of the TV screens if the demand is high. This raises an additional uncertainty about the ability of the market to absorb the increased production. How does this change the decision tree?

Figure 12.6 further modifies Figure 12.5 to include a new decision component and a chance event (see nodes 4 and 5). The decision component indicates the choice of

Figure 12.6   Edwin Electronics' Modified Decision Tree 2

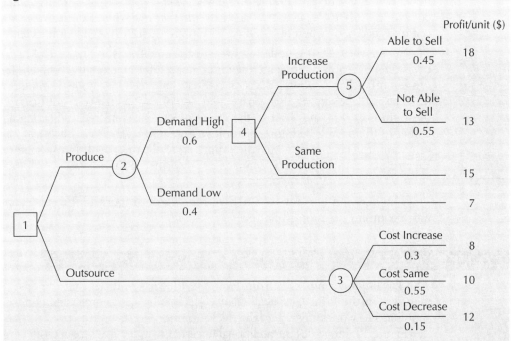

whether to increase production. If production is increased, a new chance node captures the uncertainty about Edwin's ability to sell the excess production.■

In summary, a decision tree is a graphical representation of the logical structure of a decision problem, showing the sequence of decisions and outcomes of chance events. It provides decision makers with a mechanism to structure and communicate a decision-making situation when risk is present. Decision tree analysis provides a framework with which a decision maker can deal with the uncertainty involved in a decision.

12.5 | Decision Criteria

Once a complete decision tree is structured, an analyst is in a better position to select a preferred alternative from a set of possible choices influenced by uncertain events. But how do we understand the risks associated with these uncertain events? And even if we can quantify the risks, how are choices then made? This section deals with several commonly used decision criteria for situations that involve uncertainty. Each has strengths and weaknesses that will be pointed out through a series of examples.

12.5.1 Expected Value

One criterion for selecting among alternatives when there are risky outcomes is to pick the alternative that has the highest expected value, EV, as previously presented in Example 12.3. If the units of value associated with the rightmost branches of a decision tree are measured in dollars, then this criterion may be referred to as *expected monetary value*, or EMV.

When a decision problem has been structured with a decision tree, finding the expected value of a particular decision is obtained by a procedure referred to as "rolling back" the decision tree. The procedure is as follows.

1. *Structure the problem:* Develop a decision tree representing the decision situation in question.

2. *Rollback:* Execute the **rollback procedure** (also known as **backward induction**) on the decision tree from right to left as follows:

   (a) At each chance node, compute the expected value of the possible outcomes. The resulting expected value becomes the value associated with the chance node and the branch on the left of that node (if there is one).

   (b) At each decision node, select the option with the best expected value (best may be highest value or lowest cost depending on the context). The best expected value becomes the value associated with the decision node and the branch on the left of that node (if there is one). For the option(s) not selected at this time, indicate their termination by a double-slash (//) on the corresponding branch.

   (c) Continue rolling back until the leftmost node is reached.

3. *Conclusion:* The expected value associated with the final node is the expected value of the overall decision. Tracing forward (left to right), the non-terminated decision options indicate the set of recommended decisions at each subsequent node.

EXAMPLE 12.7

Carry out a decision tree analysis on Edwin Electronics' modified tree in Figure 12.5 using the expected value criterion.

Since the decision tree is already provided, step 1 is complete. The rollback procedure described in step 2 has two phases in this case. First, the tree is rolled back to each of the chance nodes as in step 2(a) (phase 1). The expected values at nodes 2 and 3 are computed as follows:

$$EV(2) = 0.6(15) + 0.4(7) = 11.80$$

$$EV(3) = 0.3(8) + 0.55(10) + 0.15(12) = 9.70$$

Figure 12.7 shows the rollback so far.

Figure 12.7   Phase 1 of Step 2: Rolling Back to the Chance Nodes

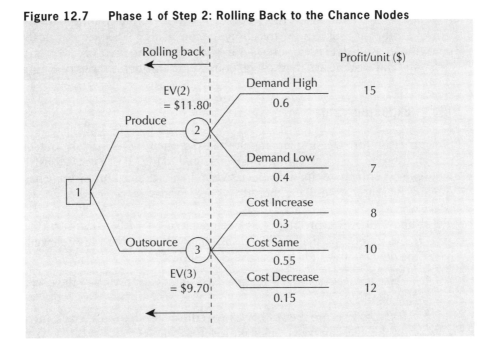

Next, the tree is further rolled back to the decision node as in step 2(b) (phase 2). The expected value at node 1 is then EV(1) = $11.80, which is equal to EV(2) since EV(2) is higher than EV(3). Figure 12.8 shows this result. As for step 3, the following conclusion is made: the expected value of the overall decision is $11.80 per unit and the recommended decision is to produce TV screens in-house.■

EXAMPLE 12.8

Perform decision tree analysis on Edwin Electronics' second modified tree in Figure 12.6.

The result of this analysis is shown in Figure 12.9. The overall expected profit for this tree is $11.95 per unit. The recommended decision is to produce TV screens in-house, and if the demand is high, the production level should be increased.■

Figure 12.8   Phase 2 of Step 2: Rolling Back to the Decision Nodes

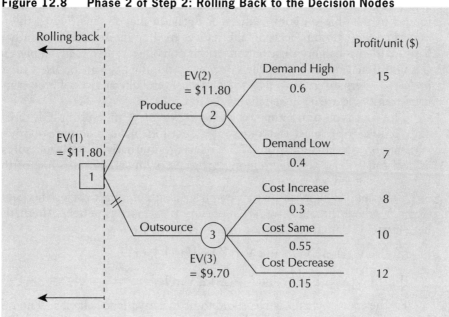

Figure 12.9   Completed Analysis for EE's Modified Decision Tree 2

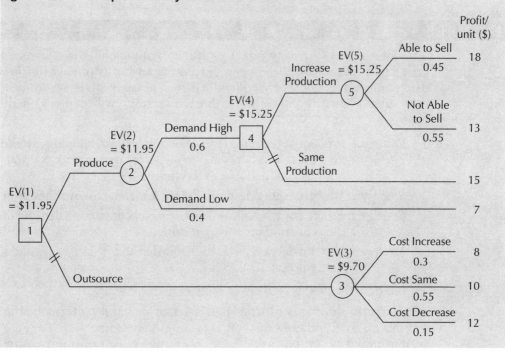

12.5.2 Dominance

Comparing several risky alternatives on the basis of the expected value criterion is straightforward, because it is easily computed and intuitive. However, expected value is only a summary measure and it does not consider the dispersion of the outcomes associated with a decision. Two decision alternatives with the same expected value may have very different ranges of possible outcome values. A decision maker sensitive to the degree of risk associated with various decision alternatives may be interested in quantifying risk on the basis of the probability distribution of the outcomes rather than simply the mean. With this additional information, the concept of dominance can be used to screen out less preferred alternatives, or to pick the best of several alternatives. Three examples of dominance reasoning are discussed in this section.

The first type of dominance reasoning involves measuring risk through the mean and variance of project outcomes. If a decision maker computes both the mean and variance of the outcomes for several decision alternatives, both measures may be used as decision criteria to select among projects, or to screen out some of the alternatives being considered.

For example, suppose that an engineer is attempting to select between two projects X and Y where the outcomes are monetary (i.e., more is better). Alternative X is said to have **mean-variance dominance** over alternative Y

(a) If $EV(X) \geq EV(Y)$ and $Var(X) < Var(Y)$, or

(b) If $EV(X) > EV(Y)$ and $Var(X) \leq Var(Y)$

Furthermore, an alternative is said to be mean-variance efficient if no other alternative has both a higher mean and a lower variance. Example 12.9 illustrates mean-variance dominance.

EXAMPLE 12.9

The Tireco Tire Company produces a line of automobile tires. Tireco is planning its production strategy for the coming year. Demand for its products varies over the year. The production planners have identified four possible strategies to meet demand, but the expected profits associated with each strategy will depend on the outcome of demand:

Strategy 1: Produce at the same level of output all year, building inventory in times of low demand and drawing on inventory during peak demand periods. Do not vary production staff levels.

Strategy 2: Vary production levels to follow demand. Use overtime as necessary.

Strategy 3: Produce at the same level all year, subcontracting demand during peak times to avoid excessive overtime.

Strategy 4: Vary production levels to follow demand by using a second shift of regular time workers as necessary.

Strategy 5: Combine strategies 3 and 4.

Marketing staff have formulated a set of demand patterns taking into account possible market conditions that might occur in the coming year. They are not sure what pattern might occur, but have made some subjective probability estimates as to the likelihood of each. Management combined these demand pattern forecasts with information on production costs, and have summarized their anticipated profits for each strategy-demand pattern as shown in Table 12.4.

The marketing department is not sure what demand pattern might occur, but they have estimated subjectively that the probability that demand takes on pattern 1 is 20%; pattern 2, 50%; and pattern 3, 30%. From these estimates, the mean and the variance associated with each strategy were calculated and are also shown in Table 12.4.

Table 12.4   Tireco Tire Company Forecast Profit Contributions (£000s)

Production Strategy	Demand Pattern			Mean	Variance
	1	2	3		
1	420	310	600	401	12 169
2	280	340	630	380	16 300
3	500	290	425	380	8775
4	600	275	390	395.5	19 812

It is natural that Tireco would want to maximize its expected profits. However, it might also like to minimize the variance of the strategy chosen at the same time so that the strategy chosen does not have a large degree of risk associated with possible outcome values. In view of these two objectives, several strategies can be removed from the above list with dominance reasoning:

1. Strategies 2 and 3 have the same mean profit, but strategy 2 has a higher variance. Strategy 2 is mean-variance dominated by strategy 3, and thus can be removed from consideration.

2. Strategy 5 has a lower mean and a higher variance when compared to strategy 1. Strategy 5 can be removed from consideration because it is mean-variance dominated by strategy 1.

3. Strategy 4 has a lower mean and higher variance when compared to strategy 1. Strategy 4 is thus mean-variance dominated by strategy 1.

Of the five strategies, only strategies 1 and 3 remain. While strategy 1 has the highest mean, it also has the highest variance, so a choice between the two cannot be made on the basis of the mean and variance alone. This set of two alternatives is thus the efficient set, because it can be reduced no further by mean-variance dominance reasoning. A choice between the two will require management to assess its willingness to trade off mean profits with the variability in profits.∎

Mean-variance analysis is useful when the mean and variance do well at capturing the distribution of possible outcomes for a decision. It is commonly used for screening investment opportunities. If more information about the distribution of outcomes is needed by a decision maker, it may be preferable to compare alternatives using dominance concepts that take into account the full distribution of outcomes.

The second type of dominance reasoning commonly used to choose between risky decision alternatives is to compare their full sets of outcomes directly. When outcome dominance exists, one or more of the alternatives can be removed from consideration. **Outcome dominance** of alternative X over alternative Y can occur in one of two ways. The first is when the worst outcome for alternative X is at least as good as the best outcome for alternative Y. Consider a slight variation of Example 12.5, in which Edwin Electronics must decide between in-house production and outsourcing. Suppose that the data for the problem is the same as earlier presented, but that the profit per unit if Edwin produces in-house and demand is low is $13 rather than $7 (refer to Figure 12.5). With this small change, inspection of the range of possible outcomes now shows that the profit per unit for the "produce" decision is better than for the "outsource" decision, for all possible outcomes. The "produce" decision is said to dominate the "outsource" decision due to outcome dominance.

The second way in which outcome dominance can occur is when one alternative is at least as preferred to another for each outcome, and is better for at least one outcome. In Example 12.9, strategy 1 has outcome dominance over strategy 5 because it has a higher profit contribution for each demand pattern.

Outcome dominance can be useful in screening out alternatives that are clearly worse than others among a set of choices. Though outcome dominance is straightforward to apply, it may not screen out or remove many alternatives.

The third common type of dominance reasoning that can be used to screen or order risky decision alternatives is stochastic dominance, as illustrated in Example 12.10.

EXAMPLE 12.10

Suppose in Example 12.5 that the profit per unit for EE when the decision to produce is taken and demand is low is $10. Figure 12.10 shows the probability distribution functions, also referred to as **risk profiles**, of the outcomes for the two decision alternatives that EE is considering.

A look at the cumulative distribution functions, also known as **cumulative risk profiles**, for the two alternatives provides further insights. This is seen in Figure 12.11.

The cumulative risk profile for the "outsource" decision either overlaps with or lies to the left of and above the cumulative risk profile of the "produce" decision. This means that for all outcomes, the probability that the "outsource" decision gives a lower profit per unit is equal to or greater than the corresponding probability for the "produce" decision. The produce strategy is said to dominate the outsource strategy according to (first-order) stochastic dominance. A more precise definition is as follows: If two decision

Figure 12.10   Illustration of Risk Profiles for Example 12.10

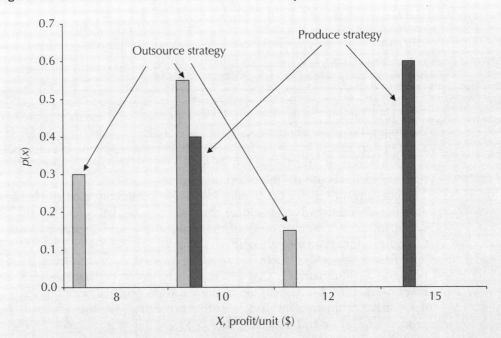

Figure 12.11   Illustration of First-Order Stochastic Dominance for Example 12.10

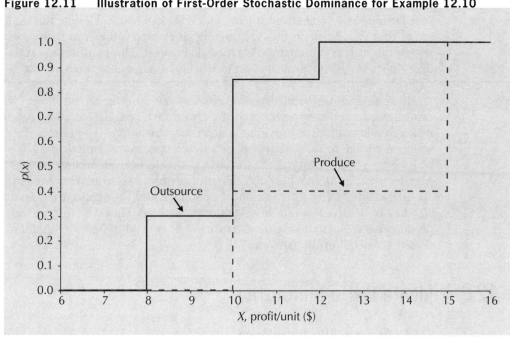

alternatives *a* and *b* have outcome cumulative distribution functions $F(x)$ and $G(x)$, respectively, then alternative *a* is said to have first order **stochastic dominance** over alternative *b* if $F(x) \geqslant G(x)$ for all x. In other words, alternative *a* is more likely to give a higher (better) outcome than alternative *b* for all possible outcomes.

While first-order stochastic dominance and outcome dominance can be used to screen alternatives, it is often the case that they are not able to provide a definitive "best" alternative. For example, the cumulative risk profiles for the original EE problem from Example 12.6 show that the cumulative risk profiles intersect (see Figure 12.12). The

Figure 12.12   Cumulative Risk Profiles for the Decision Alternatives in Example 12.6

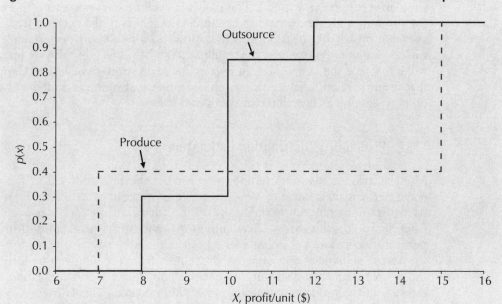

produce strategy is preferred using the EV criterion, but no definitive preference can be stated using ideas of deterministic or stochastic dominance. Despite this limitation with the use of cumulative risk profiles for ordering alternatives, they can be very useful in making statements such as "alternative A is more likely to produce a profit in excess of $1 000 000 than alternative B" or "project C is more likely to suffer a loss than project D." ■

In this section, several decision criteria were introduced and illustrated. Each has its own strengths and weaknesses, and the criteria are best viewed as approaches to decision making under risk, each providing a different perspective. The expected value criterion is widely used due to its intuitive appeal and computational simplicity. It is a powerful tool and a very good starting point for any economic analysis where uncertainty is present. Mean-variance and stochastic dominance criteria take into account more information about the probability distribution of outcomes and thus can enrich an analyst's understanding of the risk inherent in a decision alternative. However, they are not always able to produce a complete ranking of alternatives. They tend to be more useful in screening out dominated decision alternatives.

12.6 | Monte Carlo Simulation

12.6.1 Dealing With Complexity

This section deals with procedures that have been found to be effective in analyzing risk in complex decisions. Large or complicated projects may have decision trees with hundreds or thousands of terminal branches, each involving multiple sources of risk. In such situations, it may be difficult to evaluate decision alternatives with decision trees alone. For instance, the present worth computation for a project may require a number of inputs, such as the initial cost, a series of revenues and savings, operating and maintenance costs, and salvage value. When one or more of these inputs are random variables, the present worth inherits this randomness and may be very difficult to characterize.

When dealing with large and complex decision trees, a method called **Monte Carlo simulation** can be very useful. The basic idea behind the approach is to evaluate alternative decision strategies by randomly sampling branches of the decision tree, and thereby to assemble probability distributions (risk profiles) for relevant performance measures. As with any sampling procedure, the results will not be perfect, but the accuracy of the results will increase as the sample size increases. The main strength of Monte Carlo simulation is that it allows us to analyze the combined impact of multiple sources of uncertainty in order to develop an overall picture of overall risk.

12.6.2 Probability Distribution Estimation

Monte Carlo simulation attempts to construct the probability distribution of an outcome performance measure of a project (e.g., present worth) by repeatedly sampling from the input random variable probability distributions. This is a useful technique when it would otherwise be very difficult to obtain the probability distribution of the performance measure of interest for a project.

Since one basic notion of probability comes from the long-run frequency of an event, the sample frequency distribution generated by Monte Carlo simulation can be a good estimate of the probability distribution of the outcome. This is true, of course, provided that a sufficient sample size is used. Once the probability distribution is estimated, summary statistics such as the range of possible outcomes and the expected value can be analyzed to provide insight into the possible performance level for the project.

In Monte Carlo simulation, the probability distributions of the individual random variables, which are the input elements of the overall outcome, are assumed to be known in advance. By randomly sampling values for the random variables from the specified probability distributions, Monte Carlo simulation produces a sample of the overall performance measure for the project. This process can be seen as imitating the randomness in the performance measure as a result of the randomness of the project inputs. Incidentally, the name "Monte Carlo simulation" refers to the casino games at Monte Carlo, which symbolize the random behaviour. The use and practicality of Monte Carlo simulation have greatly increased with the widespread availability of application software such as spreadsheets (e.g., Excel and Lotus), spreadsheet add-ons including @Risk and Crystal Ball, and special-purpose simulation languages.

12.6.3 The Monte Carlo Simulation Approach

The following five steps are taken in developing a Monte Carlo simulation model.

1. *Analytical model:* Identify all input random variables that affect the outcome performance measure of the project in question. Develop the equation(s) for computing the outcome (denoted by Y) from a particular realization of the input random variables. This is sometimes referred to as "the deterministic version" of the model.

2. *Probability distributions:* Establish an appropriate probability distribution for each input random variable.

3. *Random sampling:* Sample a value for each input random variable from its associated probability distribution. The following is a random sampling procedure for discrete random variables:

    (a) For each discrete random variable, create a table similar to Table 12.5 containing the possible outcomes, their associated probabilities, and the corresponding random-number assignment ranges.

    (b) For each input random variable, generate a random number Z (see Close-Up 12.2). Find the range to which Z belongs from Table 12.6 and assign the appropriate outcome.

Table 12.5   Random Number Assignment Ranges

i	Outcome x_i	Probability $p(x_i)$	Random Number (Z) Assignment Range
1	x_1	$p(x_1)$	$0 \leq Z \leq p(x_1)$
2	x_2	$p(x_2)$	$p(x_1) \leq Z < p(x_1) + p(x_2)$
.	.	.	.
.	.	.	.
.	.	.	.
$m-1$	x_{m-1}	$p(x_{m-1})$	$p(x_1) + \ldots + p(x_{m-2}) \leq Z < p(x_1) + \ldots + p(x_{m-1})$
m	x_m	$p(x_m)$	$p(x_1) + \ldots + p(x_{m-1}) \leq Z < 1$

Table 12.6   Random Sampling Table

Step 3(b) (do for all random variables)		Step 3(c)
Random Number	Value of X	Value of Y
Generate Z	Assign value to X (one of x_i's) using Table 12.5 from step 3(a)	Compute Y using the analytical model from step 1

    (c) Substitute the sample values of the random variables into the expression for the outcome measure, Y, and compute the value of Y. This forms one sample point in the procedure. Table 12.6 can be the basis for a spreadsheet application of random sampling.

  4. *Repeat sampling:* Continue sampling until a sufficient sample size is obtained for the value of Y.

  5. *Summary:* Summarize the frequency distribution of the sample outcomes using a histogram. Summary statistics, like the range of possible outcomes and expected value, can also be calculated from the sample outcomes.

     One point remains to be clarified. In step 4 of Monte Carlo simulation, random sampling continues until a sufficient sample size is obtained. What is sufficient? The law of large numbers tells us that as the number of samples approaches infinity, the frequencies will converge to the true underlying probabilities. As a practical guideline, one should aim for a sample size of at least 100 in order to obtain reasonable results. One way to ensure the validity of the end result is to monitor the expected value of the frequency distribution and continue sampling until some stability appears in it. More rigorous guidelines exist for selecting the appropriate number of samples (e.g., confidence interval methods) but these are beyond the scope of this text.

N E T   V A L U E   1 2 . 1

Monte Carlo Simulation Analysis Software

Monte Carlo simulation analysis often involves a large amount of data and requires a sophisticated, multi-step analytical process with solution presentation aids such as graphics and charts. A number of commercial products are available for Monte Carlo simulation, and the internet is a good place to find out more about these products and even download a trial version of the software.

    Two examples of commercially available Monte Carlo simulation software packages are Crystal Ball (www.decisioneering.com/crystal_

ball) and @Risk (www.palisade.com). They are both designed as add-ins to spreadsheet software in order to ensure ease of use and compatibility with packages most engineers use routinely. They contain features designed to provide speedy and versatile calculation power to help answer questions such as "What is the probability that a project will be profitable?" or "How sensitive is our decision to interest rates or inflation?" There are other features to facilitate the selection of model input variables (such as a library of useful probability distributions) and to assist with the preparation and presentation of simulation results.

CLOSE-UP 12.2   Generating Random Numbers

In order to generate random numbers, a source of independent and uniformly distributed random numbers between 0 and 1 is required. These uniformly distributed random numbers can then be converted to random numbers from any probability distribution using a variety of methods, one of which is a table look-up as illustrated in Table 12.6. Today, random-number generation is easily achieved by using calculator functions or application software such as Excel or Lotus, which have built-in random-number generators for a limited number of probability distributions. Specialized commercial software such as @Risk or Crystal Ball provides a wider range of built-in random-number generators.

EXAMPLE 12.11

Pharma-Excel, a pharmaceutical company, is considering the worth of a R&D project that involves improvement of vitamin pills. Since this is a new research domain for Pharma, they are not certain about the related costs and benefits. As a part of the initial feasibility study, Pharma-Excel estimated probability distributions for the first cost and annual revenue as seen in Table 12.7, which are assumed to be independent quantities. Simulate the present worth of this project on the basis of a 10-year study period using the Monte Carlo method. Does the project seem viable? Pharma-Excel's MARR is 15% for this type of project.

Table 12.7   First Cost and Annual Revenue for Pharma-Excel's Research Project

First Cost	Probability	Annual Revenue	Probability
$1 000 000	0.2	$ 100 000	0.125
1 250 000	0.2	350 000	0.125
1 500 000	0.2	600 000	0.125
1 750 000	0.2	850 000	0.125
2 000 000	0.2	1 100 000	0.125
		1 350 000	0.125
		1 600 000	0.125
		1 850 000	0.125

As the first step, the analytical expression for the present worth of the project is developed. We use the following expression for computing the present worth of the project:

$$PW = -(\text{First Cost}) + (\text{Annual Revenue})(P/A,15\%,10)$$
$$= -(\text{First Cost}) + (\text{Annual Revenue})(5.0188)$$

The probability distributions for the first cost and annual revenue are provided in Table 12.7 (step 2). On the basis of these distributions, random numbers are assigned to particular intervals on the 0 to 1 probability scale (step 3(a)). Tables 12.8 and 12.9 summarize the random-number assignment for each random variable.

Table 12.8   Random Number Assignment Ranges for the First Cost

First Cost	Probability	Random Number (Z_1) Assignment Range
$1 000 000	0.2	$0 \leq Z_1 < 0.2$
1 250 000	0.2	$0.2 \leq Z_1 < 0.4$
1 500 000	0.2	$0.4 \leq Z_1 < 0.6$
1 750 000	0.2	$0.6 \leq Z_1 < 0.8$
2 000 000	0.2	$0.8 \leq Z_1 < 1$

Table 12.9   Random Number Assignment Ranges for the Annual Revenue

Annual Revenue	Probability	Random Number (Z_2) Assignment Range
$ 100 000	0.125	$0 \leq Z_2 < 0.125$
350 000	0.125	$0.125 \leq Z_2 < 0.25$
600 000	0.125	$0.25 \leq Z_2 < 0.375$
850 000	0.125	$0.375 \leq Z_2 < 0.5$
1 100 000	0.125	$0.5 \leq Z_2 < 0.625$
1 350 000	0.125	$0.625 \leq Z_2 < 0.75$
1 600 000	0.125	$0.75 \leq Z_2 < 0.875$
1 850 000	0.125	$0.875 \leq Z_2 < 1$

Following steps 3(b) and 3(c), the simulation results are obtained for a sample size of 200. Table 12.10 presents partial results. The frequency distribution shown in Figure 12.13 is also generated on the basis of the simulation results. Note: Each bar

Table 12.10   Partial Results for the Monte Carlo Simulation

Sample Number	Random Number (Z_1)	First Cost	Random Number (Z_2)	Annual Revenue	Present Worth
1	0.076162	$1 000 000	0.605155	$1 100 000	$4 520 680
2	0.728782	1 750 000	0.293282	600 000	1 261 280
3	0.29656	1 250 000	0.747692	1 350 000	5 525 380
4	0.940748	2 000 000	0.327516	600 000	1 011 280
5	0.384964	1 250 000	0.788017	1 600 000	6 780 080
.	.	.	.	.	.
.	.	.	.	.	.
.	.	.	.	.	.

Figure 12.13   Frequency Distribution for Pharma-Excel's Research Project

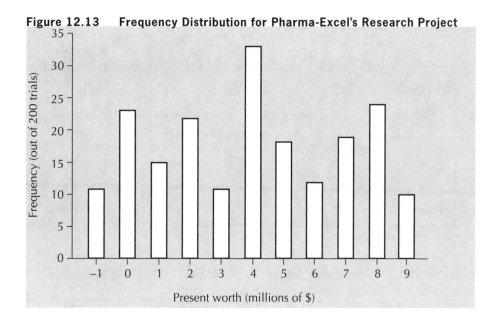

in the histogram includes all values up to that value that were not included in the previous bar.

From the simulation results, the average present worth of the research project over the 10-year study period is $3 152 437 with a maximum value of $8 284 780 and a minimum value of –$1 498 120. The project exhibited a negative present worth roughly 18% of the time. From these figures, the project seems to be viable, because it has a good chance of having a positive present worth that might be as high as $8 million.■

EXAMPLE 12.12

While considering the cogeneration plant project outlined in Example 11.1, Cogenesis Corporation wishes to determine the probability distribution of the project's present worth to better assess the probability of a negative present worth. The three random variables they wish to investigate are the initial investment, the savings in electricity costs, and the extra wood costs. Previously, management determined that the range of possible values for the initial costs was $2 800 000 to $3 300 000, the range for savings in electricity costs was $920 000 to $1 080 000 per year, and the additional wood costs were $350 000 to $400 000 per year. To the best of their knowledge, management thinks that any outcome between the lower and upper bounds for the first cost, electricity savings, and additional wood costs is equally likely. Table 12.11 shows the discrete approximation of the probability distribution functions created for the three random variables.

By randomly sampling repeatedly from each of these distributions, we can construct a probability distribution of the present worth of the project. Assuming that all other parameters are fixed at their expected scenario values, the following expression is used for computing the present worth. A portion of the Monte Carlo simulation results is shown in Table 12.12.

Table 12.11   Discrete Probability Distributions for Cogenesis' Plant Project

Initial Cost	Probability	Electricity Savings	Probability	Additional Wood Costs	Probability
$2 800 000	1/6	$ 920 000	0.2	$350 000	1/6
2 900 000	1/6	960 000	0.2	360 000	1/6
3 000 000	1/6	1 000 000	0.2	370 000	1/6
3 100 000	1/6	1 040 000	0.2	380 000	1/6
3 200 000	1/6	1 080 000	0.2	390 000	1/6
3 300 000	1/6			400 000	1/6

Table 12.12   Partial Results for the Cogenesis Monte Carlo Simulation

Initial Cost	Electricity Savings	Additional Wood Costs	Present Worth
$3 200 000	$1 080 000	$360 000	$1 635 912
2 800 000	920 000	350 000	915 502
3 000 000	960 000	380 000	790 196
2 900 000	920 000	370 000	666 114
⋮	⋮	⋮	⋮

$$PW = -(\text{First cost}) + (65\ 000 + \text{Wood costs} - \text{Electricity savings})$$
$$\times (P/A,12\%,20) - 17\ 000(P/F,12\%,10) - 35\ 000[(P/F,12\%,4)$$
$$+ (P/F,12\%,8) + (P/F,12\%,12) + (P/F,12\%,16)]$$
$$= -(\text{First cost}) + (65\ 000 + \text{Wood costs} - \text{Electricity savings})$$
$$\times (7.4694) - 17\ 000(0.32197) - 35\ 000(0.63552 + 0.40388$$
$$+ 0.25668 + 0.16312)$$
$$= -(\text{First cost}) + 7.4694(65\ 000 + \text{Wood costs}$$
$$- \text{Electricity savings}) - 56\ 545.49$$

By sampling a total of 200 times and computing a present worth for each sample, management arrived at the histogram shown in Figure 12.14. The expected present worth was $1 007 816 with a minimum of $42 032 and a maximum of $2 110 606. There were no instances where the present worth turned out to be zero or less. The probability of this project yielding a negative present worth appears to be negligible—assuming, of course, that the probability distributions for the input parameters have been specified correctly!■

Figure 12.14   Frequency Distribution for Cogenesis' Plant Project

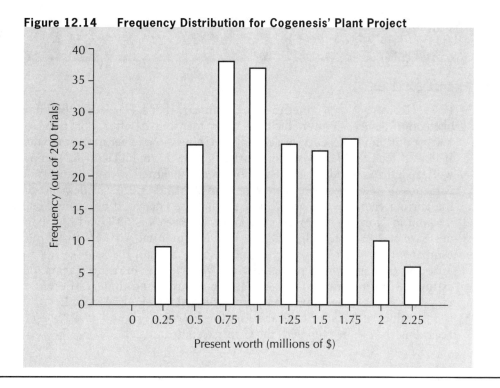

12.7 | Application Issues

A decision tree is a powerful tool for structuring decisions when one or more sources of risk are present and there is a sequence of decisions to be made. When decision trees get very large, or when there are numerous sources of uncertainty, Monte Carlo simulation is useful for approximating the probability distribution function of performance measures associated with a given decision strategy. Both methods provide a way to structure decision making under uncertainty using probability theory. By explicitly modelling the components of the decision-making process, engineers and analysts can structure a decision clearly and better understand the implications of the risk in decision making.

The major drawback of probability-based methods is that the probability distributions of each of the random variables must be specified. This can be a real challenge when the probabilities are highly subjective or there is a lack of historical or experimental data upon which to base probability assessments. Therefore, we must be aware of this grey area when we interpret the output and try to put the results of the analysis in context.

Despite this drawback, as engineers we recognize that every decision has elements of uncertainty and risk. The tools discussed in this chapter provide a framework for analyzing risk and gaining insights that would otherwise be ignored. In conjunction with the sensitivity analysis methods covered in Chapter 11, probability theory is a powerful tool for assessing project viability and the implications of the risk associated with every engineering project.

REVIEW PROBLEMS

REVIEW PROBLEM 12.1

Power Tech is a North America-based company that specializes in building power-surge protection devices. Power Tech has been focusing its efforts on the North American market until now. Recently, a deal with a Chinese manufacturing company has surfaced. If Power Tech decides to become partners with this manufacturing company, its market will expand to include Asia. It is, however, concerned with the uncertainty associated with possible changes in North American demand and Asian demand. From studying the current economy, Power Tech feels that the chance of no change or an increase in demand in North America over the next three years is 60% and the chance of demand decrease is 40%. After discussions with its potential partners in China, Power Tech estimates that Asian demand may increase (or remain the same) with a probability of 30% and decrease with a probability of 70% over the next three years. Power Tech has estimated the revenue increase that can be expected under different scenarios if it establishes the partnership; this information is shown in Table 12.13.

Conduct a decision tree analysis for Power Tech and make a recommendation regarding the partnership with the Chinese company.

Table 12.13　Expected Revenue Increase for Power Tech

	North American Demand	Asian Demand	Revenue Increase (millions of $)
	increase	increase	2
Partnership	increase	decrease	0.75
with Chinese	decrease	increase	0.5
company	decrease	decrease	−1
	increase	increase	0.75
No partnership	increase	decrease	0.5
with Chinese	decrease	increase	0.1
company	decrease	decrease	0.3

ANSWER

The result of the analysis is shown in Figure 12.15. The expected value calculations at each chance node are shown below.

The first phase of rollback (in millions of dollars):

$EV(4) = 0.3(2) + 0.7(0.75) = 1.125$
$EV(5) = 0.3(0.5) + 0.7(-1) = -0.55$
$EV(6) = 0.3(0.75) + 0.7(0.5) = 0.575$
$EV(7) = 0.3(0.1) + 0.7(0.3) = 0.24$

The second phase of rollback (in millions of dollars):

$EV(2) = 0.6(1.125) + 0.4(-0.55) = 0.455$
$EV(3) = 0.6(0.575) + 0.4(0.24) = 0.441$

Figure 12.15   Decision Tree Analysis for Power Tech (Review Problem 12.1)

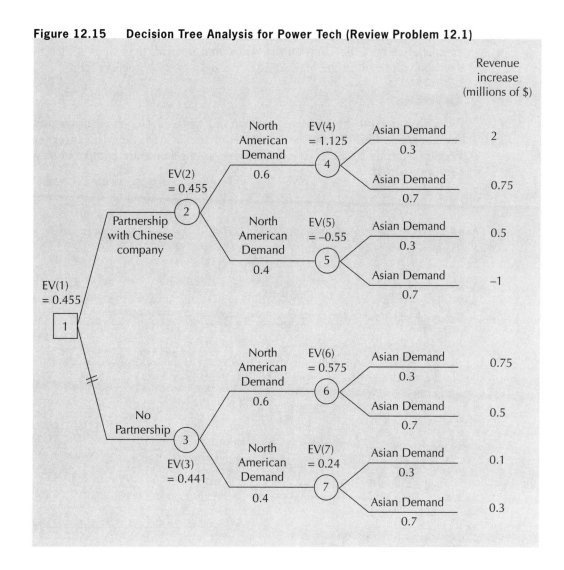

According to the expected value criterion, the partnership with the Chinese company is recommended, since the expected value for forming the partnership is higher than for not forming it. However, Power Tech should also note that these expected values have only a marginal difference. It is perhaps wise to collect more information regarding other aspects of this proposed partnership.■

REVIEW PROBLEM 12.2

A telephone company called LOTell thinks the introduction of a new internet service package for rural residential customers would give it an advantage over its competitors. However, a survey of the potential growth of rural internet home users would take at least three months. LOTell has two options at present: first, to introduce the internet package without the survey result in order to make sure that no competitors are present at the time of market entry, and second, to wait for the survey result in order to minimize the risk of failing to attract enough customers. If LOTell decides to wait for the survey result, there are three possible outcomes: the market growth is rapid (30% probability), steady (40%), or slow (30%). Depending on the survey result, LOTell may decide to introduce or not introduce the new internet service. If it decides to launch the new service after the survey, which is three months from now, then there is a 70% chance that the competitors will come up with a similar service package. What decision will result in the highest expected market share for LOTell?

ANSWER

A decision tree for the problem is shown in Figure 12.16, which also shows the results of the analysis. The recommended decision is to introduce the internet service now because it produces a higher expected gain in market share compared to waiting for the survey results.■

Figure 12.16   Decision Tree Analysis for LOTell

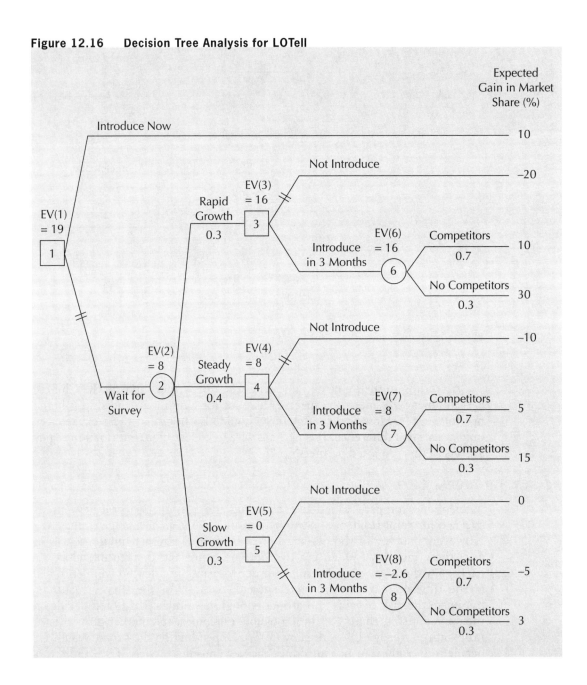

REVIEW PROBLEM 12.3

A land developer is building a new set of condominiums, which when rented will have all utilities included in the monthly rental fees paid by tenants. The developer is looking at using tankless gas water heaters as an alternative to standard gas water heaters. Tankless water heaters provide hot water without using a storage tank. Cold water passes through a pipe to the heater, and a gas burner heats the water on demand. Tankless water heaters can be cost-effective because they do not incur the standby heat loss associated with regular tank-based gas hot water heaters. Though tankless heaters cost more to install, the manufacturers claim that they can save a substantial amount on water heating bills. The first cost of a standard gas water heater is $600 and it is expected to have a service life of 10 years. A comparable tankless unit costs $1500 installed and is expected to have a service life of 20 years. The annual hot water heating costs for a small family (low demand) is approximately $480, and for a large family (high demand) $960. From past experience, the developer knows that 75% of tenants will fall into the high-demand category. The developer does not have any experience with tankless heaters, but the manufacturers indicate that energy savings, compared to a regular tank gas water heater, can be 20% to 30%. The developer has decided to model this uncertainty by a discrete distribution with a 50% probability that the savings will be 20% and a 50% probability that the savings will be 30%. The MARR is 15%.

(a) Draw a decision tree for this situation. Which of the two alternatives minimizes the expected annual worth of costs?

(b) If the developer is concerned about risk, can he find a preferred alternative using mean-variance dominance reasoning?

(c) Is one alternative preferred to the other on the basis of outcome dominance?

(d) Compare the cumulative risk profiles for the two alternatives. Does first-order stochastic dominance between the two alternatives exist?

ANSWER

(a) Figure 12.17 shows the completed decision tree for the developer. To carry out the decision analysis, the annual worth of costs for each branch of the tree must be calculated. For example, the annual worth of costs for the regular heater when demand is low is

$$AW(\text{standard heater, low demand}) = 600 \, (A/P,15\%,10) + 480$$
$$= 599.55$$

and the annual worth of costs for the tankless heater, when demand is high and savings are 20% is

$$AW(\text{tankless heater, high demand, savings}) = 1500(A/P,15\%,20) + (1-0.20)(960)$$
$$= 1007.64$$

On the basis of these annual worths, the decision tree can be "rolled back" to demonstrate that the tankless heater has expected annual costs of $869.64 and the standard heater has expected annual costs of $959.55. The tankless heater is preferred according to the EV criterion.

(b) The mean annual costs of each alternative can be taken from part (a). The variance of the annual worth of costs for the standard heater is

$$Var(\text{standard}) = 0.75 \times (1079.55 - 959.55)^2 + 0.25 \times (599.55 - 959.55)^2$$
$$= 43\ 200$$

Figure 12.17   Decision Tree for Heater

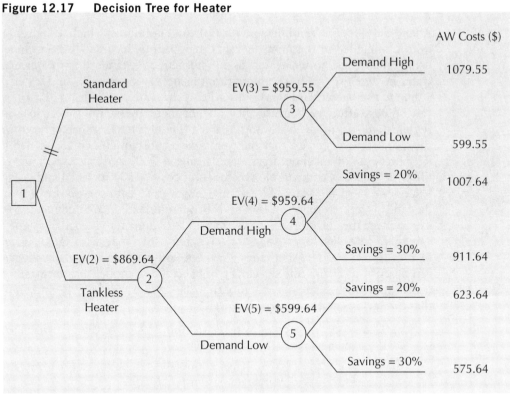

This is calculated on the basis of the fact that for the decision to purchase a standard gas hot water heater, the only uncertainty is the demand level. It will be high with probability 75%, and low with probability 25%.

The variance of the annual worth of costs for the tankless heater is

$$Var(tankless) = 0.75 \times 0.50 \times (1007.64 - 869.64)^2 + 0.75 \times 0.50 \times$$
$$(911.64 - 869.64)^2 + 0.25 \times 0.5 \times (623.64 - 869.64)^2$$
$$+ 0.25 \times 0.50 \times (575.64 - 869.64)^2$$
$$= 26\ 172$$

This is calculated on the basis of the fact that for the decision to purchase a tankless gas water heater, there are four possible outcomes. They are {demand is high, savings = 20%}, {demand is high, savings = 30%}, {demand is low, savings = 20%}, and finally {demand is low, savings = 30%}. The respective probabilities are $0.75 \times 0.50 = 0.375$, $0.75 \times 0.50 = 0.375$, $0.25 \times 0.5 = 0.125$, and $0.25 \times 0.5 = 0.125$.

The tankless heater is preferred to the standard gas water heater by mean-variance reasoning, because it has both lower expected annual worth of costs and lower variance of annual worth of costs.

(c) Neither type of outcome dominance exists. For example, the worst outcome for the tankless heater is $1007.64. This is not better than the best outcome for the standard heater ($599.55). The converse also holds in that the worst outcome for the standard heater ($1079.55) is not better than the best outcome for the tankless heater ($575.64). For the second type of outcome dominance, the possible outcomes associated with the two uncertain events must be examined (see part (b)). By directly examining the outcomes for each, it can be seen that the tankless heater is better for each outcome except when demand is low and savings are 20%. Therefore the second type of outcome dominance does not hold.

(d) Figure 12.18 provides the cumulative risk profiles for the standard and tankless heaters. The figures demonstrate that first-degree stochastic dominance cannot screen out either alternative because they overlap one another.

Figure 12.18   Cumulative Risk Profiles for Standard and Tankless Heaters

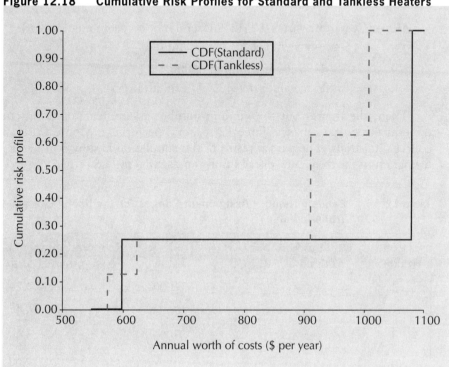

REVIEW PROBLEM 12.4

Orla is in the car rental business. The cars she uses have a useful life of two to four years. The cars can be traded in for a new car at the dealer. The trade-in values vary from €1000 to €5000 depending on the condition of the car. Orla has estimated the probability distributions for the length of useful life and the trade-in value for a typical car. Her estimates are shown in Table 12.14. Orla wants to find out the typical annual cost of

Table 12.14   Probability Distributions for the Useful Life and the Trade-In Value of a Rental Car

Useful Life (Years)	Probability	Trade-In Value	Probability
2	0.4	€1000	0.2
3	0.3	2000	0.2
4	0.3	3000	0.2
		4000	0.2
		5000	0.2

owning a car for doing this business using the Monte Carlo simulation method. Assume that the cost of a new car is €15 000 and the annual maintenance cost is €800. Her MARR is 12%.

ANSWER

Orla first comes up with the following annual cost expression for her problem:

$$
\begin{aligned}
AC &= (\text{Cost of new car})(A/P,12\%,\text{Useful life}) + (\text{Maintenance cost}) \\
&\quad - (\text{Trade-in value})(A/F,12\%,\text{Useful life}) \\
&= 15\ 000(A/P,12\%,\text{Useful life}) + 800 \\
&\quad - (\text{Trade-in value})(A/F,12\%,\text{Useful life})
\end{aligned}
$$

Then, she figures out the random-number assignment for the useful life and the trade-in value of a car (see Table 12.15), and performs a Monte Carlo simulation and collects 200 trials. The partial result of the simulation is shown in Table 12.16 and the histogram of the frequency distribution is presented in Figure 12.19.

Table 12.15   Random Number Assignment Ranges for the Useful Life and the Trade-In Value

Useful Life	Random Number (Z_1) Assignment Range	Trade-In Value	Random Number (Z_2) Assignment Range
2	$0 \leq Z_1 < 0.4$	€1000	$0 \leq Z_2 < 0.2$
3	$0.4 \leq Z_1 < 0.7$	2000	$0.2 \leq Z_2 < 0.4$
4	$0.7 \leq Z_1 < 1$	3000	$0.4 \leq Z_2 < 0.6$
		4000	$0.6 \leq Z_2 < 0.8$
		5000	$0.8 \leq Z_2 < 1$

Table 12.16   Partial Results for the Monte Carlo Simulation

Sample Number	Random Number (Z_1)	Useful Life	Random Number (Z_2)	Trade-In Value	Annual Cost
1	0.522750	3	0.325129	2000	€5652.54
2	0.809623	4	0.423085	3000	4310.81
3	0.124285	2	0.799329	4000	6988.68
4	0.104359	2	0.207269	2000	7932.08
5	0.961704	4	0.108746	1000	4729.28
.	.	.	.	.	.
.	.	.	.	.	.
.	.	.	.	.	.

Figure 12.19   Frequency Distribution for Orla's Annual Cost

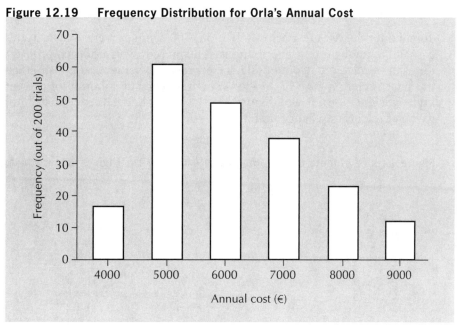

The results of the Monte Carlo simulation show that the average annual cost for each car is €5706. In the 200 samples, the annual cost ranged from €3892 to €8404.■

REVIEW PROBLEM 12.5

Betteryet Insurance (see Review Problems 11.1 to 11.3) has consulted several energy experts in order to further understand the implications of electricity and natural gas price changes on its two energy efficiency projects. Betteryet has estimated that the cost of electricity can range from $0.063 per kilowatt-hour to $0.077 per kilowatt-hour, and the price of natural gas can range from $3.35 to $3.65 per 1000 cubic feet. Both probability distributions (discrete approximations) are shown in Table 12.17. Carry out a Monte Carlo simulation to determine the probability distribution of the present worth of the two energy efficiency projects.

Table 12.17   Probability Distributions for the Cost of Electricity and the Price of Natural Gas

Electricity Cost (per kWh)	Probability	Natural Gas Price (per 1000 cubic feet)	Probability
$0.063	0.125	$3.35	1/7
0.065	0.125	3.40	1/7
0.067	0.125	3.45	1/7
0.069	0.125	3.50	1/7
0.071	0.125	3.55	1/7
0.073	0.125	3.60	1/7
0.075	0.125	3.65	1/7
0.077	0.125		

ANSWER

To conduct the Monte Carlo simulation, 300 samples were drawn from the electricity cost and natural gas cost distributions. The present worth of each project was computed for each simulated outcome using the expressions below, and a histogram of the results was constructed. Figure 12.20 shows the frequency distribution of present worth of the high-efficiency motor and Figure 12.21 shows the frequency distribution of present worth of the heat exchange unit.

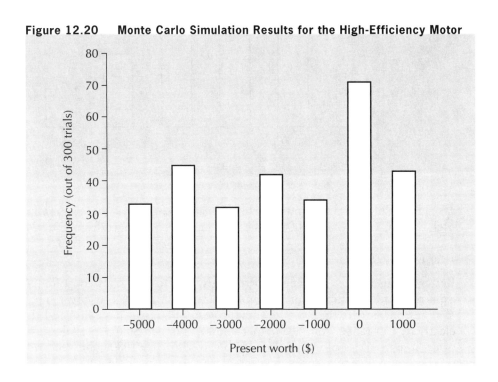

Figure 12.20   Monte Carlo Simulation Results for the High-Efficiency Motor

Figure 12.21   Monte Carlo Simulation Results for the Heat Exchange Unit

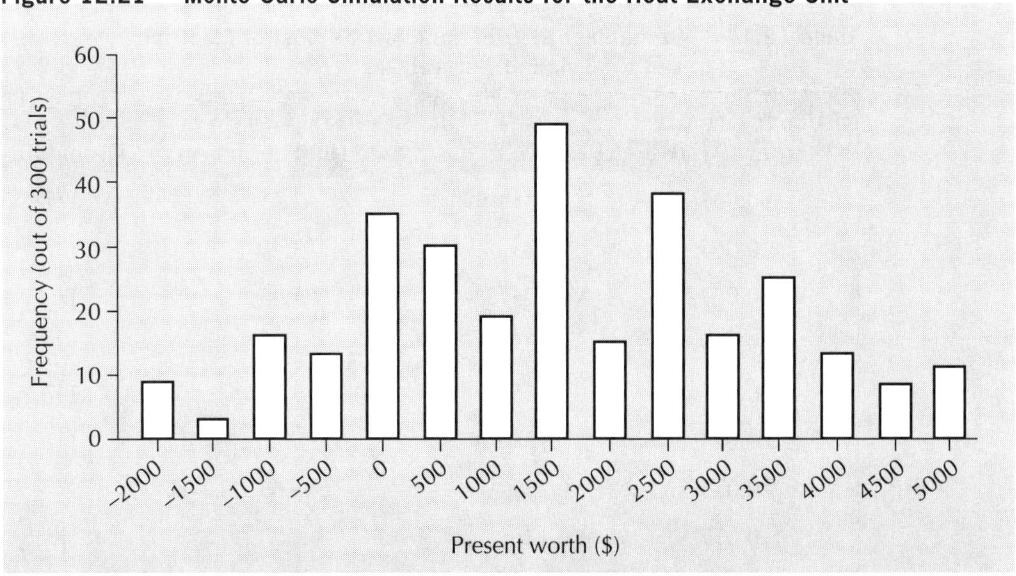

PW (High efficiency motor)

$$= -28\,000 + (P/A,10\%,10)[70\,000(\text{Electricity}) - 700]$$
$$= -28\,000 + (6.1446)[70\,000(\text{Electricity}) - 700]$$

PW(Heat exchanger)

$$= -40\,000 + (P/A,10\%,10)[29\,000(\text{Electricity})$$
$$\quad +2250(\text{Natural gas}) - 3200]$$
$$= -40\,000 + (6.1446)[29\,000(\text{Electricity}) + 2250(\text{Natural gas})$$
$$\quad - 3200]$$

It is clear from the simulations that the high-efficiency motor has roughly a 40% chance of having a positive present worth and hence is not a good investment. The heat exchange unit, on the other hand, has only a small chance of having a negative present worth, and hence looks like a better choice for Betteryet Insurance.■

SUMMARY

In this chapter, some basic probability theory was used to assist in dealing with decision making under uncertainty. Basic concepts such as random variables, probability distributions, and expected value were introduced to help understand decision tree analysis and Monte Carlo simulation analysis. A decision tree is a graphical representation of the logical structure of a decision problem in terms of the sequence of decisions and outcomes of chance events. An analysis based on a decision tree can therefore account for the sequential nature of decisions.

A variety of criteria can be used to select a preferred decision from a set of alternatives. Expected monetary value is a commonly used criterion that can form the starting point of any economic analysis. If a decision maker wants to characterize the risk associated with a decision more fully, mean-variance or stochastic dominance decision criteria can be applied.

Monte Carlo simulation is useful for estimating the probability distribution of the economic outcome of a project when there are multiple sources of uncertainty involved in a project. The probability distribution of the outcome performance measure can then be used to assess the economic viability of the project.

Engineering Economics in Action, Part 12B:
Chances Are Good

"So let me get this straight," Clem started, a few days later. "You structured this as a tree. We—Global Widgets—have the first decision whether to proceed with development of the Adaptamatic at all. Then the next node represents the probability that the Adaptamatic will have a competitor. Then there is our choice of going into production or not. Then there are the customers, which we represent as chance nodes, who can have low, medium, and high demand. Finally, there are the outcomes, each of which has a dollar value associated with it. Is that right?"

"Perfect," Naomi said.

→

"And the dollar values came from . . . ?"

"Well, some reasonable assumptions about the marketplace," Naomi replied. "We had enough information to quantify three levels of demand. Also, we figured we had as much consumer acceptance as any competitor because of our association with Powerluxe, so when a competitive product was in the marketplace, we assumed that we would have the same market share as anyone else."

"What about the probabilities?" Clem continued.

"That was harder." Bill answered this one. Naomi couldn't help but smile. A few days ago, she didn't know anything about marketing, and Bill didn't know about decision trees, but now she was answering the questions about marketing, and he was answering the ones about decision trees! Go figure.

"We had no hard information on probabilities," Bill continued. "We set it all up as a spreadsheet so that we could adjust the probabilities freely. First we put in our best guesses—subjective probabilities. We talked to Prabha up in marketing and to a bunch of people at Powerluxe, and refined that down. Finally, we tested it for a whole range of possibilities. Bottom line: chances are good that we will make a killing."

Clem looked at the decision tree and at the table Bill gave him that reported the expected value of proceeding with the Adaptamatic development under various assumptions. "Are you sure this is right? How come we make more money if we have a competitor than if we don't?"

"Good observation, Clem." Bill was beaming. This was his area of expertise. "That is interesting, isn't it? That's what's called 'building the category' in marketing. The principle is that when you have a new product, you have to spend a lot of resources educating the consumer. If you have to do that yourself, it's really costly. But a competitor can work for you by taking a share of those education costs and actually making it cheaper for you. That's what happens here."

"It was the decision tree approach that revealed this dynamic in this case," Bill continued. "By ourselves, we could lose money if demand was low, even though the expected value was fairly high. However, with a competitor on the market, we can almost guarantee making money because the efforts of our competitor reduce our marketing costs and expand the market at the same time. Combine them both in the decision tree, and we have a high expected value with almost no risk. Cool, eh?"

"Really cool." Clem glanced at Naomi with an appraising look. "Speaking of chances—chances are good that Anna's going to like this. Nice job."

PROBLEMS

For additional practice, please see the problems (with selected solutions) provided on the Student CD-ROM that accompanies this book.

12.1 An investment has possible rates of return of 7%, 10%, and 15% over five years. The probabilities of attaining these rates, estimated on the basis of the current economy, are 0.65, 0.25, and 0.1, respectively. If you have ¥10 000 to invest,

(a) What is the expected rate of return from this investment?

(b) What is the variance of the rate of return on this investment?

12.2 Rockies Adventure Wear, Inc., sells athletic and outdoor clothing through catalogue sales. Its managers want to upgrade their order-processing centre so that they have less chance of losing customers by putting them on hold. The upgrade may result in a processing capacity of 30, 40, 50, or 60 calls per hour with the probabilities of 0.2, 0.4, 0.3, and 0.1, respectively. Market research indicates that the average number of calls that Rockies may receive is 50 per hour.

(a) How many customers are expected to be lost per hour due to the lack of processing capacity?

(b) What is the variance of the loss due to the lack of processing capacity?

12.3 Power Tech builds power-surge protection devices. One of the components, a plastic moulded cover, can be produced by two automated machines, A1 and X1000. Each machine produces a number of defects with probabilities shown in the following table.

A1 No. of Defects (out of 100)	Probability	X1000 No. of Defects (out of 100)	Probability
0	0.3	0	0.25
1	0.28	1	0.33
2	0.15	2	0.26
3	0.15	3	0.1
4	0.1	4	0.05
5	0.02	5	0.01

(a) Which machine is better with regard to the expected number of defective products?

(b) If Power Tech wants to take the variance of the number of defects into account, can mean-variance dominance be used to select a preferable machine?

12.4 Lightning City is famous for having many thunderstorms during the summer months (from June to August). One of the CB Electronix factories is located in Lightning City. They have collected information, shown in the table below, regarding the number of blackouts caused by lightning.

Number of Blackouts (Per Month)	Probability (Summer Months)	Probability (Non-Summer Months)
0	0	0.45
1	0.4	0.4
2	0.25	0.15
3	0.2	0
4	0.1	0
5	0.05	0

For the first three blackouts in a month, the cost due to suspended manufacturing is $800 per blackout. For the fourth and fifth blackouts, the cost increases to $1500 per blackout. A local insurance company offers protection against lightning-related expenses. The monthly payment is $500 for complete annual coverage. Assume that the number of blackouts in any month is independent of those in any other month.

(a) What is the expected cost related to blackouts over the summer months? Over the non-summer months? Should CB consider purchasing the insurance policy ?

(b) What is the variance of costs related to blackouts over the summer months? Over the non-summer months? Can mean-variance dominance reasoning be used to decide whether to purchase the insurance policy?

12.5 A new wave-soldering machine is expected to generate monthly savings of either 800 000 rupees, 1 000 000 rupees, 1 200 000 rupees, or 1 400 000 rupees over the next two years. The manager is not sure about the likelihood of the four savings scenarios, so she assumes that they are equally likely. What is the present worth of the expected monthly savings? Use MARR of 12%, compounded monthly, for this problem.

12.6 Regional Express is a small courier service. By introducing a new computerized tracking device, they anticipate some increase in revenue, currently estimated at $2.75 per parcel. The possible new revenue ranges from $2.95 to $5.00 per parcel with probabilities shown in the table below. Assuming that Regional's monthly capacity is 60 000 parcels and the monthly operating and maintenance costs are $8000, what is the present worth of the expected revenue over 12 months? Regional's MARR is 12%, compounded monthly.

Revenue per parcel	$2.95	$3.25	$3.50	$4.00	$5.00
Probability	0.1	0.35	0.3	0.15	0.1

12.7 Katrina is thinking about buying an automobile. She figures her monthly payments will be €90 for insurance, €30 for gas, and €20 for general maintenance. The car she would like to buy may last for 4, 5, or 6 years before a major repair, with probabilities of 0.4, 0.4, and 0.2, respectively. Calculate the present worth of the monthly expenses over the expected life of the car (before a major repair). Katrina's MARR is 10%, compounded monthly.

12.8 Pharma-Excel is a pharmaceutical company. They are currently studying the feasibility of a research project that involves improvement of vitamin C pills. To examine the optimistic, expected, and pessimistic scenarios for this project, they gathered the data shown below. What is the expected annual cost of the vitamin C project? Assume Pharma-Excel's MARR is 15%. Note that the lead time is different for each scenario.

	Optimistic	Expected	Pessimistic
Research and development costs (at the end of research)	$75 000	$240 000	$500 000
Lead time to production (years)	1	2	3
Probability	0.15	0.5	0.35

12.9 Mega City Hospital is selling lottery tickets. All proceeds go to its cancer research program. Each ticket costs $100, but the campaign catchphrase promises a 1-in-1000 chance of winning the first prize. The first prize is a "dream" house, which is worth $250 000. On the basis of decision tree analysis, is buying a ticket worthwhile?

12.10 See Problem 12.9. Determine the price of a ticket so that not buying a ticket is the preferred option and determine the chance of winning so that not buying a ticket is the preferred option.

12.11 Randall at Churchill Circuits (CC) has just received an emergency order for one of CC's special-purpose circuit boards. Five are in stock at the moment. However, when they were tested last week, two were defective but were mixed up with the three good ones. There is not enough time to retest the boards before shipment to the customer. Randall can either choose one of the five boards at random to ship to the customer or he can obtain a proven non-defective one from another plant. If the customer gets a bad board, the total incremental cost to CC is $10 000. The incremental cost to CC of getting the board from another plant is $5000.

(a) What is the chance that the customer gets a bad board if Randall sends one of the five in stock?

(b) What is the expected value of the decision to send the customer one of the five boards in stock?

(c) Draw a decision tree for Randall's decision. On the basis of EV, what should he do?

12.12 St. Jacobs Cheese Factory (SJCF) is getting ready for a busy tourist season. SJCF wants to either increase production or produce the same amount as last year, depending on the demand level for the coming season. SJCF estimates the probabilities for high, medium, and low demands to be 0.4, 0.35, and 0.25, respectively, on the basis of the number of tourists forecasted by the local recreational bureau. If SJCF increases production, the expected profits corresponding to high, medium, and low demands are $750 000, $350 000, and $100 000, respectively. If SJCF does not increase production, the expected profits are $500 000, $400 000, and $200 000, respectively.

(a) Construct a decision tree for SJCF. On the basis of EV, what should SJCF do?

(b) Construct a cumulative risk profile for both decision alternatives. Does either outcome or first-order stochastic dominance exist?

12.13 LOTell, a telephone company, has two options for its new internet service package: it can introduce a combined rate for the residential phone line and the internet access or it can offer various add-on internet service rates in addition to the regular phone rate. LOTell can only afford to introduce one of the packages at this point. The expected gain in market share by introducing the internet service would likely differ for different market growth rates. LOTell has estimated that if it introduces the combined rate, it would gain 30%, 15%, and 3% of the market share with rapid, steady, or slow market growth. If it introduces the add-on rates, it gains 15%, 10%, and 5% of the market share with rapid, steady, or slow market growth.

(a) Construct a decision tree for LOTell. If it wishes to maximize expected market share growth, which package should LOTell introduce to the market now?

(b) Can either the combined rate or the add-on rate alternative be eliminated from consideration due to dominance reasoning?

12.14 Bockville Brackets (BB) uses a robot for welding small brackets onto car-frame assemblies. BB's R&D team is proposing a new design for the welding robot. The new design should provide substantial savings to BB by increasing efficiency in the robot's mobility. However, the new design is based on the latest technology, and there is some uncertainty associated with the performance level of the robot. The R&D team estimates that the new robot may exhibit high, medium, and low performance levels with the probabilities of 0.35, 0.55, and 0.05 respectively. The annual savings corresponding to high, medium, and low performance levels are $500 000, $250 000, and $150 000 respectively. The development cost of the new robot is $550 000.

(a) On the basis of a five-year study period, what is the present worth of the new robot for each performance scenario? Assume BB's MARR is 12%.

(b) Construct a decision tree. On the basis of EV, should BB approve the development of a new robot?

12.15 Refer to Review Problem 12.1. Power Tech is still considering the partnership with the Chinese manufacturing company. The analysis in Review Problem 12.1 has shown that the partnership is recommended (by a marginal difference in the expected revenue increase between the two options). Power Tech now wants to further examine the possible shipping delay and quality control problems associated with the partnership. Power Tech estimates that shipping may be delayed 40% of the time due to the distance. Independently of the shipping problem, there may be a quality problem 25% of the time due to communication difficulties and lack of close supervision by Power Tech. The payoff information is estimated as shown below. Develop a decision tree for Power Tech's shipping and quality control problems and analyze it. On the basis of EV, what is the recommendation regarding the possible partnership?

Shipping Problem	Quality Problem	Gain in Annual Profit
No shipping delay	Acceptable quality	$ 200 000
No shipping delay	Poor quality	25 000
Shipping delay	Acceptable quality	100 000
Shipping delay	Poor quality	−100 000

12.16 Refer to Problem 12.2. Rockies Adventure Wear, Inc. has upgraded its order-processing centre in order to improve the processing speed and customer access rate. Before completely switching to the upgraded system, Rockies has an option of testing it. The test will cost Rockies $50 000, which includes the testing cost and loss of business due to shutting down the business for a half-day. If Rockies does not test the system, there is a 55% chance of severe failure ($150 000 repair and loss of business costs), a 35% chance of minor failure ($35 000 repair and loss of business costs), and a 10% chance of no failure. If Rockies tests the system, the result can be favourable with the probability of 0.34, which requires no modification, and not favourable with the probability of 0.66. If the test result is not favourable, Rockies has two options: minor modification and major modification. The minor modification costs $5000 and the major modification costs $30 000. After the minor modification, there is still a 15% chance of severe failure ($150 000 costs), a 45% chance of minor failure ($35 000 costs), and a 40% chance of no failure. Finally, after the major modification, there is still a 5% chance of severe failure, a 30% chance of minor failure, and a 65% chance of no failure. What is the recommended action for Rockies, using a decision tree analysis?

12.17 Refer to Problem 12.8. As a part of Pharma-Excel's feasibility study, they want to include information on the acceptance attitude of the public toward the new vitamin C product. Regardless of the optimistic, expected, and pessimistic scenarios on research and development, there is a chance the general public may not feel comfortable with the new product because it is based on a new technology. They estimate that the likelihood of the public accepting the product (and purchasing it) is 33.3% and not accepting it is 66.7%. The expected annual profit after the research is $1 000 000 if the public accepts the new product and $200 000 if the public does not accept it.

(a) Calculate the annual worth for all possible combinations of three R&D scenarios (optimistic, expected, and pessimistic) and two scenarios on public reaction (accept or not accept). Pharma-Excel's MARR is 15%.

(b) Using the annual worth information as the payoff information, build a decision tree for Pharma's problem. Should they proceed with the development of this new vitamin C product?

12.18 Baby Bear Beads (BBB) found themselves confronting a decision problem when a packaging line suffered a major breakdown. Ross, the manager of maintenance, Rita, plant manager, and Ravi, the company president, met to discuss the problem.

Ross reported that the current line could be repaired, but the cost and result were uncertain. He estimated that for $40 000, there was a 75% chance the line would be as good as new. Otherwise, an extra $100 000 would have to be spent to achieve the same result.

Rita's studies suggested that for $90 000, the whole line might be replaced by a new piece of equipment. However, there was a 40% chance an extra $20 000 might be required to modify downstream operations to accept a slightly different package size.

Ravi, who had reviewed his sales projections, revealed that there was a 30% chance the production line would no longer be required anyway, but that this wouldn't be known until after a replacement decision was made. Rita then pointed out that there was an 80% chance the new equipment she proposed could easily be adapted to other purposes, so that the investment, including the modifications to downstream operations, could be completely recovered even if the line was no longer needed. On the other hand, the repaired packing line would have to be scrapped with essentially no recovery of the costs.

The present worth of the benefit of having the line running is $150 000. Use decision tree analysis to determine what BBB should do about the packaging line.

12.19 Refer to Review Problem 12.1. Power Tech feels comfortable about the probability estimate regarding the change in North American demand. However, they would like to examine the probability estimate for Asian demand more carefully. Perform sensitivity analysis on the probability that Asian demand increases. Try the following values, {0.1, 0.2, 0.3, 0.4, 0.5}, in which 0.3 is the base case value. Analyze the result and give a revised recommendation as to Power Tech's possible partnership.

12.20 Refer to Review Problem 12.2. LOTell is happy with the decision recommendation suggested by the previous decision tree analysis considering information on market growth. However, LOTell feels that the uncertainty in market growth is the most important factor in the overall decision regarding the introduction of the internet service package. Hence, they wish to examine the sensitivity of the probability estimates for the market growth. Answer the following questions on the basis of the decision tree developed for Review Problem 12.2.

(a) Let p_1 be the probability of rapid market growth, p_2 be the probability of steady growth, and p_3 be the probability of slow market growth. Express the expected value at Node 2, EV(2), in terms of p_1, p_2, and p_3.

(b) If EV(2) < 10, then the option to introduce the package now is preferred. Using the expression of EV(2) that was developed in part (a), graph all possible values of p_1 and p_2 that lead to the decision to introduce the package now. (You will see that p_3 from part (a) is not involved.) What can you observe from the graph regarding the values of p_1 and p_2?

12.21 Kennedy Foods Company is a producer of frozen turkeys. A new piece of freezing equipment became available in the market last month. It costs €325 000. The new equipment should increase Kennedy Foods' production efficiency, and hence its annual profit. However, the net increase in the annual profit is somewhat uncertain because it depends on the annual operating cost of the new equipment, which is uncertain at this point. Kennedy Foods estimates the possible annual revenues with the following probability distribution.

Net Increase in Annual Revenue	Probability
€25 000	0.1
30 000	0.35
35 000	0.4
40 000	0.15

(a) Express the present worth of this investment in analytical terms. Use a 10-year study period and a MARR of 15%.

(b) Show the random-number assignment ranges that can be used in Monte Carlo simulation.

(c) Carry out a Monte Carlo simulation of 100 trials. What is the expected present worth? What are the maximum and minimum PW in the sample frequency distribution? Construct a histogram of the present worth. Is it worthwhile for Kennedy Foods to purchase the new freezing equipment?

12.22 Refer back to Problem 12.3. Power Tech has decided to use the X1000 model exclusively for producing plastic moulded covers. The revenue for each non-defective unit is $0.10. For the defective units, rework costs are $0.15 per unit.

(a) Show the analytical expression for the total revenue.

(b) Using the same probability distribution as in Problem 12.3, create a table showing the random number assignment ranges for the number of defective units.

(c) Assume that the X1000 model produces moulded covers in batches of 100. Carry out a Monte Carlo simulation of 100 production runs. What is the average total revenue? What are the maximum and minimum revenue in the 100 trials? Construct a histogram of the total revenue and comment on your results.

12.23 Ron-Jing is starting her undergraduate studies in September and needs a place to live over the next four years. Her friend, Nabil, told her about his plan to buy a house and rent out part of it. She thinks it may be a good idea to buy a house too, as long as she can get a reasonable rental income and can sell the house for a good price in four years. A fair-sized house, located a 15-minute walk from the university, costs $120 000. She estimates that the net rental income, after expenses, will be $1050 per month, which seems reasonable. She is, however, concerned about the resale value. She figures that the resale value will depend on the housing market in four years. She estimates the possible resale values and their likelihoods as shown in the table below. Using the Monte Carlo method, simulate 100 trials of the present worth of investing in the house. Use a MARR of 12%. Is this a viable investment for her?

Resale Value ($)	Probability
$100 000	0.2
110 000	0.2
120 000	0.2
130 000	0.2
140 000	0.2

12.24 An oil company owns a tract of land that has good potential for containing oil. The size of the oil deposit is unknown, but from previous experience with land of similar characteristics, the geological engineers predict that an oil well will yield between 0 (a dry well) and 100 million barrels per year over a five-year period. The following probability distribution for well yield is also estimated. The cost of drilling a well is $10 million, and the profit (after deducting production costs) is $0.50 per barrel. Interest is 10% per year. Carry out a Monte Carlo simulation of 100 trials and construct a histogram of the resulting net present worth of investing in drilling a well. Comment on your results. Do you recommend drilling?

Annual Number of Barrels (in Millions)	Probability
0	0.05
10	0.1
20	0.1
30	0.1
40	0.1
50	0.1
60	0.1
70	0.1
80	0.1
90	0.1
100	0.05

 12.25 Refer to Problem 9.24. Before it purchases sonar warning devices to help trucks back up at store loading docks, St. James department store is re-examining the original estimates of two types of annual savings generated by installing the devices. St. James feels that the original estimates were somewhat optimistic, and they want to include probability information in their analysis. The table below shows their revised estimates, with probability distributions, of annual savings from faster turnaround time and reduced damage to the loading docks. The sonar system costs £220 000 and has a life of four years and a scrap value of £20 000. Carry out a Monte Carlo simulation and generate 100 random samples of the present worth of investing in the sonar system. Assume that St. James's MARR is 18% and ignore the inflation. Make a recommendation regarding the possible purchase based on the frequency distribution.

Savings From Faster Turnaround	Probability	Savings From Reduced Damage	Probability
£38 000	0.2	£24 000	0.25
41 000	0.2	26 000	0.25
44 000	0.2	28 000	0.25
47 000	0.2	30 000	0.25
50 000	0.2		

 12.26 A fabric manufacturer has been asked to extend a line of credit to a new customer, a dress manufacturer. In the past, the mill has extended credit to customers. Although most pay back the debt, some have defaulted on the payments and the fabric manufacturer has lost money. Previous experience with similar new customers indicates that 20% of customers are bad risks, 50% are average in that they will pay most of their bills, and 30% are good customers in that they will regularly pay their bills. The average length of business affiliation with bad, average, and good customers is 2, 5, and 10 years, respectively. Previous experience also indicates that, for each group, annual profits have the following probability distribution.

Bad Risk		Average Risk		Good Risk	
Annual Profit	Probability	Annual Profit	Probability	Annual Profit	Probability
$-50 000	1/7	$10 000	0.2	$20 000	1/7
-40 000	1/7	15 000	0.2	25 000	1/7
-30 000	1/7	20 000	0.2	30 000	1/7
-20 000	1/7	25 000	0.2	35 000	1/7
-10 000	1/7	30 000	0.2	40 000	1/7
0	1/7			45 000	1/7
10 000	1/7			50 000	1/7

Construct a spreadsheet with the headings as shown in the table below. Generate 100 random trials, and construct a frequency distribution of the present worth of extending a line of credit to a customer. Use a MARR of 10% per year.

Sample Number	Random Number 1	Risk Rating	Years of Business	Random Number 2	Annual Profit	PW
1						
2						
3						
4						
5						
⋮	⋮	⋮	⋮	⋮	⋮	⋮

12.27 Rockies Adventure Wear, Inc., has been selling athletic and outdoor clothing through catalogue sales. Most orders from customers are processed by phone and the rest by mail. Rockies is now considering expanding its market by introducing a web-based ordering system. The first cost of setting it up is $120 000. A market expert predicts 10 000 new customers in the first year. Each new customer generates an average of $5 of revenue for Rockies. There are, however, uncertainties regarding the possible market growth and annual operating and maintenance costs over the five years. The market may grow at a steady rate of 2%, 5%, 8%, 10%, or 15% from the initial estimate of 10 000, with each growth rate having a chance of 20%. The annual costs may be $10 000, $15 000, $20 000, $25 000, $30 000, or $35 000, and these estimates are equally likely. Rockies' MARR is 18% for this type of investment. On the basis of 100 trials generated by Monte Carlo simulation, what is the expected present worth for this project? Comment on the project's viability.

12.28 Hitomi is considering buying a new lawnmower. She has a choice of a Lawn Guy or a Clip Job. Her neighbour, Sam, looked at buying a mower himself a while ago, and gave her the following information on the two types of mowers.

	Lawn Guy	Clip Job
First cost	$350	$120
Life (years)	10	4
Annual gas	$60	$40
Annual maintenance	$30	$60
Salvage value	0	0

Due to the long life of a Lawn Guy mower and the uncertainty of future gas prices, Hitomi is reluctant to use a single estimate for its life or the annual cost of gas. As for the Clip Job, she is not sure about the annual maintenance cost, since that model has a relatively short life and it may break down easily. With help from a friend who works at a hardware store, she comes up with the probabilistic estimates shown below for the Lawn Guy's expected life and the cost of gas, and Clip Job's annual maintenance cost. Hitomi's MARR is 5%. Find the expected annual cost for each mower using the Monte Carlo simulation method. Perform at least 100 trials. Which mower is preferred?

	Lawn Guy			Clip Job	
Life	Probability	Gas	Probability	Maintenance	Probability
7	0.25	$50	0.2	$50	0.2
8	0.25	60	0.3	60	0.5
9	0.25	70	0.4	70	0.2
10	0.25	80	0.1	80	0.1

 12.29 Northern Lager Brewery (NLB), currently buys 250 000 bottle labels every year from a local label-maker. The label-maker charges NLB £0.075 per piece. A demand forecast indicates that NLB's annual demand may grow up to 400 000 in the near future, so NLB is considering making the labels themselves. If they did so, they would purchase a high-quality colour photocopier that costs £6000 and lasts for five years with no salvage value at the end of its life. The operating cost of the photocopier would be £4900 per year, including the cost of colour cartridges and special paper used for labels. NLB would also have to hire a label designer. The cost of labour is estimated to be £0.04 per label produced. On the basis of the following probabilistic estimate on the future demand, simulate 200 Monte Carlo trials of the present worth of costs for (a) continuing to purchase from the local label-maker and (b) making its own labels. NLB's MARR is 12%. Make a recommendation as to whether NLB should consider making its own labels.

Demand	Probability
200 000	0.2
250 000	0.2
300 000	0.2
350 000	0.2
400 000	0.2

12.30 Refer to Problem 11.29. Timely Testing (TT) has now established probabilistic information on the annual variable costs and annual production volume for the two testing devices offered by vendors A and B. TT would like to perform a Monte Carlo simulation analysis before deciding which of the two alternatives it should choose. First cost, annual fixed costs, and salvage value information is presented in the table below in addition to the probabilistic estimates. TT charges $3.25 per board tested. The MARR is 15%. Compare the expected annual worth and possible cost ranges for the two alternatives. What would you recommend to TT now?

Alternative A				Alternative B			
Annual Variable Costs ($/Board)	Probability	Production Volume (Boards)	Probability	Annual Variable Costs ($/Board)	Probability	Production Volume (Boards)	Probability
2.20	0.2	40 000	0.05	1.01	0.1	40 000	0.05
2.25	0.3	45 000	0.15	1.03	0.2	45 000	0.05
2.30	0.3	50 000	0.25	1.06	0.4	50 000	0.2
2.35	0.2	55 000	0.2	1.10	0.3	55 000	0.2
		60 000	0.15			60 000	0.2
		65 000	0.05			65 000	0.1
		70 000	0.05			70 000	0.1
		75 000	0.05			75 000	0.05
		80 000	0.05			80 000	0.05

	A	B
First cost ($)	200 000	350 000
Annual fixed costs ($)	16 500	35 500
Salvage value ($)	5000	20 000
Life	7	10

More Challenging Problem

12.31 A small company in Sydney, Australia, sells flavoured beverages to markets on the east and west coasts. It is contemplating two expansion options to accommodate sales growth. The first is to build a new packaging facility at site A (close to Sydney) that will supply the entire country. The other option is to build two smaller packaging plants: one at site A on the east coast, and a second at site B on the west coast. This would save them from having to transport the beverages from the east coast (where they are made) to the west coast (where they are sold). Transportation costs are $0.75 per unit.

While the marketing team is fairly sure of the market potential on the east coast, the situation on the west coast is far less certain. Current sales on the east coast are 100 000 units per year, and are 50 000 on the west coast. Marketing staff estimate that the average east coast growth rate for the product will be 2% per year, with some variation they feel is reasonably captured by a normal distribution with a standard deviation of 0.5% per year. Due to the fact that there has been a high population growth rate in the west, and also that there might be stiff competition there, the team is less certain of both profit margins and sales growth potential. They estimate that the east coast profit margin per unit will have a normal distribution with a mean of $4 and a standard deviation of $0.10, whereas the west coast sales growth rate has the potential to be larger than the east coast's but much more uncertain. A summary of data for both regions is below.

	Mean	Std. Dev.
Volume growth rate, site A (%/year)	2	0.5
Volume growth rate, site B (%/year)	5	3
Profit margin, site A ($)	4	0.1
Profit margin, site B ($)	4	3
Initial sales volume, site A	100 000	
Initial sales volume, site B	50 000	
MARR	0.15	
Transportation cost/unit, site A to site B ($)	0.75	
First cost to build at site A	$2 000 000	
First cost to build at site A and site B	$2 500 000	
BV of facilities as a percent of first cost (in 10 years)	0.3	

(a) Using the average values in the above table, what is the expected present worth of building a single facility at site A? What is the present worth of building at both site A and site B? Use a study period of 10 years and assume that at the end of the 10 years, the book value of the facilities built is 30% of the first cost. Based on expected present worth, what would your recommendation be?

(b) You would like to get a better understanding of the risks of each alternative, and have decided to conduct a Monte Carlo simulation so that you can construct a risk profile for each project. In such a simulation, what input random variables will you need to sample?

(c) Specify the analytical model you will use to evaluate the present worth of each project for a given realization of the input random variables. Using, for example, the data analysis feature of Microsoft Excel, investigate how you can generate normally distributed random variables with the mean and standard deviations shown in the table above. Conduct the Monte Carlo simulation using 300 trials to evaluate the expected present worth of the two projects. Comment on your results.

(d) Compute the variance of the present worth of each alternative. Using the concept of mean-variance dominance, what can you say about the two alternatives?

(e) Develop a risk profile for the present worth of both projects. Compare the two projects in terms of their risks. What is the risk of a negative present worth for each project?

(f) Develop a cumulative risk profile for each project, and plot them on the same graph. Does the cumulative risk profile provide any additional insight to your decision?

(g) What would your overall recommendation to management be?

MINI-CASE 12.1

Predicting Water Demand in High-Rise Buildings in Hong Kong

The city of Hong Kong has a dense population with limited land suitable for housing. As a result, building trends have been toward taller apartment units, with many in excess of 140 m. With water resource management a large concern for city planners, the design of efficient building water supply systems is very important. A key component of the design process is estimating the maximum total simultaneous water demand in such buildings so that engineers can limit the water supply failure rate to less than 1%. The challenge for city engineers is to balance the cost of the water supply system with its reliability.

A variety of methods have been used to estimate the maximum load. They typically involve using the flow rate of each water fixture, the number of water fixtures in a building, and an estimated parameter that captures the probability that all of the fixtures are in use simultaneously. These "fixture-unit" methods were initially developed in the mid-1900s, and were parameterized with empirical data from medium-rise housing. Hong Kong engineers were concerned that the methods might not be suitable for high-rise buildings.

To address this concern, the engineers conducted a survey of close to 600 households to find out the typical daily usage patterns of various water consuming appliances for each family member. Combining this with sampled flow rates and flow duration for each of the appliances, they were able to construct probability distributions for each of these sources of variation. The engineers then used this information as input to a Monte Carlo simulation, which was used to construct a cumulative risk profile for the peak water load on the system. The simulation predicted a peak load of only 50%–60% of that computed by the "fixture-unit" method, and implied a much lower cost for the high-rise buildings' water supply system. These findings were thought to be useful in updating the methodology for computing water demand in high-rise residential construction and were seen to be a step forward in balancing cost with water supply reliability.

Source: L.T. Wong and K.W. Mui (2007). "Stochastic Modelling of Water Demand by Domestic Washrooms in Residential Tower Blocks," *Water and Environment Journal*, Volume 22, Number 2, June 2008, pp. 125–130.

Discussion

As a result of technological change and the emergence of new societal needs, engineers must constantly review and update design processes. In this mini-case, increases in the height of apartment towers and the need to better manage natural resources were the driving forces behind consideration of new design parameters for load estimation for sizing water plants and piping systems. The design parameters developed years previously for low-rise or medium-rise buildings were not considered appropriate for the very high-rise housing common in Hong Kong. Also, the need to manage fresh water consumption through the use of efficient water supply systems was considered necessary.

The process of estimating overall peak water needs started with studying the individual components of an apartment water system—the typical domestic appliances one would find in an apartment. By collecting data on the water consumption of each appliance, and then by sampling appliance usage patterns by time of day, day of week, and by time of year, the engineers were able to build up a profile for water usage patterns. The sampling allowed them to establish probability distributions for the number of users served by a particular appliance, the flow rate of an appliance, its operating time and the hourly use of the appliance. Using this as their basic unit of analysis, they could then build up a model of an apartment system by specifying its size, and how the individual components interacted.

For many engineering projects, system-level behaviour can be difficult to predict, even if there is a good model of how individual components behave. For example, in this mini-case, older peak water consumption models assumed that all appliances in the same washroom could be in use simultaneously. A more realistic assumption, validated by the interview results, was that only one could be in use at a time. While the older model might have been more mathematically tractable, the model developed in this mini-case allowed for more complex and realistic interactions. As a result, the engineers were able to produce a predictive model that met the needs of technological change and societal need.

Questions

1. Many projects can be implemented without an initial simulation study. Name a few examples, and explain why decisions about their acceptability can be made without a simulation study.

2. What are the characteristics of projects where simulation makes sense? Provide several examples of this type of project.

3. Simulations can help with evaluating different design alternatives, including the possibility of pinpointing potential problems. Can you think of other benefits provided by a simulation?

4. The output of a single simulation trial is a random variable. What is the problem with using only one trial for the purposes of decision making?

5. Ji-Ye is evaluating a proposal to build a plant that will cost 元70 000 000. A consultant has offered her a simulation that was developed to model the performance of a similar plant in the United States. If she wanted to use the American simulation, what considerations would she need to make for its application in a Chinese setting? How might she judge how much to spend on the simulation?

CHAPTER

13

Qualitative Considerations and Multiple Criteria

Engineering Economics in Action, Part 13A: Don't Box Them In

Naomi and Bill Astad were seated in Naomi's office. "OK," said Naomi. "Now that we know we can handle the demand, it's time to work on the design, right? What is the best design?" She was referring to the self-adjusting vacuum cleaner project for Powerluxe that she and Bill had been working on for several months.

"Probably the best way to find out," Bill answered, "will be to get the information from interviews with small groups of consumers."

"All right," said Naomi. "We have to know what to ask them. I guess the most important step for us is to define the relevant characteristics of vacuums."

"I agree," Bill responded. "We couldn't get meaningful responses about choices if we left out some important aspect of vacuums like suction power. One way to get the relevant characteristics will be to talk to people who have designed vacuums before, and probably to vacuum cleaner salespeople, too." They both smiled at the humorous prospect of seeking out vacuum cleaner salespeople, instead of trying to avoid them.

"We're going to need some technical people on the team," Naomi said. "We will have to develop a set of technically feasible possibilities."

"Exactly," Bill replied. "Moreover, we need to have working models of the feasible types. That is, we can't just ask questions about attributes in the abstract. Most people would have a hard time inferring actual performance from numbers about weight or suction power, for example. Also, consumers are not directly interested in these measurements. They don't care what the vacuum weighs. They care about what it takes to move it around and go up and down stairs. This depends on several aspects of the cleaner. It includes weight, but also the way the cleaner is balanced and the size of the wheels."

"That makes sense," Naomi said. "I assume that we would want to structure the interviews to make use of some form of MCDM approach."

"Huh?" Bill said.

13.1 | Introduction

Most of this book has been concerned with making decisions based on a single economic measure such as present worth, annual worth, or internal rate of return. This is natural, since many of the decisions that are made by an individual, and most that are made by businesses, have the financial impact of a project as a primary consideration. However, rarely are costs and benefits the only consideration in evaluating a project. Sometimes other considerations are paramount.

For decisions made by and for an individual, cost may be relatively unimportant. One individual may buy vegetables on the basis of their freshness, regardless of the cost. A dress or suit may be purchased because it is fashionable or attractive. A car may be chosen for its comfort and not its cost.

Traditionally, firms were different from individuals in this way. It was felt that all decisions for a firm *should* be made on the basis of the costs and benefits as measured in money (even if they sometimes were not, in practice), since the firm's survival depended solely on being financially competitive.

Society has changed, however. Companies now make decisions that apparently involve factors that are very difficult to measure in monetary terms. Money spent by firms on charities and good causes provides a benefit in image that is very hard to quantify. Resource companies that demonstrate a concern for the environment incur costs with no clear financial benefit. Companies that provide benefits for employees beyond statute or collective agreement norms gain something that is hard to measure.

The fact that firms are making decisions on the basis of criteria other than only money most individuals would hail as a good thing. It seems to be a good thing for the companies, too, since those that do so tend to be successful. However, it can make the process of decision making more difficult, because there is no longer a single measure of value.

Money has the convenient feature that, in general, more is better. For example, of several mutually exclusive projects (of identical service lives), the one with the highest present worth is the best choice. People prefer a higher salary to a lower one. However, if there are reasons to make a choice other than just the cost, things get somewhat more difficult. For example, which is better: the project with the higher present worth that involves clear-cutting a forest, or the one with lower present worth that preserves the forest? Does a high salary compensate for working for a company that does business with a totalitarian government?

Although such considerations have had particular influence in recent years, the problem of including both qualitative *and* quantitative criteria in engineering decisions has always been present. This leads to the question of how a decision maker deals with multiple objectives, be they quantitative or qualitative. There are three basic approaches to the problem.

1. Model and analyze the costs alone. Leave the other considerations to be dealt with on the basis of experience and managerial judgment. In other words, consider the problem in two stages. First, treat it as if cost were the only important criterion. Subsequently, make a decision based on the refined cost information—the economic analysis—and all other considerations. The benefit of this approach is its simplicity; the methods for analyzing costs are well established and defensible. The liability is that errors can be made, since humans have only a limited ability to process information. A bad decision can be made, and, moreover, it can be hard to explain why a particular decision was made.

2. Convert other criteria to money, and then treat the problem as a cost-minimization or profit-maximization problem. Before environmental issues were recognized as being so important, the major criterion that was not easily converted to money was human health and safety. Elaborate schemes were developed to measure the cost of a lost life or injury so that good economic decisions were made. For example, one method was to estimate the money that a worker would have made if he or she had not been injured. With an estimate like this, the cost of a project in lives and injuries could be compared with the profits obtained. A benefit of this approach is that it does take non-monetary criteria into account. A drawback is the difficult and politically sensitive task of determining the cost of a human life or the cost of cutting down a 300-year-old tree.

3. Use a **multi-criterion decision making (MCDM)** approach. There are several MCDM methods that explicitly consider multiple criteria and guide a decision maker to superior or optimal choices. The benefit of MCDM is that all important criteria can be explicitly taken into account in an appropriate manner. The main drawback is that MCDM methods take time and effort to use.

In recent years and in many circumstances, looking at only the monetary costs and benefits of projects has become inappropriate. Consequently, considerable attention has been focused on how best to make a choice under competing criteria. The first two approaches listed above still have validity in some circumstances; in particular, when non-monetary criteria are relatively unimportant, it makes sense to look at costs alone. However, it is necessary to use an MCDM method of some sort in much of engineering decision making today.

In this chapter, we focus on three useful MCDM approaches. The first, *efficiency*, permits the identification of a subset of superior alternatives when there are multiple criteria. The second approach, *decision matrixes*, is a version of multi-attribute utility theory (MAUT) that is widely practised. The third, the *analytic hierarchy process (AHP)*, is a relatively new but popular MAUT approach. It should be noted that all of these methods make assumptions about the tradeoffs among criteria that may not be suitable in particular cases. They should not be applied blindly, but critically and with a strong dose of common sense.

13.2 | Efficiency

When dealing with a single criterion like cost, it is usually clear which alternatives are better than others. The rule for a present worth analysis of mutually exclusive alternatives (with identical service lives) is, for example, that the highest present worth alternative is best.

All criteria can be measured in some way. The scale might be continuous, such as "weight in kilograms" or "distance from home," or discrete, such as "number of doors" or "operators needed." The measurement might be subjective, such as a rating of "excellent," "very good," "good," "fair," and "poor," or conform to an objective physical property such as voltage or luminescence.

Once measured, the value of the alternative can be established with respect to that criterion. It may be that the smaller or lower measurement is better, as is often the case with cost, or that the higher is better, as with a criterion such as "lives saved." Sometimes a target is desired, for example, a target weight or room temperature. In this case, the criterion could be adjusted to be the distance from the target, with a shorter distance being better.

Consequently, given one criterion, we can recognize which of several alternatives is best. However, once there are multiple criteria, the problem is more difficult. This is because an alternative can be highly valued with respect to one criterion and lowly valued with respect to another.

EXAMPLE 13.1

Brisbane Meats will be replacing its effluent treatment system. It has evaluated several alternatives, shown in Figure 13.1. Two criteria were considered: present worth and discharge purity. Which alternatives can be eliminated from further consideration?

Consider alternatives A and E in Figure 13.1. Alternative E *dominates* alternative A because it is less costly and it provides purer discharge. If these were the only criteria to consider in making a choice, one would always choose E over A. Similarly, one can eliminate F, B, and H, all of which are dominated by D and other alternatives.

Now consider alternative G, which has the same cost as E but has poor discharge purity. One would still always choose E over G, since, for the purity criterion, E is better at the same price.

Three alternatives now remain: E, D, and C. E is cheapest, but provides the least pure output of the three. C is the most expensive, but provides the greatest purity, while D is in the middle. Certainly none of these dominates the others. There is a natural tendency to focus on D, since it seems to balance the two criteria, but this really depends on the relative importance of the criteria to the decision maker. For example, if cost were very important, E might be the best choice, since the difference in purity between E and C may be considered relatively small.■

Figure 13.1   Selecting an Effluent Treatment

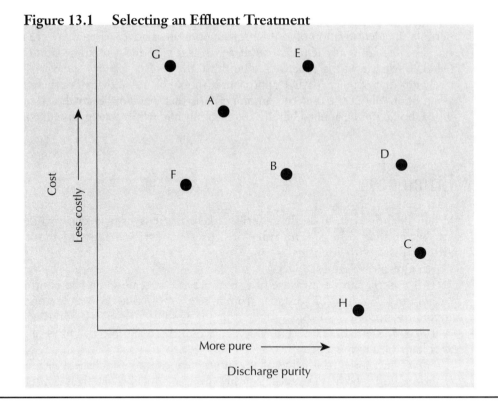

Decisions that involve only two criteria can be simplified graphically as done for Example 13.1, but when there are more than two criteria graphical methods become more difficult.

An alternative is **efficient** if no other alternative is valued as at least equal to it for all criteria and preferred to it for at least one. If an alternative is not efficient it is **inefficient**; this is the same as a **dominated** alternative.

EXAMPLE 13.2

Skiven is evaluating surveillance cameras for a security system. The criteria he is taking into account are price, weight, picture clarity, and low-light performance. The details for the 10 models are shown in Table 13.1. Skiven wants a camera with low cost, low weight, a high score for picture clarity, and a high score for low-light performance. Which models can be eliminated from further consideration?

To determine the efficient alternatives, the following algorithm can be used.

1. Order the alternatives according to one criterion, the *index criterion*. The cameras for Example 13.2 are already ordered by cost, so cost can be the index criterion.

2. Start with the second most preferred alternative for the index criterion. Call this the *candidate alternative*.

3. Compare the candidate alternative with each of the alternatives that are more preferred for the index criterion. (For the first candidate alternative, there is only one.)

4. If any alternative equals or exceeds the candidate for all criteria, and exceeds it for at least one, the candidate is dominated, and can be eliminated from further

consideration. If no alternative equals or exceeds the candidate for all criteria and exceeds it for at least one, the candidate is efficient.

5. The next most preferred alternative for the index criterion becomes the new candidate; go to step 3. Stop if there are no more alternatives to consider.

The algorithm for Example 13.2 starts by comparing camera 2 against camera 1. It can be seen that camera 2 is better for weight and picture clarity, although worse for low-light performance and cost, so it is not dominated. Looking at camera 3, it is equal to camera 1 for picture clarity, and worse than camera 1 for all other characteristics. It is dominated, since to avoid domination it would have to be better than camera 1 for at least one criterion. Moreover, we need not consider it for the remainder of the algorithm. We then compare camera 4 with only cameras 1 and 2. Since it is better in weight than the other two, it is not dominated, and we continue to camera 5. Camera 5 is dominated in comparison with camera 4, since it is worse in all respects.

Table 13.1   Surveillance Camera Characteristics

Camera	Price ($)	Weight (Grams)	Picture Clarity (10-Point Scale)	Low-Light Performance (10-Point Scale)
1	230	900	3	6
2	243	640	5	4
3	274	910	3	5
4	313	433	5	7
5	365	450	2	4
6	415	330	6	6
7	418	552	7	5
8	565	440	3	6
9	590	630	7	4
10	765	255	9	5

Carrying through in this manner shows that camera 3 is dominated by camera 1, camera 5 is dominated by camera 4, camera 8 is dominated by camera 6, and camera 9 is dominated by camera 7. The set of efficient alternatives consists of cameras 1, 2, 4, 6, 7, and 10. This set clearly includes the best choice, since there is no reason to choose a dominated alternative.■

Usually there is more than one alternative in the efficient set. Sometimes reducing the number of alternatives to be considered makes the problem easier to solve through intuition or judgment, but usually it is desirable to have some clear method for selecting a single alternative. One popular method is to use decision matrixes.

13.3 | Decision Matrixes

Usually not all of the criteria that can be identified for a decision problem are equally important. Often cost is the most important criterion, but in some cases another criterion, safety, for example, might be most important. As suggested with Example 13.1, the choice of the most important criterion will have a direct effect on which alternative is best.

One approach to choosing the best alternative is to put numerical weights on the criteria. For example, if cost were most important, it would have a high weight, while a less important criterion might be given a low weight. If criteria are evaluated according to a scale that can be used directly as a measure of preference, then the weights and preference measures can be combined mathematically to determine a best alternative. This approach is called **multi-attribute utility theory (MAUT)**.

Many different specific techniques for making decisions are based on MAUT. This section deals with decision matrixes, which are commonly used in engineering studies. The subsequent section reviews the analytic hierarchy process, a MAUT method of increasing popularity.

In a **decision matrix**, the rows of the matrix represent the criteria and the columns the alternatives. There is an extra column for the weights of the criteria. The cells of the matrix (other than the criteria weights) contain an evaluation of the alternatives on a scale from 0 to 10, where 0 is worst and 10 is best. The weights are chosen so that they sum to 10.

The following algorithm can be used:

1. Give a weight to each criterion to express its relative importance: the higher the weight, the more important the criterion. Choose the weight values so that they sum to 10.

2. For each alternative, give a rating from 0 to 10 of how well it meets each criterion. A rating of 0 is given to the worst possible fulfillment of the criterion and 10 to the best possible.

3. For each alternative, multiply each rating by the corresponding criterion weight, and sum to give an overall score.

4. The alternative with the highest score is best. The value of the score can be interpreted as the percentage of an ideal solution achieved by the alternative being evaluated.

5. Carry out some sensitivity analysis with respect to weights or rating estimates to verify the indicated decision or to determine under which conditions different choices are made.

EXAMPLE 13.3

Skiven is evaluating surveillance cameras for a security system. The criteria he is taking into account, in order of importance for him, are low-light performance, picture clarity, weight, and price. The details for the six efficient models are shown in Table 13.2. Which model is best?

In order to follow the steps given above, we need to determine the criteria weights. It is usually fairly easy for a decision maker to determine which criteria are more important than others, but generally more difficult to specify particular weights. There exist many formal methods for establishing such weights in a rigorous way, but in practice, estimating weights on the basis of careful consideration or a discussion with the decision maker is sufficient. Recall that a sensitivity analysis forms part of the overall decision process, and this compensates somewhat for the imprecision of the weights.

Skiven suggests that weights of 1, 1.5, 3.5, and 4 for price, weight, picture clarity, and low-light performance, respectively, are appropriate weights for this problem. These weights are listed as the second column of Table 13.3.

Table 13.2   Efficient Set of Surveillance Camera Alternatives

Camera	Price ($)	Weight (Grams)	Picture Clarity (10-Point Scale)	Low-Light Performance (10-Point Scale)
1	230	900	3	6
2	243	640	5	4
4	313	433	5	7
6	415	330	6	6
7	418	552	7	5
10	765	255	9	5

Table 13.3   Decision Matrix for Example 13.3

Criterion	Criterion Weight	Alternatives					
		1	2	4	6	7	10
Price	1.0	10.0	9.8	8.4	6.5	6.5	0.0
Weight	1.5	0.0	4.0	7.2	8.8	5.4	10.0
Clarity	3.5	3.0	5.0	5.0	6.0	7.0	9.0
Low-light performance	4.0	6.0	4.0	7.0	6.0	5.0	5.0
Score	10.0	44.5	49.3	64.8	64.8	59.1	66.5

The ratings for each alternative for picture clarity and low-light performance are already on a scale from 0 to 10, so those ratings can be used directly. To select ratings for the price and weight, two different measures could be used.

1. *Normalization:* The rating r for the least preferred alternative (α) is 0 and the most preferred (β) is 10. For each remaining measure (γ) the rating r can be determined as

$$r = 10 \times \frac{\gamma - \alpha}{\beta - \alpha}$$

For this problem, the rating of alternative 6 for price would be

$$r_{6,price} = 10 \times \frac{415 - 765}{230 - 765} = 6.54$$

The advantage of normalization is that it provides a mathematical basis for the rating evaluations. One disadvantage is that the rating may not reflect the value as perceived by the decision maker. A second disadvantage is that it may overrate the best alternative and underrate the worst, since these are set to the extreme values. A third disadvantage is that the addition or deletion of a single alternative (the one with the highest or lowest evaluation for a criterion) will change the entire set of ratings.

2. *Subjective evaluation:* Ask the decision maker to rate the alternatives on the 0 to 10 scale. For example, asked to rate alternative 6 for cost, Skiven might give it a 7. The advantages of subjective evaluation include that it is relatively immune to changes in the alternative set, and that it may be more accurate since it includes

perceptions of worth that cannot be directly calculated from the criteria measures. Its main disadvantage is that people often make mistakes and give inconsistent evaluations.

For the ratings shown in Table 13.3, the normalization process was used. The overall score is then calculated by summing for each alternative the rating for a criterion multiplied by the weighting for that criterion. From Table 13.3, the total score for alternative 1 is calculated as

$$1 \times 10 + 1.5 \times 0 + 3.5 \times 3 + 4 \times 6 = 44.5$$

It can be seen in Table 13.3 that the highest score is for alternative 10. This means essentially that the greatest total benefit is achieved if alternative 10 is taken.

Also note that a "perfect" alternative, that is, one that rated 10 on every criterion, would have a total score of 100. Thus the 66.5 score for alternative 10 means that it is only about 66.5% of the score of a perfect alternative. The practice of making weights sum to 10 and rating the alternatives on a scale from 0 to 10 is done specifically so that the resulting score can be interpreted as a percentage of the ideal; if this is not desired, any relative weights or rating scale can be used.

Alternative 10 is the best choice for the particular weights and ratings given, but some sensitivity analysis should be done to verify its robustness. There are several ways to do this sensitivity analysis, but the most sensible is to vary the weights of the criteria to see how the results change. This is easy to do when a spreadsheet is being used to calculate the scores.

Table 13.4 shows a range of criteria weights and the corresponding alternative scores. It can be seen that cameras 4 and 6 also can be identified as best in some of the criteria weight possibilities. For the final recommendation, it may be necessary to review these results with Skiven to let him determine which of the weight possibilities are most appropriate for him.■

As has been seen in Example 13.3, the decision matrix approach structures information about multiple objectives of the problem. An additive utility model permits the calculation of an overall score for each alternative. A comparison of the scores permits the best one to be selected. Doing a sensitivity analysis may reveal promising alternatives from relatively small changes in the alternative weight assumptions.

Table 13.4   Sensitivity Analysis for the Surveillance Camera

Criterion	Criterion Weights						
Price	1	1	1	1	2	2.5	1
Weight	1.5	2	1	2	2	2.5	1
Picture clarity	3.5	3	3	2	2	2.5	4
Low-light performance	4	4	5	5	4	2.5	4
Alternative	**Alternative Scores**						
Camera 1	44.5	43.0	49.0	46.0	50.0	47.5	46.0
Camera 2	49.3	48.8	48.8	47.8	53.6	57.0	49.8
Camera 4	64.8	65.9	**65.7**	**67.9**	**69.4**	67.2	63.7
Camera 6	64.8	66.2	63.4	66.2	66.8	**68.4**	63.4
Camera 7	59.1	58.3	57.9	56.3	57.8	59.7	59.9
Camera 10	**66.5**	**67.7**	62.0	63.6	58.0	60.0	**66.0**

13.4 | The Analytic Hierarchy Process

The **analytic hierarchy process (AHP)** is also a MAUT approach. It offers two features beyond what is done in decision matrixes. First, it provides a mechanism for structuring the problem that is particularly useful for large, complex decisions. Second, it provides a better method for establishing the criteria weights.

N E T   V A L U E   1 3 . 1

AHP Software

The analytic hierarchy process (AHP) is effective for structuring and analyzing complex, multi-attribute decision-making problems. It has a wide range of applications including vendor selection, risk assessment, strategic planning, resource allocation, and human resources management.

AHP has such broad industrial application areas that software has been developed to support its use. Expert Choice™ software has gained acceptance as a useful tool for companies to help

them deal with decision-making situations that might otherwise be too complex to structure and solve.

The Expert Choice website (www. expertchoice.com) provides several case studies describing the software's use for a number of large international companies, reference materials for AHP, and access to a trial version of the product.

AHP is somewhat more complicated to carry out than decision matrixes. In order to describe the procedure, we first list the basic steps. Example 13.4, which follows the list of steps, explains in more detail the operations at each step. The basic steps of AHP are as follows.

1. Identify the decision to be made, called the **goal**. Structure the goal, criteria, and alternatives into a hierarchy, as illustrated in Figure 13.2. The criteria could be more than one level (not illustrated in Figure 13.2) to provide additional structure to very complex problems.

Figure 13.2   AHP Hierarchy

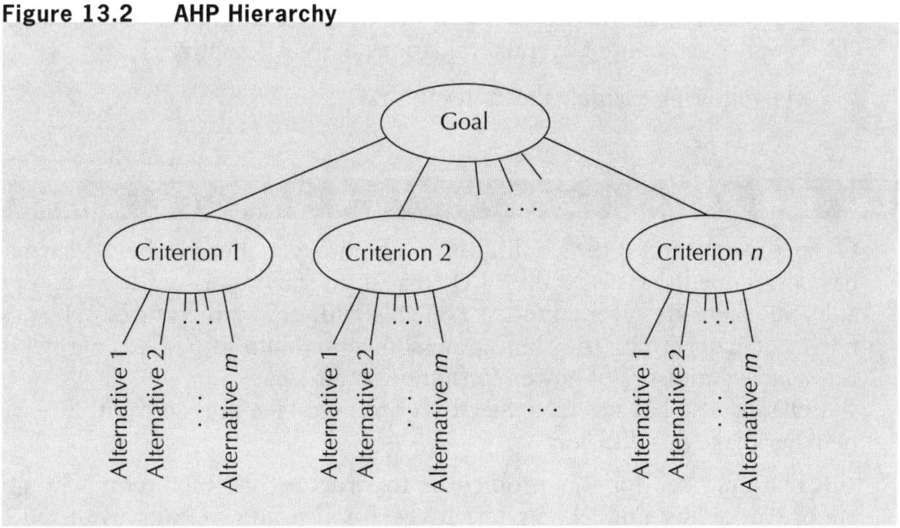

Table 13.5   The AHP Value Scale for Comparison of Two Alternatives or Two Criteria (A and B)

Value	Interpretation
1/9 = 0.111	Extreme preference/importance of B over A
1/7 = 0.1429	Very strong preference/importance of B over A
1/5 = 0.2	Strong preference/importance of B over A
1/3 = 0.333	Moderate preference/importance of B over A
1	Equal preference/importance of A and B
3	Moderate preference/importance of A over B
5	Strong preference/importance of A over B
7	Very strong preference/importance of A over B
9	Extreme preference/importance of A over B
Intermediate values	*For more detail between above values*

2. Perform pairwise comparisons for alternatives. **Pairwise comparison** is an evaluation of the importance or preference of a pair of alternatives. Comparisons are made for all possible pairs of alternatives *with respect to each criterion*. This is done by giving each pair of alternatives a value according to Table 13.5 for their relationship for each criterion. These values are placed in a **pairwise comparison matrix (PCM)**.

3. **Priority weights** for the alternatives are calculated by normalizing the elements of the PCM and averaging the row entries. (The columns of priority weights together form a *priority matrix*.)

4. Perform pairwise comparisons for criteria. As in step 2, all pairs of criteria are compared using the AHP value scale (Table 13.5). A PCM is determined, as in step 3, and priority weights are calculated for the criteria.

5. Alternative priority weights are multiplied by the corresponding criteria priority weights and summed to give an overall alternative ranking. (That is, the priority matrix of alternatives is multiplied by the column of criteria priority weights to give a column of overall evaluations of the alternatives.)

The following example illustrates this process.

EXAMPLE 13.4

Oksana is examining the cooling of a laboratory at Beaconsfield Pharmaceuticals. She has determined that a 12 000 BTU (British thermal unit) per hour cooling unit is suitable, but there are several models available with different features. The available quantitative data concerning the choices are shown in Table 13.6. The energy efficiency rating is a standard measure of power consumption efficiency.

Oksana also has several subjective criteria to take into account. She will use AHP to help her make this decision.

The first step for this problem is to structure the hierarchy. The goal is clear: to choose a cooling unit. The alternatives are also known, and are listed in Table 13.6.

Table 13.6   Cooling Unit Features

Model	Price	Energy Efficiency Rating
1	$640	9.5
2	$600	9.1
3	$959	10.0
4	$480	9.0
5	$460	9.0

After some consideration, Oksana concludes that the following are critical to her consideration:

1. Cost
2. Energy consumption
3. Loudness
4. Perceived comfort

The resulting hierarchy is illustrated in Figure 13.3.

The second step is to construct a PCM for the alternatives with respect to each criterion. For illustration purposes, we will do this step for the criterion "perceived comfort" only.

Oksana first considers two of the alternatives only, say, 1 and 2, with respect to "perceived comfort." She gives the preferred one (which is alternative 1) a rating from the scale shown in Table 13.5. In this case, Oksana judges that alternative 1 is moderately better than 2; the rating is a 3. This appears in the PCM shown in Figure 13.4 in row 1, column 2, corresponding to alternative 1 and alternative 2. Correspondingly, the reciprocal, 1/3, is put in row 2, column 1. This can be interpreted loosely as indicating that alternative 1 is three times as desirable as alternative 2 for perceived comfort, and correspondingly alternative 2 is 1/3 as desirable as alternative 1.

As another example, consider the comparison of alternatives 3 and 4. Alternative 4 is strongly preferred to 3 in "perceived comfort," so a 5 appears in row 4, column 3, and

Figure 13.3   The AHP Hierarchy for Example 13.4

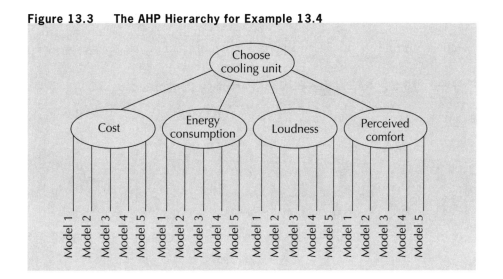

Figure 13.4   PCM for Perceived Comfort

$$
\begin{bmatrix}
1 & 3 & 5 & 1 & 3 \\
\frac{1}{3} & 1 & 3 & \frac{1}{3} & 1 \\
\frac{1}{5} & \frac{1}{3} & 1 & \frac{1}{5} & \frac{1}{3} \\
1 & 3 & 5 & 1 & 3 \\
\frac{1}{3} & 1 & 3 & \frac{1}{3} & 1
\end{bmatrix}
$$

1/5 appears in row 3, column 4. Similar comparisons of all pairs complete the PCM in Figure 13.4 for "perceived comfort." In summary, row 1 of Figure 13.4 shows the results of comparing alternative 1 with alternatives 1, 2, 3, 4, and 5. Row 2 shows the results of comparing alternative 2 with alternatives 1, 2, 3, 4, and 5, and so on. Note that an alternative is equally preferred to itself, so all main diagonal entries are 1. PCMs for the other three criteria are developed in exactly the same manner.

The next step is to determine priority weights for the alternatives. This is done by first normalizing the columns of the PCM and then averaging the rows. To normalize the columns, sum the column entries, and divide each entry by this sum. To average the rows, sum the rows (after normalizing) and divide by the number of entries per row.

For example, the sum of column 1 of the PCM in Figure 13.4 is 2.866. Then the normalized entries for the column will be 1/2.866, 0.333/2.866, etc. The complete normalized PCM for "perceived comfort" is shown in Figure 13.5. The priority weights are then calculated as the average of each row of the normalized PCM, and are also illustrated in Figure 13.5.

A similar process can be carried out for the other three criteria. The four columns (one for each criterion) of priority weights form a priority matrix, shown in Figure 13.6. The first column of this matrix consists of the priority weights for cost, the second for energy efficiency, the third for loudness, and the fourth for perceived comfort.

The next step is to construct a PCM for the criteria themselves. This is done in the same manner as for the alternatives for each criterion, except that now one rates the criteria in pairwise comparisons with each other. The PCM Oksana creates is illustrated as Figure 13.7, with the rows and columns in the following order: cost, energy consumption, loudness, and perceived comfort. For example, energy consumption is

Figure 13.5   Normalized PCM for Perceived Comfort

	Normalized PCM				**Average**
0.349	0.360	0.294	0.349	0.360	0.342
0.116	0.120	0.176	0.116	0.120	0.130
0.070	0.040	0.059	0.070	0.040	0.056
0.349	0.360	0.294	0.349	0.360	0.342
0.116	0.120	0.176	0.116	0.120	0.130

Figure 13.6   Priority Matrix for Example 13.4

$$\begin{bmatrix} 0.90 & 0.256 & 0.033 & 0.342 \\ 0.114 & 0.230 & 0.468 & 0.130 \\ 0.031 & 0.338 & 0.282 & 0.056 \\ 0.383 & 0.088 & 0.086 & 0.342 \\ 0.383 & 0.088 & 0.131 & 0.130 \end{bmatrix}$$

Figure 13.7   PCM for Goal

$$\begin{bmatrix} 1 & \frac{1}{3} & 5 & 1 \\ 3 & 1 & 7 & 3 \\ \frac{1}{5} & \frac{1}{7} & 1 & \frac{1}{5} \\ 1 & \frac{1}{3} & 5 & 1 \end{bmatrix}$$

moderately more important than energy cost. Thus, there is a 3 in row 2, column 1 and a 1/3 in row 1, column 2. The normalized PCM and row averages are shown in Figure 13.8.

The order of the rows and columns of Figures 13.7 and 13.8 is: cost, energy efficiency, loudness, and perceived comfort. Thus, the criterion with the highest priority rating is noise, at 0.524, then cost and perceived comfort identical at 0.212, and finally energy efficiency last at 0.053.

The final stage of the process consists of determining an overall score for each alternative. Note that the entire process of AHP has essentially led to the development of a decision matrix: the priority ratings for the criteria are the weights, while the priority ratings for the alternatives are the ratings of the alternatives for the criteria. Consequently, the final score is determined by multiplying each alternative priority rating by the appropriate criterion rating and then summing.

This can also be viewed as matrix multiplication of the priority matrixes for the alternatives by the column of priority weights of the criteria, as shown in Figure 13.9. The interpretation of the column vector on the right in Figure 13.9 is a ranking of the

Figure 13.8   Normalized PCM and Average Values for Goal

Average

$$\begin{bmatrix} 0.192 & 0.184 & 0.278 & 0.192 \\ 0.577 & 0.553 & 0.389 & 0.577 \\ 0.038 & 0.079 & 0.056 & 0.038 \\ 0.192 & 0.184 & 0.278 & 0.192 \end{bmatrix} \quad \begin{bmatrix} 0.212 \\ 0.524 \\ 0.053 \\ 0.212 \end{bmatrix}$$

Figure 13.9   Final Alternative Scores for Example 13.4

$$
\begin{bmatrix}
0.090 & 0.256 & 0.033 & 0.342 \\
0.114 & 0.230 & 0.468 & 0.130 \\
0.031 & 0.338 & 0.282 & 0.056 \\
0.383 & 0.088 & 0.086 & 0.342 \\
0.383 & 0.088 & 0.131 & 0.130
\end{bmatrix}
\times
\begin{bmatrix}
0.212 \\
0.524 \\
0.053 \\
0.212
\end{bmatrix}
=
\begin{bmatrix}
0.227 \\
0.197 \\
0.211 \\
0.204 \\
0.162
\end{bmatrix}
$$

alternatives. The best alternative is number 1, followed by number 3, 4, and 2; number 5 is the worst.

In conclusion, the best cooling unit is model 1. Oksana should buy this one for the laboratory. ■

13.5 | The Consistency Ratio for AHP

The subjective evaluation of the PCMs can be inconsistent. For example, Joe can say that alternative 1 is five times as important as alternative 2, and alternative 2 is five times as important as alternative 3, but then claim that alternative 1 is only twice as important as alternative 3. Or he might even say alternative 1 is less important than alternative 3.

The fact that the construction of PCMs includes redundant information is useful because it helps get a good estimate of the best rating for the alternative. However, a check has to be made that the decision maker is being consistent.

A measure called the **consistency ratio** (to measure the consistency of the reported comparisons) can be calculated for any PCM. The consistency ratio ranges from 0 (perfect consistency) to 1 (no consistency). A consistency ratio of 0.1 or less is considered acceptable in practice. The calculation of the consistency ratio is briefly reviewed in Appendix 13A.

REVIEW PROBLEMS

REVIEW PROBLEM 13.1

Warmtrex makes thermostat controls for baseboard heaters. As part of the control manufacturing process, a two-centimetre steel diaphragm is fitted to a steel cup. The diaphragm is used to open a safety switch rapidly to avoid arcing across the contacts. Currently the cup is seam-welded, which is both expensive and a source of quality problems. The company wants to explore the use of adhesives to replace the welding process. Table 13.7 lists the ones examined, along with various properties for each.

High-temperature resistance is desirable, as are tensile bond strength and pressure resistance. Fast curing speeds are desirable to reduce work-in-progress inventory storage costs, and, of course, cheaper material costs are important.

Warmtrex wants to select a single adhesive type for comparison experiments against the current seam-welding method.

Table 13.7   Possible Adhesives for Thermostat Control

Adhesive	Maximum Temperature (°C)	Tensile Bond Strength (kPa)	Pressure Resistance (kPa)	Curing Speed	Cost
Acrylic	106	21 000	3738	Medium	Cheap
Silicone	200	3 150	560	Slow	Medium
Cyanoacrylate	250	3 500	630	Fast	Cheap
Methacrylate A	225	28 000	4984	Slow	Expensive
Methacrylate B	225	7 000	1246	Medium	Expensive

(a) Are any of the listed adhesives in Table 13.7 inefficient?

(b) Discussions with management indicate that the criteria can be weighted as follows:

Temperature resistance	1.5
Bond strength	1.5
Pressure resistance	2.5
Curing speed	3.5
Cost	1.0
Total	10.0

Create a decision matrix for this problem using the above weights. Normalize the data in Table 13.7 to estimate the ratings of each alternative for each criterion. Use only the efficient alternatives. Use the maximum temperature figures to measure temperature resistance. For curing speed, set fast as 8, medium as 5, and slow as 2, while for cost, set expensive as 2, medium as 5, and cheap as 10. Under these conditions, which is the recommended adhesive?

(c) The analytic hierarchy process (AHP) was performed for this problem. The hierarchy is shown in Figure 13.10. The priority matrix in Figure 13.11

Figure 13.10   The AHP Hierarchy for Review Problem 13.1(c)

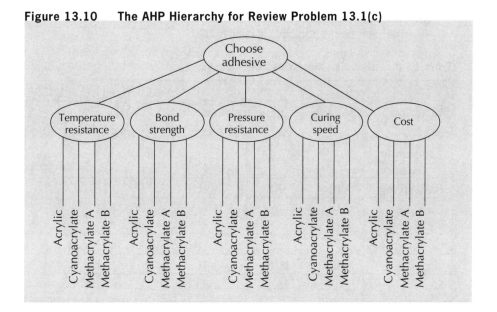

Figure 13.11 Priority Matrix for Review Problem 13.1(c)

$$\begin{bmatrix} 0.095 & 0.368 & 0.392 & 0.213 & 0.341 \\ 0.331 & 0.077 & 0.067 & 0.502 & 0.278 \\ 0.287 & 0.445 & 0.438 & 0.061 & 0.188 \\ 0.287 & 0.110 & 0.103 & 0.224 & 0.193 \end{bmatrix}$$

Figure 13.12 PCM for Goal for Review Problem 13.1(c)

$$\begin{bmatrix} 1 & 1 & \frac{1}{2} & \frac{1}{3} & 2 \\ 1 & 1 & \frac{1}{2} & \frac{1}{3} & 2 \\ 2 & 2 & 1 & \frac{1}{2} & 3 \\ 3 & 3 & 2 & 1 & 4 \\ \frac{1}{2} & \frac{1}{2} & \frac{1}{3} & \frac{1}{4} & 1 \end{bmatrix}$$

represents the results from PCMs calculated for the different criteria. The rows of Figure 13.11 correspond to the alternatives acrylic, cyanoacrylate, methacrylate A, and methacrylate B, respectively, while the columns correspond to temperature resistance, bond strength, pressure resistance, curing speed, and cost, respectively. A PCM for the goal is shown in Figure 13.12, with the criteria in the same order as for Figure 13.11. With this information, what is the best adhesive to recommend for the experiment?

ANSWER

(a) It can be observed that silicone is dominated by cyanoacrylate in all criteria, and therefore is inefficient.

(b) As shown in Table 13.8, under the weighting and rating conditions specified, the methacrylate A adhesive is clearly best.

Table 13.8 Decision Matrix of Adhesives

Criterion	Criterion Weight	Alternatives			
		Acrylic	Cyanoacrylate	Methacrylate A	Methacrylate B
Temperature resistance	1.5	0.0	10.0	8.3	8.3
Bond strength	1.5	7.1	0.0	10.0	1.4
Pressure resistance	2.5	7.1	0.0	10.0	1.4
Curing speed	3.5	5.0	8.0	2.0	5.0
Cost	1.0	10.0	10.0	2.0	2.0
Score	10.0	56.1	53.0	61.4	37.6

Figure 13.13   Normalized PCM for Goal for Review Problem 13.1(c)

		Normalized PCM			Average
0.133	0.133	0.115	0.138	0.167	0.137
0.133	0.133	0.115	0.138	0.167	0.137
0.267	0.267	0.231	0.207	0.250	0.244
0.4	0.4	0.462	0.414	0.333	0.402
0.067	0.067	0.077	0.130	0.083	0.079

(c) First we have to normalize the PCM for the goal, and then average to get the criteria weights, as shown in Figure 13.13.

Then we multiply the priority matrix by the average criteria weights, as illustrated in Figure 13.14.

Since the second alternative in Figure 13.14 has the highest net weight, it is preferred. The recommended adhesive using AHP is the cyanoacrylate. This disagrees with the result obtained using decision matrixes.

Figure 13.14   Calculating Alternative Weights for Review Problem 13.1(c)

$$
\begin{bmatrix}
0.095 & 0.368 & 0.392 & 0.213 & 0.341 \\
0.331 & 0.077 & 0.067 & 0.502 & 0.278 \\
0.287 & 0.445 & 0.438 & 0.061 & 0.188 \\
0.287 & 0.110 & 0.103 & 0.224 & 0.193
\end{bmatrix}
\times
\begin{bmatrix}
0.137 \\
0.137 \\
0.244 \\
0.402 \\
0.079
\end{bmatrix}
=
\begin{bmatrix}
0.272 \\
0.296 \\
0.247 \\
0.185
\end{bmatrix}
$$

SUMMARY

In this chapter, three approaches for dealing explicitly with multiple criteria were presented. The first, *efficiency*, allows the identification of alternatives that are not dominated by others. An alternative is dominated if there is another alternative at least as good with respect to all criteria, and better in at least one.

The second approach, *decision matrixes*, is in wide usage. Decision matrixes are a multi-attribute utility theory (MAUT) method in which criteria are subjectively weighted and then multiplied by subjectively evaluated criteria ratings to give an overall score. The weights sum to 10 and the criteria ratings are on a scale from 0 to 10, resulting in an overall score that could range from 0 to 100. The alternative with the highest score is best, and the value of the score can be considered a percentage of an "ideal" alternative.

The third approach presented in this chapter is the *analytic hierarchy process*. AHP is also a MAUT method. Pairwise comparisons are used to extract criterion weights in a more rigorous manner than for decision matrixes. Pairwise comparisons similarly are used to rate alternatives. Multiplying criterion weights and alternative ratings gives an overall evaluation for each alternative.

Engineering Economics in Action, Part 13B:
Moving On

Three months later Naomi was seated in Anna Kulkowski's office. The Powerluxe project report had been submitted two days before.

"Naomi, the work you and Bill did and your report are first-rate," Anna said. "We're going to start negotiations with Powerluxe to bring the Adaptamatic line of vacuum cleaners to market. I had no idea anybody could make such clear recommendations on such a complex problem. Congratulations on a good job."

Naomi thought back on all of the people who had helped her in her almost two years at Global Widgets: how Clem had taught her practical problem solving; how Dave had shown her the ropes; how Terry had helped her realize the benefits of attention to detail. Bill had shown her how the real world mixed engineering with marketing, business, and government. Anna, too, had shown her how to manage people. "Thank you very much, Ms. Kulkowski," Naomi responded, a small break in her voice betraying her emotion.

"Do you enjoy this kind of work?" Anna asked.

"Yes, I do," Naomi replied. "It's exciting to see how the engineering relates to everything else."

"I think we have a new long-term assignment for you," Anna said. "This is just the first step for Global Widgets in developing new products. We have a first-rate team of engineers. We want to make better use of them. Ed Burns is going to head up a product development group. He read your report on the Adaptamatic line of vacuum cleaners and was quite impressed. He and I would like you in that development group. We need someone who understands the engineering and can relate it to markets. What do you say?"

"I'm in!" Naomi had a big grin on her face.

A few days later, Naomi answered the phone.

"Hey, Naomi, it's Terry. I hear you got promoted!"

"Hi, Terry. Nice to hear from you. Well, the money's about the same, but it sure will be interesting. How are things with you?" Naomi had fond memories of working with Terry.

"Well, I graduate next month, and I have a job. Guess where?"

Naomi knew exactly where, Clem had told her about the interviews. It wasn't really fair to the other candidates—Clem had decided to hire Terry as soon as he had applied. "Here? Really?"

PROBLEMS

For additional practice, please see the problems (with selected solutions) provided on the Student CD-ROM that accompanies this book.

13.1 The table on the next page shows information about alternative choices. Criteria C and E are to be minimized, while all the rest are to be maximized. Which of the alternatives can be eliminated from further consideration?

	Criteria				
Alternative	A	B	C	D	E
1	340	5	11	1.2	1
2	570	8	22	3.3	1
3	410	9	22	3.2	2
4	120	4	36	0.9	3
5	122	1	46	1.3	2
6	345	8	47	0.6	3
7	119	4	57	1.1	2
8	554	2	89	2.1	3
9	317	9	117	0.9	1
10	129	5	165	1.5	3

13.2 The following is a partially completed pairwise comparison matrix. Complete it. What are the corresponding priority weights?

$$\begin{bmatrix} - & \frac{1}{2} & - & \frac{1}{9} \\ - & - & - & - \\ 4 & 2 & - & \frac{1}{2} \\ - & 4 & - & - \end{bmatrix}$$

 13.3 A large city's transit commission is considering building a new subway line. Twelve alternatives are being considered. All of the alternatives are shown in the tables, with their criteria values. The relevant criteria are:

C1   Population and jobs served per kilometre

C2   Projected daily traffic per kilometre

C3   Capital cost per kilometre (in millions of dollars)

C4   IRR

C5   Structural effect on urbanization

It is desirable to have high population served, high traffic, low capital cost per kilometre, and a high IRR. Criterion C5 concerns the benefits for urban growth caused by the subway location, and is measured on a scale from 0 to 10, with the higher values being preferred.

	Alternative Subway Lines					
Criterion	L1	L2	L3	L4	L5	L6
C1	81 900	31 800	11 500	31 100	23 000	16 100
C2	25 500	11 600	7 100	10 500	10 200	3 500
C3	65	45	29	35	40	10
C4	8.6	6.3	4.5	14.1	13	11.8
C5	3	7	4	6	5	4

	Alternative Subway Lines					
Criterion	**L7**	**L8**	**L9**	**L10**	**L11**	**L12**
C1	13200	28 200	36 500	24 400	18 400	13 900
C2	3 700	7 400	10 300	7 100	4 700	3 100
C3	32	30	13.2	40	43	25
C4	3.9	6	3	3.7	3.7	5.8
C5	9	10	6	9	5	4

(a) Which of these alternatives are efficient?

(b) Establish ratings for each of the efficient alternatives from part (a) for a decision matrix through normalization. The weights of the five criteria are C1: 1.5, C2: 2, C3: 2.5, C4: 3, C5: 1. Construct a decision matrix and determine the best subway route.

(c) If the criteria weights were C1: 2, C2: 2, C3: 2, C4: 2, C5: 2, would the recommended alternative be different?

13.4 Selective Steel is considering buying a new CNC punch press. They put high value on the reliability of this equipment, since it will be central to their production process. Speed and quality are also important to them, but not as important as reliability. As well as examining these factors, the company will make sure that the equipment will be easily adaptable to changes in production.

Construct a decision matrix for the alternatives listed and appropriate criteria. Select appropriate weightings for the criteria (state any assumptions), and determine the preferred alternative. Do a reasonable sensitivity analysis to determine the conditions under which the choice of alternative may change. A spreadsheet program should be used.

The alternatives are as follows:

1. Name: Accumate Plasmapress
   Cost: $428 600
   Service: Average
   Reliability: Average
   Speed: High
   Quality: Good
   Flexibility: Excellent
   Size: Average
   Other: None

2. Name: Weissman Model 4560
   Cost: $383 765
   Service: Below average
   Reliability: Average
   Speed: High
   Quality: Fair
   Flexibility: Not good
   Size: Average
   Other: Many tool stations (desirable), but small tools (not desirable)

3. Name: A. D. Hockley Model 661-84
    Cost: $533 725
    Service: Untested
    Reliability: Average
    Speed: Slow
    Quality: Very good
    Flexibility: Very good
    Size: Very compact
    Other: Small turret (not desirable)

4. Name: Frammit Manu-Centre 1500/45
    Cost: $393 000
    Service: Average
    Reliability: Below average
    Speed: Average
    Quality: Average
    Flexibility: Average
    Size: Average
    Other: Has perforating and coining feature (very desirable) Poor torch design causes too rapid wear (undesirable)

5. Name: Frammit Manu-Centre Lasertool 1250/30/1500
    Cost: $340 056
    Service: Average
    Reliability: Average
    Speed: Exceptionally fast
    Quality: Exceptionally good
    Flexibility: Average
    Size: Undesirably small
    Other: Cannot handle heavy-gauge metal (not desirable)

6. Name: Boxcab 3025/12P CNC
    Cost: $405 232
    Service: Untested
    Reliability: Excellent
    Speed: Average
    Quality: Very good
    Flexibility: Average
    Size: Average
    Other: None

13.5 Complete each of the following pairwise comparison matrixes:

(a)
$$\begin{bmatrix} - & 1 & - & - \\ - & - & - & 2 \\ \frac{1}{5} & 3 & - & - \\ 3 & - & \frac{1}{2} & - \end{bmatrix}$$

(b)

$$
\begin{bmatrix}
- & 9 & - & 1 & 3 \\
- & - & 3 & - & - \\
\frac{1}{7} & - & - & \frac{1}{4} & 2 \\
- & 6 & - & - & - \\
- & 1 & - & 2 & -
\end{bmatrix}
$$

13.6 For each of the PCMs in Problem 13.5, compute priority weights for the alternatives.

13.7 (a) Francis has several job opportunities for his co-op work term. He would like a job with good pay that is close to home, contributes to his engineering studies, and is with a smaller company. Which of the opportunities listed below should be removed from further consideration?

Job		Pay	Home	Studies	Size
			Criteria		
1.	Spinoff Consulting	1700	2	3	5
2.	Nub Automotive	1600	5	3	500
3.	Soutel	2200	80	4	150
4.	Turbine Hydro	1800	100	3	3000
5.	Fitzsimon Associates	1700	100	1	20
6.	General Auto	2000	150	2	2500
7.	Ring Casper	2200	250	5	300
8.	Jones Mines	2700	500	3	20
9.	Resources, Inc.	2700	2000	2	40

Pay: Monthly salary in euros

Home: Distance from home in kilometres

Studies: Contribution to engineering studies, 0 = none, 5 = a lot

Size: Number of employees at that location

(b) Francis feels that the following weights can represent the importance of his four criteria:

Pay:	4.0
Home:	2.5
Studies:	2.0
Size:	1.5

Using normalization to establish the ratings, which job is best?

Problems 13.8 to 13.23 are based on the following situation.

John is considering the selection of a consultant to provide ongoing support for a computer system. John wishes to contract with one consultant only, based on the criteria of cost, reliability, familiarity with equipment, location, and quality. Cost is measured by the quoted daily rate. Reliability and quality are measured on a qualitative scale based on discussions with references and interviews with the consultants. Familiarity with equipment has three possibilities: none, some, or much. Location is measured by distance from the consultant's office to the plant site in kilometres.

The specific data for the consultants are as follows:

	Consultants				
Criterion	**A**	**B**	**C**	**D**	**E**
Cost	$500	$500	$450	$600	$400
Reliability	Good	Good	Excellent	Good	Fair
Familiarity	None	Some	Some	Much	Some
Location	3	1	5	2	1
Quality	Excellent	Good	Fair	Excellent	Good

13.8 Is any choice of consultant inefficient?

13.9 John can assign weights to the criteria as follows:

Criterion	**Weight**
Cost	2
Reliability	3
Familiarity	1
Location	1
Quality	3

Using the following tables to convert qualitative evaluations to numbers and using the formula for determining ratings by normalization, construct a decision matrix for choosing a consultant. Which consultant is best?

For Reliability and Quantity	
Description	**Value**
Excellent	5
Very Good	4
Good	3
Fair	2
Poor	1

For Familiarity	
Description	**Value**
None	1
Some	2
Much	3

13.10 Draw an AHP hierarchy for this problem.

13.11 John considers a cost difference of $100 or more to be of strong importance, and a difference of $50 to be of moderate importance. Construct a PCM for the criterion *cost*.

13.12 John considers the difference between good reliability and fair reliability to be of moderate importance, and the difference between good and excellent to be of strong importance. He considers the difference between fair and excellent to be of very strong importance. Construct a PCM for the criterion *reliability*.

13.13 John considers the difference between no familiarity with the computing equipment and some familiarity to be of strong importance, and the difference between none and much familiarity to be of extreme importance. He considers the difference between some and much to be of strong importance. Construct a PCM for the criterion *familiarity*.

13.14 John considers each kilometre of distance worth two units on the AHP value scale. For example, the value of the location of consultant D over consultant C is $(5 - 2) \times 2 = 6$, and a 6 would be placed in the fourth row, third column, of the PCM for the criterion *location*. Construct the complete PCM for the criterion *location*.

13.15 John considers the difference between fair quality and good quality to be of strong importance, and the difference between good and excellent to be of very strong importance. He considers the difference between fair and excellent to be of extreme importance. Construct a PCM for the criterion *quality*.

13.16 Calculate the priority weights for the criterion *cost* from the PCM constructed in Problem 13.11.

13.17 Calculate the priority weights for the criterion *reliability* from the PCM constructed in Problem 13.12.

13.18 Calculate the priority weights for the criterion *familiarity* from the PCM constructed in Problem 13.13.

13.19 Calculate the priority weights for the criterion *location* from the PCM constructed in Problem 13.14.

13.20 Calculate the priority weights for the criterion *quality* from the PCM constructed in Problem 13.15.

13.21 Construct the priority matrix from the answers to Problems 13.16 to 13.20.

13.22 A partially completed PCM for the criteria is shown below. Complete the PCM and calculate the priority weights for the criteria.

$$
\begin{bmatrix}
- & \frac{1}{3} & - & 3 & - \\
- & - & - & - & 1 \\
\frac{1}{3} & \frac{1}{7} & - & 1 & - \\
- & \frac{1}{6} & - & - & - \\
1 & - & 7 & 7 & -
\end{bmatrix}
$$

13.23 Determine the overall score for each alternative. Which consultant is best?

Problems 13.24 to 13.31 are based on the following case.

Fabian has several job opportunities for his co-op work term. The pay is in dollars per month. The distance from home is in kilometres. Relevance to studies is on a five-point scale, 5 meaning very relevant to studies and 0 meaning not relevant at all. The company size refers to the number of employees at the job location.

In general, Fabian wants a job with good pay that is close to home, contributes to his engineering studies, and is with a smaller company.

	Job	Pay	Criteria		
			Distance From Home	Relevance to Studies	Company Size
1.	Spinoff Consulting	1700	2	3	5
2.	Soutel	2200	80	4	150
3.	Ring Casper	2200	250	5	300
4.	Jones Mines	2700	500	3	20

13.24 Draw an AHP hierarchy for this problem.

13.25 (a) Complete the PCM below for *pay*.

$$
\begin{bmatrix}
- & \frac{1}{5} & \frac{1}{5} & \frac{1}{7} \\
- & - & 1 & \frac{1}{5} \\
- & - & - & \frac{1}{5} \\
- & - & - & -
\end{bmatrix}
$$

(b) What are the priority weights for *pay*?

13.26 (a) Complete the PCM below for *distance from home*.

$$
\begin{bmatrix}
- & - & - & - \\
\frac{1}{5} & - & - & - \\
\frac{1}{7} & \frac{1}{5} & - & - \\
\frac{1}{9} & \frac{1}{7} & \frac{1}{2} & -
\end{bmatrix}
$$

(b) What are the priority weights for *distance from home*?

13.27 (a) Complete the PCM below for *studies*.

$$
\begin{bmatrix}
- & - & \frac{1}{3} & - \\
2 & - & - & 2 \\
- & 2 & - & - \\
1 & - & \frac{1}{3} & -
\end{bmatrix}
$$

(b) What are the priority weights for *studies*?

13.28 (a) Complete the PCM below for *size*.

$$
\begin{bmatrix}
- & - & - & 1 \\
\frac{1}{7} & - & - & \frac{1}{7} \\
\frac{1}{9} & \frac{1}{2} & - & \frac{1}{9} \\
- & - & - & -
\end{bmatrix}
$$

(b) What are the priority weights for *size*?

13.29 Form a priority matrix for the PCMs in Problems 13.25 to 13.28.

13.30 Given the following PCM for the criteria, calculate the priority weights for the criteria.

$$
\begin{bmatrix}
1 & 3 & 5 & 5 \\
\frac{1}{3} & 1 & 3 & 3 \\
\frac{1}{5} & \frac{1}{3} & 1 & 2 \\
\frac{1}{5} & \frac{1}{3} & \frac{1}{2} & 1
\end{bmatrix}
$$

13.31 Using the results of Problems 13.29 and 13.30, calculate the priority weights for the goal. Which job is Fabian's best choice? His second best choice?

More Challenging Problem

13.32 Consider a case of two alternatives being evaluated on the basis of two criteria. An AHP analysis reveals that alternative 2 is the clearly preferred choice over alternative 1. However, a new possibility, alternative 3, is brought to your attention. Alternative 3 has almost exactly the same characteristics on the two criteria under consideration as alternative 2. When alternative 3 is added to your AHP analysis, is it possible that alternative 1 could become the preferred choice? Prove your answer. (*Hint:* Use the following simple example for the two-alternative case, and then add a third alternative that is identical to the second one.)

PCM for criterion 1:
$$
\begin{bmatrix}
1 & \frac{1}{3} \\
3 & 1
\end{bmatrix}
$$

PCM for criterion 2:
$$
\begin{bmatrix}
1 & 3 \\
\frac{1}{3} & 1
\end{bmatrix}
$$

PCM for goal:
$$
\begin{bmatrix}
1 & 3 \\
\frac{1}{3} & 1
\end{bmatrix}
$$

MINI-CASE 13.1

Northwind Stoneware

Northwind Stoneware makes consumer stoneware products. Stoneware is fired in a kiln, which is an enclosure made of a porous brick having heating elements designed to raise the internal temperature to over 1200°C. Clay items such as stoneware can be hardened by firing them, following a particular temperature pattern called a firing curve.

Quality and cost problems led Northwind to examine better ways to control the firing curve. Alternatives available to them included

1. Direct human control (the temperature sensitivity of the human eye cannot be matched by any automatic control)

2. Use of a KilnSitter, a mechanical switch that shuts off the kiln at a preset temperature

3. Use of a pyrometer, an electrical instrument for measuring heat, with a programmable controller

4. Use of a pyrometer and a computer

The criteria used to determine the best choice included installation cost, effectiveness, reliability, energy savings, maintenance costs, and other applications.

A decision matrix evaluation method was used. Under a variety of criteria weights, the use of a pyrometer and a computer was the recommended choice. The exception was when installation cost was given overwhelming weight; in this case the use of a KilnSitter was recommended.

Discussion

Real life is always more complicated than any model used for analysis. In reading about a case in which decisions are made, the complexity of the real decision process is always hidden, because the very process of describing the situation is itself a model. In describing something, choices are made about what is important and worth describing and what is unimportant and not worth describing. In real life, one has to go through a process of separating from a great mass of information exactly what is important and what is not important.

The process of solving the quality and cost problems for Northwind Stoneware first involved collecting a great deal of information about the manufacturing process. Many people were interviewed, production was observed, and technical details of stoneware chemistry were researched. Market analysis was required to identify possible solutions to the process. Several person-weeks of work were required to gather the information to create the decision matrix mentioned above.

We often concentrate on the mechanics of using a decision tool instead of the mechanics of getting good data to use as input to the tool. In many cases, getting good data is 90% of the solution.

Questions

Think of an everyday decision situation that you have been involved in recently: it might be deciding which novel to buy, which apartment to rent, or which movie to go to.

1. Write down a one-page description of the decision. Include where you were, what choices you had, what you chose to do, and why.

2. Think about how you decided what to write for Question 1. Was it easy to know what to include in your description, and what not to include?

3. Can you immediately identify alternatives and criteria, as formally defined in this chapter, from your one-page description? If not, why not?

4. Construct a formal decision matrix model for your problem, augmenting the alternatives and criteria if necessary. Fill in the weights and ratings. Comment on any difficulties you have coming up with the weights and ratings.

5. Calculate the overall score for your decision matrix. Did you actually choose the one with the highest score? If not, why not?

Appendix 13A   Calculating the Consistency Ratio for AHP

The consistency ratio (*CR*) for AHP provides a measure of the ability of a decision maker to report preferences over alternatives or criteria. Calculating the *CR* can be done without understanding the concepts underlying it, but some background can be helpful. In this appendix, a brief overview of the basis for the *CR* is given, but the main purpose is to present an algorithm for calculating the *CR*. For the background information, some understanding of linear algebra is desirable, but the algorithm for calculating the *CR* can be easily followed without this background information.

Recall that in AHP a pairwise comparison matrix (PCM) is developed by comparing, for example, the alternatives with respect to one criterion. If alternative x is compared with y and found to be twice as preferred, and y is compared with z and is three times as preferred, then it is easy to deduce that x should be six times as preferred as z. However, in AHP every pairwise comparison is made giving several judgments about the same relationship. Humans will not necessarily be perfectly consistent in reporting preferences; one of the strengths of AHP is that, by getting redundant preference information, the quality of the information is improved.

Observe that if a decision maker were perfectly consistent, the columns of a PCM would be multiples of each other. They would differ in scale because for each column n, the nth row is fixed as 1, but the relative values would be constant.

Recall that an eigenvalue λ is one of n solutions to the equation

$$\mathbf{Aw} = \lambda\mathbf{w}$$

where $\mathbf{A}$ is a square $n \times n$ matrix and $\mathbf{w}$ is an $n \times 1$ eigenvector.

There is one eigenvector corresponding to each eigenvalue λ. A PCM is unusual in that, in addition to being a square matrix, all of the entries are positive, the corresponding entries across the main diagonal are reciprocals, and there are 1s along the main diagonal. It can be shown that, in this case, if the decision maker is perfectly consistent, there will be a single non-zero eigenvalue, and $n - 1$ eigenvalues of value 0.

In practice, a PCM is not perfectly consistent. However, assuming that the error is relatively small, there should be one large eigenvalue λ_{max} and $n - 1$ small ones. Further, it can be shown that, with small inconsistencies, $\lambda_{max} > n$.

The *CR* is developed from the difference between λ_{max} and n. First, this difference is divided by $n - 1$, effectively distributing it over the other, supposedly zero, eigenvalues. This gives the consistency index (CI).

$$CI = \frac{\lambda_{max} - n}{n - 1}$$

Second, the *CI* is divided by the *CI* of a random matrix of the same size, called a *random index (RI)*, to form the consistency ratio (*CR*). The *RI* for matrixes of up to 10 rows and columns is shown in Table 13A.1; these were developed by averaging the *CI*s for hundreds of randomly generated matrixes.

$$CR = \frac{CI}{RI}$$

The idea is that, if the *PCM* were completely random, we would expect the *CI* to be equal to the *RI*. Thus, a random *PCM* would have a *CR* of 1. However, a consistent *PCM*, having a *CI* of 0, would also have a *CR* of 0. *CR*s of between 0 and 1 indicate

Table 13A.1 Random Indexes for Various Sizes of Matrices

Size ($n \times n$)	Random Index
2	0
3	0.58
4	0.90
5	1.12
6	1.24
7	1.32
8	1.41
9	1.45
10	1.49

how consistent or random the PCM is, in a manner that is independent of the size of the matrix.

The *CR* is actually a well designed statistical measure of the deviation of a particular PCM from perfect consistency. As a common rule, an upper limit of 0.1 is usually used. Thus, if $CR \leq 0.1$, the PCM is *acceptably* consistent. If the $CR > 0.1$, the PCM should be re-evaluated by the decision maker.

This background can be used to construct an algorithm for calculating the *CR* for any PCM. Essentially, one determines λ_{max} and its associated eigenvector, called the **principal eigenvector**. The general algorithm for a PCM **A** of size $n \times n$ is as follows:

1. Find **w**, the eigenvector associated with λ_{max}, from normalizing the following:

$$\mathbf{w} = \lim_{k \to \infty} \frac{\mathbf{A}^k \mathbf{e}}{n}$$

where **e** is a column vector $[111 \ldots 111]^T$.

In other words, form a sequence of powers of the matrix **A**. Normalize each of the resulting matrixes by dividing each element of **A** by the sum of the elements from the corresponding column. Form **w** by summing each row of normalized **A** and dividing this by n. Eventually, **w** will not noticeably change from one power of **A** to the next-higher power. This value of **w** is the desired principal eigenvector.

In the main part of this chapter, we calculated the priority weights for a PCM. That procedure was an approximate method of determining **w**; the priority weights and the eigenvector associated with λ_{max} are the same thing. The procedure mentioned here is more accurate, but more difficult to compute and time-consuming.

2. Since **w** is a vector, λ_{max} can be found by solving $\mathbf{Aw} = \lambda_{max}\mathbf{w}$ for any element of **w**, or equivalently, any row of **A**. Thus, for any i, compute

$$\lambda_{max} = \frac{\sum_{j} (a_{ij} \times w_j)}{w_i}$$

3. Calculate the *CI* from

$$CI = \frac{\lambda_{\max} - n}{n - 1}$$

4. With reference to Table 13A.1, calculate the CR from

$$CR = \frac{CI}{RI}$$

If $CR \leq 0.1$, the *PCM* is acceptably consistent.

EXAMPLE 13A.1

Does the PCM of Figure 13.4 meet the requirement for a consistency ratio less than or equal to 0.1?

The PCM of Figure 13.4 is reproduced as Figure 13A.1, with successive powers of the original matrix, normalized versions of the powers, and the calculated **w**. The calculations were done using a spreadsheet program; most popular spreadsheet programs can automatically find powers of matrixes.

It can be seen that **w** converges very quickly to a stable set of values. Normally, the less consistent a PCM is, the longer it will take to converge. Also, note that the elements of **w** are very close to the priority weights that were calculated for this PCM in Section 13.4.

To determine $\lambda_{\max}$, we must now solve $\mathbf{Aw} = \lambda_{\max}\mathbf{w}$ for one row of **A**. Selecting the first row of **A** results in the following expression:

$$[1\ 3\ 5\ 1\ 3] \begin{bmatrix} 0.343 \\ 0.129 \\ 0.055 \\ 0.343 \\ 0.129 \end{bmatrix} = \lambda_{\max} \times 0.343$$

Or, equivalently,

$$\lambda_{\max} = \frac{(1 \times 0.343) + (3 \times 0.129) + (5 \times 0.055) + (1 \times 0.343) + (3 \times 0.129)}{0.343}$$

$$= 5.0583$$

The consistency index can then be calculated from

$$CI = \frac{\lambda_{\max} - n}{n - 1}$$

$$= \frac{5.0583 - 5}{5 - 1}$$

$$= 0.0146$$

Figure 13A.1   Calculating the Principal Eigenvector w

	A						Normalized						w
	1	3	5	1	3		0.349	0.36	0.29	0.349	0.36		0.342
	0.333	1	3	0.333	1		0.116	0.12	0.18	0.116	0.12		0.130
A^1	0.2	0.333	1	0.2	0.333		0.07	0.04	0.06	0.07	0.04		0.056
	1	3	5	1	3		0.349	0.36	0.29	0.349	0.36		0.342
	0.333	1	3	0.33	1		0.116	0.12	0.18	0.116	0.12		0.130
	5	13.67	33	5	13.667		0.34	0.346	0.35	0.34	0.346		0.343
	1.933	5	13.3	1.93	5		0.132	0.126	0.13	0.132	0.126		0.129
A^2	0.822	2.2	5	0.82	2.2		0.056	0.056	0.05	0.056	0.056		0.055
	5	13.67	33	5	13.667		0.34	0.346	0.35	0.34	0.346		0.343
	1.933	5	13.3	1.93	5		0.132	0.126	0.13	0.132	0.126		0.129
	25.711	68.33	165	25.7	68.333		0.343	0.343	0.34	0.343	0.343		0.343
	9.667	25.71	61.7	9.67	25.711		0.129	0.129	0.13	0.129	0.129		01.29
A^3	4.111	1	26.4	4.11	11		0.055	0.055	0.06	0.055	0.055		0.055
	25.711	68.33	165	25.7	68.333		0.343	0.343	0.34	0.343	0.343		0.343
	9.667	25.71	61.7	9.67	25.711		0.129	0.129	0.13	0.129	0.129		01.29
	129.98	345.9	832	130	345.93		0.343	0.343	0.34	0.343	0.343		0.349
	48.807	130	313	48.8	129.98		0.129	0.129	0.13	0.129	0.129		0.129
A^4	20.84	55.47	134	20.8	55.474		0.055	0.055	0.06	0.055	0.055		0.055
	129.98	345.9	832	130	345.93		0.343	0.343	0.34	0.343	0.343		0.349
	48.807	130	313	48.8	129.98		0.129	0.129	0.13	0.129	0.129		0.129
	657	1749	4207	657	1749.1		0.343	0.343	0.34	0.343	0.343		0.343
	246.79	657	1581	247	657		0.129	0.129	0.13	0.129	0.129		0.129
A^5	105.37	280.5	675	105	280.5		0.055	0.055	0.06	0.055	0.055		0.055
	657	1749	4207	657	1749.1		0.343	0.343	0.34	0.343	0.343		0.343
	246.79	657	1581	247	657		0.129	0.129	0.13	0.129	0.129		0.129

As seen in Table 13A.1, the *RI* for a matrix of five rows and columns is 1.12. We can then calculate the consistency ratio as

$$CR = \frac{CI}{RI} = \frac{0.0146}{1.12} = 0.013$$

Clearly, the *CR* is very much less than 0.1. It can thus be concluded that the original PCM is acceptably consistent.■

PROBLEMS FOR APPENDIX 13A

 13A.1 Does the following PCM meet the requirement for a consistency ratio less than or equal to 0.1? What are the values of the consistency index and the consistency ratio?

$$
\begin{bmatrix}
1 & \frac{1}{2} & \frac{1}{4} & \frac{1}{9} \\
2 & 1 & \frac{1}{2} & \frac{1}{4} \\
4 & 3 & 1 & 1 \\
9 & 7 & 1 & 1
\end{bmatrix}
$$

 13A.2 Does the following PCM meet the requirement for a consistency ratio less than or equal to 0.1? What are the values of the consistency index and the consistency ratio?

$$
\begin{bmatrix}
1 & \frac{1}{2} & \frac{1}{4} & \frac{1}{9} \\
2 & 1 & \frac{1}{2} & \frac{1}{4} \\
4 & 3 & 1 & 1 \\
9 & 7 & 1 & 1
\end{bmatrix}
$$

 13A.3 Does the following PCM meet the requirement for a consistency ratio less than or equal to 0.1? What are the values of the consistency index and the consistency ratio? How does an accurate evaluation of the principal eigenvector for this PCM compare with the priority weights calculated in Problem 13.30?

$$
\begin{bmatrix}
1 & 3 & 5 & 5 \\
\frac{1}{3} & 1 & 3 & 3 \\
\frac{1}{5} & \frac{1}{3} & 1 & 2 \\
\frac{1}{5} & \frac{1}{3} & \frac{1}{2} & 1
\end{bmatrix}
$$

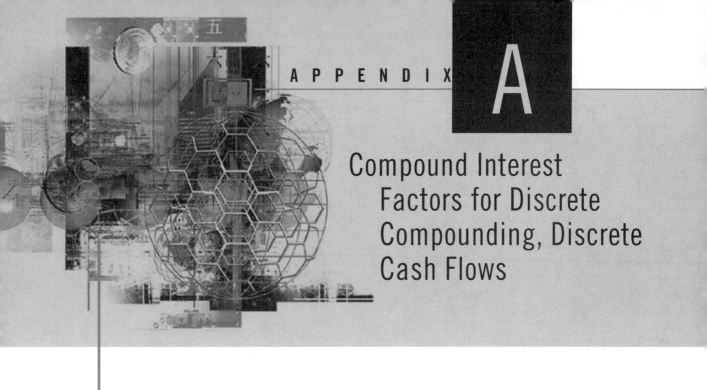

APPENDIX A

Compound Interest Factors for Discrete Compounding, Discrete Cash Flows

$i = 0.5\%$ Discrete Compounding, Discrete Cash Flows

	SINGLE PAYMENT		UNIFORM SERIES				Arithmetic Gradient Series Factor
	Compound Amount Factor	Present Worth Factor	Sinking Fund Factor	Uniform Series Factor	Capital Recovery Factor	Series Present Worth Factor	
N	(F/P,i,N)	(P/F,i,N)	(A/F,i,N)	(F/A,i,N)	(A/P,i,N)	(P/A,i,N)	(A/G,i,N)
1	1.0050	0.99502	1.0000	1.0000	1.0050	0.99502	0.00000
2	1.0100	0.99007	0.49875	2.0050	0.50375	1.9851	0.49875
3	1.0151	0.98515	0.33167	3.0150	0.33667	2.9702	0.99667
4	1.0202	0.98025	0.24813	4.0301	0.25313	3.9505	1.4938
5	1.0253	0.97537	0.19801	5.0503	0.20301	4.9259	1.9900
6	1.0304	0.97052	0.16460	6.0755	0.16960	5.8964	2.4855
7	1.0355	0.96569	0.14073	7.1059	0.14573	6.8621	2.9801
8	1.0407	0.96089	0.12283	8.1414	0.12783	7.8230	3.4738
9	1.0459	0.95610	0.10891	9.1821	0.11391	8.7791	3.9668
10	1.0511	0.95135	0.09777	10.228	0.10277	9.7304	4.4589
11	1.0564	0.94661	0.08866	11.279	0.09366	10.677	4.9501
12	1.0617	0.94191	0.08107	12.336	0.08607	11.619	5.4406
13	1.0670	0.93722	0.07464	13.397	0.07964	12.556	5.9302
14	1.0723	0.93256	0.06914	14.464	0.07414	13.489	6.4190
15	1.0777	0.92792	0.06436	15.537	0.06936	14.417	6.9069
16	1.0831	0.92330	0.06019	16.614	0.06519	15.340	7.3940
17	1.0885	0.91871	0.05651	17.697	0.06151	16.259	7.8803
18	1.0939	0.91414	0.05323	18.786	0.05823	17.173	8.3658
19	1.0994	0.90959	0.05030	19.880	0.05530	18.082	8.8504
20	1.1049	0.90506	0.04767	20.979	0.05267	18.987	9.3342
21	1.1104	0.90056	0.04528	22.084	0.05028	19.888	9.8172
22	1.1160	0.89608	0.04311	23.194	0.04811	20.784	10.299
23	1.1216	0.89162	0.04113	24.310	0.04613	21.676	10.781
24	1.1272	0.88719	0.03932	25.432	0.04432	22.563	11.261
25	1.1328	0.88277	0.03765	26.559	0.04265	23.446	11.741
26	1.1385	0.87838	0.03611	27.692	0.04111	24.324	12.220
27	1.1442	0.87401	0.03469	28.830	0.03969	25.198	12.698
28	1.1499	0.86966	0.03336	29.975	0.03836	26.068	13.175
29	1.1556	0.86533	0.03213	31.124	0.03713	26.933	13.651
30	1.1614	0.86103	0.03098	32.280	0.03598	27.794	14.126
31	1.1672	0.85675	0.02990	33.441	0.03490	28.651	14.601
32	1.1730	0.85248	0.02889	34.609	0.03389	29.503	15.075
33	1.1789	0.84824	0.02795	35.782	0.03295	30.352	15.548
34	1.1848	0.84402	0.02706	36.961	0.03206	31.196	16.020
35	1.1907	0.83982	0.02622	38.145	0.03122	32.035	16.492
40	1.2208	0.81914	0.02265	44.159	0.02765	36.172	18.836
45	1.2516	0.79896	0.01987	50.324	0.02487	40.207	21.159
50	1.2832	0.77929	0.01765	56.645	0.02265	44.143	23.462
55	1.3156	0.76009	0.01584	63.126	0.02084	47.981	25.745
60	1.3489	0.74137	0.01433	69.770	0.01933	51.726	28.006
65	1.3829	0.72311	0.01306	76.582	0.01806	55.377	30.247
70	1.4178	0.70530	0.01197	83.566	0.01697	58.939	32.468
75	1.4536	0.68793	0.01102	90.727	0.01602	62.414	34.668
80	1.4903	0.67099	0.01020	98.068	0.01520	65.802	36.847
85	1.5280	0.65446	0.00947	105.59	0.01447	69.108	39.006
90	1.5666	0.63834	0.00883	113.31	0.01383	72.331	41.145
95	1.6061	0.62262	0.00825	121.22	0.01325	75.476	43.263
100	1.6467	0.60729	0.00773	129.33	0.01273	78.543	45.361

$i = 1\%$ **Discrete Compounding, Discrete Cash Flows**

	SINGLE PAYMENT		UNIFORM SERIES				Arithmetic Gradient Series Factor
	Compound Amount Factor	Present Worth Factor	Sinking Fund Factor	Uniform Series Factor	Capital Recovery Factor	Series Present Worth Factor	
N	(F/P,i,N)	(P/F,i,N)	(A/F,i,N)	(F/A,i,N)	(A/P,i,N)	(P/A,i,N)	(A/G,i,N)
1	1.0100	0.99010	1.0000	1.0000	1.0100	0.99010	0.00000
2	1.0201	0.98030	0.49751	2.0100	0.50751	1.9704	0.49751
3	1.0303	0.97059	0.33002	3.0301	0.34002	2.9410	0.99337
4	1.0406	0.96098	0.24628	4.0604	0.25628	3.9020	1.4876
5	1.0510	0.95147	0.19604	5.1010	0.20604	4.8534	1.9801
6	1.0615	0.94205	0.16255	6.1520	0.17255	5.7955	2.4710
7	1.0721	0.93272	0.13863	7.2135	0.14863	6.7282	2.9602
8	1.0829	0.92348	0.12069	8.2857	0.13069	7.6517	3.4478
9	1.0937	0.91434	0.10674	9.3685	0.11674	8.5660	3.9337
10	1.1046	0.90529	0.09558	10.462	0.10558	9.4713	4.4179
11	1.1157	0.89632	0.08645	11.567	0.09645	10.368	4.9005
12	1.1268	0.88745	0.07885	12.683	0.08885	11.255	5.3815
13	1.1381	0.87866	0.07241	13.809	0.08241	12.134	5.8607
14	1.1495	0.86996	0.06690	14.947	0.07690	13.004	6.3384
15	1.1610	0.86135	0.06212	16.097	0.07212	13.865	6.8143
16	1.1726	0.85282	0.05794	17.258	0.06794	14.718	7.2886
17	1.1843	0.84438	0.05426	18.430	0.06426	15.562	7.7613
18	1.1961	0.83602	0.05098	19.615	0.06098	16.398	8.2323
19	1.2081	0.82774	0.04805	20.811	0.05805	17.226	8.7017
20	1.2202	0.81954	0.04542	22.019	0.05542	18.046	9.1694
21	1.2324	0.81143	0.04303	23.239	0.05303	18.857	9.6354
22	1.2447	0.80340	0.04086	24.472	0.05086	19.660	10.100
23	1.2572	0.79544	0.03889	25.716	0.04889	20.456	10.563
24	1.2697	0.78757	0.03707	26.973	0.04707	21.243	11.024
25	1.2824	0.77977	0.03541	28.243	0.04541	22.023	11.483
26	1.2953	0.77205	0.03387	29.526	0.04387	22.795	11.941
27	1.3082	0.76440	0.03245	30.821	0.04245	23.560	12.397
28	1.3213	0.75684	0.03112	32.129	0.04112	24.316	12.852
29	1.3345	0.74934	0.02990	33.450	0.03990	25.066	13.304
30	1.3478	0.74192	0.02875	34.785	0.03875	25.808	13.756
31	1.3613	0.73458	0.02768	36.133	0.03768	26.542	14.205
32	1.3749	0.72730	0.02667	37.494	0.03667	27.270	14.653
33	1.3887	0.72010	0.02573	38.869	0.03573	27.990	15.099
34	1.4026	0.71297	0.02484	40.258	0.03484	28.703	15.544
35	1.4166	0.70591	0.02400	41.660	0.03400	29.409	15.987
40	1.4889	0.67165	0.02046	48.886	0.03046	32.835	18.178
45	1.5648	0.63905	0.01771	56.481	0.02771	36.095	20.327
50	1.6446	0.60804	0.01551	64.463	0.02551	39.196	22.436
55	1.7285	0.57853	0.01373	72.852	0.02373	42.147	24.505
60	1.8167	0.55045	0.01224	81.670	0.02224	44.955	26.533
65	1.9094	0.52373	0.01100	90.937	0.02100	47.627	28.522
70	2.0068	0.49831	0.00993	100.68	0.01993	50.169	30.470
75	2.1091	0.47413	0.00902	110.91	0.01902	52.587	32.379
80	2.2167	0.45112	0.00822	121.67	0.01822	54.888	34.249
85	2.3298	0.42922	0.00752	132.98	0.01752	57.078	36.080
90	2.4486	0.40839	0.00690	144.86	0.01690	59.161	37.872
95	2.5735	0.38857	0.00636	157.35	0.01636	61.143	39.626
100	2.7048	0.36971	0.00587	170.48	0.01587	63.029	41.343

$i = 1.5\%$ **Discrete Compounding, Discrete Cash Flows**

	SINGLE PAYMENT		UNIFORM SERIES				Arithmetic Gradient Series Factor
	Compound Amount Factor	Present Worth Factor	Sinking Fund Factor	Uniform Series Factor	Capital Recovery Factor	Series Present Worth Factor	
N	(F/P,i,N)	(P/F,i,N)	(A/F,i,N)	(F/A,i,N)	(A/P,i,N)	(P/A,i,N)	(A/G,i,N)
1	1.0150	0.98522	1.0000	1.0000	1.0150	0.98522	0.00000
2	1.0302	0.97066	0.49628	2.0150	0.51128	1.9559	0.49628
3	1.0457	0.95632	0.32838	3.0452	0.34338	2.9122	0.99007
4	1.0614	0.94218	0.24444	4.0909	0.25944	3.8544	1.4814
5	1.0773	0.92826	0.19409	5.1523	0.20909	4.7826	1.9702
6	1.0934	0.91454	0.16053	6.2296	0.17553	5.6972	2.4566
7	1.1098	0.90103	0.13656	7.3230	0.15156	6.5982	2.9405
8	1.1265	0.88771	0.11858	8.4328	0.13358	7.4859	3.4219
9	1.1434	0.87459	0.10461	9.5593	0.11961	8.3605	3.9008
10	1.1605	0.86167	0.09343	10.703	0.10843	9.2222	4.3772
11	1.1779	0.84893	0.08429	11.863	0.09929	10.071	4.8512
12	1.1956	0.83639	0.07668	13.041	0.09168	10.908	5.3227
13	1.2136	0.82403	0.07024	14.237	0.08524	11.732	5.7917
14	1.2318	0.81185	0.06472	15.450	0.07972	12.543	6.2582
15	1.2502	0.79985	0.05994	16.682	0.07494	13.343	6.7223
16	1.2690	0.78803	0.05577	17.932	0.07077	14.131	7.1839
17	1.2880	0.77639	0.05208	19.201	0.06708	14.908	7.6431
18	1.3073	0.76491	0.04881	20.489	0.06381	15.673	8.0997
19	1.3270	0.75361	0.04588	21.797	0.06088	16.426	8.5539
20	1.3469	0.74247	0.04325	23.124	0.05825	17.169	9.0057
21	1.3671	0.73150	0.04087	24.471	0.05587	17.900	9.4550
22	1.3876	0.72069	0.03870	25.838	0.05370	18.621	9.9018
23	1.4084	0.71004	0.03673	27.225	0.05173	19.331	10.346
24	1.4295	0.69954	0.03492	28.634	0.04992	20.030	10.788
25	1.4509	0.68921	0.03326	30.063	0.04826	20.720	11.228
26	1.4727	0.67902	0.03173	31.514	0.04673	21.399	11.665
27	1.4948	0.66899	0.03032	32.987	0.04532	22.068	12.099
28	1.5172	0.65910	0.02900	34.481	0.04400	22.727	12.531
29	1.5400	0.64936	0.02778	35.999	0.04278	23.376	12.961
30	1.5631	0.63976	0.02664	37.539	0.04164	24.016	13.388
31	1.5865	0.63031	0.02557	39.102	0.04057	24.646	13.813
32	1.6103	0.62099	0.02458	40.688	0.03958	25.267	14.236
33	1.6345	0.61182	0.02364	42.299	0.03864	25.879	14.656
34	1.6590	0.60277	0.02276	43.933	0.03776	26.482	15.073
35	1.6839	0.59387	0.02193	45.592	0.03693	27.076	15.488
40	1.8140	0.55126	0.01843	54.268	0.03343	29.916	17.528
45	1.9542	0.51171	0.01572	63.614	0.03072	32.552	19.507
50	2.1052	0.47500	0.01357	73.683	0.02857	35.000	21.428
55	2.2679	0.44093	0.01183	84.530	0.02683	37.271	23.289
60	2.4432	0.40930	0.01039	96.215	0.02539	39.380	25.093
65	2.6320	0.37993	0.00919	108.80	0.02419	41.338	26.839
70	2.8355	0.35268	0.00817	122.36	0.02317	43.155	28.529
75	3.0546	0.32738	0.00730	136.97	0.02230	44.842	30.163
80	3.2907	0.30389	0.00655	152.71	0.02155	46.407	31.742
85	3.5450	0.28209	0.00589	169.67	0.02089	47.861	33.268
90	3.8189	0.26185	0.00532	187.93	0.02032	49.210	34.740
95	4.1141	0.24307	0.00482	207.61	0.01982	50.462	36.160
100	4.4320	0.22563	0.00437	228.80	0.01937	51.625	37.530

$i = 2\%$ Discrete Compounding, Discrete Cash Flows

	SINGLE PAYMENT		UNIFORM SERIES				Arithmetic Gradient Series Factor
	Compound Amount Factor	Present Worth Factor	Sinking Fund Factor	Uniform Series Factor	Capital Recovery Factor	Series Present Worth Factor	
N	(F/P,i,N)	(P/F,i,N)	(A/F,i,N)	(F/A,i,N)	(A/P,i,N)	(P/A,i,N)	(A/G,i,N)
1	1.0200	0.98039	1.0000	1.0000	1.0200	0.98039	0.00000
2	1.0404	0.96117	0.49505	2.0200	0.51505	1.9416	0.49505
3	1.0612	0.94232	0.32675	3.0604	0.34675	2.8839	0.98680
4	1.0824	0.92385	0.24262	4.1216	0.26262	3.8077	1.4752
5	1.1041	0.90573	0.19216	5.2040	0.21216	4.7135	1.9604
6	1.1262	0.88797	0.15853	6.3081	0.17853	5.6014	2.4423
7	1.1487	0.87056	0.13451	7.4343	0.15451	6.4720	2.9208
8	1.1717	0.85349	0.11651	8.5830	0.13651	7.3255	3.3961
9	1.1951	0.83676	0.10252	9.7546	0.12252	8.1622	3.8681
10	1.2190	0.82035	0.09133	10.950	0.11133	8.9826	4.3367
11	1.2434	0.80426	0.08218	12.169	0.10218	9.787	4.8021
12	1.2682	0.78849	0.07456	13.412	0.09456	10.575	5.2642
13	1.2936	0.77303	0.06812	14.680	0.08812	11.348	5.7231
14	1.3195	0.75788	0.06260	15.974	0.08260	12.106	6.1786
15	1.3459	0.74301	0.05783	17.293	0.07783	12.849	6.6309
16	1.3728	0.72845	0.05365	18.639	0.07365	13.578	7.0799
17	1.4002	0.71416	0.04997	20.012	0.06997	14.292	7.5256
18	1.4282	0.70016	0.04670	21.412	0.06670	14.992	7.9681
19	1.4568	0.68643	0.04378	22.841	0.06378	15.678	8.4073
20	1.4859	0.67297	0.04116	24.297	0.06116	16.351	8.8433
21	1.5157	0.65978	0.03878	25.783	0.05878	17.011	9.2760
22	1.5460	0.64684	0.03663	27.299	0.05663	17.658	9.7050
23	1.5769	0.63416	0.03467	28.845	0.05467	18.292	10.1320
24	1.6084	0.62172	0.03287	30.422	0.05287	18.914	10.5550
25	1.6406	0.60953	0.03122	32.030	0.05122	19.523	10.9740
26	1.6734	0.59758	0.02970	33.671	0.04970	20.121	11.391
27	1.7069	0.58586	0.02829	35.344	0.04829	20.707	11.804
28	1.7410	0.57437	0.02699	37.051	0.04699	21.281	12.214
29	1.7758	0.56311	0.02578	38.792	0.04578	21.844	12.621
30	1.8114	0.55207	0.02465	40.568	0.04465	22.396	13.025
31	1.8476	0.54125	0.02360	42.379	0.04360	22.938	13.426
32	1.8845	0.53063	0.02261	44.227	0.04261	23.468	13.823
33	1.9222	0.52023	0.02169	46.112	0.04169	23.989	14.217
34	1.9607	0.51003	0.02082	48.034	0.04082	24.499	14.608
35	1.9999	0.50003	0.02000	49.994	0.04000	24.999	14.996
40	2.2080	0.45289	0.01656	60.402	0.03656	27.355	16.889
45	2.4379	0.41020	0.01391	71.893	0.03391	29.490	18.703
50	2.6916	0.37153	0.01182	84.579	0.03182	31.424	20.442
55	2.9717	0.33650	0.01014	98.587	0.03014	33.175	22.106
60	3.2810	0.30478	0.00877	114.05	0.02877	34.761	23.696
65	3.6225	0.27605	0.00763	131.13	0.02763	36.197	25.215
70	3.9996	0.25003	0.00667	149.98	0.02667	37.499	26.663
75	4.4158	0.22646	0.00586	170.79	0.02586	38.677	28.043
80	4.8754	0.20511	0.00516	193.77	0.02516	39.745	29.357
85	5.3829	0.18577	0.00456	219.14	0.02456	40.711	30.606
90	5.9431	0.16826	0.00405	247.16	0.02405	41.587	31.793
95	6.5617	0.15240	0.00360	278.08	0.02360	42.380	32.919
100	7.2446	0.13803	0.00320	312.23	0.02320	43.098	33.986

$i = 3\%$ **Discrete Compounding, Discrete Cash Flows**

	SINGLE PAYMENT		UNIFORM SERIES				Arithmetic Gradient Series Factor
	Compound Amount Factor	Present Worth Factor	Sinking Fund Factor	Uniform Series Factor	Capital Recovery Factor	Series Present Worth Factor	
N	$(F/P,i,N)$	$(P/F,i,N)$	$(A/F,i,N)$	$(F/A,i,N)$	$(A/P,i,N)$	$(P/A,i,N)$	$(A/G,i,N)$
1	1.0300	0.97087	1.0000	1.0000	1.0300	0.97087	0.00000
2	1.0609	0.94260	0.49261	2.0300	0.52261	1.9135	0.49261
3	1.0927	0.91514	0.32353	3.0909	0.35353	2.8286	0.98030
4	1.1255	0.88849	0.23903	4.1836	0.26903	3.7171	1.4631
5	1.1593	0.86261	0.18835	5.3091	0.21835	4.5797	1.9409
6	1.1941	0.83748	0.15460	6.4684	0.18460	5.4172	2.4138
7	1.2299	0.81309	0.13051	7.6625	0.16051	6.2303	2.8819
8	1.2668	0.78941	0.11246	8.8923	0.14246	7.0197	3.3450
9	1.3048	0.76642	0.09843	10.159	0.12843	7.7861	3.8032
10	1.3439	0.74409	0.08723	11.464	0.11723	8.5302	4.2565
11	1.3842	0.72242	0.07808	12.808	0.10808	9.2526	4.7049
12	1.4258	0.70138	0.07046	14.192	0.10046	9.9540	5.1485
13	1.4685	0.68095	0.06403	15.618	0.09403	10.635	5.5872
14	1.5126	0.66112	0.05853	17.086	0.08853	11.296	6.0210
15	1.5580	0.64186	0.05377	18.599	0.08377	11.938	6.4500
16	1.6047	0.62317	0.04961	20.157	0.07961	12.561	6.8742
17	1.6528	0.60502	0.04595	21.762	0.07595	13.166	7.2936
18	1.7024	0.58739	0.04271	23.414	0.07271	13.754	7.7081
19	1.7535	0.57029	0.03981	25.117	0.06981	14.324	8.1179
20	1.8061	0.55368	0.03722	26.870	0.06722	14.877	8.5229
21	1.8603	0.53755	0.03487	28.676	0.06487	15.415	8.9231
22	1.9161	0.52189	0.03275	30.537	0.06275	15.937	9.3186
23	1.9736	0.50669	0.03081	32.453	0.06081	16.444	9.7093
24	2.0328	0.49193	0.02905	34.426	0.05905	16.936	10.095
25	2.0938	0.47761	0.02743	36.459	0.05743	17.413	10.477
26	2.1566	0.46369	0.02594	38.553	0.05594	17.877	10.853
27	2.2213	0.45019	0.02456	40.710	0.05456	18.327	11.226
28	2.2879	0.43708	0.02329	42.931	0.05329	18.764	11.593
29	2.3566	0.42435	0.02211	45.219	0.05211	19.188	11.956
30	2.4273	0.41199	0.02102	47.575	0.05102	19.600	12.314
31	2.5001	0.39999	0.02000	50.003	0.05000	20.000	12.668
32	2.5751	0.38834	0.01905	52.503	0.04905	20.389	13.017
33	2.6523	0.37703	0.01816	55.078	0.04816	20.766	13.362
34	2.7319	0.36604	0.01732	57.730	0.04732	21.132	13.702
35	2.8139	0.35538	0.01654	60.462	0.04654	21.487	14.037
40	3.2620	0.30656	0.01326	75.401	0.04326	23.115	15.650
45	3.7816	0.26444	0.01079	92.720	0.04079	24.519	17.156
50	4.3839	0.22811	0.00887	112.80	0.03887	25.730	18.558
55	5.0821	0.19677	0.00735	136.07	0.03735	26.774	19.860
60	5.8916	0.16973	0.00613	163.05	0.03613	27.676	21.067
65	6.8300	0.14641	0.00515	194.33	0.03515	28.453	22.184
70	7.9178	0.12630	0.00434	230.59	0.03434	29.123	23.215
75	9.1789	0.10895	0.00367	272.63	0.03367	29.702	24.163
80	10.641	0.09398	0.00311	321.36	0.03311	30.201	25.035
85	12.336	0.08107	0.00265	377.86	0.03265	30.631	25.835
90	14.300	0.06993	0.00226	443.35	0.03226	31.002	26.567
95	16.578	0.06032	0.00193	519.27	0.03193	31.323	27.235
100	19.219	0.05203	0.00165	607.29	0.03165	31.599	27.844

i = 4%   Discrete Compounding, Discrete Cash Flows

	SINGLE PAYMENT		UNIFORM SERIES				Arithmetic Gradient Series Factor
	Compound Amount Factor	Present Worth Factor	Sinking Fund Factor	Uniform Series Factor	Capital Recovery Factor	Series Present Worth Factor	
N	(F/P,i,N)	(P/F,i,N)	(A/F,i,N)	(F/A,i,N)	(A/P,i,N)	(P/A,i,N)	(A/G,i,N)
1	1.0400	0.96154	1.0000	1.0000	1.0400	0.96154	0.00000
2	1.0816	0.92456	0.49020	2.0400	0.53020	1.8861	0.49020
3	1.1249	0.88900	0.32035	3.1216	0.36035	2.7751	0.97386
4	1.1699	0.85480	0.23549	4.2465	0.27549	3.6299	1.4510
5	1.2167	0.82193	0.18463	5.4163	0.22463	4.4518	1.9216
6	1.2653	0.79031	0.15076	6.6330	0.19076	5.2421	2.3857
7	1.3159	0.75992	0.12661	7.8983	0.16661	6.0021	2.8433
8	1.3686	0.73069	0.10853	9.2142	0.14853	6.7327	3.2944
9	1.4233	0.70259	0.09449	10.583	0.13449	7.4353	3.7391
10	1.4802	0.67556	0.08329	12.006	0.12329	8.1109	4.1773
11	1.5395	0.64958	0.07415	13.486	0.11415	8.7605	4.6090
12	1.6010	0.62460	0.06655	15.026	0.10655	9.3851	5.0343
13	1.6651	0.60057	0.06014	16.627	0.10014	9.9856	5.4533
14	1.7317	0.57748	0.05467	18.292	0.09467	10.563	5.8659
15	1.8009	0.55526	0.04994	20.024	0.08994	11.118	6.2721
16	1.8730	0.53391	0.04582	21.825	0.08582	11.652	6.6720
17	1.9479	0.51337	0.04220	23.698	0.08220	12.166	7.0656
18	2.0258	0.49363	0.03899	25.645	0.07899	12.659	7.4530
19	2.1068	0.47464	0.03614	27.671	0.07614	13.134	7.8342
20	2.1911	0.45639	0.03358	29.778	0.07358	13.590	8.2091
21	2.2788	0.43883	0.03128	31.969	0.07128	14.029	8.5779
22	2.3699	0.42196	0.02920	34.248	0.06920	14.451	8.9407
23	2.4647	0.40573	0.02731	36.618	0.06731	14.857	9.2973
24	2.5633	0.39012	0.02559	39.083	0.06559	15.247	9.6479
25	2.6658	0.37512	0.02401	41.646	0.06401	15.622	9.9925
26	2.7725	0.36069	0.02257	44.312	0.06257	15.983	10.331
27	2.8834	0.34682	0.02124	47.084	0.06124	16.330	10.664
28	2.9987	0.33348	0.02001	49.968	0.06001	16.663	10.991
29	3.1187	0.32065	0.01888	52.966	0.05888	16.984	11.312
30	3.2434	0.30832	0.01783	56.085	0.05783	17.292	11.627
31	3.3731	0.29646	0.01686	59.328	0.05686	17.588	11.937
32	3.5081	0.28506	0.01595	62.701	0.05595	17.874	12.241
33	3.6484	0.27409	0.01510	66.210	0.05510	18.148	12.540
34	3.7943	0.26355	0.01431	69.858	0.05431	18.411	12.832
35	3.9461	0.25342	0.01358	73.652	0.05358	18.665	13.120
40	4.8010	0.20829	0.01052	95.026	0.05052	19.793	14.477
45	5.8412	0.17120	0.00826	121.03	0.04826	20.720	15.705
50	7.1067	0.14071	0.00655	152.67	0.04655	21.482	16.812
55	8.6464	0.11566	0.00523	191.16	0.04523	22.109	17.807
60	10.520	0.09506	0.00420	237.99	0.04420	22.623	18.697
65	12.799	0.07813	0.00339	294.97	0.04339	23.047	19.491
70	15.572	0.06422	0.00275	364.29	0.04275	23.395	20.196
75	18.945	0.05278	0.00223	448.63	0.04223	23.680	20.821
80	23.050	0.04338	0.00181	551.24	0.04181	23.915	21.372
85	28.044	0.03566	0.00148	676.09	0.04148	24.109	21.857
90	34.119	0.02931	0.00121	827.98	0.04121	24.267	22.283
95	41.511	0.02409	0.00099	1012.8	0.04099	24.398	22.655
100	50.505	0.01980	0.00081	1237.6	0.04081	24.505	22.980

$i = 5\%$ **Discrete Compounding, Discrete Cash Flows**

	SINGLE PAYMENT		UNIFORM SERIES				Arithmetic Gradient Series Factor
	Compound Amount Factor	Present Worth Factor	Sinking Fund Factor	Uniform Series Factor	Capital Recovery Factor	Series Present Worth Factor	
N	(F/P,i,N)	(P/F,i,N)	(A/F,i,N)	(F/A,i,N)	(A/P,i,N)	(P/A,i,N)	(A/G,i,N)
1	1.0500	0.95238	1.0000	1.0000	1.0500	0.95238	0.00000
2	1.1025	0.90703	0.48780	2.0500	0.53780	1.8594	0.48780
3	1.1576	0.86384	0.31721	3.1525	0.36721	2.7232	0.96749
4	1.2155	0.82270	0.23201	4.3101	0.28201	3.5460	1.4391
5	1.2763	0.78353	0.18097	5.5256	0.23097	4.3295	1.9025
6	1.3401	0.74622	0.14702	6.8019	0.19702	5.0757	2.3579
7	1.4071	0.71068	0.12282	8.1420	0.17282	5.7864	2.8052
8	1.4775	0.67684	0.10472	9.5491	0.15472	6.4632	3.2445
9	1.5513	0.64461	0.09069	11.027	0.14069	7.1078	3.6758
10	1.6289	0.61391	0.07950	12.578	0.12950	7.7217	4.0991
11	1.7103	0.58468	0.07039	14.207	0.12039	8.3064	4.5144
12	1.7959	0.55684	0.06283	15.917	0.11283	8.8633	4.9219
13	1.8856	0.53032	0.05646	17.713	0.10646	9.3936	5.3215
14	1.9799	0.50507	0.05102	19.599	0.10102	9.8986	5.7133
15	2.0789	0.48102	0.04634	21.579	0.09634	10.380	6.0973
16	2.1829	0.45811	0.04227	23.657	0.09227	10.838	6.4736
17	2.2920	0.43630	0.03870	25.840	0.08870	11.274	6.8423
18	2.4066	0.41552	0.03555	28.132	0.08555	11.690	7.2034
19	2.5270	0.39573	0.03275	30.539	0.08275	12.085	7.5569
20	2.6533	0.37689	0.03024	33.066	0.08024	12.462	7.9030
21	2.7860	0.35894	0.02800	35.719	0.07800	12.821	8.2416
22	2.9253	0.34185	0.02597	38.505	0.07597	13.163	8.5730
23	3.0715	0.32557	0.02414	41.430	0.07414	13.489	8.8971
24	3.2251	0.31007	0.02247	44.502	0.07247	13.799	9.2140
25	3.3864	0.29530	0.02095	47.727	0.07095	14.094	9.5238
26	3.5557	0.28124	0.01956	51.113	0.06956	14.375	9.8266
27	3.7335	0.26785	0.01829	54.669	0.06829	14.643	10.122
28	3.9201	0.25509	0.01712	58.403	0.06712	14.898	10.411
29	4.1161	0.24295	0.01605	62.323	0.06605	15.141	10.694
30	4.3219	0.23138	0.01505	66.439	0.06505	15.372	10.969
31	4.5380	0.22036	0.01413	70.761	0.06413	15.593	11.238
32	4.7649	0.20987	0.01328	75.299	0.06328	15.803	11.501
33	5.0032	0.19987	0.01249	80.064	0.06249	16.003	11.757
34	5.2533	0.19035	0.01176	85.067	0.06176	16.193	12.006
35	5.5160	0.18129	0.01107	90.320	0.06107	16.374	12.250
40	7.0400	0.14205	0.00828	120.80	0.05828	17.159	13.377
45	8.9850	0.11130	0.00626	159.70	0.05626	17.774	14.364
50	11.467	0.08720	0.00478	209.35	0.05478	18.256	15.223
55	14.636	0.06833	0.00367	272.71	0.05367	18.633	15.966
60	18.679	0.05354	0.00283	353.58	0.05283	18.929	16.606
65	23.840	0.04195	0.00219	456.80	0.05219	19.161	17.154
70	30.426	0.03287	0.00170	588.53	0.05170	19.343	17.621
75	38.833	0.02575	0.00132	756.65	0.05132	19.485	18.018
80	49.561	0.02018	0.00103	971.23	0.05103	19.596	18.353
85	63.254	0.01581	0.00080	1245.1	0.05080	19.684	18.635
90	80.730	0.01239	0.00063	1594.6	0.05063	19.752	18.871
95	103.03	0.00971	0.00049	2040.7	0.05049	19.806	19.069
100	131.50	0.00760	0.00038	2610.0	0.05038	19.848	19.234

$i = 6\%$ Discrete Compounding, Discrete Cash Flows

	SINGLE PAYMENT		UNIFORM SERIES				Arithmetic Gradient Series Factor
	Compound Amount Factor	Present Worth Factor	Sinking Fund Factor	Uniform Series Factor	Capital Recovery Factor	Series Present Worth Factor	
N	(F/P,i,N)	(P/F,i,N)	(A/F,i,N)	(F/A,i,N)	(A/P,i,N)	(P/A,i,N)	(A/G,i,N)
1	1.0600	0.94340	1.0000	1.0000	1.0600	0.94340	0.00000
2	1.1236	0.89000	0.48544	2.0600	0.54544	1.8334	0.48544
3	1.1910	0.83962	0.31411	3.1836	0.37411	2.6730	0.96118
4	1.2625	0.79209	0.22859	4.3746	0.28859	3.4651	1.4272
5	1.3382	0.74726	0.17740	5.6371	0.23740	4.2124	1.8836
6	1.4185	0.70496	0.14336	6.9753	0.20336	4.9173	2.3304
7	1.5036	0.66506	0.11914	8.3938	0.17914	5.5824	2.7676
8	1.5938	0.62741	0.10104	9.8975	0.16104	6.2098	3.1952
9	1.6895	0.59190	0.08702	11.491	0.14702	6.8017	3.6133
10	1.7908	0.55839	0.07587	13.181	0.13587	7.3601	4.0220
11	1.8983	0.52679	0.06679	14.972	0.12679	7.8869	4.4213
12	2.0122	0.49697	0.05928	16.870	0.11928	8.3838	4.8113
13	2.1329	0.46884	0.05296	18.882	0.11296	8.8527	5.1920
14	2.2609	0.44230	0.04758	21.015	0.10758	9.2950	5.5635
15	2.3966	0.41727	0.04296	23.276	0.10296	9.7122	5.9260
16	2.5404	0.39365	0.03895	25.673	0.09895	10.106	6.2794
17	2.6928	0.37136	0.03544	28.213	0.09544	10.477	6.6240
18	2.8543	0.35034	0.03236	30.906	0.09236	10.828	6.9597
19	3.0256	0.33051	0.02962	33.760	0.08962	11.158	7.2867
20	3.2071	0.31180	0.02718	36.786	0.08718	11.470	7.6051
21	3.3996	0.29416	0.02500	39.993	0.08500	11.764	7.9151
22	3.6035	0.27751	0.02305	43.392	0.08305	12.042	8.2166
23	3.8197	0.26180	0.02128	46.996	0.08128	12.303	8.5099
24	4.0489	0.24698	0.01968	50.816	0.07968	12.550	8.7951
25	4.2919	0.23300	0.01823	54.865	0.07823	12.783	9.0722
26	4.5494	0.21981	0.01690	59.156	0.07690	13.003	9.3414
27	4.8223	0.20737	0.01570	63.706	0.07570	13.211	9.6029
28	5.1117	0.19563	0.01459	68.528	0.07459	13.406	9.8568
29	5.4184	0.18456	0.01358	73.640	0.07358	13.591	10.103
30	5.7435	0.17411	0.01265	79.058	0.07265	13.765	10.342
31	6.0881	0.16425	0.01179	84.802	0.07179	13.929	10.574
32	6.4534	0.15496	0.01100	90.890	0.07100	14.084	10.799
33	6.8406	0.14619	0.01027	97.343	0.07027	14.230	11.017
34	7.2510	0.13791	0.00960	104.18	0.06960	14.368	11.228
35	7.6861	0.13011	0.00897	111.43	0.06897	14.498	11.432
40	10.286	0.09722	0.00646	154.76	0.06646	15.046	12.359
45	13.765	0.07265	0.00470	212.74	0.06470	15.456	13.141
50	18.420	0.05429	0.00344	290.34	0.06344	15.762	13.796
55	24.650	0.04057	0.00254	394.17	0.06254	15.991	14.341
60	32.988	0.03031	0.00188	533.13	0.06188	16.161	14.791
65	44.145	0.02265	0.00139	719.08	0.06139	16.289	15.160
70	59.076	0.01693	0.00103	967.93	0.06103	16.385	15.461
75	79.057	0.01265	0.00077	1300.9	0.06077	16.456	15.706
80	105.80	0.00945	0.00057	1746.6	0.06057	16.509	15.903
85	141.58	0.00706	0.00043	2343.0	0.06043	16.549	16.062
90	189.46	0.00528	0.00032	3141.1	0.06032	16.579	16.189
95	253.55	0.00394	0.00024	4209.1	0.06024	16.601	16.290
100	339.30	0.00295	0.00018	5638.4	0.06018	16.618	16.371

$i = 7\%$ Discrete Compounding, Discrete Cash Flows

	SINGLE PAYMENT		UNIFORM SERIES				Arithmetic Gradient Series Factor
	Compound Amount Factor	Present Worth Factor	Sinking Fund Factor	Uniform Series Factor	Capital Recovery Factor	Series Present Worth Factor	
N	(F/P,i,N)	(P/F,i,N)	(A/F,i,N)	(F/A,i,N)	(A/P,i,N)	(P/A,i,N)	(A/G,i,N)
1	1.0700	0.93458	1.0000	1.0000	1.0700	0.93458	0.00000
2	1.1449	0.87344	0.48309	2.0700	0.55309	1.8080	0.48309
3	1.2250	0.81630	0.31105	3.2149	0.38105	2.6243	0.95493
4	1.3108	0.76290	0.22523	4.4399	0.29523	3.3872	1.4155
5	1.4026	0.71299	0.17389	5.7507	0.24389	4.1002	1.8650
6	1.5007	0.66634	0.13980	7.1533	0.20980	4.7665	2.3032
7	1.6058	0.62275	0.11555	8.6540	0.18555	5.3893	2.7304
8	1.7182	0.58201	0.09747	10.260	0.16747	5.9713	3.1465
9	1.8385	0.54393	0.08349	11.978	0.15349	6.5152	3.5517
10	1.9672	0.50835	0.07238	13.816	0.14238	7.0236	3.9461
11	2.1049	0.47509	0.06336	15.784	0.13336	7.4987	4.3296
12	2.2522	0.44401	0.05590	17.888	0.12590	7.9427	4.7025
13	2.4098	0.41496	0.04965	20.141	0.11965	8.3577	5.0648
14	2.5785	0.38782	0.04434	22.550	0.11434	8.7455	5.4167
15	2.7590	0.36245	0.03979	25.129	0.10979	9.1079	5.7583
16	2.9522	0.33873	0.03586	27.888	0.10586	9.4466	6.0897
17	3.1588	0.31657	0.03243	30.840	0.10243	9.7632	6.4110
18	3.3799	0.29586	0.02941	33.999	0.09941	10.059	6.7225
19	3.6165	0.27651	0.02675	37.379	0.09675	10.336	7.0242
20	3.8697	0.25842	0.02439	40.995	0.09439	10.594	7.3163
21	4.1406	0.24151	0.02229	44.865	0.09229	10.836	7.5990
22	4.4304	0.22571	0.02041	49.006	0.09041	11.061	7.8725
23	4.7405	0.21095	0.01871	53.436	0.08871	11.272	8.1369
24	5.0724	0.19715	0.01719	58.177	0.08719	11.469	8.3923
25	5.4274	0.18425	0.01581	63.249	0.08581	11.654	8.6391
26	5.8074	0.17220	0.01456	68.676	0.08456	11.826	8.8773
27	6.2139	0.16093	0.01343	74.484	0.08343	11.987	9.1072
28	6.6488	0.15040	0.01239	80.698	0.08239	12.137	9.3289
29	7.1143	0.14056	0.01145	87.347	0.08145	12.278	9.5427
30	7.6123	0.13137	0.01059	94.461	0.08059	12.409	9.7487
31	8.1451	0.12277	0.00980	102.07	0.07980	12.532	9.9471
32	8.7153	0.11474	0.00907	110.22	0.07907	12.647	10.138
33	9.3253	0.10723	0.00841	118.93	0.07841	12.754	10.322
34	9.9781	0.10022	0.00780	128.26	0.07780	12.854	10.499
35	10.677	0.09366	0.00723	138.24	0.07723	12.948	10.669
40	14.974	0.06678	0.00501	199.64	0.07501	13.332	11.423
45	21.002	0.04761	0.00350	285.75	0.07350	13.606	12.036
50	29.457	0.03395	0.00246	406.53	0.07246	13.801	12.529
55	41.315	0.02420	0.00174	575.93	0.07174	13.940	12.921
60	57.946	0.01726	0.00123	813.52	0.07123	14.039	13.232
65	81.273	0.01230	0.00087	1146.8	0.07087	14.110	13.476
70	113.99	0.00877	0.00062	1614.1	0.07062	14.160	13.666
75	159.88	0.00625	0.00044	2269.7	0.07044	14.196	13.814
80	224.23	0.00446	0.00031	3189.1	0.07031	14.222	13.927
85	314.50	0.00318	0.00022	4478.6	0.07022	14.240	14.015
90	441.10	0.00227	0.00016	6287.2	0.07016	14.253	14.081
95	618.67	0.00162	0.00011	8823.9	0.07011	14.263	14.132
100	867.72	0.00115	0.00008	12 382.0	0.07008	14.269	14.170

$i = 8\%$ Discrete Compounding, Discrete Cash Flows

	SINGLE PAYMENT		UNIFORM SERIES				Arithmetic Gradient Series Factor
	Compound Amount Factor	Present Worth Factor	Sinking Fund Factor	Uniform Series Factor	Capital Recovery Factor	Series Present Worth Factor	
N	(F/P,i,N)	(P/F,i,N)	(A/F,i,N)	(F/A,i,N)	(A/P,i,N)	(P/A,i,N)	(A/G,i,N)
1	1.0800	0.92593	1.0000	1.0000	1.0800	0.92593	0.00000
2	1.1664	0.85734	0.48077	2.0800	0.56077	1.7833	0.48077
3	1.2597	0.79383	0.30803	3.2464	0.38803	2.5771	0.94874
4	1.3605	0.73503	0.22192	4.5061	0.30192	3.3121	1.4040
5	1.4693	0.68058	0.17046	5.8666	0.25046	3.9927	1.8465
6	1.5869	0.63017	0.13632	7.3359	0.21632	4.6229	2.2763
7	1.7138	0.58349	0.11207	8.9228	0.19207	5.2064	2.6937
8	1.8509	0.54027	0.09401	10.637	0.17401	5.7466	3.0985
9	1.9990	0.50025	0.08008	12.488	0.16008	6.2469	3.4910
10	2.1589	0.46319	0.06903	14.487	0.14903	6.7101	3.8713
11	2.3316	0.42888	0.06008	16.645	0.14008	7.1390	4.2395
12	2.5182	0.39711	0.05270	18.977	0.13270	7.5361	4.5957
13	2.7196	0.36770	0.04652	21.495	0.12652	7.9038	4.9402
14	2.9372	0.34046	0.04130	24.215	0.12130	8.2442	5.2731
15	3.1722	0.31524	0.03683	27.152	0.11683	8.5595	5.5945
16	3.4259	0.29189	0.03298	30.324	0.11298	8.8514	5.9046
17	3.7000	0.27027	0.02963	33.750	0.10963	9.1216	6.2037
18	3.9960	0.25025	0.02670	37.450	0.10670	9.3719	6.4920
19	4.3157	0.23171	0.02413	41.446	0.10413	9.6036	6.7697
20	4.6610	0.21455	0.02185	45.762	0.10185	9.8181	7.0369
21	5.0338	0.19866	0.01983	50.423	0.09983	10.017	7.2940
22	5.4365	0.18394	0.01803	55.457	0.09803	10.201	7.5412
23	5.8715	0.17032	0.01642	60.893	0.09642	10.371	7.7786
24	6.3412	0.15770	0.01498	66.765	0.09498	10.529	8.0066
25	6.8485	0.14602	0.01368	73.106	0.09368	10.675	8.2254
26	7.3964	0.13520	0.01251	79.954	0.09251	10.810	8.4352
27	7.9881	0.12519	0.01145	87.351	0.09145	10.935	8.6363
28	8.6271	0.11591	0.01049	95.339	0.09049	11.051	8.8289
29	9.3173	0.10733	0.00962	103.97	0.08962	11.158	9.0133
30	10.063	0.09938	0.00883	113.28	0.08883	11.258	9.1897
31	10.868	0.09202	0.00811	123.35	0.08811	11.350	9.3584
32	11.737	0.08520	0.00745	134.21	0.08745	11.435	9.5197
33	12.676	0.07889	0.00685	145.95	0.08685	11.514	9.6737
34	13.690	0.07305	0.00630	158.63	0.08630	11.587	9.8208
35	14.785	0.06763	0.00580	172.32	0.08580	11.655	9.9611
40	21.725	0.04603	0.00386	259.06	0.08386	11.925	10.570
45	31.920	0.03133	0.00259	386.51	0.08259	12.108	11.045
50	46.902	0.02132	0.00174	573.77	0.08174	12.233	11.411
55	68.914	0.01451	0.00118	848.92	0.08118	12.319	11.690
60	101.26	0.00988	0.00080	1253.2	0.08080	12.377	11.902
65	148.78	0.00672	0.00054	1847.2	0.08054	12.416	12.060
70	218.61	0.00457	0.00037	2720.1	0.08037	12.443	12.178
75	321.20	0.00311	0.00025	4002.6	0.08025	12.461	12.266
80	471.95	0.00212	0.00017	5886.9	0.08017	12.474	12.330
85	693.46	0.00144	0.00012	8655.7	0.08012	12.482	12.377
90	1018.9	0.00098	0.00008	12 724.0	0.08008	12.488	12.412
95	1497.1	0.00067	0.00005	18 702.0	0.08005	12.492	12.437
100	2199.8	0.00045	0.00004	27 485.0	0.08004	12.494	12.455

$i = 9\%$ **Discrete Compounding, Discrete Cash Flows**

	SINGLE PAYMENT		UNIFORM SERIES				Arithmetic Gradient Series Factor
	Compound Amount Factor	Present Worth Factor	Sinking Fund Factor	Uniform Series Factor	Capital Recovery Factor	Series Present Worth Factor	
N	(F/P,i,N)	(P/F,i,N)	(A/F,i,N)	(F/A,i,N)	(A/P,i,N)	(P/A,i,N)	(A/G,i,N)
1	1.0900	0.91743	1.0000	1.0000	1.0900	0.91743	0.00000
2	1.1881	0.84168	0.47847	2.0900	0.56847	1.7591	0.47847
3	1.2950	0.77218	0.30505	3.2781	0.39505	2.5313	0.94262
4	1.4116	0.70843	0.21867	4.5731	0.30867	3.2397	1.3925
5	1.5386	0.64993	0.16709	5.9847	0.25709	3.8897	1.8282
6	1.6771	0.59627	0.13292	7.5233	0.22292	4.4859	2.2498
7	1.8280	0.54703	0.10869	9.2004	0.19869	5.0330	2.6574
8	1.9926	0.50187	0.09067	11.028	0.18067	5.5348	3.0512
9	2.1719	0.46043	0.07680	13.021	0.16680	5.9952	3.4312
10	2.3674	0.42241	0.06582	15.193	0.15582	6.4177	3.7978
11	2.5804	0.38753	0.05695	17.560	0.14695	6.8052	4.1510
12	2.8127	0.35553	0.04965	20.141	0.13965	7.1607	4.4910
13	3.0658	0.32618	0.04357	22.953	0.13357	7.4869	4.8182
14	3.3417	0.29925	0.03843	26.019	0.12843	7.7862	5.1326
15	3.6425	0.27454	0.03406	29.361	0.12406	8.0607	5.4346
16	3.9703	0.25187	0.03030	33.003	0.12030	8.3126	5.7245
17	4.3276	0.23107	0.02705	36.974	0.11705	8.5436	6.0024
18	4.7171	0.21199	0.02421	41.301	0.11421	8.7556	6.2687
19	5.1417	0.19449	0.02173	46.018	0.11173	8.9501	6.5236
20	5.6044	0.17843	0.01955	51.160	0.10955	9.1285	6.7674
21	6.1088	0.16370	0.01762	56.765	0.10762	9.2922	7.0006
22	6.6586	0.15018	0.01590	62.873	0.10590	9.4424	7.2232
23	7.2579	0.13778	0.01438	69.532	0.10438	9.5802	7.4357
24	7.9111	0.12640	0.01302	76.790	0.10302	9.7066	7.6384
25	8.6231	0.11597	0.01181	84.701	0.10181	9.8226	7.8316
26	9.3992	0.10639	0.01072	93.324	0.10072	9.9290	8.0156
27	10.245	0.09761	0.00973	102.72	0.09973	10.027	8.1906
28	11.167	0.08955	0.00885	112.97	0.09885	10.116	8.3571
29	12.172	0.08215	0.00806	124.14	0.09806	10.198	8.5154
30	13.268	0.07537	0.00734	136.31	0.09734	10.274	8.6657
31	14.462	0.06915	0.00669	149.58	0.09669	10.343	8.8083
32	15.763	0.06344	0.00610	164.04	0.09610	10.406	8.9436
33	17.182	0.05820	0.00556	179.80	0.09556	10.464	9.0718
34	18.728	0.05339	0.00508	196.98	0.09508	10.518	9.1933
35	20.414	0.04899	0.00464	215.71	0.09464	10.567	9.3083
40	31.409	0.03184	0.00296	337.88	0.09296	10.757	9.7957
45	48.327	0.02069	0.00190	525.86	0.09190	10.881	10.160
50	74.358	0.01345	0.00123	815.08	0.09123	10.962	10.430
55	114.41	0.00874	0.00079	1260.1	0.09079	11.014	10.626
60	176.03	0.00568	0.00051	1944.8	0.09051	11.048	10.768
65	270.85	0.00369	0.00033	2998.3	0.09033	11.070	10.870
70	416.73	0.00240	0.00022	4619.2	0.09022	11.084	10.943
75	641.19	0.00156	0.00014	7113.2	0.09014	11.094	10.994
80	986.55	0.00101	0.00009	10951.0	0.09009	11.100	11.030
85	1517.9	0.00066	0.00006	16855.0	0.09006	11.104	11.055
90	2335.5	0.00043	0.00004	25939.0	0.09004	11.106	11.073
95	3593.5	0.00028	0.00003	39917.0	0.09003	11.108	11.085
100	5529.0	0.00018	0.00002	61423.0	0.09002	11.109	11.093

$i = 10\%$ Discrete Compounding, Discrete Cash Flows

	SINGLE PAYMENT		UNIFORM SERIES				Arithmetic Gradient Series Factor
	Compound Amount Factor	Present Worth Factor	Sinking Fund Factor	Uniform Series Factor	Capital Recovery Factor	Series Present Worth Factor	
N	$(F/P,i,N)$	$(P/F,i,N)$	$(A/F,i,N)$	$(F/A,i,N)$	$(A/P,i,N)$	$(P/A,i,N)$	$(A/G,i,N)$
1	1.1000	0.90909	1.0000	1.0000	1.1000	0.90909	0.00000
2	1.2100	0.82645	0.47619	2.1000	0.57619	1.7355	0.47619
3	1.3310	0.75131	0.30211	3.3100	0.40211	2.4869	0.93656
4	1.4641	0.68301	0.21547	4.6410	0.31547	3.1699	1.3812
5	1.6105	0.62092	0.16380	6.1051	0.26380	3.7908	1.8101
6	1.7716	0.56447	0.12961	7.7156	0.22961	4.3553	2.2236
7	1.9487	0.51316	0.10541	9.4872	0.20541	4.8684	2.6216
8	2.1436	0.46651	0.08744	11.436	0.18744	5.3349	3.0045
9	2.3579	0.42410	0.07364	13.579	0.17364	5.7590	3.3724
10	2.5937	0.38554	0.06275	15.937	0.16275	6.1446	3.7255
11	2.8531	0.35049	0.05396	18.531	0.15396	6.4951	4.0641
12	3.1384	0.31863	0.04676	21.384	0.14676	6.8137	4.3884
13	3.4523	0.28966	0.04078	24.523	0.14078	7.1034	4.6988
14	3.7975	0.26333	0.03575	27.975	0.13575	7.3667	4.9955
15	4.1772	0.23939	0.03147	31.772	0.13147	7.6061	5.2789
16	4.5950	0.21763	0.02782	35.950	0.12782	7.8237	5.5493
17	5.0545	0.19784	0.02466	40.545	0.12466	8.0216	5.8071
18	5.5599	0.17986	0.02193	45.599	0.12193	8.2014	6.0526
19	6.1159	0.16351	0.01955	51.159	0.11955	8.3649	6.2861
20	6.7275	0.14864	0.01746	57.275	0.11746	8.5136	6.5081
21	7.4002	0.13513	0.01562	64.002	0.11562	8.6487	6.7189
22	8.1403	0.12285	0.01401	71.403	0.11401	8.7715	6.9189
23	8.9543	0.11168	0.01257	79.543	0.11257	8.8832	7.1085
24	9.8497	0.10153	0.01130	88.497	0.11130	8.9847	7.2881
25	10.835	0.09230	0.01017	98.347	0.11017	9.0770	7.4580
26	11.918	0.08391	0.00916	109.18	0.10916	9.1609	7.6186
27	13.110	0.07628	0.00826	121.10	0.10826	9.2372	7.7704
28	14.421	0.06934	0.00745	134.21	0.10745	9.3066	7.9137
29	15.863	0.06304	0.00673	148.63	0.10673	9.3696	8.0489
30	17.449	0.05731	0.00608	164.49	0.10608	9.4269	8.1762
31	19.194	0.05210	0.00550	181.94	0.10550	9.4790	8.2962
32	21.114	0.04736	0.00497	201.14	0.10497	9.5264	8.4091
33	23.225	0.04306	0.00450	222.25	0.10450	9.5694	8.5152
34	25.548	0.03914	0.00407	245.48	0.10407	9.6086	8.6149
35	28.102	0.03558	0.00369	271.02	0.10369	9.6442	8.7086
40	45.259	0.02209	0.00226	442.59	0.10226	9.7791	9.0962
45	72.890	0.01372	0.00139	718.90	0.10139	9.8628	9.3740
50	117.39	0.00852	0.00086	1163.9	0.10086	9.9148	9.5704
55	189.06	0.00529	0.00053	1880.6	0.10053	9.9471	9.7075
60	304.48	0.00328	0.00033	3034.8	0.10033	9.9672	9.8023
65	490.37	0.00204	0.00020	4893.7	0.10020	9.9796	9.8672
70	789.75	0.00127	0.00013	7887.5	0.10013	9.9873	9.9113
75	1271.9	0.00079	0.00008	12 709.0	0.10008	9.9921	9.9410

$i = 11\%$ Discrete Compounding, Discrete Cash Flows

	SINGLE PAYMENT		UNIFORM SERIES				Arithmetic Gradient Series Factor
	Compound Amount Factor	Present Worth Factor	Sinking Fund Factor	Uniform Series Factor	Capital Recovery Factor	Series Present Worth Factor	
N	$(F/P,i,N)$	$(P/F,i,N)$	$(A/F,i,N)$	$(F/A,i,N)$	$(A/P,i,N)$	$(P/A,i,N)$	$(A/G,i,N)$
1	1.1100	0.90090	1.0000	1.0000	1.1100	0.90090	0.00000
2	1.2321	0.81162	0.47393	2.1100	0.58393	1.7125	0.47393
3	1.3676	0.73119	0.29921	3.3421	0.40921	2.4437	0.93055
4	1.5181	0.65873	0.21233	4.7097	0.32233	3.1024	1.3700
5	1.6851	0.59345	0.16057	6.2278	0.27057	3.6959	1.7923
6	1.8704	0.53464	0.12638	7.9129	0.23638	4.2305	2.1976
7	2.0762	0.48166	0.10222	9.783	0.21222	4.7122	2.5863
8	2.3045	0.43393	0.08432	11.859	0.19432	5.1461	2.9585
9	2.5580	0.39092	0.07060	14.164	0.18060	5.5370	3.3144
10	2.8394	0.35218	0.05980	16.722	0.16980	5.8892	3.6544
11	3.1518	0.31728	0.05112	19.561	0.16112	6.2065	3.9788
12	3.4985	0.28584	0.04403	22.713	0.15403	6.4924	4.2879
13	3.8833	0.25751	0.03815	26.212	0.14815	6.7499	4.5822
14	4.3104	0.23199	0.03323	30.095	0.14323	6.9819	4.8619
15	4.7846	0.20900	0.02907	34.405	0.13907	7.1909	5.1275
16	5.3109	0.18829	0.02552	39.190	0.13552	7.3792	5.3794
17	5.8951	0.16963	0.02247	44.501	0.13247	7.5488	5.6180
18	6.5436	0.15282	0.01984	50.396	0.12984	7.7016	5.8439
19	7.2633	0.13768	0.01756	56.939	0.12756	7.8393	6.0574
20	8.0623	0.12403	0.01558	64.203	0.12558	7.9633	6.2590
21	8.949	0.11174	0.01384	72.265	0.12384	8.0751	6.4491
22	9.934	0.10067	0.01231	81.214	0.12231	8.1757	6.6283
23	11.026	0.09069	0.01097	91.15	0.12097	8.2664	6.7969
24	12.239	0.08170	0.00979	102.17	0.11979	8.3481	6.9555
25	13.585	0.07361	0.00874	114.41	0.11874	8.4217	7.1045
26	15.080	0.06631	0.00781	128.00	0.11781	8.4881	7.2443
27	16.739	0.05974	0.00699	143.08	0.11699	8.5478	7.3754
28	18.580	0.05382	0.00626	159.82	0.11626	8.6016	7.4982
29	20.624	0.04849	0.00561	178.40	0.11561	8.6501	7.6131
30	22.892	0.04368	0.00502	199.02	0.11502	8.6938	7.7206
31	25.410	0.03935	0.00451	221.91	0.11451	8.7331	7.8210
32	28.206	0.03545	0.00404	247.32	0.11404	8.7686	7.9147
33	31.308	0.03194	0.00363	275.53	0.11363	8.8005	8.0021
34	34.752	0.02878	0.00326	306.84	0.11326	8.8293	8.0836
35	38.575	0.02592	0.00293	341.59	0.11293	8.8552	8.1594
40	65.001	0.01538	0.00172	581.83	0.11172	8.9511	8.4659
45	109.53	0.00913	0.00101	986.6	0.11101	9.0079	8.6763
50	184.56	0.00542	0.00060	1668.8	0.11060	9.0417	8.8185
55	311.00	0.00322	0.00035	2818.2	0.11035	9.0617	8.9135

$i = 12\%$ Discrete Compounding, Discrete Cash Flows

	SINGLE PAYMENT		UNIFORM SERIES				Arithmetic Gradient Series Factor
	Compound Amount Factor	Present Worth Factor	Sinking Fund Factor	Uniform Series Factor	Capital Recovery Factor	Series Present Worth Factor	Arithmetic Gradient Series Factor
N	(F/P,i,N)	(P/F,i,N)	(A/F,i,N)	(F/A,i,N)	(A/P,i,N)	(P/A,i,N)	(A/G,i,N)
1	1.1200	0.89286	1.0000	1.0000	1.1200	0.89286	0.00000
2	1.2544	0.79719	0.47170	2.1200	0.59170	1.6901	0.47170
3	1.4049	0.71178	0.29635	3.3744	0.41635	2.4018	0.92461
4	1.5735	0.63552	0.20923	4.7793	0.32923	3.0373	1.3589
5	1.7623	0.56743	0.15741	6.3528	0.27741	3.6048	1.7746
6	1.9738	0.50663	0.12323	8.1152	0.24323	4.1114	2.1720
7	2.2107	0.45235	0.09912	10.089	0.21912	4.5638	2.5515
8	2.4760	0.40388	0.08130	12.300	0.20130	4.9676	2.9131
9	2.7731	0.36061	0.06768	14.776	0.18768	5.3282	3.2574
10	3.1058	0.32197	0.05698	17.549	0.17698	5.6502	3.5847
11	3.4785	0.28748	0.04842	20.655	0.16842	5.9377	3.8953
12	3.8960	0.25668	0.04144	24.133	0.16144	6.1944	4.1897
13	4.3635	0.22917	0.03568	28.029	0.15568	6.4235	4.4683
14	4.8871	0.20462	0.03087	32.393	0.15087	6.6282	4.7317
15	5.4736	0.18270	0.02682	37.280	0.14682	6.8109	4.9803
16	6.1304	0.16312	0.02339	42.753	0.14339	6.9740	5.2147
17	6.8660	0.14564	0.02046	48.884	0.14046	7.1196	5.4353
18	7.6900	0.13004	0.01794	55.750	0.13794	7.2497	5.6427
19	8.6128	0.11611	0.01576	63.440	0.13576	7.3658	5.8375
20	9.6463	0.10367	0.01388	72.052	0.13388	7.4694	6.0202
21	10.804	0.09256	0.01224	81.699	0.13224	7.5620	6.1913
22	12.100	0.08264	0.01081	92.503	0.13081	7.6446	6.3514
23	13.552	0.07379	0.00956	104.60	0.12956	7.7184	6.5010
24	15.179	0.06588	0.00846	118.16	0.12846	7.7843	6.6406
25	17.000	0.05882	0.00750	133.33	0.12750	7.8431	6.7708
26	19.040	0.05252	0.00665	150.33	0.12665	7.8957	6.8921
27	21.325	0.04689	0.00590	169.37	0.12590	7.9426	7.0049
28	23.884	0.04187	0.00524	190.70	0.12524	7.9844	7.1098
29	26.750	0.03738	0.00466	214.58	0.12466	8.0218	7.2071
30	29.960	0.03338	0.00414	241.33	0.12414	8.0552	7.2974
31	33.555	0.02980	0.00369	271.29	0.12369	8.0850	7.3811
32	37.582	0.02661	0.00328	304.85	0.12328	8.1116	7.4586
33	42.092	0.02376	0.00292	342.43	0.12292	8.1354	7.5302
34	47.143	0.02121	0.00260	384.52	0.12260	8.1566	7.5965
35	52.800	0.01894	0.00232	431.66	0.12232	8.1755	7.6577
40	93.051	0.01075	0.00130	767.09	0.12130	8.2438	7.8988
45	163.99	0.00610	0.00074	1358.2	0.12074	8.2825	8.0572
50	289.00	0.00346	0.00042	2400.0	0.12042	8.3045	8.1597
55	509.32	0.00196	0.00024	4236.0	0.12024	8.3170	8.2251

$i = 13\%$ Discrete Compounding, Discrete Cash Flows

	SINGLE PAYMENT		UNIFORM SERIES				Arithmetic Gradient Series Factor
	Compound Amount Factor	Present Worth Factor	Sinking Fund Factor	Uniform Series Factor	Capital Recovery Factor	Series Present Worth Factor	
N	(F/P,i,N)	(P/F,i,N)	(A/F,i,N)	(F/A,i,N)	(A/P,i,N)	(P/A,i,N)	(A/G,i,N)
1	1.1300	0.88496	1.0000	1.0000	1.1300	0.88496	0.00000
2	1.2769	0.78315	0.46948	2.1300	0.59948	1.6681	0.46948
3	1.4429	0.69305	0.29352	3.4069	0.42352	2.3612	0.91872
4	1.6305	0.61332	0.20619	4.8498	0.33619	2.9745	1.3479
5	1.8424	0.54276	0.15431	6.4803	0.28431	3.5172	1.7571
6	2.0820	0.48032	0.12015	8.3227	0.25015	3.9975	2.1468
7	2.3526	0.42506	0.09611	10.405	0.22611	4.4226	2.5171
8	2.6584	0.37616	0.07839	12.757	0.20839	4.7988	2.8685
9	3.0040	0.33288	0.06487	15.416	0.19487	5.1317	3.2014
10	3.3946	0.29459	0.05429	18.420	0.18429	5.4262	3.5162
11	3.8359	0.26070	0.04584	21.814	0.17584	5.6869	3.8134
12	4.3345	0.23071	0.03899	25.650	0.16899	5.9176	4.0936
13	4.8980	0.20416	0.03335	29.985	0.16335	6.1218	4.3573
14	5.5348	0.18068	0.02867	34.883	0.15867	6.3025	4.6050
15	6.2543	0.15989	0.02474	40.417	0.15474	6.4624	4.8375
16	7.0673	0.14150	0.02143	46.672	0.15143	6.6039	5.0552
17	7.9861	0.12522	0.01861	53.739	0.14861	6.7291	5.2589
18	9.0243	0.11081	0.01620	61.725	0.14620	6.8399	5.4491
19	10.197	0.09806	0.01413	70.749	0.14413	6.9380	5.6265
20	11.523	0.08678	0.01235	80.947	0.14235	7.0248	5.7917
21	13.021	0.07680	0.01081	92.470	0.14081	7.1016	5.9454
22	14.714	0.06796	0.00948	105.49	0.13948	7.1695	6.0881
23	16.627	0.06014	0.00832	120.20	0.13832	7.2297	6.2205
24	18.788	0.05323	0.00731	136.83	0.13731	7.2829	6.3431
25	21.231	0.04710	0.00643	155.62	0.13643	7.3300	6.4566
26	23.991	0.04168	0.00565	176.85	0.13565	7.3717	6.5614
27	27.109	0.03689	0.00498	200.84	0.13498	7.4086	6.6582
28	30.633	0.03264	0.00439	227.95	0.13439	7.4412	6.7474
29	34.616	0.02889	0.00387	258.58	0.13387	7.4701	6.8296
30	39.116	0.02557	0.00341	293.20	0.13341	7.4957	6.9052
31	44.201	0.02262	0.00301	332.32	0.13301	7.5183	6.9747
32	49.947	0.02002	0.00266	376.52	0.13266	7.5383	7.0385
33	56.440	0.01772	0.00234	426.46	0.13234	7.5560	7.0971
34	63.777	0.01568	0.00207	482.90	0.13207	7.5717	7.1507
35	72.069	0.01388	0.00183	546.68	0.13183	7.5856	7.1998
40	132.78	0.00753	0.00099	1013.7	0.13099	7.6344	7.3888
45	244.64	0.00409	0.00053	1874.2	0.13053	7.6609	7.5076
50	450.74	0.00222	0.00029	3459.5	0.13029	7.6752	7.5811
55	830.45	0.00120	0.00016	6380.4	0.13016	7.6830	7.6260

$i = 14\%$ Discrete Compounding, Discrete Cash Flows

	SINGLE PAYMENT		UNIFORM SERIES				Arithmetic Gradient Series Factor
	Compound Amount Factor	Present Worth Factor	Sinking Fund Factor	Uniform Series Factor	Capital Recovery Factor	Series Present Worth Factor	
N	(F/P,i,N)	(P/F,i,N)	(A/F,i,N)	(F/A,i,N)	(A/P,i,N)	(P/A,i,N)	(A/G,i,N)
1	1.1400	0.87719	1.0000	1.0000	1.1400	0.87719	0.00000
2	1.2996	0.76947	0.46729	2.1400	0.60729	1.6467	0.46729
3	1.4815	0.67497	0.29073	3.4396	0.43073	2.3216	0.91290
4	1.6890	0.59208	0.20320	4.9211	0.34320	2.9137	1.3370
5	1.9254	0.51937	0.15128	6.6101	0.29128	3.4331	1.7399
6	2.1950	0.45559	0.11716	8.5355	0.25716	3.8887	2.1218
7	2.5023	0.39964	0.09319	10.730	0.23319	4.2883	2.4832
8	2.8526	0.35056	0.07557	13.233	0.21557	4.6389	2.8246
9	3.2519	0.30751	0.06217	16.085	0.20217	4.9464	3.1463
10	3.7072	0.26974	0.05171	19.337	0.19171	5.2161	3.4490
11	4.2262	0.23662	0.04339	23.045	0.18339	5.4527	3.7333
12	4.8179	0.20756	0.03667	27.271	0.17667	5.6603	3.9998
13	5.4924	0.18207	0.03116	32.089	0.17116	5.8424	4.2491
14	6.2613	0.15971	0.02661	37.581	0.16661	6.0021	4.4819
15	7.1379	0.14010	0.02281	43.842	0.16281	6.1422	4.6990
16	8.1372	0.12289	0.01962	50.980	0.15962	6.2651	4.9011
17	9.2765	0.10780	0.01692	59.118	0.15692	6.3729	5.0888
18	10.575	0.09456	0.01462	68.394	0.15462	6.4674	5.2630
19	12.056	0.08295	0.01266	78.969	0.15266	6.5504	5.4243
20	13.743	0.07276	0.01099	91.025	0.15099	6.6231	5.5734
21	15.668	0.06383	0.00954	104.77	0.14954	6.6870	5.7111
22	17.861	0.05599	0.00830	120.44	0.14830	6.7429	5.8381
23	20.362	0.04911	0.00723	138.30	0.14723	6.7921	5.9549
24	23.212	0.04308	0.00630	158.66	0.14630	6.8351	6.0624
25	26.462	0.03779	0.00550	181.87	0.14550	6.8729	6.1610
26	30.167	0.03315	0.00480	208.33	0.14480	6.9061	6.2514
27	34.390	0.02908	0.00419	238.50	0.14419	6.9352	6.3342
28	39.204	0.02551	0.00366	272.89	0.14366	6.9607	6.4100
29	44.693	0.02237	0.00320	312.09	0.14320	6.9830	6.4791
30	50.950	0.01963	0.00280	356.79	0.14280	7.0027	6.5423
31	58.083	0.01722	0.00245	407.74	0.14245	7.0199	6.5998
32	66.215	0.01510	0.00215	465.82	0.14215	7.0350	6.6522
33	75.485	0.01325	0.00188	532.04	0.14188	7.0482	6.6998
34	86.053	0.01162	0.00165	607.52	0.14165	7.0599	6.7431
35	98.100	0.01019	0.00144	693.57	0.14144	7.0700	6.7824
40	188.88	0.00529	0.00075	1342.0	0.14075	7.1050	6.9300
45	363.68	0.00275	0.00039	2590.6	0.14039	7.1232	7.0188
50	700.23	0.00143	0.00020	4994.5	0.14020	7.1327	7.0714
55	1348.2	0.00074	0.00010	9623.1	0.14010	7.1376	7.1020

$i = 15\%$ Discrete Compounding, Discrete Cash Flows

	SINGLE PAYMENT		UNIFORM SERIES				Arithmetic Gradient Series Factor
	Compound Amount Factor	Present Worth Factor	Sinking Fund Factor	Uniform Series Factor	Capital Recovery Factor	Series Present Worth Factor	
N	(F/P,i,N)	(P/F,i,N)	(A/F,i,N)	(F/A,i,N)	(A/P,i,N)	(P/A,i,N)	(A/G,i,N)
1	1.1500	0.86957	1.0000	1.0000	1.1500	0.86957	0.00000
2	1.3225	0.75614	0.46512	2.1500	0.61512	1.6257	0.46512
3	1.5209	0.65752	0.28798	3.4725	0.43798	2.2832	0.90713
4	1.7490	0.57175	0.20027	4.9934	0.35027	2.8550	1.3263
5	2.0114	0.49718	0.14832	6.7424	0.29832	3.3522	1.7228
6	2.3131	0.43233	0.11424	8.7537	0.26424	3.7845	2.0972
7	2.6600	0.37594	0.09036	11.067	0.24036	4.1604	2.4498
8	3.0590	0.32690	0.07285	13.727	0.22285	4.4873	2.7813
9	3.5179	0.28426	0.05957	16.786	0.20957	4.7716	3.0922
10	4.0456	0.24718	0.04925	20.304	0.19925	5.0188	3.3832
11	4.6524	0.21494	0.04107	24.349	0.19107	5.2337	3.6549
12	5.3503	0.18691	0.03448	29.002	0.18448	5.4206	3.9082
13	6.1528	0.16253	0.02911	34.352	0.17911	5.5831	4.1438
14	7.0757	0.14133	0.02469	40.505	0.17469	5.7245	4.3624
15	8.1371	0.12289	0.02102	47.580	0.17102	5.8474	4.5650
16	9.3576	0.10686	0.01795	55.717	0.16795	5.9542	4.7522
17	10.761	0.09293	0.01537	65.075	0.16537	6.0472	4.9251
18	12.375	0.08081	0.01319	75.836	0.16319	6.1280	5.0843
19	14.232	0.07027	0.01134	88.212	0.16134	6.1982	5.2307
20	16.367	0.06110	0.00976	102.44	0.15976	6.2593	5.3651
21	18.822	0.05313	0.00842	118.81	0.15842	6.3125	5.4883
22	21.645	0.04620	0.00727	137.63	0.15727	6.3587	5.6010
23	24.891	0.04017	0.00628	159.28	0.15628	6.3988	5.7040
24	28.625	0.03493	0.00543	184.17	0.15543	6.4338	5.7979
25	32.919	0.03038	0.00470	212.79	0.15470	6.4641	5.8834
26	37.857	0.02642	0.00407	245.71	0.15407	6.4906	5.9612
27	43.535	0.02297	0.00353	283.57	0.15353	6.5135	6.0319
28	50.066	0.01997	0.00306	327.10	0.15306	6.5335	6.0960
29	57.575	0.01737	0.00265	377.17	0.15265	6.5509	6.1541
30	66.212	0.01510	0.00230	434.75	0.15230	6.5660	6.2066
31	76.144	0.01313	0.00200	500.96	0.15200	6.5791	6.2541
32	87.565	0.01142	0.00173	577.10	0.15173	6.5905	6.2970
33	100.70	0.00993	0.00150	664.67	0.15150	6.6005	6.3357
34	115.80	0.00864	0.00131	765.37	0.15131	6.6091	6.3705
35	133.18	0.00751	0.00113	881.17	0.15113	6.6166	6.4019
40	267.86	0.00373	0.00056	1779.1	0.15056	6.6418	6.5168
45	538.77	0.00186	0.00028	3585.1	0.15028	6.6543	6.5830
50	1083.7	0.00092	0.00014	7217.7	0.15014	6.6605	6.6205
55	2179.6	0.00046	0.00007	14524.0	0.15007	6.6636	6.6414

$i = 20\%$ Discrete Compounding, Discrete Cash Flows

	SINGLE PAYMENT		UNIFORM SERIES				Arithmetic Gradient Series Factor
	Compound Amount Factor	Present Worth Factor	Sinking Fund Factor	Uniform Series Factor	Capital Recovery Factor	Series Present Worth Factor	
N	(F/P,i,N)	(P/F,i,N)	(A/F,i,N)	(F/A,i,N)	(A/P,i,N)	(P/A,i,N)	(A/G,i,N)
1	1.2000	0.83333	1.0000	1.0000	1.2000	0.83333	0.00000
2	1.4400	0.69444	0.45455	2.2000	0.65455	1.5278	0.45455
3	1.7280	0.57870	0.27473	3.6400	0.47473	2.1065	0.87912
4	2.0736	0.48225	0.18629	5.3680	0.38629	2.5887	1.2742
5	2.4883	0.40188	0.13438	7.4416	0.33438	2.9906	1.6405
6	2.9860	0.33490	0.10071	9.9299	0.30071	3.3255	1.9788
7	3.5832	0.27908	0.07742	12.916	0.27742	3.6046	2.2902
8	4.2998	0.23257	0.06061	16.499	0.26061	3.8372	2.5756
9	5.1598	0.19381	0.04808	20.799	0.24808	4.0310	2.8364
10	6.1917	0.16151	0.03852	25.959	0.23852	4.1925	3.0739
11	7.4301	0.13459	0.03110	32.150	0.23110	4.3271	3.2893
12	8.9161	0.11216	0.02526	39.581	0.22526	4.4392	3.4841
13	10.699	0.09346	0.02062	48.497	0.22062	4.5327	3.6597
14	12.839	0.07789	0.01689	59.196	0.21689	4.6106	3.8175
15	15.407	0.06491	0.01388	72.035	0.21388	4.6755	3.9588
16	18.488	0.05409	0.01144	87.442	0.21144	4.7296	4.0851
17	22.186	0.04507	0.00944	105.93	0.20944	4.7746	4.1976
18	26.623	0.03756	0.00781	128.12	0.20781	4.8122	4.2975
19	31.948	0.03130	0.00646	154.74	0.20646	4.8435	4.3861
20	38.338	0.02608	0.00536	186.69	0.20536	4.8696	4.4643
21	46.005	0.02174	0.00444	225.03	0.20444	4.8913	4.5334
22	55.206	0.01811	0.00369	271.03	0.20369	4.9094	4.5941
23	66.247	0.01509	0.00307	326.24	0.20307	4.9245	4.6475
24	79.497	0.01258	0.00255	392.48	0.20255	4.9371	4.6943
25	95.396	0.01048	0.00212	471.98	0.20212	4.9476	4.7352
26	114.48	0.00874	0.00176	567.38	0.20176	4.9563	4.7709
27	137.37	0.00728	0.00147	681.85	0.20147	4.9636	4.8020
28	164.84	0.00607	0.00122	819.22	0.20122	4.9697	4.8291
29	197.81	0.00506	0.00102	984.07	0.20102	4.9747	4.8527
30	237.38	0.00421	0.00085	1181.9	0.20085	4.9789	4.8731
31	284.85	0.00351	0.00070	1419.3	0.20070	4.9824	4.8908
32	341.82	0.00293	0.00059	1704.1	0.20059	4.9854	4.9061
33	410.19	0.00244	0.00049	2045.9	0.20049	4.9878	4.9194
34	492.22	0.00203	0.00041	2456.1	0.20041	4.9898	4.9308
35	590.67	0.00169	0.00034	2948.3	0.20034	4.9915	4.9406

i = 25%   Discrete Compounding, Discrete Cash Flows

	SINGLE PAYMENT		UNIFORM SERIES				Arithmetic Gradient Series Factor
	Compound Amount Factor	Present Worth Factor	Sinking Fund Factor	Uniform Series Factor	Capital Recovery Factor	Series Present Worth Factor	
N	(F/P,i,N)	(P/F,i,N)	(A/F,i,N)	(F/A,i,N)	(A/P,i,N)	(P/A,i,N)	(A/G,i,N)
1	1.2500	0.80000	1.0000	1.0000	1.2500	0.80000	0.00000
2	1.5625	0.64000	0.44444	2.2500	0.69444	1.4400	0.44444
3	1.9531	0.51200	0.26230	3.8125	0.51230	1.9520	0.85246
4	2.4414	0.40960	0.17344	5.7656	0.42344	2.3616	1.2249
5	3.0518	0.32768	0.12185	8.2070	0.37185	2.6893	1.5631
6	3.8147	0.26214	0.08882	11.259	0.33882	2.9514	1.8683
7	4.7684	0.20972	0.06634	15.073	0.31634	3.1611	2.1424
8	5.9605	0.16777	0.05040	19.842	0.30040	3.3289	2.3872
9	7.4506	0.13422	0.03876	25.802	0.28876	3.4631	2.6048
10	9.3132	0.10737	0.03007	33.253	0.28007	3.5705	2.7971
11	11.642	0.08590	0.02349	42.566	0.27349	3.6564	2.9663
12	14.552	0.06872	0.01845	54.208	0.26845	3.7251	3.1145
13	18.190	0.05498	0.01454	68.760	0.26454	3.7801	3.2437
14	22.737	0.04398	0.01150	86.949	0.26150	3.8241	3.3559
15	28.422	0.03518	0.00912	109.69	0.25912	3.8593	3.4530
16	35.527	0.02815	0.00724	138.11	0.25724	3.8874	3.5366
17	44.409	0.02252	0.00576	173.64	0.25576	3.9099	3.6084
18	55.511	0.01801	0.00459	218.04	0.25459	3.9279	3.6698
19	69.389	0.01441	0.00366	273.56	0.25366	3.9424	3.7222
20	86.736	0.01153	0.00292	342.94	0.25292	3.9539	3.7667
21	108.42	0.00922	0.00233	429.68	0.25233	3.9631	3.8045
22	135.53	0.00738	0.00186	538.10	0.25186	3.9705	3.8365
23	169.41	0.00590	0.00148	673.63	0.25148	3.9764	3.8634
24	211.76	0.00472	0.00119	843.03	0.25119	3.9811	3.8861
25	264.70	0.00378	0.00095	1054.8	0.25095	3.9849	3.9052
26	330.87	0.00302	0.00076	1319.5	0.25076	3.9879	3.9212
27	413.59	0.00242	0.00061	1650.4	0.25061	3.9903	3.9346
28	516.99	0.00193	0.00048	2064.0	0.25048	3.9923	3.9457
29	646.23	0.00155	0.00039	2580.9	0.25039	3.9938	3.9551
30	807.79	0.00124	0.00031	3227.2	0.25031	3.9950	3.9628
31	1009.7	0.00099	0.00025	4035.0	0.25025	3.9960	3.9693
32	1262.2	0.00079	0.00020	5044.7	0.25020	3.9968	3.9746
33	1577.7	0.00063	0.00016	6306.9	0.25016	3.9975	3.9791
34	1972.2	0.00051	0.00013	7884.6	0.25013	3.9980	3.9828
35	2465.2	0.00041	0.00010	9856.8	0.25010	3.9984	3.9858

$i = 30\%$ Discrete Compounding, Discrete Cash Flows

	SINGLE PAYMENT		UNIFORM SERIES				Arithmetic Gradient Series Factor
	Compound Amount Factor	Present Worth Factor	Sinking Fund Factor	Uniform Series Factor	Capital Recovery Factor	Series Present Worth Factor	
N	(F/P,i,N)	(P/F,i,N)	(A/F,i,N)	(F/A,i,N)	(A/P,i,N)	(P/A,i,N)	(A/G,i,N)
1	1.3000	0.76923	1.0000	1.0000	1.3000	0.76923	0.00000
2	1.6900	0.59172	0.43478	2.3000	0.73478	1.3609	0.43478
3	2.1970	0.45517	0.25063	3.9900	0.55063	1.8161	0.82707
4	2.8561	0.35013	0.16163	6.1870	0.46163	2.1662	1.1783
5	3.7129	0.26933	0.11058	9.0431	0.41058	2.4356	1.4903
6	4.8268	0.20718	0.07839	12.756	0.37839	2.6427	1.7654
7	6.2749	0.15937	0.05687	17.583	0.35687	2.8021	2.0063
8	8.1573	0.12259	0.04192	23.858	0.34192	2.9247	2.2156
9	10.604	0.09430	0.03124	32.015	0.33124	3.0190	2.3963
10	13.786	0.07254	0.02346	42.619	0.32346	3.0915	2.5512
11	17.922	0.05580	0.01773	56.405	0.31773	3.1473	2.6833
12	23.298	0.04292	0.01345	74.327	0.31345	3.1903	2.7952
13	30.288	0.03302	0.01024	97.625	0.31024	3.2233	2.8895
14	39.374	0.02540	0.00782	127.91	0.30782	3.2487	2.9685
15	51.186	0.01954	0.00598	167.29	0.30598	3.2682	3.0344
16	66.542	0.01503	0.00458	218.47	0.30458	3.2832	3.0892
17	86.504	0.01156	0.00351	285.01	0.30351	3.2948	3.1345
18	112.46	0.00889	0.00269	371.52	0.30269	3.3037	3.1718
19	146.19	0.00684	0.00207	483.97	0.30207	3.3105	3.2025
20	190.05	0.00526	0.00159	630.17	0.30159	3.3158	3.2275
21	247.06	0.00405	0.00122	820.22	0.30122	3.3198	3.2480
22	321.18	0.00311	0.00094	1067.3	0.30094	3.3230	3.2646
23	417.54	0.00239	0.00072	1388.5	0.30072	3.3254	3.2781
24	542.80	0.00184	0.00055	1806.0	0.30055	3.3272	3.2890
25	705.64	0.00142	0.00043	2348.8	0.30043	3.3286	3.2979
26	917.33	0.00109	0.00033	3054.4	0.30033	3.3297	3.3050
27	1192.5	0.00084	0.00025	3971.8	0.30025	3.3305	3.3107
28	1550.3	0.00065	0.00019	5164.3	0.30019	3.3312	3.3153
29	2015.4	0.00050	0.00015	6714.6	0.30015	3.3317	3.3189
30	2620.0	0.00038	0.00011	8730.0	0.30011	3.3321	3.3219
31	3406.0	0.00029	0.00009	11 350.0	0.30009	3.3324	3.3242
32	4427.8	0.00023	0.00007	14 756.0	0.30007	3.3326	3.3261
33	5756.1	0.00017	0.00005	19 184.0	0.30005	3.3328	3.3276
34	7483.0	0.00013	0.00004	24 940.0	0.30004	3.3329	3.3288
35	9727.9	0.00010	0.00003	32 423.0	0.30003	3.3330	3.3297

$i = 40\%$ Discrete Compounding, Discrete Cash Flows

	SINGLE PAYMENT		UNIFORM SERIES				Arithmetic Gradient Series Factor
	Compound Amount Factor	Present Worth Factor	Sinking Fund Factor	Uniform Series Factor	Capital Recovery Factor	Series Present Worth Factor	
N	(F/P,i,N)	(P/F,i,N)	(A/F,i,N)	(F/A,i,N)	(A/P,i,N)	(P/A,i,N)	(A/G,i,N)
1	1.4000	0.71429	1.0000	1.0000	1.4000	0.71429	0.00000
2	1.9600	0.51020	0.41667	2.4000	0.81667	1.2245	0.41667
3	2.7440	0.36443	0.22936	4.3600	0.62936	1.5889	0.77982
4	3.8416	0.26031	0.14077	7.1040	0.54077	1.8492	1.0923
5	5.3782	0.18593	0.09136	10.946	0.49136	2.0352	1.3580
6	7.5295	0.13281	0.06126	16.324	0.46126	2.1680	1.5811
7	10.541	0.09486	0.04192	23.853	0.44192	2.2628	1.7664
8	14.758	0.06776	0.02907	34.395	0.42907	2.3306	1.9185
9	20.661	0.04840	0.02034	49.153	0.42034	2.3790	2.0422
10	28.925	0.03457	0.01432	69.814	0.41432	2.4136	2.1419
11	40.496	0.02469	0.01013	98.739	0.41013	2.4383	2.2215
12	56.694	0.01764	0.00718	139.23	0.40718	2.4559	2.2845
13	79.371	0.01260	0.00510	195.93	0.40510	2.4685	2.3341
14	111.12	0.00900	0.00363	275.30	0.40363	2.4775	2.3729
15	155.57	0.00643	0.00259	386.42	0.40259	2.4839	2.4030
16	217.80	0.00459	0.00185	541.99	0.40185	2.4885	2.4262
17	304.91	0.00328	0.00132	759.78	0.40132	2.4918	2.4441
18	426.88	0.00234	0.00094	1064.70	0.40094	2.4941	2.4577
19	597.63	0.00167	0.00067	1491.58	0.40067	2.4958	2.4682
20	836.68	0.00120	0.00048	2089.21	0.40048	2.4970	2.4761
21	1171.36	0.00085	0.00034	2925.89	0.40034	2.4979	2.4821
22	1639.90	0.00061	0.00024	4097.24	0.40024	2.4985	2.4866
23	2295.86	0.00044	0.00017	5737.14	0.40017	2.4989	2.4900
24	3214.20	0.00031	0.00012	8033.00	0.40012	2.4992	2.4925
25	4499.88	0.00022	0.00009	11 247.0	0.40009	2.4994	2.4944
26	6299.83	0.00016	0.00006	15 747.0	0.40006	2.4996	2.4959
27	8819.76	0.00011	0.00005	22 047.0	0.40005	2.4997	2.4969
28	12 348.0	0.00008	0.00003	30 867.0	0.40003	2.4998	2.4977
29	17 287.0	0.00006	0.00002	43 214.0	0.40002	2.4999	2.4983
30	24 201.0	0.00004	0.00002	60 501.0	0.40002	2.4999	2.4988
31	33 882.0	0.00003	0.00001	84 703.0	0.40001	2.4999	2.4991
32	47 435.0	0.00002	0.00001	118 585.0	0.40001	2.4999	2.4993
33	66 409.0	0.00002	0.00001	166 019.0	0.40001	2.5000	2.4995
34	92 972.0	0.00001	0.00000	232 428.0	0.40000	2.5000	2.4996
35	130 161.0	0.00001	0.00000	325 400.0	0.40000	2.5000	2.4997

i = 50%   Discrete Compounding, Discrete Cash Flows

	SINGLE PAYMENT		UNIFORM SERIES				Arithmetic Gradient Series Factor
	Compound Amount Factor	Present Worth Factor	Sinking Fund Factor	Uniform Series Factor	Capital Recovery Factor	Series Present Worth Factor	
N	(F/P,i,N)	(P/F,i,N)	(A/F,i,N)	(F/A,i,N)	(A/P,i,N)	(P/A,i,N)	(A/G,i,N)
1	1.5000	0.66667	1.0000	1.0000	1.5000	0.66667	0.00000
2	2.2500	0.44444	0.40000	2.5000	0.90000	1.1111	0.40000
3	3.3750	0.29630	0.21053	4.7500	0.71053	1.4074	0.73684
4	5.0625	0.19753	0.12308	8.1250	0.62308	1.6049	1.0154
5	7.5938	0.13169	0.07583	13.1875	0.57583	1.7366	1.2417
6	11.3906	0.08779	0.04812	20.781	0.54812	1.8244	1.4226
7	17.0859	0.05853	0.03108	32.172	0.53108	1.8829	1.5648
8	25.6289	0.03902	0.02030	49.258	0.52030	1.9220	1.6752
9	38.443	0.02601	0.01335	74.887	0.51335	1.9480	1.7596
10	57.665	0.01734	0.00882	113.330	0.50882	1.9653	1.8235
11	86.498	0.01156	0.00585	170.995	0.50585	1.9769	1.8713
12	129.746	0.00771	0.00388	257.493	0.50388	1.9846	1.9068
13	194.620	0.00514	0.00258	387.239	0.50258	1.9897	1.9329
14	291.929	0.00343	0.00172	581.86	0.50172	1.9931	1.9519
15	437.894	0.00228	0.00114	873.79	0.50114	1.9954	1.9657
16	656.841	0.00152	0.00076	1311.68	0.50076	1.9970	1.9756
17	985.261	0.00101	0.00051	1968.52	0.50051	1.9980	1.9827
18	1477.89	0.00068	0.00034	2953.78	0.50034	1.9986	1.9878
19	2216.84	0.00045	0.00023	4431.68	0.50023	1.9991	1.9914
20	3325.26	0.00030	0.00015	6648.51	0.50015	1.9994	1.9940
21	4987.89	0.00020	0.00010	9973.77	0.50010	1.9996	1.9958
22	7481.83	0.00013	0.00007	14962.0	0.50007	1.9997	1.9971
23	11223.0	0.00009	0.00004	22443.0	0.50004	1.9998	1.9980
24	16834.0	0.00006	0.00003	33666.0	0.50003	1.9999	1.9986
25	25251.0	0.00004	0.00002	50500.0	0.50002	1.9999	1.9990
26	37877.0	0.00003	0.00001	75752.0	0.50001	1.9999	1.9993
27	56815.0	0.00002	0.00001	113628.0	0.50001	2.0000	1.9995
28	85223.0	0.00001	0.00001	170443.0	0.50001	2.0000	1.9997
29	127834.0	0.00001	0.00000	255666.0	0.50000	2.0000	1.9998
30	191751.0	0.00001	0.00000	383500.0	0.50000	2.0000	1.9998
31	287627.0	0.00000	0.00000	575251.0	0.50000	2.0000	1.9999
32	431440.0	0.00000	0.00000	862878.0	0.50000	2.0000	1.9999
33	647160.0	0.00000	0.00000	1294318.0	0.50000	2.0000	1.9999
34	970740.0	0.00000	0.00000	1941477.0	0.50000	2.0000	2.0000
35	1456110.0	0.00000	0.00000	2912217.0	0.50000	2.0000	2.0000

Compound Interest Factors for Continuous Compounding, Discrete Cash Flows

$r = 1\%$ Continuous Compounding, Discrete Cash Flows

	SINGLE PAYMENT		UNIFORM SERIES				Arithmetic Gradient Series Factor
	Compound Amount Factor	Present Worth Factor	Sinking Fund Factor	Uniform Series Factor	Capital Recovery Factor	Series Present Worth Factor	
N	(F/P,r,N)	(P/F,r,N)	(A/F,r,N)	(F/A,r,N)	(A/P,r,N)	(P/A,r,N)	(A/G,r,N)
1	1.0101	0.99005	1.0000	1.0000	1.0101	0.99005	0.00000
2	1.0202	0.98020	0.49750	2.0101	0.50755	1.97025	0.49750
3	1.0305	0.97045	0.33001	3.0303	0.34006	2.94069	0.99333
4	1.0408	0.96079	0.24626	4.0607	0.25631	3.90148	1.48750
5	1.0513	0.95123	0.19602	5.1015	0.20607	4.85271	1.98000
6	1.0618	0.94176	0.16253	6.1528	0.17258	5.79448	2.47084
7	1.0725	0.93239	0.13861	7.2146	0.14866	6.72687	2.96000
8	1.0833	0.92312	0.12067	8.2871	0.13072	7.64999	3.44751
9	1.0942	0.91393	0.10672	9.3704	0.11677	8.56392	3.93334
10	1.1052	0.90484	0.09556	10.4646	0.10561	9.46876	4.41751
11	1.1163	0.89583	0.08643	11.5698	0.09648	10.36459	4.90002
12	1.1275	0.88692	0.07883	12.6860	0.08888	11.25151	5.38086
13	1.1388	0.87810	0.07239	13.8135	0.08244	12.12961	5.86004
14	1.1503	0.86936	0.06688	14.9524	0.07693	12.99896	6.33755
15	1.1618	0.86071	0.06210	16.1026	0.07215	13.85967	6.81340
16	1.1735	0.85214	0.05792	17.2645	0.06797	14.71182	7.28759
17	1.1853	0.84366	0.05424	18.4380	0.06429	15.55548	7.76012
18	1.1972	0.83527	0.05096	19.6233	0.06101	16.39075	8.23098
19	1.2092	0.82696	0.04803	20.8205	0.05808	17.21771	8.70018
20	1.2214	0.81873	0.04539	22.0298	0.05544	18.03644	9.16772
21	1.2337	0.81058	0.04301	23.2512	0.05306	18.84703	9.63360
22	1.2461	0.80252	0.04084	24.4848	0.05089	19.64954	10.09782
23	1.2586	0.79453	0.03886	25.7309	0.04891	20.44408	10.56039
24	1.2712	0.78663	0.03705	26.9895	0.04710	21.23071	11.02129
25	1.2840	0.77880	0.03538	28.2608	0.04543	22.00951	11.48054
26	1.2969	0.77105	0.03385	29.5448	0.04390	22.78056	11.93813
27	1.3100	0.76338	0.03242	30.8417	0.04247	23.54394	12.39407
28	1.3231	0.75578	0.03110	32.1517	0.04115	24.29972	12.84835
29	1.3364	0.74826	0.02987	33.4748	0.03992	25.04798	13.30098
30	1.3499	0.74082	0.02873	34.8112	0.03878	25.78880	13.75196
31	1.3634	0.73345	0.02765	36.1611	0.03770	26.52225	14.20128
32	1.3771	0.72615	0.02665	37.5245	0.03670	27.24840	14.64895
33	1.3910	0.71892	0.02571	38.9017	0.03576	27.96732	15.09498
34	1.4049	0.71177	0.02482	40.2926	0.03487	28.67909	15.53935
35	1.4191	0.70469	0.02398	41.6976	0.03403	29.38378	15.98208
40	1.4918	0.67032	0.02043	48.9370	0.03048	32.80343	18.17104
45	1.5683	0.63763	0.01768	56.5475	0.02773	36.05630	20.31900
50	1.6487	0.60653	0.01549	64.5483	0.02554	39.15053	22.42613
55	1.7333	0.57695	0.01371	72.9593	0.02376	42.09385	24.49262
60	1.8221	0.54881	0.01222	81.8015	0.02227	44.89362	26.51868
65	1.9155	0.52205	0.01098	91.0971	0.02103	47.55684	28.50455
70	2.0138	0.49659	0.00991	100.869	0.01996	50.09018	30.45046
75	2.1170	0.47237	0.00900	111.142	0.01905	52.49997	32.35670
80	2.2255	0.44933	0.00820	121.942	0.01825	54.79223	34.22354
85	2.3396	0.42741	0.00750	133.296	0.01755	56.97269	36.05128
90	2.4596	0.40657	0.00689	145.232	0.01694	59.04681	37.84024
95	2.5857	0.38674	0.00634	157.779	0.01639	61.01978	39.59075
100	2.7183	0.36788	0.00585	170.970	0.01582	63.21206	41.30316

$r = 2\%$ Continuous Compounding, Discrete Cash Flows

	SINGLE PAYMENT		UNIFORM SERIES				Arithmetic Gradient Series Factor
	Compound Amount Factor	Present Worth Factor	Sinking Fund Factor	Uniform Series Factor	Capital Recovery Factor	Series Present Worth Factor	
N	(F/P,r,N)	(P/F,r,N)	(A/F,r,N)	(F/A,r,N)	(A/P,r,N)	(P/A,r,N)	(A/G,r,N)
1	1.0202	0.98020	1.0000	1.0000	1.0202	0.98020	0.00000
2	1.0408	0.96079	0.49500	2.0202	0.51520	1.94099	0.49500
3	1.0618	0.94176	0.32669	3.0610	0.34689	2.88275	0.98667
4	1.0833	0.92312	0.24255	4.1228	0.26275	3.80587	1.47500
5	1.1052	0.90484	0.19208	5.2061	0.21228	4.71071	1.96001
6	1.1275	0.88692	0.15845	6.3113	0.17865	5.59763	2.44168
7	1.1503	0.86936	0.13443	7.4388	0.15463	6.46699	2.92003
8	1.1735	0.85214	0.11643	8.5891	0.13663	7.31913	3.39505
9	1.1972	0.83527	0.10243	9.7626	0.12263	8.15440	3.86674
10	1.2214	0.81873	0.09124	10.9598	0.11144	8.97313	4.33511
11	1.2461	0.80252	0.08209	12.1812	0.10230	9.77565	4.80016
12	1.2712	0.78663	0.07448	13.4273	0.09468	10.56228	5.26190
13	1.2969	0.77105	0.06803	14.6985	0.08824	11.33333	5.72032
14	1.3231	0.75578	0.06252	15.9955	0.08272	12.08911	6.17543
15	1.3499	0.74082	0.05774	17.3186	0.07794	12.82993	6.62723
16	1.3771	0.72615	0.05357	18.6685	0.07377	13.55608	7.07573
17	1.4049	0.71177	0.04989	20.0456	0.07009	14.26785	7.52093
18	1.4333	0.69768	0.04662	21.4505	0.06682	14.96553	7.96283
19	1.4623	0.68386	0.04370	22.8839	0.06390	15.64939	8.40144
20	1.4918	0.67032	0.04107	24.3461	0.06128	16.31971	8.83677
21	1.5220	0.65705	0.03870	25.8380	0.05890	16.97675	9.26882
22	1.5527	0.64404	0.03655	27.3599	0.05675	17.62079	9.69759
23	1.5841	0.63128	0.03459	28.9126	0.05479	18.25207	10.12309
24	1.6161	0.61878	0.03279	30.4967	0.05299	18.87086	10.54533
25	1.6487	0.60653	0.03114	32.1128	0.05134	19.47739	10.96431
26	1.6820	0.59452	0.02962	33.7615	0.04982	20.07191	11.38005
27	1.7160	0.58275	0.02821	35.4435	0.04842	20.65466	11.79253
28	1.7507	0.57121	0.02691	37.1595	0.04711	21.22587	12.20178
29	1.7860	0.55990	0.02570	38.9102	0.04590	21.78576	12.60780
30	1.8221	0.54881	0.02457	40.6963	0.04477	22.33458	13.01059
31	1.8589	0.53794	0.02352	42.5184	0.04372	22.87252	13.41017
32	1.8965	0.52729	0.02253	44.3773	0.04274	23.39981	13.80654
33	1.9348	0.51685	0.02161	46.2738	0.04181	23.91666	14.19971
34	1.9739	0.50662	0.02074	48.2086	0.04094	24.42328	14.58969
35	2.0138	0.49659	0.01993	50.1824	0.04013	24.91987	14.97648
40	2.2255	0.44933	0.01648	60.6663	0.03668	27.25913	16.86302
45	2.4596	0.40657	0.01384	72.2528	0.03404	29.37579	18.67137
50	2.7183	0.36788	0.01176	85.0578	0.03196	31.29102	20.40283
55	3.0042	0.33287	0.01008	99.2096	0.03028	33.02399	22.05883
60	3.3201	0.30119	0.00871	114.850	0.02891	34.59205	23.64090
65	3.6693	0.27253	0.00757	132.135	0.02777	36.01089	25.15068
70	4.0552	0.24660	0.00661	151.237	0.02681	37.29471	26.58991
75	4.4817	0.22313	0.00580	172.349	0.02600	38.45635	27.96040
80	4.9530	0.20190	0.00511	195.682	0.02531	39.50745	29.26404
85	5.4739	0.18268	0.00452	221.468	0.02472	40.45853	30.50278
90	6.0496	0.16530	0.00400	249.966	0.02420	41.31910	31.67864
95	6.6859	0.14957	0.00355	281.461	0.02375	42.09777	32.79365
100	7.3891	0.13534	0.00316	316.269	0.02336	42.80234	33.84990

$r = 3\%$ **Continuous Compounding, Discrete Cash Flows**

	SINGLE PAYMENT		UNIFORM SERIES				Arithmetic Gradient Series Factor
	Compound Amount Factor	Present Worth Factor	Sinking Fund Factor	Uniform Series Factor	Capital Recovery Factor	Series Present Worth Factor	
N	(F/P,r,N)	(P/F,r,N)	(A/F,r,N)	(F/A,r,N)	(A/P,r,N)	(P/A,r,N)	(A/G,r,N)
1	1.0305	0.97045	1.0000	1.0000	1.0305	0.97045	0.00000
2	1.0618	0.94176	0.49250	2.0305	0.52296	1.91221	0.49250
3	1.0942	0.91393	0.32338	3.0923	0.35384	2.82614	0.98000
4	1.1275	0.88692	0.23886	4.1865	0.26932	3.71306	1.46251
5	1.1618	0.86071	0.18818	5.3140	0.21864	4.57377	1.94002
6	1.1972	0.83527	0.15442	6.4758	0.18488	5.40904	2.41255
7	1.2337	0.81058	0.13033	7.6730	0.16078	6.21962	2.88009
8	1.2712	0.78663	0.11228	8.9067	0.14273	7.00625	3.34265
9	1.3100	0.76338	0.09825	10.1779	0.12871	7.76963	3.80025
10	1.3499	0.74082	0.08705	11.4879	0.11750	8.51045	4.25287
11	1.3910	0.71892	0.07790	12.8378	0.10835	9.22937	4.70055
12	1.4333	0.69768	0.07028	14.2287	0.10073	9.92705	5.14328
13	1.4770	0.67706	0.06385	15.6621	0.09430	10.60411	5.58107
14	1.5220	0.65705	0.05835	17.1390	0.08880	11.26115	6.01393
15	1.5683	0.63763	0.05359	18.6610	0.08404	11.89878	6.44189
16	1.6161	0.61878	0.04943	20.2293	0.07989	12.51756	6.86494
17	1.6653	0.60050	0.04578	21.8454	0.07623	13.11806	7.28311
18	1.7160	0.58275	0.04253	23.5107	0.07299	13.70081	7.69641
19	1.7683	0.56553	0.03964	25.2267	0.07010	14.26633	8.10485
20	1.8221	0.54881	0.03704	26.9950	0.06750	14.81515	8.50845
21	1.8776	0.53259	0.03470	28.8171	0.06516	15.34774	8.90722
22	1.9348	0.51685	0.03258	30.6947	0.06303	15.86459	9.30119
23	1.9937	0.50158	0.03065	32.6295	0.06110	16.36617	9.69038
24	2.0544	0.48675	0.02888	34.6232	0.05934	16.85292	10.07479
25	2.1170	0.47237	0.02726	36.6776	0.05772	17.32528	10.45445
26	2.1815	0.45841	0.02578	38.7946	0.05623	17.78369	10.82939
27	2.2479	0.44486	0.02440	40.9761	0.05486	18.22855	11.19962
28	2.3164	0.43171	0.02314	43.2240	0.05359	18.66026	11.56517
29	2.3869	0.41895	0.02196	45.5404	0.05241	19.07921	11.92605
30	2.4596	0.40657	0.02086	47.9273	0.05132	19.48578	12.28230
31	2.5345	0.39455	0.01985	50.3869	0.05030	19.88033	12.63393
32	2.6117	0.38289	0.01890	52.9214	0.04935	20.26323	12.98098
33	2.6912	0.37158	0.01801	55.5331	0.04846	20.63480	13.32346
34	2.7732	0.36059	0.01717	58.2243	0.04763	20.99540	13.66140
35	2.8577	0.34994	0.01639	60.9975	0.04685	21.34534	13.99484
40	3.3201	0.30119	0.01313	76.1830	0.04358	22.94587	15.59532
45	3.8574	0.25924	0.01066	93.8259	0.04111	24.32346	17.08739
50	4.4817	0.22313	0.00875	114.324	0.03920	25.50917	18.47499
55	5.2070	0.19205	0.00724	138.140	0.03769	26.52971	19.76232
60	6.0496	0.16530	0.00603	165.809	0.03649	27.40811	20.95382
65	7.0287	0.14227	0.00505	197.957	0.03551	28.16415	22.05405
70	8.1662	0.12246	0.00425	235.307	0.03470	28.81487	23.06771
75	9.4877	0.10540	0.00359	278.702	0.03404	29.37496	23.99955
80	11.0232	0.09072	0.00304	329.119	0.03349	29.85703	24.85433
85	12.8071	0.07808	0.00258	387.696	0.03303	30.27196	25.63678
90	14.8797	0.06721	0.00219	455.753	0.03265	30.62908	26.35156
95	17.2878	0.05784	0.00187	534.823	0.03232	30.93647	27.00324
100	20.0855	0.04979	0.00160	626.690	0.03205	31.20103	27.59626

r = 4%   Continuous Compounding, Discrete Cash Flows

	SINGLE PAYMENT		UNIFORM SERIES				Arithmetic Gradient Series Factor
	Compound Amount Factor	Present Worth Factor	Sinking Fund Factor	Uniform Series Factor	Capital Recovery Factor	Series Present Worth Factor	
N	(F/P,r,N)	(P/F,r,N)	(A/F,r,N)	(F/A,r,N)	(A/P,r,N)	(P/A,r,N)	(A/G,r,N)
1	1.0408	0.96079	1.0000	1.0000	1.0408	0.96079	0.00000
2	1.0833	0.92312	0.49000	2.0408	0.53081	1.88391	0.49000
3	1.1275	0.88692	0.32009	3.1241	0.36090	2.77083	0.97334
4	1.1735	0.85214	0.23521	4.2516	0.27602	3.62297	1.45002
5	1.2214	0.81873	0.18433	5.4251	0.22514	4.44170	1.92006
6	1.2712	0.78663	0.15045	6.6465	0.19127	5.22833	2.38345
7	1.3231	0.75578	0.12630	7.9178	0.16711	5.98411	2.84021
8	1.3771	0.72615	0.10821	9.2409	0.14903	6.71026	3.29036
9	1.4333	0.69768	0.09418	10.6180	0.13499	7.40794	3.73391
10	1.4918	0.67032	0.08298	12.0513	0.12379	8.07826	4.17089
11	1.5527	0.64404	0.07384	13.5432	0.11465	8.72229	4.60130
12	1.6161	0.61878	0.06624	15.0959	0.10705	9.34108	5.02517
13	1.6820	0.59452	0.05984	16.7120	0.10065	9.93560	5.44252
14	1.7507	0.57121	0.05437	18.3940	0.09518	10.50681	5.85339
15	1.8221	0.54881	0.04964	20.1447	0.09045	11.05562	6.25780
16	1.8965	0.52729	0.04552	21.9668	0.08633	11.58291	6.65577
17	1.9739	0.50662	0.04191	23.8633	0.08272	12.08953	7.04734
18	2.0544	0.48675	0.03870	25.8371	0.07951	12.57628	7.43255
19	2.1383	0.46767	0.03585	27.8916	0.07666	13.04395	7.81143
20	2.2255	0.44933	0.03330	30.0298	0.07411	13.49328	8.18401
21	2.3164	0.43171	0.03100	32.2554	0.07181	13.92499	8.55034
22	2.4109	0.41478	0.02893	34.5717	0.06974	14.33977	8.91045
23	2.5093	0.39852	0.02704	36.9826	0.06785	14.73829	9.26438
24	2.6117	0.38289	0.02532	39.4919	0.06613	15.12118	9.61219
25	2.7183	0.36788	0.02375	42.1036	0.06456	15.48906	9.95392
26	2.8292	0.35345	0.02231	44.8219	0.06312	15.84252	10.28960
27	2.9447	0.33960	0.02099	47.6511	0.06180	16.18211	10.61930
28	3.0649	0.32628	0.01976	50.5958	0.06058	16.50839	10.94305
29	3.1899	0.31349	0.01864	53.6607	0.05945	16.82188	11.26092
30	3.3201	0.30119	0.01759	56.8506	0.05840	17.12307	11.57295
31	3.4556	0.28938	0.01662	60.1707	0.05743	17.41246	11.87920
32	3.5966	0.27804	0.01572	63.6263	0.05653	17.69049	12.17971
33	3.7434	0.26714	0.01488	67.2230	0.05569	17.95763	12.47456
34	3.8962	0.25666	0.01409	70.9664	0.05490	18.21429	12.76379
35	4.0552	0.24660	0.01336	74.8626	0.05417	18.46089	13.04745
40	4.9530	0.20190	0.01032	96.8625	0.05113	19.55620	14.38452
45	6.0496	0.16530	0.00808	123.733	0.04889	20.45296	15.59182
50	7.3891	0.13534	0.00639	156.553	0.04720	21.18717	16.67745
55	9.0250	0.11080	0.00509	196.640	0.04590	21.78829	17.64976
60	11.0232	0.09072	0.00407	245.601	0.04488	22.28044	18.51721
65	13.4637	0.07427	0.00327	305.403	0.04409	22.68338	19.28820
70	16.4446	0.06081	0.00264	378.445	0.04345	23.01328	19.97102
75	20.0855	0.04979	0.00214	467.659	0.04295	23.28338	20.57366
80	24.5325	0.04076	0.00173	576.625	0.04255	23.50452	21.10378
85	29.9641	0.03337	0.00141	709.717	0.04222	23.68558	21.56867
90	36.5982	0.02732	0.00115	872.275	0.04196	23.83381	21.97512
95	44.7012	0.02237	0.00093	1070.82	0.04174	23.95517	22.32948
100	54.5982	0.01832	0.00076	1313.33	0.04157	24.05454	22.63760

$r = 5\%$ **Continuous Compounding, Discrete Cash Flows**

	SINGLE PAYMENT		UNIFORM SERIES				Arithmetic Gradient Series Factor
	Compound Amount Factor	Present Worth Factor	Sinking Fund Factor	Uniform Series Factor	Capital Recovery Factor	Series Present Worth Factor	
N	$(F/P,r,N)$	$(P/F,r,N)$	$(A/F,r,N)$	$(F/A,r,N)$	$(A/P,r,N)$	$(P/A,r,N)$	$(A/G,r,N)$
1	1.0513	0.95123	1.0000	1.0000	1.0513	0.95123	0.00000
2	1.1052	0.90484	0.48750	2.0513	0.53877	1.85607	0.48750
3	1.1618	0.86071	0.31681	3.1564	0.36808	2.71677	0.96668
4	1.2214	0.81873	0.23157	4.3183	0.28284	3.53551	1.43754
5	1.2840	0.77880	0.18052	5.5397	0.23179	4.31431	1.90011
6	1.3499	0.74082	0.14655	6.8237	0.19782	5.05512	2.35439
7	1.4191	0.70469	0.12235	8.1736	0.17362	5.75981	2.80042
8	1.4918	0.67032	0.10425	9.5926	0.15552	6.43013	3.23821
9	1.5683	0.63763	0.09022	11.0845	0.14149	7.06776	3.66780
10	1.6487	0.60653	0.07903	12.6528	0.13031	7.67429	4.08923
11	1.7333	0.57695	0.06992	14.3015	0.12119	8.25124	4.50252
12	1.8221	0.54881	0.06236	16.0347	0.11364	8.80005	4.90774
13	1.9155	0.52205	0.05600	17.8569	0.10727	9.32210	5.30491
14	2.0138	0.49659	0.05058	19.7724	0.10185	9.81868	5.69409
15	2.1170	0.47237	0.04590	21.7862	0.09717	10.29105	6.07534
16	2.2255	0.44933	0.04184	23.9032	0.09311	10.74038	6.44871
17	2.3396	0.42741	0.03827	26.1287	0.08954	11.16779	6.81425
18	2.4596	0.40657	0.03513	28.4683	0.08640	11.57436	7.17205
19	2.5857	0.38674	0.03233	30.9279	0.08360	11.96111	7.52215
20	2.7183	0.36788	0.02984	33.5137	0.08111	12.32898	7.86463
21	2.8577	0.34994	0.02760	36.2319	0.07887	12.67892	8.19957
22	3.0042	0.33287	0.02558	39.0896	0.07685	13.01179	8.52703
23	3.1582	0.31664	0.02376	42.0938	0.07503	13.32843	8.84710
24	3.3201	0.30119	0.02210	45.2519	0.07337	13.62962	9.15986
25	3.4903	0.28650	0.02059	48.5721	0.07186	13.91613	9.46539
26	3.6693	0.27253	0.01921	52.0624	0.07048	14.18866	9.76377
27	3.8574	0.25924	0.01794	55.7317	0.06921	14.44790	10.05510
28	4.0552	0.24660	0.01678	59.5891	0.06805	14.69450	10.33946
29	4.2631	0.23457	0.01571	63.6443	0.06698	14.92907	10.61695
30	4.4817	0.22313	0.01473	67.9074	0.06600	15.15220	10.88766
31	4.7115	0.21225	0.01381	72.3891	0.06509	15.36445	11.15168
32	4.9530	0.20190	0.01297	77.1006	0.06424	15.56634	11.40912
33	5.2070	0.19205	0.01219	82.0536	0.06346	15.75839	11.66006
34	5.4739	0.18268	0.01146	87.2606	0.06273	15.94108	11.90461
35	5.7546	0.17377	0.01078	92.7346	0.06205	16.11485	12.14288
40	7.3891	0.13534	0.00802	124.613	0.05930	16.86456	13.24346
45	9.4877	0.10540	0.00604	165.546	0.05731	17.44844	14.20240
50	12.1825	0.08208	0.00458	218.105	0.05586	17.90317	15.03289
55	15.6426	0.06393	0.00350	285.592	0.05477	18.25731	15.74801
60	20.0855	0.04979	0.00269	372.247	0.05396	18.53311	16.36042
65	25.7903	0.03877	0.00207	483.515	0.05334	18.74791	16.88218
70	33.1155	0.03020	0.00160	626.385	0.05287	18.91519	17.32453
75	42.5211	0.02352	0.00123	809.834	0.05251	19.04547	17.69786
80	54.5982	0.01832	0.00096	1045.39	0.05223	19.14694	18.01158
85	70.1054	0.01426	0.00074	1347.84	0.05201	19.22595	18.27416
90	90.0171	0.01111	0.00058	1736.20	0.05185	19.28749	18.49313
95	115.584	0.00865	0.00045	2234.87	0.05172	19.33542	18.67508
100	148.413	0.00674	0.00035	2875.17	0.05162	19.37275	18.82580

$r = 6\%$ **Continuous Compounding, Discrete Cash Flows**

	SINGLE PAYMENT		UNIFORM SERIES				Arithmetic Gradient Series Factor
	Compound Amount Factor	Present Worth Factor	Sinking Fund Factor	Uniform Series Factor	Capital Recovery Factor	Series Present Worth Factor	
N	(F/P,r,N)	(P/F,r,N)	(A/F,r,N)	(F/A,r,N)	(A/P,r,N)	(P/A,r,N)	(A/G,r,N)
1	1.0618	0.94176	1.0000	1.0000	1.0618	0.94176	0.00000
2	1.1275	0.88692	0.48500	2.0618	0.54684	1.82868	0.48500
3	1.1972	0.83527	0.31355	3.1893	0.37538	2.66396	0.96002
4	1.2712	0.78663	0.22797	4.3866	0.28981	3.45058	1.42508
5	1.3499	0.74082	0.17675	5.6578	0.23858	4.19140	1.88019
6	1.4333	0.69768	0.14270	7.0077	0.20454	4.88908	2.32539
7	1.5220	0.65705	0.11847	8.4410	0.18031	5.54612	2.76072
8	1.6161	0.61878	0.10037	9.9629	0.16221	6.16491	3.18622
9	1.7160	0.58275	0.08636	11.5790	0.14820	6.74766	3.60195
10	1.8221	0.54881	0.07522	13.2950	0.13705	7.29647	4.00797
11	1.9348	0.51685	0.06615	15.1171	0.12799	7.81332	4.40435
12	2.0544	0.48675	0.05864	17.0519	0.12048	8.30007	4.79114
13	2.1815	0.45841	0.05234	19.1064	0.11418	8.75848	5.16845
14	2.3164	0.43171	0.04698	21.2878	0.10881	9.19019	5.53633
15	2.4596	0.40657	0.04237	23.6042	0.10420	9.59676	5.89490
16	2.6117	0.38289	0.03837	26.0638	0.10020	9.97965	6.24424
17	2.7732	0.36059	0.03487	28.6755	0.09671	10.34025	6.58445
18	2.9447	0.33960	0.03180	31.4487	0.09363	10.67984	6.91564
19	3.1268	0.31982	0.02908	34.3934	0.09091	10.99966	7.23792
20	3.3201	0.30119	0.02665	37.5202	0.08849	11.30085	7.55141
21	3.5254	0.28365	0.02449	40.8403	0.08632	11.58451	7.85622
22	3.7434	0.26714	0.02254	44.3657	0.08438	11.85164	8.15248
23	3.9749	0.25158	0.02079	48.1091	0.08262	12.10322	8.44032
24	4.2207	0.23693	0.01920	52.0840	0.08104	12.34015	8.71986
25	4.4817	0.22313	0.01776	56.3047	0.07960	12.56328	8.99124
26	4.7588	0.21014	0.01645	60.7864	0.07829	12.77342	9.25460
27	5.0531	0.19790	0.01526	65.5452	0.07709	12.97131	9.51008
28	5.3656	0.18637	0.01416	70.5983	0.07600	13.15769	9.75782
29	5.6973	0.17552	0.01316	75.9639	0.07500	13.33321	9.99796
30	6.0496	0.16530	0.01225	81.6612	0.07408	13.49851	10.23066
31	6.4237	0.15567	0.01140	87.7109	0.07324	13.65418	10.45605
32	6.8210	0.14661	0.01062	94.1346	0.07246	13.80079	10.67429
33	7.2427	0.13807	0.00991	100.956	0.07174	13.93886	10.88553
34	7.6906	0.13003	0.00924	108.198	0.07108	14.06889	11.08992
35	8.1662	0.12246	0.00863	115.889	0.07047	14.19134	11.28761
40	11.0232	0.09072	0.00617	162.091	0.06801	14.70461	12.18092
45	14.8797	0.06721	0.00446	224.458	0.06629	15.08484	12.92953
50	20.0855	0.04979	0.00324	308.645	0.06508	15.36653	13.55188
55	27.1126	0.03688	0.00237	422.285	0.06420	15.57520	14.06541
60	36.5982	0.02732	0.00174	575.683	0.06357	15.72980	14.48619
65	49.4024	0.02024	0.00128	782.748	0.06311	15.84432	14.82876
70	66.6863	0.01500	0.00094	1062.26	0.06278	15.92916	15.10600
75	90.0171	0.01111	0.00069	1439.56	0.06253	15.99202	15.32913
80	121.510	0.00823	0.00051	1948.85	0.06235	16.03858	15.50782
85	164.022	0.00610	0.00038	2636.34	0.06222	16.07307	15.65026
90	221.406	0.00452	0.00028	3564.34	0.06212	16.09863	15.76333
95	298.867	0.00335	0.00021	4817.01	0.06204	16.11756	15.85273
100	403.429	0.00248	0.00015	6507.94	0.06199	16.13158	15.92318

$r = 7\%$　　　　　　　　　　　　　　　　**Continuous Compounding, Discrete Cash Flows**

	SINGLE PAYMENT		UNIFORM SERIES				Arithmetic Gradient Series Factor
	Compound Amount Factor	Present Worth Factor	Sinking Fund Factor	Uniform Series Factor	Capital Recovery Factor	Series Present Worth Factor	
N	(F/P,r,N)	(P/F,r,N)	(A/F,r,N)	(F/A,r,N)	(A/P,r,N)	(P/A,r,N)	(A/G,r,N)
1	1.0725	0.93239	1.0000	1.0000	1.0725	0.93239	0.00000
2	1.1503	0.86936	0.48251	2.0725	0.55502	1.80175	0.48251
3	1.2337	0.81058	0.31029	3.2228	0.38280	2.61234	0.95337
4	1.3231	0.75578	0.22439	4.4565	0.29690	3.36812	1.41262
5	1.4191	0.70469	0.17302	5.7796	0.24553	4.07281	1.86030
6	1.5220	0.65705	0.13891	7.1987	0.21142	4.72985	2.29645
7	1.6323	0.61263	0.11467	8.7206	0.18718	5.34248	2.72114
8	1.7507	0.57121	0.09659	10.3529	0.16910	5.91369	3.13444
9	1.8776	0.53259	0.08262	12.1036	0.15513	6.44628	3.53643
10	2.0138	0.49659	0.07152	13.9812	0.14403	6.94287	3.92721
11	2.1598	0.46301	0.06252	15.9950	0.13503	7.40588	4.30688
12	2.3164	0.43171	0.05508	18.1547	0.12759	7.83759	4.67555
13	2.4843	0.40252	0.04885	20.4711	0.12136	8.24012	5.03334
14	2.6645	0.37531	0.04356	22.9554	0.11607	8.61543	5.38039
15	2.8577	0.34994	0.03903	25.6199	0.11154	8.96536	5.71683
16	3.0649	0.32628	0.03512	28.4775	0.10762	9.29164	6.04282
17	3.2871	0.30422	0.03170	31.5424	0.10421	9.59587	6.35849
18	3.5254	0.28365	0.02871	34.8295	0.10122	9.87952	6.66402
19	3.7810	0.26448	0.02607	38.3549	0.09858	10.14400	6.95958
20	4.0552	0.24660	0.02373	42.1359	0.09624	10.39059	7.24533
21	4.3492	0.22993	0.02165	46.1911	0.09416	10.62052	7.52146
22	4.6646	0.21438	0.01979	50.5404	0.09229	10.83490	7.78815
23	5.0028	0.19989	0.01811	55.2050	0.09062	11.03479	8.04559
24	5.3656	0.18637	0.01661	60.2078	0.08912	11.22116	8.29397
25	5.7546	0.17377	0.01525	65.5733	0.08776	11.39494	8.53348
26	6.1719	0.16203	0.01402	71.3279	0.08653	11.55696	8.76434
27	6.6194	0.15107	0.01290	77.4998	0.08541	11.70803	8.98674
28	7.0993	0.14086	0.01189	84.1192	0.08440	11.84889	9.20088
29	7.6141	0.13134	0.01096	91.2185	0.08347	11.98023	9.40697
30	8.1662	0.12246	0.01012	98.8326	0.08263	12.10268	9.60521
31	8.7583	0.11418	0.00935	106.999	0.08185	12.21686	9.79582
32	9.3933	0.10646	0.00864	115.757	0.08115	12.32332	9.97900
33	10.0744	0.09926	0.00799	125.150	0.08050	12.42258	10.15495
34	10.8049	0.09255	0.00740	135.225	0.07990	12.51513	10.32389
35	11.5883	0.08629	0.00685	146.030	0.07936	12.60143	10.48603
40	16.4446	0.06081	0.00469	213.006	0.07720	12.95288	11.20165
45	23.3361	0.04285	0.00325	308.049	0.07575	13.20055	11.77687
50	33.1155	0.03020	0.00226	442.922	0.07477	13.37508	12.23466
55	46.9931	0.02128	0.00158	634.315	0.07408	13.49807	12.59571
60	66.6863	0.01500	0.00110	905.916	0.07361	13.58473	12.87812
65	94.6324	0.01057	0.00077	1291.34	0.07328	13.64581	13.09734
70	134.290	0.00745	0.00054	1838.27	0.07305	13.68885	13.26638
75	190.566	0.00525	0.00038	2614.41	0.07289	13.71918	13.39591
80	270.426	0.00370	0.00027	3715.81	0.07278	13.74055	13.49462
85	383.753	0.00261	0.00019	5278.76	0.07270	13.75561	13.56947
90	544.572	0.00184	0.00013	7496.70	0.07264	13.76622	13.62598
95	772.784	0.00129	0.00009	10 644.0	0.07260	13.77370	13.66846
100	1096.63	0.00091	0.00007	15 110.0	0.07257	13.77897	13.70028

r = 8%   Continuous Compounding, Discrete Cash Flows

	SINGLE PAYMENT		UNIFORM SERIES				Arithmetic Gradient Series Factor
	Compound Amount Factor	Present Worth Factor	Sinking Fund Factor	Uniform Series Factor	Capital Recovery Factor	Series Present Worth Factor	
N	(*F/P,r,N*)	(*P/F,r,N*)	(*A/F,r,N*)	(*F/A,r,N*)	(*A/P,r,N*)	(*P/A,r,N*)	(*A/G,r,N*)
1	1.0833	0.92312	1.0000	1.0000	1.0833	0.92312	0.00000
2	1.1735	0.85214	0.48001	2.0833	0.56330	1.77526	0.48001
3	1.2712	0.78663	0.30705	3.2568	0.39034	2.56189	0.94672
4	1.3771	0.72615	0.22085	4.5280	0.30413	3.28804	1.40018
5	1.4918	0.67032	0.16934	5.9052	0.25263	3.95836	1.84044
6	1.6161	0.61878	0.13519	7.3970	0.21848	4.57714	2.26758
7	1.7507	0.57121	0.11095	9.0131	0.19424	5.14835	2.68169
8	1.8965	0.52729	0.09290	10.7637	0.17619	5.67564	3.08288
9	2.0544	0.48675	0.07899	12.6602	0.16227	6.16239	3.47127
10	2.2255	0.44933	0.06796	14.7147	0.15125	6.61172	3.84700
11	2.4109	0.41478	0.05903	16.9402	0.14232	7.02651	4.21022
12	2.6117	0.38289	0.05168	19.3511	0.13496	7.40940	4.56110
13	2.8292	0.35345	0.04553	21.9628	0.12882	7.76285	4.89980
14	3.0649	0.32628	0.04034	24.7920	0.12362	8.08913	5.22653
15	3.3201	0.30119	0.03590	27.8569	0.11918	8.39033	5.54147
16	3.5966	0.27804	0.03207	31.1770	0.11536	8.66836	5.84486
17	3.8962	0.25666	0.02876	34.7736	0.11204	8.92503	6.13689
18	4.2207	0.23693	0.02586	38.6698	0.10915	9.16195	6.41781
19	4.5722	0.21871	0.02332	42.8905	0.10660	9.38067	6.68785
20	4.9530	0.20190	0.02107	47.4627	0.10436	9.58256	6.94726
21	5.3656	0.18637	0.01908	52.4158	0.10237	9.76894	7.19628
22	5.8124	0.17204	0.01731	57.7813	0.10059	9.94098	7.43518
23	6.2965	0.15882	0.01572	63.5938	0.09901	10.09980	7.66421
24	6.8210	0.14661	0.01431	69.8903	0.09760	10.24641	7.88363
25	7.3891	0.13534	0.01304	76.7113	0.09632	10.38174	8.09372
26	8.0045	0.12493	0.01189	84.1003	0.09518	10.50667	8.29475
27	8.6711	0.11533	0.01086	92.1048	0.09414	10.62200	8.48698
28	9.3933	0.10646	0.00992	100.776	0.09321	10.72845	8.67068
29	10.1757	0.09827	0.00908	110.169	0.09236	10.82673	8.84614
30	11.0232	0.09072	0.00831	120.345	0.09160	10.91745	9.01360
31	11.9413	0.08374	0.00761	131.368	0.09090	11.00119	9.17336
32	12.9358	0.07730	0.00698	143.309	0.09026	11.07849	9.32566
33	14.0132	0.07136	0.00640	156.245	0.08969	11.14986	9.47078
34	15.1803	0.06587	0.00587	170.258	0.08916	11.21573	9.60898
35	16.4446	0.06081	0.00539	185.439	0.08868	11.27654	9.74051
40	24.5325	0.04076	0.00354	282.547	0.08683	11.51725	10.30689
45	36.5982	0.02732	0.00234	427.416	0.08563	11.67860	10.74256
50	54.5982	0.01832	0.00155	643.535	0.08484	11.78676	11.07380
55	81.4509	0.01228	0.00104	965.947	0.08432	11.85926	11.32302
60	121.510	0.00823	0.00069	1446.93	0.08398	11.90785	11.50878
65	181.272	0.00552	0.00046	2164.47	0.08375	11.94043	11.64610
70	270.426	0.00370	0.00031	3234.91	0.08360	11.96227	11.74685
75	403.429	0.00248	0.00021	4831.83	0.08349	11.97690	11.82030
80	601.845	0.00166	0.00014	7214.15	0.08343	11.98672	11.87352
85	897.847	0.00111	0.00009	10 768.0	0.08338	11.99329	11.91189
90	1339.43	0.00075	0.00006	16 070.0	0.08335	11.99770	11.93942
95	1998.20	0.00050	0.00004	23 980.0	0.08333	12.00066	11.95910
100	2980.96	0.00034	0.00003	35 779.0	0.08332	12.00264	11.97311

$r = 9\%$ **Continuous Compounding, Discrete Cash Flows**

	SINGLE PAYMENT		UNIFORM SERIES				Arithmetic Gradient Series Factor
	Compound Amount Factor	Present Worth Factor	Sinking Fund Factor	Uniform Series Factor	Capital Recovery Factor	Series Present Worth Factor	
N	(F/P,r,N)	(P/F,r,N)	(A/F,r,N)	(F/A,r,N)	(A/P,r,N)	(P/A,r,N)	(A/G,r,N)
1	1.0942	0.91393	1.0000	1.0000	1.0942	0.91393	0.00000
2	1.1972	0.83527	0.47752	2.0942	0.57169	1.74920	0.47752
3	1.3100	0.76338	0.30382	3.2914	0.39800	2.51258	0.94008
4	1.4333	0.69768	0.21733	4.6014	0.31150	3.21026	1.38776
5	1.5683	0.63763	0.16571	6.0347	0.25988	3.84789	1.82063
6	1.7160	0.58275	0.13153	7.6030	0.22570	4.43063	2.23880
7	1.8776	0.53259	0.10731	9.3190	0.20148	4.96323	2.64241
8	2.0544	0.48675	0.08931	11.1966	0.18349	5.44998	3.03160
9	2.2479	0.44486	0.07547	13.2510	0.16964	5.89484	3.40654
10	2.4596	0.40657	0.06452	15.4990	0.15869	6.30141	3.76743
11	2.6912	0.37158	0.05568	17.9586	0.14986	6.67298	4.11449
12	2.9447	0.33960	0.04843	20.6498	0.14260	7.01258	4.44793
13	3.2220	0.31037	0.04238	23.5945	0.13656	7.32294	4.76801
14	3.5254	0.28365	0.03729	26.8165	0.13146	7.60660	5.07498
15	3.8574	0.25924	0.03296	30.3419	0.12713	7.86584	5.36913
16	4.2207	0.23693	0.02924	34.1993	0.12341	8.10277	5.65074
17	4.6182	0.21654	0.02603	38.4200	0.12020	8.31930	5.92011
18	5.0531	0.19790	0.02324	43.0382	0.11741	8.51720	6.17755
19	5.5290	0.18087	0.02079	48.0913	0.11497	8.69807	6.42339
20	6.0496	0.16530	0.01865	53.6202	0.11282	8.86337	6.65794
21	6.6194	0.15107	0.01676	59.6699	0.11093	9.01444	6.88154
22	7.2427	0.13807	0.01509	66.2893	0.10926	9.15251	7.09452
23	7.9248	0.12619	0.01360	73.5320	0.10777	9.27869	7.29723
24	8.6711	0.11533	0.01228	81.4568	0.10645	9.39402	7.49000
25	9.4877	0.10540	0.01110	90.1280	0.10527	9.49942	7.67318
26	10.3812	0.09633	0.01004	99.6157	0.10421	9.59574	7.84712
27	11.3589	0.08804	0.00909	109.997	0.10327	9.68378	8.01215
28	12.4286	0.08046	0.00824	121.356	0.10241	9.76424	8.16862
29	13.5991	0.07353	0.00747	133.784	0.10165	9.83778	8.31685
30	14.8797	0.06721	0.00679	147.383	0.10096	9.90498	8.45719
31	16.2810	0.06142	0.00616	162.263	0.10034	9.96640	8.58995
32	17.8143	0.05613	0.00560	178.544	0.09978	10.02254	8.71547
33	19.4919	0.05130	0.00509	196.358	0.09927	10.07384	8.83405
34	21.3276	0.04689	0.00463	215.850	0.09881	10.12073	8.94600
35	23.3361	0.04285	0.00422	237.178	0.09839	10.16358	9.05164
40	36.5982	0.02732	0.00265	378.004	0.09682	10.32847	9.49496
45	57.3975	0.01742	0.00167	598.863	0.09584	10.43361	9.82070
50	90.0171	0.01111	0.00106	945.238	0.09523	10.50065	10.05692
55	141.175	0.00708	0.00067	1488.46	0.09485	10.54339	10.22624
60	221.406	0.00452	0.00043	2340.41	0.09460	10.57065	10.34639
65	347.234	0.00288	0.00027	3676.53	0.09445	10.58803	10.43088
70	544.572	0.00184	0.00017	5771.98	0.09435	10.59911	10.48983
75	854.059	0.00117	0.00011	9058.30	0.09428	10.60618	10.53069
80	1339.43	0.00075	0.00007	14212.0	0.09424	10.61068	10.55884
85	2100.65	0.00048	0.00004	22295.0	0.09422	10.61356	10.57813
90	3294.47	0.00030	0.00003	34972.0	0.09420	10.61539	10.59128
95	5166.75	0.00019	0.00002	54853.0	0.09419	10.61655	10.60022
100	8103.08	0.00012	0.00001	86033.0	0.09419	10.61730	10.60627

$r = 10\%$ Continuous Compounding, Discrete Cash Flows

	SINGLE PAYMENT		UNIFORM SERIES				Arithmetic Gradient Series Factor
	Compound Amount Factor	Present Worth Factor	Sinking Fund Factor	Uniform Series Factor	Capital Recovery Factor	Series Present Worth Factor	
N	(F/P,r,N)	(P/F,r,N)	(A/F,r,N)	(F/A,r,N)	(A/P,r,N)	(P/A,r,N)	(A/G,r,N)
1	1.1052	0.90484	1.0000	1.0000	1.1052	0.90484	0.00000
2	1.2214	0.81873	0.47502	2.1052	0.58019	1.72357	0.47502
3	1.3499	0.74082	0.30061	3.3266	0.40578	2.46439	0.93344
4	1.4918	0.67032	0.21384	4.6764	0.31901	3.13471	1.37535
5	1.6487	0.60653	0.16212	6.1683	0.26729	3.74124	1.80086
6	1.8221	0.54881	0.12793	7.8170	0.23310	4.29005	2.21012
7	2.0138	0.49659	0.10374	9.6391	0.20892	4.78663	2.60329
8	2.2255	0.44933	0.08582	11.6528	0.19099	5.23596	2.98060
9	2.4596	0.40657	0.07205	13.8784	0.17723	5.64253	3.34227
10	2.7183	0.36788	0.06121	16.3380	0.16638	6.01041	3.68856
11	3.0042	0.33287	0.05248	19.0563	0.15765	6.34328	4.01976
12	3.3201	0.30119	0.04533	22.0604	0.15050	6.64448	4.33618
13	3.6693	0.27253	0.03940	25.3806	0.14457	6.91701	4.63814
14	4.0552	0.24660	0.03442	29.0499	0.13959	7.16361	4.92598
15	4.4817	0.22313	0.03021	33.1051	0.13538	7.38674	5.20008
16	4.9530	0.20190	0.02661	37.5867	0.13178	7.58863	5.46081
17	5.4739	0.18268	0.02351	42.5398	0.12868	7.77132	5.70856
18	6.0496	0.16530	0.02083	48.0137	0.12600	7.93662	5.94373
19	6.6859	0.14957	0.01850	54.0634	0.12367	8.08618	6.16673
20	7.3891	0.13534	0.01646	60.7493	0.12163	8.22152	6.37798
21	8.1662	0.12246	0.01468	68.1383	0.11985	8.34398	6.57790
22	9.0250	0.11080	0.01311	76.3045	0.11828	8.45478	6.76690
23	9.9742	0.10026	0.01172	85.3295	0.11689	8.55504	6.94542
24	11.0232	0.09072	0.01049	95.3037	0.11566	8.64576	7.11388
25	12.1825	0.08208	0.00940	106.327	0.11458	8.72784	7.27269
26	13.4637	0.07427	0.00844	118.509	0.11361	8.80211	7.42228
27	14.8797	0.06721	0.00758	131.973	0.11275	8.86932	7.56305
28	16.4446	0.06081	0.00681	146.853	0.11198	8.93013	7.69541
29	18.1741	0.05502	0.00612	163.297	0.11129	8.98515	7.81975
30	20.0855	0.04979	0.00551	181.472	0.11068	9.03494	7.93646
31	22.1980	0.04505	0.00496	201.557	0.11013	9.07999	8.04593
32	24.5325	0.04076	0.00447	223.755	0.10964	9.12075	8.14851
33	27.1126	0.03688	0.00403	248.288	0.10920	9.15763	8.24458
34	29.9641	0.03337	0.00363	275.400	0.10880	9.19101	8.33446
35	33.1155	0.03020	0.00327	305.364	0.10845	9.22121	8.41851
40	54.5982	0.01832	0.00196	509.629	0.10713	9.33418	8.76204
45	90.0171	0.01111	0.00118	846.404	0.10635	9.40270	9.00281
50	148.413	0.00674	0.00071	1401.65	0.10588	9.44427	9.16915
55	244.692	0.00409	0.00043	2317.10	0.10560	9.46947	9.28264
60	403.429	0.00248	0.00026	3826.43	0.10543	9.48476	9.35924
65	665.142	0.00150	0.00016	6314.88	0.10533	9.49404	9.41046
70	1096.63	0.00091	0.00010	10 418.0	0.10527	9.49966	9.44444
75	1808.04	0.00055	0.00006	17 182.0	0.10523	9.50307	9.46683

$r = 11\%$ Continuous Compounding, Discrete Cash Flows

	SINGLE PAYMENT		UNIFORM SERIES				Arithmetic Gradient Series Factor
	Compound Amount Factor	Present Worth Factor	Sinking Fund Factor	Uniform Series Factor	Capital Recovery Factor	Series Present Worth Factor	
N	$(F/P,r,N)$	$(P/F,r,N)$	$(A/F,r,N)$	$(F/A,r,N)$	$(A/P,r,N)$	$(P/A,r,N)$	$(A/G,r,N)$
1	1.1163	0.89583	1.0000	1.0000	1.1163	0.89583	0.00000
2	1.2461	0.80252	0.47253	2.1163	0.58881	1.69835	0.47253
3	1.3910	0.71892	0.29741	3.3624	0.41369	2.41728	0.92681
4	1.5527	0.64404	0.21038	4.7533	0.32666	3.06131	1.36297
5	1.7333	0.57695	0.15858	6.3060	0.27486	3.63826	1.78115
6	1.9348	0.51685	0.12439	8.0393	0.24067	4.15511	2.18154
7	2.1598	0.46301	0.10026	9.9741	0.21654	4.61813	2.56437
8	2.4109	0.41478	0.08241	12.1338	0.19869	5.03291	2.92993
9	2.6912	0.37158	0.06875	14.5447	0.18503	5.40449	3.27852
10	3.0042	0.33287	0.05802	17.2360	0.17430	5.73736	3.61047
11	3.3535	0.29820	0.04941	20.2401	0.16568	6.03556	3.92615
12	3.7434	0.26714	0.04238	23.5936	0.15866	6.30269	4.22597
13	4.1787	0.23931	0.03658	27.3370	0.15286	6.54200	4.51035
14	4.6646	0.21438	0.03173	31.5157	0.14801	6.75638	4.77973
15	5.2070	0.19205	0.02764	36.1803	0.14392	6.94843	5.03457
16	5.8124	0.17204	0.02416	41.3873	0.14044	7.12048	5.27536
17	6.4883	0.15412	0.02119	47.1998	0.13746	7.27460	5.50257
18	7.2427	0.13807	0.01863	53.6881	0.13490	7.41267	5.71673
19	8.0849	0.12369	0.01641	60.9308	0.13269	7.53636	5.91832
20	9.0250	0.11080	0.01449	69.0157	0.13077	7.64716	6.10787
21	10.0744	0.09926	0.01281	78.0407	0.12909	7.74642	6.28588
22	11.2459	0.08892	0.01135	88.1151	0.12763	7.83534	6.45287
23	12.5535	0.07966	0.01006	99.3610	0.12634	7.91500	6.60934
24	14.0132	0.07136	0.00894	111.915	0.12521	7.98636	6.75579
25	15.6426	0.06393	0.00794	125.928	0.12422	8.05029	6.89273
26	17.4615	0.05727	0.00706	141.570	0.12334	8.10756	7.02063
27	19.4919	0.05130	0.00629	159.032	0.12257	8.15886	7.13998
28	21.7584	0.04596	0.00560	178.524	0.12188	8.20482	7.25122
29	24.2884	0.04117	0.00499	200.282	0.12127	8.24599	7.35482
30	27.1126	0.03688	0.00445	224.571	0.12073	8.28288	7.45120
31	30.2652	0.03304	0.00397	251.683	0.12025	8.31592	7.54080
32	33.7844	0.02960	0.00355	281.949	0.11982	8.34552	7.62400
33	37.7128	0.02652	0.00317	315.733	0.11945	8.37203	7.70121
34	42.0980	0.02375	0.00283	353.446	0.11911	8.39579	7.77278
35	46.9931	0.02128	0.00253	395.544	0.11881	8.41707	7.83909
40	81.4509	0.01228	0.00145	691.883	0.11772	8.49449	8.10288
45	141.175	0.00708	0.00083	1205.52	0.11711	8.53916	8.27905
50	244.692	0.00409	0.00048	2095.77	0.11676	8.56493	8.39490
55	424.113	0.00236	0.00027	3638.80	0.11655	8.57980	8.47009

$r = 12\%$ Continuous Compounding, Discrete Cash Flows

	SINGLE PAYMENT		UNIFORM SERIES				Arithmetic Gradient Series Factor
	Compound Amount Factor	Present Worth Factor	Sinking Fund Factor	Uniform Series Factor	Capital Recovery Factor	Series Present Worth Factor	
N	(F/P,r,N)	(P/F,r,N)	(A/F,r,N)	(F/A,r,N)	(A/P,r,N)	(P/A,r,N)	(A/G,r,N)
1	1.1275	0.88692	1.0000	1.0000	1.1275	0.88692	0.00000
2	1.2712	0.78663	0.47004	2.1275	0.59753	1.67355	0.47004
3	1.4333	0.69768	0.29423	3.3987	0.42172	2.37122	0.92019
4	1.6161	0.61878	0.20695	4.8321	0.33445	2.99001	1.35061
5	1.8221	0.54881	0.15508	6.4481	0.28258	3.53882	1.76148
6	2.0544	0.48675	0.12092	8.2703	0.24841	4.02557	2.15307
7	2.3164	0.43171	0.09686	10.3247	0.22435	4.45728	2.52566
8	2.6117	0.38289	0.07911	12.6411	0.20660	4.84018	2.87962
9	2.9447	0.33960	0.06556	15.2528	0.19306	5.17977	3.21532
10	3.3201	0.30119	0.05495	18.1974	0.18245	5.48097	3.53320
11	3.7434	0.26714	0.04647	21.5176	0.17397	5.74810	3.83374
12	4.2207	0.23693	0.03959	25.2610	0.16708	5.98503	4.11743
13	4.7588	0.21014	0.03392	29.4817	0.16142	6.19516	4.38480
14	5.3656	0.18637	0.02921	34.2405	0.15670	6.38154	4.63641
15	6.0496	0.16530	0.02525	39.6061	0.15275	6.54684	4.87283
16	6.8210	0.14661	0.02190	45.6557	0.14940	6.69344	5.09464
17	7.6906	0.13003	0.01906	52.4767	0.14655	6.82347	5.30246
18	8.6711	0.11533	0.01662	60.1673	0.14412	6.93880	5.49687
19	9.7767	0.10228	0.01453	68.8384	0.14202	7.04108	5.67850
20	11.0232	0.09072	0.01272	78.6151	0.14022	7.13180	5.84796
21	12.4286	0.08046	0.01116	89.6383	0.13865	7.21226	6.00583
22	14.0132	0.07136	0.00980	102.067	0.13729	7.28362	6.15274
23	15.7998	0.06329	0.00861	116.080	0.13611	7.34691	6.28926
24	17.8143	0.05613	0.00758	131.880	0.13508	7.40305	6.41597
25	20.0855	0.04979	0.00668	149.694	0.13418	7.45283	6.53344
26	22.6464	0.04416	0.00589	169.780	0.13339	7.49699	6.64221
27	25.5337	0.03916	0.00520	192.426	0.13269	7.53616	6.74280
28	28.7892	0.03474	0.00459	217.960	0.13208	7.57089	6.83574
29	32.4597	0.03081	0.00405	246.749	0.13155	7.60170	6.92152
30	36.5982	0.02732	0.00358	279.209	0.13108	7.62902	7.00059
31	41.2644	0.02423	0.00317	315.807	0.13066	7.65326	7.07342
32	46.5255	0.02149	0.00280	357.071	0.13030	7.67475	7.14043
33	52.4573	0.01906	0.00248	403.597	0.12997	7.69381	7.20202
34	59.1455	0.01691	0.00219	456.054	0.12969	7.71072	7.25859
35	66.6863	0.01500	0.00194	515.200	0.12944	7.72572	7.31050
40	121.510	0.00823	0.00106	945.203	0.12855	7.77878	7.51141
45	221.406	0.00452	0.00058	1728.72	0.12808	7.80791	7.63916
50	403.429	0.00248	0.00032	3156.38	0.12781	7.82389	7.71909
55	735.095	0.00136	0.00017	5757.75	0.12767	7.83266	7.76841

$r = 13\%$ Continuous Compounding, Discrete Cash Flows

	SINGLE PAYMENT		UNIFORM SERIES				Arithmetic Gradient Series Factor
	Compound Amount Factor	Present Worth Factor	Sinking Fund Factor	Uniform Series Factor	Capital Recovery Factor	Series Present Worth Factor	
N	$(F/P,r,N)$	$(P/F,r,N)$	$(A/F,r,N)$	$(F/A,r,N)$	$(A/P,r,N)$	$(P/A,r,N)$	$(A/G,r,N)$
1	1.1388	0.87810	1.0000	1.0000	1.1388	0.87810	0.00000
2	1.2969	0.77105	0.46755	2.1388	0.60637	1.64915	0.46755
3	1.4770	0.67706	0.29106	3.4358	0.42988	2.32620	0.91358
4	1.6820	0.59452	0.20355	4.9127	0.34238	2.92072	1.33827
5	1.9155	0.52205	0.15164	6.5948	0.29046	3.44277	1.74189
6	2.1815	0.45841	0.11750	8.5103	0.25633	3.90118	2.12473
7	2.4843	0.40252	0.09353	10.6918	0.23236	4.30370	2.48718
8	2.8292	0.35345	0.07589	13.1761	0.21472	4.65716	2.82968
9	3.2220	0.31037	0.06248	16.0053	0.20131	4.96752	3.15272
10	3.6693	0.27253	0.05201	19.2273	0.19084	5.24005	3.45683
11	4.1787	0.23931	0.04367	22.8966	0.18250	5.47936	3.74260
12	4.7588	0.21014	0.03693	27.0753	0.17576	5.68950	4.01065
13	5.4195	0.18452	0.03141	31.8341	0.17024	5.87402	4.26162
14	6.1719	0.16203	0.02684	37.2536	0.16567	6.03604	4.49618
15	7.0287	0.14227	0.02303	43.4255	0.16186	6.17832	4.71503
16	8.0045	0.12493	0.01982	50.4542	0.15865	6.30325	4.91888
17	9.1157	0.10970	0.01711	58.4586	0.15593	6.41295	5.10844
18	10.3812	0.09633	0.01480	67.5743	0.15363	6.50928	5.28441
19	11.8224	0.08458	0.01283	77.9556	0.15166	6.59386	5.44753
20	13.4637	0.07427	0.01114	89.7780	0.14997	6.66814	5.59848
21	15.3329	0.06522	0.00969	103.242	0.14851	6.73335	5.73798
22	17.4615	0.05727	0.00843	118.575	0.14726	6.79062	5.86669
23	19.8857	0.05029	0.00735	136.036	0.14618	6.84091	5.98528
24	22.6464	0.04416	0.00641	155.922	0.14524	6.88507	6.09441
25	25.7903	0.03877	0.00560	178.568	0.14443	6.92384	6.19468
26	29.3708	0.03405	0.00489	204.359	0.14372	6.95789	6.28670
27	33.4483	0.02990	0.00428	233.729	0.14311	6.98779	6.37104
28	38.0918	0.02625	0.00374	267.178	0.14257	7.01404	6.44825
29	43.3801	0.02305	0.00328	305.269	0.14210	7.03709	6.51885
30	49.4024	0.02024	0.00287	348.650	0.14170	7.05733	6.58333
31	56.2609	0.01777	0.00251	398.052	0.14134	7.07511	6.64216
32	64.0715	0.01561	0.00220	454.313	0.14103	7.09071	6.69578
33	72.9665	0.01370	0.00193	518.384	0.14076	7.10442	6.74459
34	83.0963	0.01203	0.00169	591.351	0.14052	7.11645	6.78899
35	94.6324	0.01057	0.00148	674.447	0.14031	7.12702	6.82934
40	181.272	0.00552	0.00077	1298.53	0.13960	7.16340	6.98125
45	347.234	0.00288	0.00040	2493.97	0.13923	7.18239	7.07317
50	665.142	0.00150	0.00021	4783.90	0.13904	7.19231	7.12785
55	1274.11	0.00078	0.00011	9170.36	0.13894	7.19748	7.15994

$r = 14\%$ Continuous Compounding, Discrete Cash Flows

	SINGLE PAYMENT		UNIFORM SERIES				Arithmetic Gradient Series Factor
	Compound Amount Factor	Present Worth Factor	Sinking Fund Factor	Uniform Series Factor	Capital Recovery Factor	Series Present Worth Factor	
N	$(F/P,r,N)$	$(P/F,r,N)$	$(A/F,r,N)$	$(F/A,r,N)$	$(A/P,r,N)$	$(P/A,r,N)$	$(A/G,r,N)$
1	1.1503	0.86936	1.0000	1.0000	1.1503	0.86936	0.00000
2	1.3231	0.75578	0.46506	2.1503	0.61533	1.62514	0.46506
3	1.5220	0.65705	0.28790	3.4734	0.43818	2.28219	0.90697
4	1.7507	0.57121	0.20019	4.9954	0.35046	2.85340	1.32596
5	2.0138	0.49659	0.14824	6.7460	0.29851	3.34998	1.72235
6	2.3164	0.43171	0.11416	8.7598	0.26443	3.78169	2.09652
7	2.6645	0.37531	0.09028	11.0762	0.24056	4.15700	2.44894
8	3.0649	0.32628	0.07278	13.7406	0.22305	4.48328	2.78015
9	3.5254	0.28365	0.05950	16.8055	0.20978	4.76694	3.09076
10	4.0552	0.24660	0.04919	20.3309	0.19946	5.01354	3.38141
11	4.6646	0.21438	0.04101	24.3861	0.19128	5.22792	3.65282
12	5.3656	0.18637	0.03442	29.0507	0.18470	5.41429	3.90573
13	6.1719	0.16203	0.02906	34.4162	0.17933	5.57632	4.14092
14	7.0993	0.14086	0.02464	40.5881	0.17491	5.71717	4.35918
15	8.1662	0.12246	0.02097	47.6874	0.17124	5.83963	4.56135
16	9.3933	0.10646	0.01790	55.8536	0.16818	5.94609	4.74824
17	10.8049	0.09255	0.01533	65.2469	0.16560	6.03864	4.92069
18	12.4286	0.08046	0.01315	76.0518	0.16342	6.11910	5.07952
19	14.2963	0.06995	0.01130	88.4804	0.16158	6.18905	5.22555
20	16.4446	0.06081	0.00973	102.777	0.16000	6.24986	5.35957
21	18.9158	0.05287	0.00839	119.221	0.15866	6.30272	5.48237
22	21.7584	0.04596	0.00724	138.137	0.15751	6.34868	5.59471
23	25.0281	0.03996	0.00625	159.896	0.15653	6.38864	5.69731
24	28.7892	0.03474	0.00541	184.924	0.15568	6.42337	5.79087
25	33.1155	0.03020	0.00468	213.713	0.15495	6.45357	5.87608
26	38.0918	0.02625	0.00405	246.828	0.15433	6.47982	5.95356
27	43.8160	0.02282	0.00351	284.920	0.15378	6.50265	6.02392
28	50.4004	0.01984	0.00304	328.736	0.15332	6.52249	6.08772
29	57.9743	0.01725	0.00264	379.137	0.15291	6.53974	6.14552
30	66.6863	0.01500	0.00229	437.111	0.15256	6.55473	6.19780
31	76.7075	0.01304	0.00198	503.797	0.15226	6.56777	6.24505
32	88.2347	0.01133	0.00172	580.505	0.15200	6.57910	6.28769
33	101.494	0.00985	0.00150	668.740	0.15177	6.58895	6.32614
34	116.746	0.00857	0.00130	770.234	0.15157	6.59752	6.36077
35	134.290	0.00745	0.00113	886.980	0.15140	6.60497	6.39193
40	270.426	0.00370	0.00056	1792.90	0.15083	6.62991	6.50606
45	544.572	0.00184	0.00028	3617.21	0.15055	6.64230	6.57173
50	1096.63	0.00091	0.00014	7290.91	0.15041	6.64845	6.60888
55	2208.35	0.00045	0.00007	14689.0	0.15034	6.65151	6.62960

$r = 15\%$ Continuous Compounding, Discrete Cash Flows

	SINGLE PAYMENT		UNIFORM SERIES				Arithmetic Gradient Series Factor
	Compound Amount Factor	Present Worth Factor	Sinking Fund Factor	Uniform Series Factor	Capital Recovery Factor	Series Present Worth Factor	
N	(F/P,r,N)	(P/F,r,N)	(A/F,r,N)	(F/A,r,N)	(A/P,r,N)	(P/A,r,N)	(A/G,r,N)
1	1.1618	0.86071	1.0000	1.0000	1.1618	0.86071	0.00000
2	1.3499	0.74082	0.46257	2.1618	0.62440	1.60153	0.46257
3	1.5683	0.63763	0.28476	3.5117	0.44660	2.23915	0.90037
4	1.8221	0.54881	0.19685	5.0800	0.35868	2.78797	1.31369
5	2.1170	0.47237	0.14488	6.9021	0.30672	3.26033	1.70289
6	2.4596	0.40657	0.11088	9.0191	0.27271	3.66690	2.06846
7	2.8577	0.34994	0.08712	11.4787	0.24895	4.01684	2.41096
8	3.3201	0.30119	0.06975	14.3364	0.23159	4.31803	2.73106
9	3.8574	0.25924	0.05664	17.6565	0.21847	4.57727	3.02947
10	4.4817	0.22313	0.04648	21.5139	0.20832	4.80040	3.30699
11	5.2070	0.19205	0.03847	25.9956	0.20030	4.99245	3.56446
12	6.0496	0.16530	0.03205	31.2026	0.19388	5.15775	3.80276
13	7.0287	0.14227	0.02684	37.2522	0.18868	5.30003	4.02281
14	8.1662	0.12246	0.02258	44.2809	0.18442	5.42248	4.22554
15	9.4877	0.10540	0.01907	52.4471	0.18090	5.52788	4.41191
16	11.0232	0.09072	0.01615	61.9348	0.17798	5.61860	4.58286
17	12.8071	0.07808	0.01371	72.9580	0.17554	5.69668	4.73935
18	14.8797	0.06721	0.01166	85.7651	0.17349	5.76389	4.88231
19	17.2878	0.05784	0.00994	100.645	0.17177	5.82173	5.01264
20	20.0855	0.04979	0.00848	117.933	0.17031	5.87152	5.13125
21	23.3361	0.04285	0.00725	138.018	0.16908	5.91437	5.23898
22	27.1126	0.03688	0.00620	161.354	0.16803	5.95125	5.33666
23	31.5004	0.03175	0.00531	188.467	0.16714	5.98300	5.42507
24	36.5982	0.02732	0.00455	219.967	0.16638	6.01032	5.50497
25	42.5211	0.02352	0.00390	256.565	0.16573	6.03384	5.57706
26	49.4024	0.02024	0.00334	299.087	0.16518	6.05408	5.64200
27	57.3975	0.01742	0.00287	348.489	0.16470	6.07151	5.70042
28	66.6863	0.01500	0.00246	405.886	0.16430	6.08650	5.75289
29	77.4785	0.01291	0.00212	472.573	0.16395	6.09941	5.79997
30	90.0171	0.01111	0.00182	550.051	0.16365	6.11052	5.84215
31	104.585	0.00956	0.00156	640.068	0.16340	6.12008	5.87989
32	121.510	0.00823	0.00134	744.653	0.16318	6.12831	5.91362
33	141.175	0.00708	0.00115	866.164	0.16299	6.13539	5.94374
34	164.022	0.00610	0.00099	1007.34	0.16283	6.14149	5.97060
35	190.566	0.00525	0.00085	1171.36	0.16269	6.14674	5.99453

r = 20%   **Continuous Compounding, Discrete Cash Flows**

	SINGLE PAYMENT		UNIFORM SERIES				Arithmetic Gradient Series Factor
	Compound Amount Factor	Present Worth Factor	Sinking Fund Factor	Uniform Series Factor	Capital Recovery Factor	Series Present Worth Factor	
N	(F/P,r,N)	(P/F,r,N)	(A/F,r,N)	(F/A,r,N)	(A/P,r,N)	(P/A,r,N)	(A/G,r,N)
1	1.2214	0.81873	1.0000	1.0000	1.2214	0.81873	0.00000
2	1.4918	0.67032	0.45017	2.2214	0.67157	1.48905	0.45017
3	1.8221	0.54881	0.26931	3.7132	0.49071	2.03786	0.86755
4	2.2255	0.44933	0.18066	5.5353	0.40206	2.48719	1.25279
5	2.7183	0.36788	0.12885	7.7609	0.35025	2.85507	1.60677
6	3.3201	0.30119	0.09543	10.4792	0.31683	3.15627	1.93058
7	4.0552	0.24660	0.07247	13.7993	0.29387	3.40286	2.22548
8	4.9530	0.20190	0.05601	17.8545	0.27741	3.60476	2.49289
9	6.0496	0.16530	0.04385	22.8075	0.26525	3.77006	2.73435
10	7.3891	0.13534	0.03465	28.8572	0.25606	3.90539	2.95148
11	9.0250	0.11080	0.02759	36.2462	0.24899	4.01620	3.14594
12	11.0232	0.09072	0.02209	45.2712	0.24349	4.10691	3.31943
13	13.4637	0.07427	0.01776	56.2944	0.23917	4.18119	3.47363
14	16.4446	0.06081	0.01434	69.7581	0.23574	4.24200	3.61019
15	20.0855	0.04979	0.01160	86.2028	0.23300	4.29178	3.73072
16	24.5325	0.04076	0.00941	106.288	0.23081	4.33255	3.83675
17	29.9641	0.03337	0.00764	130.821	0.22905	4.36592	3.92972
18	36.5982	0.02732	0.00622	160.785	0.22762	4.39324	4.01101
19	44.7012	0.02237	0.00507	197.383	0.22647	4.41561	4.08188
20	54.5982	0.01832	0.00413	242.084	0.22553	4.43393	4.14351
21	66.6863	0.01500	0.00337	296.683	0.22477	4.44893	4.19695
22	81.4509	0.01228	0.00275	363.369	0.22415	4.46120	4.24320
23	99.4843	0.01005	0.00225	444.820	0.22365	4.47125	4.28312
24	121.510	0.00823	0.00184	544.304	0.22324	4.47948	4.31750
25	148.413	0.00674	0.00150	665.814	0.22290	4.48622	4.34706
26	181.272	0.00552	0.00123	814.228	0.22263	4.49174	4.37243
27	221.406	0.00452	0.00100	995.500	0.22241	4.49626	4.39415
28	270.426	0.00370	0.00082	1216.91	0.22222	4.49995	4.41273
29	330.300	0.00303	0.00067	1487.33	0.22208	4.50298	4.42859
30	403.429	0.00248	0.00055	1817.63	0.22195	4.50546	4.44211
31	492.749	0.00203	0.00045	2221.06	0.22185	4.50749	4.45362
32	601.845	0.00166	0.00037	2713.81	0.22177	4.50915	4.46340
33	735.095	0.00136	0.00030	3315.66	0.22170	4.51051	4.47170
34	897.847	0.00111	0.00025	4050.75	0.22165	4.51163	4.47874
35	1096.63	0.00091	0.00020	4948.60	0.22160	4.51254	4.48471

$r = 25\%$ **Continuous Compounding, Discrete Cash Flows**

	SINGLE PAYMENT		UNIFORM SERIES				Arithmetic Gradient Series Factor
	Compound Amount Factor	Present Worth Factor	Sinking Fund Factor	Uniform Series Factor	Capital Recovery Factor	Series Present Worth Factor	
N	$(F/P,r,N)$	$(P/F,r,N)$	$(A/F,r,N)$	$(F/A,r,N)$	$(A/P,r,N)$	$(P/A,r,N)$	$(A/G,r,N)$
1	1.2840	0.77880	1.0000	1.0000	1.2840	0.77880	0.00000
2	1.6487	0.60653	0.43782	2.2840	0.72185	1.38533	0.43782
3	2.1170	0.47237	0.25428	3.9327	0.53830	1.85770	0.83505
4	2.7183	0.36788	0.16530	6.0497	0.44932	2.22558	1.19290
5	3.4903	0.28650	0.11405	8.7680	0.39808	2.51208	1.51306
6	4.4817	0.22313	0.08158	12.2584	0.36560	2.73521	1.79751
7	5.7546	0.17377	0.05974	16.7401	0.34376	2.90899	2.04855
8	7.3891	0.13534	0.04445	22.4947	0.32848	3.04432	2.26867
9	9.4877	0.10540	0.03346	29.8837	0.31749	3.14972	2.46046
10	12.1825	0.08208	0.02540	39.3715	0.30942	3.23181	2.62656
11	15.6426	0.06393	0.01940	51.5539	0.30342	3.29573	2.76958
12	20.0855	0.04979	0.01488	67.1966	0.29891	3.34552	2.89206
13	25.7903	0.03877	0.01146	87.2821	0.29548	3.38429	2.99641
14	33.1155	0.03020	0.00884	113.072	0.29287	3.41449	3.08488
15	42.5211	0.02352	0.00684	146.188	0.29087	3.43801	3.15955
16	54.5982	0.01832	0.00530	188.709	0.28932	3.45633	3.22229
17	70.1054	0.01426	0.00411	243.307	0.28814	3.47059	3.27481
18	90.0171	0.01111	0.00319	313.413	0.28722	3.48170	3.31860
19	115.584	0.00865	0.00248	403.430	0.28650	3.49035	3.35499
20	148.413	0.00674	0.00193	519.014	0.28595	3.49709	3.38514
21	190.566	0.00525	0.00150	667.427	0.28552	3.50234	3.41003
22	244.692	0.00409	0.00117	857.993	0.28519	3.50642	3.43053
23	314.191	0.00318	0.00091	1102.69	0.28493	3.50961	3.44737
24	403.429	0.00248	0.00071	1416.88	0.28473	3.51208	3.46117
25	518.013	0.00193	0.00055	1820.30	0.28457	3.51401	3.47246
26	665.142	0.00150	0.00043	2338.32	0.28445	3.51552	3.48166
27	854.059	0.00117	0.00033	3003.46	0.28436	3.51669	3.48916
28	1096.63	0.00091	0.00026	3857.52	0.28428	3.51760	3.49526
29	1408.10	0.00071	0.00020	4954.15	0.28423	3.51831	3.50020
30	1808.04	0.00055	0.00016	6362.26	0.28418	3.51886	3.50421
31	2321.57	0.00043	0.00012	8170.30	0.28415	3.51930	3.50745
32	2980.96	0.00034	0.00010	10 492.0	0.28412	3.51963	3.51007
33	3827.63	0.00026	0.00007	13 473.0	0.28410	3.51989	3.51219
34	4914.77	0.00020	0.00006	17 300.0	0.28408	3.52010	3.51389
35	6310.69	0.00016	0.00005	22 215.0	0.28407	3.52025	3.51526

Compound Interest Factors for Continuous Compounding, Continuous Compounding Periods

$r = 1\%$ **Continuous Compounding, Continuous Compounding Periods**

T	Sinking Fund Factor $(A/F,r,T)$	Uniform Series Factor $(F/A,r,T)$	Capital Recovery Factor $(A/P,r,T)$	Series Present Worth Factor $(P/A,r,T)$
1	0.99501	1.0050	1.0050	0.99502
2	0.49502	2.0201	0.50502	1.9801
3	0.32836	3.0455	0.33836	2.9554
4	0.24503	4.0811	0.25503	3.9211
5	0.19504	5.1271	0.20504	4.8771
6	0.16172	6.1837	0.17172	5.8235
7	0.13792	7.2508	0.14792	6.7606
8	0.12007	8.3287	0.13007	7.6884
9	0.10619	9.4174	0.11619	8.6069
10	0.09508	10.5171	0.10508	9.5163
11	0.08600	11.6278	0.09600	10.4166
12	0.07843	12.7497	0.08843	11.3080
13	0.07203	13.8828	0.08203	12.1905
14	0.06655	15.0274	0.07655	13.0642
15	0.06179	16.1834	0.07179	13.9292
16	0.05763	17.3511	0.06763	14.7856
17	0.05397	18.5305	0.06397	15.6335
18	0.05071	19.7217	0.06071	16.4730
19	0.04779	20.9250	0.05779	17.3041
20	0.04517	22.1403	0.05517	18.1269
21	0.04279	23.3678	0.05279	18.9416
22	0.04064	24.6077	0.05064	19.7481
23	0.03867	25.8600	0.04867	20.5466
24	0.03687	27.1249	0.04687	21.3372
25	0.03521	28.4025	0.04521	22.1199
26	0.03368	29.6930	0.04368	22.8948
27	0.03226	30.9964	0.04226	23.6621
28	0.03095	32.3130	0.04095	24.4216
29	0.02972	33.6427	0.03972	25.1736
30	0.02858	34.9859	0.03858	25.9182
31	0.02752	36.3425	0.03752	26.6553
32	0.02652	37.7128	0.03652	27.3851
33	0.02558	39.0968	0.03558	28.1076
34	0.02469	40.4948	0.03469	28.8230
35	0.02386	41.9068	0.03386	29.5312
40	0.02033	49.1825	0.03033	32.9680
45	0.01760	56.8312	0.02760	36.2372
50	0.01541	64.8721	0.02541	39.3469
55	0.01364	73.3253	0.02364	42.3050
60	0.01216	82.2119	0.02216	45.1188
65	0.01092	91.5541	0.02092	47.7954
70	0.00986	101.375	0.01986	50.3415
75	0.00895	111.700	0.01895	52.7633
80	0.00816	122.554	0.01816	55.0671
85	0.00746	133.965	0.01746	57.2585
90	0.00685	145.960	0.01685	59.3430
95	0.00631	158.571	0.01631	61.3259
100	0.00582	171.828	0.01582	63.2121

$r = 2\%$ **Continuous Compounding, Continuous Compounding Periods**

T	Sinking Fund Factor $(A/F,r,T)$	Uniform Series Factor $(F/A,r,T)$	Capital Recovery Factor $(A/P,r,T)$	Series Present Worth Factor $(P/A,r,T)$
1	0.99003	1.0101	1.0100	0.99007
2	0.49007	2.0405	0.51007	1.9605
3	0.32343	3.0918	0.34343	2.9118
4	0.24013	4.1644	0.26013	3.8442
5	0.19017	5.2585	0.21017	4.7581
6	0.15687	6.3748	0.17687	5.6540
7	0.13309	7.5137	0.15309	6.5321
8	0.11527	8.6755	0.13527	7.3928
9	0.10141	9.8609	0.12141	8.2365
10	0.09033	11.0701	0.11033	9.0635
11	0.08128	12.3038	0.10128	9.8741
12	0.07373	13.5625	0.09373	10.6686
13	0.06736	14.8465	0.08736	11.4474
14	0.06189	16.1565	0.08189	12.2108
15	0.05717	17.4929	0.07717	12.9591
16	0.05303	18.8564	0.07303	13.6925
17	0.04939	20.2474	0.06939	14.4115
18	0.04615	21.6665	0.06615	15.1162
19	0.04326	23.1142	0.06326	15.8069
20	0.04066	24.5912	0.06066	16.4840
21	0.03832	26.0981	0.05832	17.1477
22	0.03619	27.6354	0.05619	17.7982
23	0.03424	29.2037	0.05424	18.4358
24	0.03246	30.8037	0.05246	19.0608
25	0.03083	32.4361	0.05083	19.6735
26	0.02932	34.1014	0.04932	20.2740
27	0.02793	35.8003	0.04793	20.8626
28	0.02664	37.5336	0.04664	21.4395
29	0.02544	39.3019	0.04544	22.0051
30	0.02433	41.1059	0.04433	22.5594
31	0.02328	42.9464	0.04328	23.1028
32	0.02231	44.8240	0.04231	23.6354
33	0.02140	46.7396	0.04140	24.1574
34	0.02054	48.6939	0.04054	24.6692
35	0.01973	50.6876	0.03973	25.1707
40	0.01632	61.2770	0.03632	27.5336
45	0.01370	72.9802	0.03370	29.6715
50	0.01164	85.9141	0.03164	31.6060
55	0.00998	100.208	0.02998	33.3564
60	0.00862	116.006	0.02862	34.9403
65	0.00749	133.465	0.02749	36.3734
70	0.00655	152.760	0.02655	37.6702
75	0.00574	174.084	0.02574	38.8435
80	0.00506	197.652	0.02506	39.9052
85	0.00447	223.697	0.02447	40.8658
90	0.00396	252.482	0.02396	41.7351
95	0.00352	284.295	0.02352	42.5216
100	0.00313	319.453	0.02313	43.2332

$r = 3\%$ **Continuous Compounding, Continuous Compounding Periods**

T	Sinking Fund Factor $(A/F,r,T)$	Uniform Series Factor $(F/A,r,T)$	Capital Recovery Factor $(A/P,r,T)$	Series Present Worth Factor $(P/A,r,T)$
1	0.98507	1.0152	1.0151	0.98515
2	0.48515	2.0612	0.51515	1.9412
3	0.31856	3.1391	0.34856	2.8690
4	0.23530	4.2499	0.26530	3.7693
5	0.18537	5.3945	0.21537	4.6431
6	0.15212	6.5739	0.18212	5.4910
7	0.12838	7.7893	0.15838	6.3139
8	0.11060	9.0416	0.14060	7.1124
9	0.09679	10.3321	0.12679	7.8874
10	0.08575	11.6620	0.11575	8.6394
11	0.07673	13.0323	0.10673	9.3692
12	0.06923	14.4443	0.09923	10.0775
13	0.06290	15.8994	0.09290	10.7648
14	0.05748	17.3987	0.08748	11.4318
15	0.05279	18.9437	0.08279	12.0791
16	0.04870	20.5358	0.07870	12.7072
17	0.04509	22.1764	0.07509	13.3168
18	0.04190	23.8669	0.07190	13.9084
19	0.03905	25.6089	0.06905	14.4825
20	0.03649	27.4040	0.06649	15.0396
21	0.03418	29.2537	0.06418	15.5803
22	0.03209	31.1597	0.06209	16.1050
23	0.03019	33.1239	0.06019	16.6141
24	0.02845	35.1478	0.05845	17.1083
25	0.02686	37.2333	0.05686	17.5878
26	0.02539	39.3824	0.05539	18.0531
27	0.02404	41.5969	0.05404	18.5047
28	0.02279	43.8789	0.05279	18.9430
29	0.02163	46.2304	0.05163	19.3683
30	0.02055	48.6534	0.05055	19.7810
31	0.01955	51.1503	0.04955	20.1815
32	0.01861	53.7232	0.04861	20.5702
33	0.01774	56.3745	0.04774	20.9474
34	0.01692	59.1065	0.04692	21.3135
35	0.01615	61.9217	0.04615	21.6687
40	0.01293	77.3372	0.04293	23.2935
45	0.01050	95.2475	0.04050	24.6920
50	0.00862	116.056	0.03862	25.8957
55	0.00713	140.233	0.03713	26.9317
60	0.00594	168.322	0.03594	27.8234
65	0.00498	200.956	0.03498	28.5909
70	0.00419	238.872	0.03419	29.2515
75	0.00353	282.925	0.03353	29.8200
80	0.00299	334.106	0.03299	30.3094
85	0.00254	393.570	0.03254	30.7306
90	0.00216	462.658	0.03216	31.0931
95	0.00184	542.926	0.03184	31.4052
100	0.00157	636.185	0.03157	31.6738

$r = 4\%$ — Continuous Compounding, Continuous Compounding Periods

T	Sinking Fund Factor $(A/F,r,T)$	Uniform Series Factor $(F/A,r,T)$	Capital Recovery Factor $(A/P,r,T)$	Series Present Worth Factor $(P/A,r,T)$
1	0.98013	1.0203	1.0201	0.98026
2	0.48027	2.0822	0.52027	1.9221
3	0.31373	3.1874	0.35373	2.8270
4	0.23053	4.3378	0.27053	3.6964
5	0.18067	5.5351	0.22067	4.5317
6	0.14747	6.7812	0.18747	5.3343
7	0.12379	8.0782	0.16379	6.1054
8	0.10606	9.4282	0.14606	6.8463
9	0.09231	10.8332	0.13231	7.5581
10	0.08133	12.2956	0.12133	8.2420
11	0.07237	13.8177	0.11237	8.8991
12	0.06493	15.4019	0.10493	9.5304
13	0.05865	17.0507	0.09865	10.1370
14	0.05329	18.7668	0.09329	10.7198
15	0.04865	20.5530	0.08865	11.2797
16	0.04462	22.4120	0.08462	11.8177
17	0.04107	24.3469	0.08107	12.3346
18	0.03794	26.3608	0.07794	12.8312
19	0.03514	28.4569	0.07514	13.3083
20	0.03264	30.6385	0.07264	13.7668
21	0.03039	32.9092	0.07039	14.2072
22	0.02835	35.2725	0.06835	14.6304
23	0.02650	37.7323	0.06650	15.0370
24	0.02482	40.2924	0.06482	15.4277
25	0.02328	42.9570	0.06328	15.8030
26	0.02187	45.7304	0.06187	16.1636
27	0.02057	48.6170	0.06057	16.5101
28	0.01937	51.6214	0.05937	16.8430
29	0.01827	54.7483	0.05827	17.1628
30	0.01724	58.0029	0.05724	17.4701
31	0.01629	61.3903	0.05629	17.7654
32	0.01540	64.9160	0.05540	18.0491
33	0.01458	68.5855	0.05458	18.3216
34	0.01381	72.4048	0.05381	18.5835
35	0.01309	76.3800	0.05309	18.8351
40	0.01012	98.8258	0.05012	19.9526
45	0.00792	126.241	0.04792	20.8675
50	0.00626	159.726	0.04626	21.6166
55	0.00498	200.625	0.04498	22.2299
60	0.00399	250.579	0.04399	22.7321
65	0.00321	311.593	0.04321	23.1432
70	0.00259	386.116	0.04259	23.4797
75	0.00210	477.138	0.04210	23.7553
80	0.00170	588.313	0.04170	23.9809
85	0.00138	724.103	0.04138	24.1657
90	0.00112	889.956	0.04112	24.3169
95	0.00092	1092.53	0.04092	24.4407
100	0.00075	1339.95	0.04075	24.5421

$r = 5\%$ **Continuous Compounding, Continuous Compounding Periods**

T	Sinking Fund Factor $(A/F,r,T)$	Uniform Series Factor $(F/A,r,T)$	Capital Recovery Factor $(A/P,r,T)$	Series Present Worth Factor $(P/A,r,T)$
1	0.97521	1.0254	1.0252	0.97541
2	0.47542	2.1034	0.52542	1.9033
3	0.30896	3.2367	0.35896	2.7858
4	0.22583	4.4281	0.27583	3.6254
5	0.17604	5.6805	0.22604	4.4240
6	0.14291	6.9972	0.19291	5.1836
7	0.11931	8.3814	0.16931	5.9062
8	0.10166	9.8365	0.15166	6.5936
9	0.08798	11.3662	0.13798	7.2474
10	0.07707	12.9744	0.12707	7.8694
11	0.06819	14.6651	0.11819	8.4610
12	0.06082	16.4424	0.11082	9.0238
13	0.05461	18.3108	0.10461	9.5591
14	0.04932	20.2751	0.09932	10.0683
15	0.04476	22.3400	0.09476	10.5527
16	0.04080	24.5108	0.09080	11.0134
17	0.03732	26.7929	0.08732	11.4517
18	0.03426	29.1921	0.08426	11.8686
19	0.03153	31.7142	0.08153	12.2652
20	0.02910	34.3656	0.07910	12.6424
21	0.02692	37.1530	0.07692	13.0012
22	0.02495	40.0833	0.07495	13.3426
23	0.02317	43.1639	0.07317	13.6673
24	0.02155	46.4023	0.07155	13.9761
25	0.02008	49.8069	0.07008	14.2699
26	0.01873	53.3859	0.06873	14.5494
27	0.01750	57.1485	0.06750	14.8152
28	0.01637	61.1040	0.06637	15.0681
29	0.01532	65.2623	0.06532	15.3086
30	0.01436	69.6338	0.06436	15.5374
31	0.01347	74.2294	0.06347	15.7550
32	0.01265	79.0606	0.06265	15.9621
33	0.01189	84.1396	0.06189	16.1590
34	0.01118	89.4789	0.06118	16.3463
35	0.01052	95.0921	0.06052	16.5245
40	0.00783	127.781	0.05783	17.2933
45	0.00589	169.755	0.05589	17.8920
50	0.00447	223.650	0.05447	18.3583
55	0.00341	292.853	0.05341	18.7214
60	0.00262	381.711	0.05262	19.0043
65	0.00202	495.807	0.05202	19.2245
70	0.00156	642.309	0.05156	19.3961
75	0.00120	830.422	0.05120	19.5296
80	0.00093	1071.96	0.05093	19.6337
85	0.00072	1382.11	0.05072	19.7147
90	0.00056	1780.34	0.05056	19.7778
95	0.00044	2291.69	0.05044	19.8270
100	0.00034	2948.26	0.05034	19.8652

$r = 6\%$ **Continuous Compounding, Continuous Compounding Periods**

T	Sinking Fund Factor $(A/F,r,T)$	Uniform Series Factor $(F/A,r,T)$	Capital Recovery Factor $(A/P,r,T)$	Series Present Worth Factor $(P/A,r,T)$
1	0.97030	1.0306	1.0303	0.97059
2	0.47060	2.1249	0.53060	1.8847
3	0.30423	3.2870	0.36423	2.7455
4	0.22120	4.5208	0.28120	3.5562
5	0.17150	5.8310	0.23150	4.3197
6	0.13846	7.2222	0.19846	5.0387
7	0.11495	8.6994	0.17495	5.7159
8	0.09739	10.2679	0.15739	6.3536
9	0.08380	11.9334	0.14380	6.9542
10	0.07298	13.7020	0.13298	7.5198
11	0.06419	15.5799	0.12419	8.0525
12	0.05690	17.5739	0.11690	8.5541
13	0.05078	19.6912	0.11078	9.0266
14	0.04558	21.9394	0.10558	9.4715
15	0.04111	24.3267	0.10111	9.8905
16	0.03723	26.8616	0.09723	10.2851
17	0.03384	29.5532	0.09384	10.6568
18	0.03085	32.4113	0.09085	11.0067
19	0.02821	35.4461	0.08821	11.3363
20	0.02586	38.6686	0.08586	11.6468
21	0.02376	42.0904	0.08376	11.9391
22	0.02187	45.7237	0.08187	12.2144
23	0.02017	49.5817	0.08017	12.4737
24	0.01863	53.6783	0.07863	12.7179
25	0.01723	58.0282	0.07723	12.9478
26	0.01596	62.6470	0.07596	13.1644
27	0.01480	67.5515	0.07480	13.3684
28	0.01374	72.7593	0.07374	13.5604
29	0.01277	78.2891	0.07277	13.7413
30	0.01188	84.1608	0.07188	13.9117
31	0.01106	90.3956	0.07106	14.0721
32	0.01031	97.0160	0.07031	14.2232
33	0.00961	104.046	0.06961	14.3655
34	0.00897	111.510	0.06897	14.4995
35	0.00837	119.436	0.06837	14.6257
40	0.00599	167.053	0.06599	15.1547
45	0.00432	231.329	0.06432	15.5466
50	0.00314	318.092	0.06314	15.8369
55	0.00230	435.211	0.06230	16.0519
60	0.00169	593.304	0.06169	16.2113
65	0.00124	806.707	0.06124	16.3293
70	0.00091	1094.77	0.06091	16.4167
75	0.00067	1483.62	0.06067	16.4815
80	0.00050	2008.51	0.06050	16.5295
85	0.00037	2717.03	0.06037	16.5651
90	0.00027	3673.44	0.06027	16.5914
95	0.00020	4964.46	0.06020	16.6109
100	0.00015	6707.15	0.06015	16.6254

r = 7%   **Continuous Compounding, Continuous Compounding Periods**

	Sinking Fund Factor	Uniform Series Factor	Capital Recovery Factor	Series Present Worth Factor
T	(A/F,r,T)	(F/A,r,T)	(A/P,r,T)	(P/A,r,T)
1	0.96541	1.0358	1.0354	0.96580
2	0.46582	2.1468	0.53582	1.8663
3	0.29956	3.3383	0.36956	2.7059
4	0.21663	4.6161	0.28663	3.4888
5	0.16704	5.9867	0.23704	4.2187
6	0.13411	7.4566	0.20411	4.8993
7	0.11070	9.0331	0.18070	5.5339
8	0.09325	10.7239	0.16325	6.1256
9	0.07976	12.5373	0.14976	6.6773
10	0.06905	14.4822	0.13905	7.1916
11	0.06036	16.5681	0.13036	7.6712
12	0.05318	18.8052	0.12318	8.1184
13	0.04716	21.2046	0.11716	8.5354
14	0.04206	23.7779	0.11206	8.9241
15	0.03768	26.5379	0.10768	9.2866
16	0.03390	29.4979	0.10390	9.6246
17	0.03061	32.6726	0.10061	9.9397
18	0.02772	36.0774	0.09772	10.2335
19	0.02517	39.7292	0.09517	10.5075
20	0.02291	43.6457	0.09291	10.7629
21	0.02090	47.8462	0.09090	11.0011
22	0.01910	52.3513	0.08910	11.2231
23	0.01749	57.1830	0.08749	11.4302
24	0.01603	62.3651	0.08603	11.6232
25	0.01472	67.9229	0.08472	11.8032
26	0.01353	73.8837	0.08353	11.9711
27	0.01246	80.2767	0.08246	12.1275
28	0.01148	87.1332	0.08148	12.2735
29	0.01058	94.4869	0.08058	12.4095
30	0.00977	102.374	0.07977	12.5363
31	0.00902	110.833	0.07902	12.6546
32	0.00834	119.905	0.07834	12.7649
33	0.00771	129.635	0.07771	12.8677
34	0.00714	140.070	0.07714	12.9636
35	0.00661	151.262	0.07661	13.0529
40	0.00453	220.638	0.07453	13.4170
45	0.00313	319.087	0.07313	13.6735
50	0.00218	458.792	0.07218	13.8543
55	0.00152	657.044	0.07152	13.9817
60	0.00107	938.376	0.07107	14.0715
65	0.00075	1337.61	0.07075	14.1348
70	0.00053	1904.14	0.07053	14.1793
75	0.00037	2708.09	0.07037	14.2107
80	0.00026	3848.95	0.07026	14.2329
85	0.00018	5467.90	0.07018	14.2485
90	0.00013	7765.31	0.07013	14.2595
95	0.00009	11 025.0	0.07009	14.2672
100	0.00006	15 652.0	0.07006	14.2727

$r = 8\%$ **Continuous Compounding, Continuous Compounding Periods**

	Sinking Fund Factor	Uniform Series Factor	Capital Recovery Factor	Series Present Worth Factor
T	$(A/F,r,T)$	$(F/A,r,T)$	$(A/P,r,T)$	$(P/A,r,T)$
1	0.96053	1.0411	1.0405	0.96105
2	0.46107	2.1689	0.54107	1.8482
3	0.29493	3.3906	0.37493	2.6672
4	0.21213	4.7141	0.29213	3.4231
5	0.16266	6.1478	0.24266	4.1210
6	0.12985	7.7009	0.20985	4.7652
7	0.10657	9.3834	0.18657	5.3599
8	0.08924	11.2060	0.16924	5.9088
9	0.07587	13.1804	0.15587	6.4156
10	0.06528	15.3193	0.14528	6.8834
11	0.05670	17.6362	0.13670	7.3152
12	0.04964	20.1462	0.12964	7.7138
13	0.04373	22.8652	0.12373	8.0818
14	0.03874	25.8107	0.11874	8.4215
15	0.03448	29.0015	0.11448	8.7351
16	0.03081	32.4580	0.11081	9.0245
17	0.02762	36.2024	0.10762	9.2917
18	0.02484	40.2587	0.10484	9.5384
19	0.02240	44.6528	0.10240	9.7661
20	0.02024	49.4129	0.10024	9.9763
21	0.01833	54.5694	0.09833	10.1703
22	0.01662	60.1555	0.09662	10.3494
23	0.01510	66.2067	0.09510	10.5148
24	0.01374	72.7620	0.09374	10.6674
25	0.01252	79.8632	0.09252	10.8083
26	0.01142	87.5559	0.09142	10.9384
27	0.01043	95.8892	0.09043	11.0584
28	0.00953	104.917	0.08953	11.1693
29	0.00872	114.696	0.08872	11.2716
30	0.00798	125.290	0.08798	11.3660
31	0.00731	136.766	0.08731	11.4532
32	0.00670	149.198	0.08670	11.5337
33	0.00615	162.665	0.08615	11.6080
34	0.00564	177.254	0.08564	11.6766
35	0.00518	193.058	0.08518	11.7399
40	0.00340	294.157	0.08340	11.9905
45	0.00225	444.978	0.08225	12.1585
50	0.00149	669.977	0.08149	12.2711
55	0.00099	1005.64	0.08099	12.3465
60	0.00066	1506.38	0.08066	12.3971
65	0.00044	2253.40	0.08044	12.4310
70	0.00030	3367.83	0.08030	12.4538
75	0.00020	5030.36	0.08020	12.4690
80	0.00013	7510.56	0.08013	12.4792
85	0.00009	11 211.0	0.08009	12.4861
90	0.00006	16 730.0	0.08006	12.4907
95	0.00004	24 965.0	0.08004	12.4937
100	0.00003	37 249.0	0.08003	12.4958

$r = 9\%$ Continuous Compounding, Continuous Compounding Periods

	Sinking Fund Factor	Uniform Series Factor	Capital Recovery Factor	Series Present Worth Factor
T	$(A/F,r,T)$	$(F/A,r,T)$	$(A/P,r,T)$	$(P/A,r,T)$
1	0.95567	1.0464	1.0457	0.95632
2	0.45635	2.1913	0.54635	1.8303
3	0.29036	3.4440	0.38036	2.6291
4	0.20769	4.8148	0.29769	3.3592
5	0.15836	6.3146	0.24836	4.0264
6	0.12570	7.9556	0.21570	4.6361
7	0.10255	9.7512	0.19255	5.1934
8	0.08535	11.7159	0.17535	5.7028
9	0.07212	13.8656	0.16212	6.1682
10	0.06166	16.2178	0.15166	6.5937
11	0.05322	18.7915	0.14322	6.9825
12	0.04628	21.6076	0.13628	7.3378
13	0.04050	24.6888	0.13050	7.6626
14	0.03564	28.0602	0.12564	7.9594
15	0.03150	31.7492	0.12150	8.2307
16	0.02794	35.7855	0.11794	8.4786
17	0.02487	40.2020	0.11487	8.7052
18	0.02221	45.0343	0.11221	8.9122
19	0.01987	50.3218	0.10987	9.1015
20	0.01782	56.1072	0.10782	9.2745
21	0.01602	62.4374	0.10602	9.4325
22	0.01442	69.3638	0.10442	9.5770
23	0.01300	76.9425	0.10300	9.7090
24	0.01173	85.2349	0.10173	9.8297
25	0.01060	94.3082	0.10060	9.9400
26	0.00959	104.236	0.09959	10.0408
27	0.00869	115.099	0.09869	10.1329
28	0.00787	126.984	0.09787	10.2171
29	0.00714	139.989	0.09714	10.2941
30	0.00648	154.219	0.09648	10.3644
31	0.00589	169.789	0.09589	10.4287
32	0.00535	186.825	0.09535	10.4874
33	0.00487	205.466	0.09487	10.5411
34	0.00443	225.862	0.09443	10.5901
35	0.00403	248.178	0.09403	10.6350
40	0.00253	395.536	0.09253	10.8075
45	0.00160	626.638	0.09160	10.9175
50	0.00101	989.079	0.09101	10.9877
55	0.00064	1557.50	0.09064	11.0324
60	0.00041	2448.96	0.09041	11.0609
65	0.00026	3847.05	0.09026	11.0791
70	0.00017	6039.69	0.09017	11.0907
75	0.00011	9478.43	0.09011	11.0981
80	0.00007	14 871.0	0.09007	11.1028
85	0.00004	23 329.0	0.09004	11.1058
90	0.00003	36 594.0	0.09003	11.1077
95	0.00002	57 397.0	0.09002	11.1090
100	0.00001	90 023.0	0.09001	11.1097

$r = 10\%$ Continuous Compounding, Continuous Compounding Periods

T	Sinking Fund Factor $(A/F,r,T)$	Uniform Series Factor $(F/A,r,T)$	Capital Recovery Factor $(A/P,r,T)$	Series Present Worth Factor $(P/A,r,T)$
1	0.95083	1.0517	1.0508	0.95163
2	0.45167	2.2140	0.55167	1.8127
3	0.28583	3.4986	0.38583	2.5918
4	0.20332	4.9182	0.30332	3.2968
5	0.15415	6.4872	0.25415	3.9347
6	0.12164	8.2212	0.22164	4.5119
7	0.09864	10.1375	0.19864	5.0341
8	0.08160	12.2554	0.18160	5.5067
9	0.06851	14.5960	0.16851	5.9343
10	0.05820	17.1828	0.15820	6.3212
11	0.04990	20.0417	0.14990	6.6713
12	0.04310	23.2012	0.14310	6.9881
13	0.03746	26.6930	0.13746	7.2747
14	0.03273	30.5520	0.13273	7.5340
15	0.02872	34.8169	0.12872	7.7687
16	0.02530	39.5303	0.12530	7.9810
17	0.02235	44.7395	0.12235	8.1732
18	0.01980	50.4965	0.11980	8.3470
19	0.01759	56.8589	0.11759	8.5043
20	0.01565	63.8906	0.11565	8.6466
21	0.01395	71.6617	0.11395	8.7754
22	0.01246	80.2501	0.11246	8.8920
23	0.01114	89.7418	0.11114	8.9974
24	0.00998	100.232	0.10998	9.0928
25	0.00894	111.825	0.10894	9.1792
26	0.00802	124.637	0.10802	9.2573
27	0.00720	138.797	0.10720	9.3279
28	0.00647	154.446	0.10647	9.3919
29	0.00582	171.741	0.10582	9.4498
30	0.00524	190.855	0.10524	9.5021
31	0.00472	211.980	0.10472	9.5495
32	0.00425	235.325	0.10425	9.5924
33	0.00383	261.126	0.10383	9.6312
34	0.00345	289.641	0.10345	9.6663
35	0.00311	321.155	0.10311	9.6980
40	0.00187	535.982	0.10187	9.8168
45	0.00112	890.171	0.10112	9.8889
50	0.00068	1474.13	0.10068	9.9326
55	0.00041	2436.92	0.10041	9.9591
60	0.00025	4024.29	0.10025	9.9752
65	0.00015	6641.42	0.10015	9.9850
70	0.00009	10 956.0	0.10009	9.9909
75	0.00006	18 070.0	0.10006	9.9945
80	0.00003	29 800.0	0.10003	9.9966
85	0.00002	49 138.0	0.10002	9.9980
90	0.00001	81 021.0	0.10001	9.9988
95	0.00001	133 587.0	0.10001	9.9993
100	0.00000	220 255.0	0.10000	9.9995

$r = 11\%$ Continuous Compounding, Continuous Compounding Periods

T	Sinking Fund Factor $(A/F,r,T)$	Uniform Series Factor $(F/A,r,T)$	Capital Recovery Factor $(A/P,r,T)$	Series Present Worth Factor $(P/A,r,T)$
1	0.94601	1.0571	1.0560	0.94696
2	0.44702	2.2371	0.55702	1.7953
3	0.28135	3.5543	0.39135	2.5552
4	0.19902	5.0246	0.30902	3.2360
5	0.15002	6.6659	0.26002	3.8459
6	0.11767	8.4981	0.22767	4.3923
7	0.09485	10.5433	0.20485	4.8817
8	0.07796	12.8264	0.18796	5.3202
9	0.06504	15.3749	0.17504	5.7129
10	0.05489	18.2197	0.16489	6.0648
11	0.04674	21.3953	0.15674	6.3800
12	0.04010	24.9402	0.15010	6.6624
13	0.03461	28.8973	0.14461	6.9154
14	0.03002	33.3145	0.14002	7.1420
15	0.02615	38.2453	0.13615	7.3450
16	0.02286	43.7494	0.13286	7.5269
17	0.02004	49.8936	0.13004	7.6898
18	0.01762	56.7522	0.12762	7.8357
19	0.01553	64.4083	0.12553	7.9665
20	0.01371	72.9547	0.12371	8.0836
21	0.01212	82.4948	0.12212	8.1885
22	0.01074	93.1442	0.12074	8.2825
23	0.00952	105.032	0.11952	8.3667
24	0.00845	118.302	0.11845	8.4422
25	0.00751	133.115	0.11751	8.5097
26	0.00668	149.650	0.11668	8.5703
27	0.00595	168.108	0.11595	8.6245
28	0.00530	188.713	0.11530	8.6731
29	0.00472	211.713	0.11472	8.7166
30	0.00421	237.388	0.11421	8.7556
31	0.00376	266.048	0.11376	8.7905
32	0.00336	298.040	0.11336	8.8218
33	0.00300	333.753	0.11300	8.8499
34	0.00268	373.618	0.11268	8.8750
35	0.00239	418.119	0.11239	8.8975
40	0.00137	731.372	0.11137	8.9793
45	0.00078	1274.32	0.11078	9.0265
50	0.00045	2215.38	0.11045	9.0538
55	0.00026	3846.48	0.11026	9.0695
60	0.00015	6 674.0	0.11015	9.0785
65	0.00009	11 574.0	0.11009	9.0838
70	0.00005	20 067.0	0.11005	9.0868
75	0.00003	34 788.0	0.11003	9.0885
80	0.00002	60302.218	0.11002	9.0895
85	0.00001	104525.668	0.11001	9.0901
90	0.00001	181176.095	0.11001	9.0905
95	0.00000	314030.679	0.11000	9.0906
100	0.00000	544301.288	0.11000	9.0908

$r = 12\%$ **Continuous Compounding, Continuous Compounding Periods**

T	Sinking Fund Factor $(A/F,r,T)$	Uniform Series Factor $(F/A,r,T)$	Capital Recovery Factor $(A/P,r,T)$	Series Present Worth Factor $(P/A,r,T)$
1	0.94120	1.0625	1.0612	0.94233
2	0.44240	2.2604	0.56240	1.7781
3	0.27693	3.6111	0.39693	2.5194
4	0.19478	5.1340	0.31478	3.1768
5	0.14596	6.8510	0.26596	3.7599
6	0.11381	8.7869	0.23381	4.2771
7	0.09116	10.9697	0.21116	4.7357
8	0.07446	13.4308	0.19446	5.1426
9	0.06171	16.2057	0.18171	5.5034
10	0.05172	19.3343	0.17172	5.8234
11	0.04374	22.8618	0.16374	6.1072
12	0.03726	26.8391	0.15726	6.3589
13	0.03192	31.3235	0.15192	6.5822
14	0.02749	36.3796	0.14749	6.7802
15	0.02376	42.0804	0.14376	6.9558
16	0.02062	48.5080	0.14062	7.1116
17	0.01794	55.7551	0.13794	7.2498
18	0.01564	63.9261	0.13564	7.3723
19	0.01367	73.1390	0.13367	7.4810
20	0.01197	83.5265	0.13197	7.5774
21	0.01050	95.2383	0.13050	7.6628
22	0.00922	108.443	0.12922	7.7387
23	0.00811	123.332	0.12811	7.8059
24	0.00714	140.119	0.12714	7.8655
25	0.00629	159.046	0.12629	7.9184
26	0.00554	180.386	0.12554	7.9654
27	0.00489	204.448	0.12489	8.0070
28	0.00432	231.577	0.12432	8.0439
29	0.00381	262.164	0.12381	8.0766
30	0.00337	296.652	0.12337	8.1056
31	0.00298	335.537	0.12298	8.1314
32	0.00264	379.379	0.12264	8.1542
33	0.00233	428.811	0.12233	8.1745
34	0.00206	484.546	0.12206	8.1924
35	0.00183	547.386	0.12183	8.2084
40	0.00100	1004.25	0.12100	8.2648
45	0.00054	1836.72	0.12054	8.2957
50	0.00030	3353.57	0.12030	8.3127
55	0.00016	6117.46	0.12016	8.3220
60	0.00009	11 154.0	0.12009	8.3271
65	0.00005	20 330.0	0.12005	8.3299
70	0.00003	37 051.0	0.12003	8.3315
75	0.00001	67 517.0	0.12001	8.3323
80	0.00001	123 032.0	0.12001	8.3328
85	0.00000	224 185.0	0.12000	8.3330
90	0.00000	408 498.0	0.12000	8.3332
95	0.00000	744 339.0	0.12000	8.3332
100	0.00000	1 356 282.0	0.12000	8.3333

$r = 13\%$ **Continuous Compounding, Continuous Compounding Periods**

T	Sinking Fund Factor $(A/F,r,T)$	Uniform Series Factor $(F/A,r,T)$	Capital Recovery Factor $(A/P,r,T)$	Series Present Worth Factor $(P/A,r,T)$
1	0.93641	1.0679	1.0664	0.93773
2	0.43781	2.2841	0.56781	1.7611
3	0.27255	3.6691	0.40255	2.4842
4	0.19061	5.2464	0.32061	3.1191
5	0.14199	7.0426	0.27199	3.6766
6	0.11003	9.0882	0.24003	4.1661
7	0.08758	11.4179	0.21758	4.5960
8	0.07107	14.0709	0.20107	4.9734
9	0.05851	17.0923	0.18851	5.3049
10	0.04870	20.5331	0.17870	5.5959
11	0.04090	24.4515	0.17090	5.8515
12	0.03459	28.9140	0.16459	6.0759
13	0.02942	33.9960	0.15942	6.2729
14	0.02514	39.7835	0.15514	6.4460
15	0.02156	46.3745	0.15156	6.5979
16	0.01856	53.8805	0.14856	6.7313
17	0.01602	62.4286	0.14602	6.8485
18	0.01386	72.1634	0.14386	6.9513
19	0.01201	83.2496	0.14201	7.0417
20	0.01043	95.8749	0.14043	7.1210
21	0.00907	110.253	0.13907	7.1906
22	0.00790	126.627	0.13790	7.2518
23	0.00688	145.274	0.13688	7.3055
24	0.00601	166.511	0.13601	7.3526
25	0.00524	190.695	0.13524	7.3940
26	0.00458	218.237	0.13458	7.4304
27	0.00401	249.602	0.13401	7.4623
28	0.00350	285.322	0.13350	7.4904
29	0.00307	326.000	0.13307	7.5150
30	0.00269	372.327	0.13269	7.5366
31	0.00235	425.084	0.13235	7.5556
32	0.00206	485.166	0.13206	7.5722
33	0.00181	553.588	0.13181	7.5869
34	0.00158	631.510	0.13158	7.5997
35	0.00139	720.249	0.13139	7.6110
40	0.00072	1386.71	0.13072	7.6499
45	0.00038	2663.34	0.13038	7.6702
50	0.00020	5108.78	0.13020	7.6807
55	0.00010	9793.12	0.13010	7.6863
60	0.00005	18 766.0	0.13005	7.6892
65	0.00003	35 954.0	0.13003	7.6907
70	0.00001	68 879.0	0.13001	7.6914
75	0.00001	131 948.0	0.13001	7.6919
80	0.00000	252 759.0	0.13000	7.6921
85	0.00000	484 177.0	0.13000	7.6922
90	0.00000	927 467.0	0.13000	7.6922
95	0.00000	1 776 608.0	0.13000	7.6923
100	0.00000	3 403 172.0	0.13000	7.6923

$r = 14\%$ Continuous Compounding, Continuous Compounding Periods

T	Sinking Fund Factor $(A/F,r,T)$	Uniform Series Factor $(F/A,r,T)$	Capital Recovery Factor $(A/P,r,T)$	Series Present Worth Factor $(P/A,r,T)$
1	0.93163	1.0734	1.0716	0.93316
2	0.43326	2.3081	0.57326	1.7444
3	0.26822	3.7283	0.40822	2.4497
4	0.18650	5.3619	0.32650	3.0628
5	0.13810	7.2411	0.27810	3.5958
6	0.10635	9.4026	0.24635	4.0592
7	0.08411	11.8890	0.22411	4.4621
8	0.06780	14.7490	0.20780	4.8123
9	0.05544	18.0387	0.19544	5.1168
10	0.04582	21.8229	0.18582	5.3815
11	0.03820	26.1756	0.17820	5.6116
12	0.03207	31.1825	0.17207	5.8116
13	0.02707	36.9418	0.16707	5.9855
14	0.02295	43.5666	0.16295	6.1367
15	0.01954	51.1869	0.15954	6.2682
16	0.01668	59.9524	0.15668	6.3824
17	0.01428	70.0350	0.15428	6.4818
18	0.01225	81.6328	0.15225	6.5681
19	0.01053	94.9735	0.15053	6.6432
20	0.00906	110.319	0.14906	6.7085
21	0.00781	127.970	0.14781	6.7652
22	0.00674	148.274	0.14674	6.8146
23	0.00583	171.629	0.14583	6.8575
24	0.00504	198.494	0.14504	6.8947
25	0.00436	229.396	0.14436	6.9272
26	0.00377	264.942	0.14377	6.9553
27	0.00327	305.829	0.14327	6.9798
28	0.00283	352.860	0.14283	7.0011
29	0.00246	406.959	0.14246	7.0196
30	0.00213	469.188	0.14213	7.0357
31	0.00185	540.768	0.14185	7.0497
32	0.00160	623.105	0.14160	7.0619
33	0.00139	717.815	0.14139	7.0725
34	0.00121	826.757	0.14121	7.0817
35	0.00105	952.070	0.14105	7.0897
40	0.00052	1924.47	0.14052	7.1164
45	0.00026	3882.66	0.14026	7.1297
50	0.00013	7825.95	0.14013	7.1363
55	0.00006	15 767.0	0.14006	7.1396
60	0.00003	31 758.0	0.14003	7.1413
65	0.00002	63 959.0	0.14002	7.1421
70	0.00001	128 805.0	0.14001	7.1425
75	0.00000	259 389.0	0.14000	7.1427
80	0.00000	522 353.0	0.14000	7.1428
85	0.00000	1 051 897.0	0.14000	7.1428
90	0.00000	2 118 268.0	0.14000	7.1428
95	0.00000	4 265 676.0	0.14000	7.1428
100	0.00000	8 590 023.0	0.14000	7.1429

$r = 15\%$ Continuous Compounding, Continuous Compounding Periods

T	Sinking Fund Factor $(A/F,r,T)$	Uniform Series Factor $(F/A,r,T)$	Capital Recovery Factor $(A/P,r,T)$	Series Present Worth Factor $(P/A,r,T)$
1	0.92687	1.0789	1.0769	0.92861
2	0.42874	2.3324	0.57874	1.7279
3	0.26394	3.7887	0.41394	2.4158
4	0.18246	5.4808	0.33246	3.0079
5	0.13429	7.4467	0.28429	3.5176
6	0.10277	9.7307	0.25277	3.9562
7	0.08075	12.3843	0.23075	4.3337
8	0.06465	15.4674	0.21465	4.6587
9	0.05249	19.0495	0.20249	4.9384
10	0.04308	23.2113	0.19308	5.1791
11	0.03566	28.0465	0.18566	5.3863
12	0.02971	33.6643	0.17971	5.5647
13	0.02488	40.1913	0.17488	5.7182
14	0.02093	47.7745	0.17093	5.8503
15	0.01767	56.5849	0.16767	5.9640
16	0.01497	66.8212	0.16497	6.0619
17	0.01270	78.7140	0.16270	6.1461
18	0.01081	92.5315	0.16081	6.2186
19	0.00921	108.585	0.15921	6.2810
20	0.00786	127.237	0.15786	6.3348
21	0.00672	148.907	0.15672	6.3810
22	0.00574	174.084	0.15574	6.4208
23	0.00492	203.336	0.15492	6.4550
24	0.00421	237.322	0.15421	6.4845
25	0.00361	276.807	0.15361	6.5099
26	0.00310	322.683	0.15310	6.5317
27	0.00266	375.983	0.15266	6.5505
28	0.00228	437.909	0.15228	6.5667
29	0.00196	509.856	0.15196	6.5806
30	0.00169	593.448	0.15169	6.5926
31	0.00145	690.567	0.15145	6.6029
32	0.00124	803.403	0.15124	6.6118
33	0.00107	934.500	0.15107	6.6194
34	0.00092	1086.81	0.15092	6.6260
35	0.00079	1263.78	0.15079	6.6317
40	0.00037	2682.86	0.15037	6.6501
45	0.00018	5687.06	0.15018	6.6589
50	0.00008	12 047.0	0.15008	6.6630
55	0.00004	25 511.0	0.15004	6.6649
60	0.00002	54 014.0	0.15002	6.6658
65	0.00001	114 355.0	0.15001	6.6663
70	0.00000	242 097.0	0.15000	6.6665
75	0.00000	512 526.0	0.15000	6.6666

$r = 20\%$ **Continuous Compounding, Continuous Compounding Periods**

T	Sinking Fund Factor $(A/F,r,T)$	Uniform Series Factor $(F/A,r,T)$	Capital Recovery Factor $(A/P,r,T)$	Series Present Worth Factor $(P/A,r,T)$
1	0.90333	1.1070	1.1033	0.90635
2	0.40665	2.4591	0.60665	1.6484
3	0.24327	4.1106	0.44327	2.2559
4	0.16319	6.1277	0.36319	2.7534
5	0.11640	8.5914	0.31640	3.1606
6	0.08620	11.6006	0.28620	3.4940
7	0.06546	15.2760	0.26546	3.7670
8	0.05059	19.7652	0.25059	3.9905
9	0.03961	25.2482	0.23961	4.1735
10	0.03130	31.9453	0.23130	4.3233
11	0.02492	40.1251	0.22492	4.4460
12	0.01995	50.1159	0.21995	4.5464
13	0.01605	62.3187	0.21605	4.6286
14	0.01295	77.2232	0.21295	4.6959
15	0.01048	95.4277	0.21048	4.7511
16	0.00850	117.663	0.20850	4.7962
17	0.00691	144.821	0.20691	4.8331
18	0.00562	177.991	0.20562	4.8634
19	0.00458	218.506	0.20458	4.8881
20	0.00373	267.991	0.20373	4.9084
21	0.00304	328.432	0.20304	4.9250
22	0.00249	402.254	0.20249	4.9386
23	0.00203	492.422	0.20203	4.9497
24	0.00166	602.552	0.20166	4.9589
25	0.00136	737.066	0.20136	4.9663
26	0.00111	901.361	0.20111	4.9724
27	0.00091	1102.03	0.20091	4.9774
28	0.00074	1347.13	0.20074	4.9815
29	0.00061	1646.50	0.20061	4.9849
30	0.00050	2012.14	0.20050	4.9876
31	0.00041	2458.75	0.20041	4.9899
32	0.00033	3004.23	0.20033	4.9917
33	0.00027	3670.48	0.20027	4.9932
34	0.00022	4484.24	0.20022	4.9944
35	0.00018	5478.17	0.20018	4.9954
40	0.00007	14 900.0	0.20007	4.9983
45	0.00002	40 510.0	0.20002	4.9994
50	0.00001	110 127.0	0.20001	4.9998
55	0.00000	299 366.0	0.20000	4.9999
60	0.00000	813 769.0	0.20000	5.0000
65	0.00000	2 212 062.0	0.20000	5.0000
70	0.00000	6 013 016.0	0.20000	5.0000
75	0.00000	16 345 082.0	0.20000	5.0000

$r = 25\%$ Continuous Compounding, Continuous Compounding Periods

T	Sinking Fund Factor $(A/F,r,T)$	Uniform Series Factor $(F/A,r,T)$	Capital Recovery Factor $(A/P,r,T)$	Series Present Worth Factor $(P/A,r,T)$
1	0.88020	1.1361	1.1302	0.88480
2	0.38537	2.5949	0.63537	1.5739
3	0.22381	4.4680	0.47381	2.1105
4	0.14549	6.8731	0.39549	2.5285
5	0.10039	9.9614	0.35039	2.8540
6	0.07180	13.9268	0.32180	3.1075
7	0.05258	19.0184	0.30258	3.3049
8	0.03913	25.5562	0.28913	3.4587
9	0.02945	33.9509	0.27945	3.5784
10	0.02236	44.7300	0.27236	3.6717
11	0.01707	58.5705	0.26707	3.7443
12	0.01310	76.3421	0.26310	3.8009
13	0.01008	99.1614	0.26008	3.8449
14	0.00778	128.462	0.25778	3.8792
15	0.00602	166.084	0.25602	3.9059
16	0.00466	214.393	0.25466	3.9267
17	0.00362	276.422	0.25362	3.9429
18	0.00281	356.069	0.25281	3.9556
19	0.00218	458.337	0.25218	3.9654
20	0.00170	589.653	0.25170	3.9730
21	0.00132	758.265	0.25132	3.9790
22	0.00103	974.768	0.25103	3.9837
23	0.00080	1252.76	0.25080	3.9873
24	0.00062	1609.72	0.25062	3.9901
25	0.00048	2068.05	0.25048	3.9923
26	0.00038	2656.57	0.25038	3.9940
27	0.00029	3412.24	0.25029	3.9953
28	0.00023	4382.53	0.25023	3.9964
29	0.00018	5628.42	0.25018	3.9972
30	0.00014	7228.17	0.25014	3.9978
31	0.00011	9282.29	0.25011	3.9983
32	0.00008	11 920.0	0.25008	3.9987
33	0.00007	15 307.0	0.25007	3.9990
34	0.00005	19 655.0	0.25005	3.9992
35	0.00004	25 239.0	0.25004	3.9994
40	0.00001	88 102.0	0.25001	3.9998
45	0.00000	307 516.0	0.25000	3.9999
50	0.00000	1 073 345.0	0.25000	4.0000
55	0.00000	3 746 353.0	0.25000	4.0000

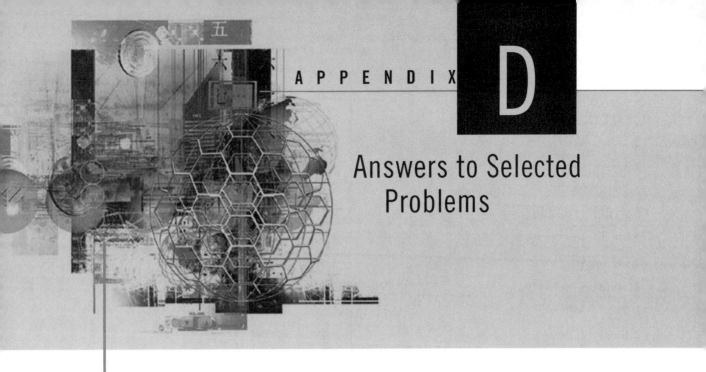

D

Answers to Selected Problems

CHAPTER 1

CHAPTER 2

2.1	$120
2.3	$5000
2.5	8%
2.7(a)	1210 yuan
2.9	18.75%
2.11	$29 719
2.13 (a)	5 years
2.15 (a)	$6728
2.17 (a)	26.6%
2.19	5%
2.21	665 270 rupees
2.23	Brand 2 about $51 less
2.25	$i_{e \text{ continuous}}$ = 8.318%
2.27	$i_{e \text{ weekly}}$ = 5.65%
2.29	0.5%
2.31	2140 yuan
2.37	Decisional equivalence holds
2.41(a)	105 months
2.43 (b)	Lost 60 rupees by locking in

CHAPTER 3

3.3	R317.22
3.5	20.3%
3.7	£74 790
3.9	$18 466 per year
3.11	$94.13
3.13	$2664
3.15	11.7%
3.17	5.8 years
3.19 (b)	$26.44 per week
3.21	162.5 MW
3.23	£3.98
3.25	$34 616
3.27	£3339
3.31	¥3 598 109
3.33	$8 013 275
3.35	$74 514
3.37	No, P = $3 670 261
3.39	$85.9 million
3.41	€122 316
3.43	27 months
3.45	Rs 21 098
3.47	$3 086 287
3.49	$257 143
3.51	No more than $504
3.53	Up to $4587
3A.1	Purchase cost > savings by $1701
3A.3	$8353

CHAPTER 4

4.1	BC, CD
4.5	A, AB, BC, ABC, BCD
4.7 (a)	–£2164
4.9 (a)	18.1%
4.11	Second offer, PW = $137 000
4.13	Earthen dam, PW = $396 038
4.15	No, PW = –$16
4.17	Hydraulic press, AW = $24 716
4.19	Plastic liner, PW = €1 100 000
4.21	XJ3, PW = –$6565
4.23	No, AC(car) = £10 126
4.25	3.71 years
4.27	About 20%
4.29	1344 yuan
4.31 (a)	T, PW = $96 664
4.33	T, AW = $26 154

4.35 (a) Only B, AW = ¥15 746

4.39 Landfill site, AW = $125 351

4.43 Curtains, payback period = 1.67 years

CHAPTER 5

5.1 (a) 9.2%

5.3 12.4%

5.5 B and D

5.7 (b) 7.58%

5.9 New machine

5.11 21.7%

5.13 3.19%

5.15 Approximate ERR > 30%

5.17 (c) Clip Job

5.19 Used refrigerator

5.21 (b) 2.46%

5.23 (b) 2

5.25 E

5.27 $119

5.29 A and C

5.31 B

5.33 Payback period

5.35 IRR or present worth

5.37 Payback period

5.39 Annual worth

5A.1 (d) No, 12.6%

5A.3 (e) No, 21.3%

CHAPTER 6

6.1 (a) Functional loss

6.3 Use-related loss

6.5 Time-related loss

6.7 (b) $7714

6.9 (b) £5806

6.13 12%

6.15 A, PW = –$22 321

6.17 (a) $312.50

6.19 Fryer 1, AW = $37 668

6.21 $25 000

6.23 Acid test = 0.9

6.25 Current assets, fixed assets, depreciation, income taxes, net income (operations), and net income

6.27 Equity ratio = 0.296

6.29 Total assets = $46 500, total equity = $15 450, net income = $2700

6.31 e.g., current = 0.9, equity = 0.39, ROA = 2%

6.33 For 2005, total assets = $3900, total equity = $1350, net income = $450

CHAPTER 7

7.1 $10 920

7.3 No

7.5 (c) 6 years

7.7 Every 3 years

7.9 4.7 years

7.11 Move robot immediately

7.13 Yes

7.15 (a) Yes

7.17 (c) $12 879

7.19 Replace immediately

7.21 (b) Defender two years, followed by Challenger 2

7.23 Replace the computer now, and keep the replacement for three years

7.25 (a) Replace old pump after four years

7.27 No

7.29 Challenger EAC is $39 452 with an economic life of five years

7.31 Overhaul old cutter and replace after four years

7.33 (a) A1 = 4 years, B1 = 7 years

7.35 (a) A3 = 12 years, B3 = 4 years

CHAPTER 8

8.1 (a) 8.4%

8.3 (a) 0.3448

8.5 No

8.7 6.6%, yes

8.9 Free machine

8.11 $3478, $5798

8.13 $8026

8.15 9.87%

8.17 9.14%

8.19 10.75%

8.21 T

8.23 T: 20.69

8.25 T

8.27 £7136

8.29 2

8.31 £618

8.33 £199 998

8.35 T

8.37 T

8.39 T: 36.85

8.41 (a) 5-year

8.43 −$25 961

8.45 $2284

8.47 $177 002

8.49 X

8.51 X

8.53 T: 29.96

8.55 (d) 9, 25%

8.57 $12 962

8.59 Free machine

8.61 −$1681

8.63 $198 590

8.65 T

8.67 T

8.69 T: 34.57%

CHAPTER 9

9.1 (a) Actual

9.3 (a) $292

9.5 (a) 14.5%

9.7 (b) 2.9%

9.9 67 years

9.11 14.3%

9.13 $14 683

9.15 11 216 million rubles

9.17 (a) 7.75%

9.21 (b) −$125 532

9.23 (a) About 16%

9.25 (b) $243 547

9.27 (c) 2.13%

9.29 (b) $998 for f = 2%

CHAPTER 10

10.1 (a) 1.55

10.1 (b) 2

10.3 (c) 0

10.5 (a) RUB1599 per day

10.11 $212 500

10.15 (a) 8 580 000 yuan per year

10.17 (d) 1333 rubles per day

10.19 (a) River route

10.23 (b) 1.07

10.25 (a) 1.08, 1.1

10.27 (a) 0.75

10.29 (a) Tennis courts, BCR = 1.18

10.31   Yes, BCR = 1.66

10.37 (c) $34 510

10.39 (a) 2 lanes

CHAPTER 11

11.3 (a) Break-even analysis for multiple projects

11.5   PW(base) = $34 617

11.7   PW(base) = €6 636 000, most sensitive to the first cost

11.9 (a) AW(base) = $176 620, most sensitive to the first cost

11.11 (a) 179 deliveries per month

11.13 (a) $141 045, outside of the likely values

11.15 (b) 47 375 boards per year

11.17 $17 535

11.19 (a) $4146

11.19 (b) Break-even operating hours = 671

11.21   Break-even maintenance cost $61.50

11.23   Break-even interest rate 11.2%

11.25   Lease if tax rate is below 27%

CHAPTER 12

12.1 (a) ¥855

12.3 (b) X1000, Var(defects) = 1.36/100 units^2

12.5   Rs 23 367 000

12.7   €6384

12.9   E(buy ticket) = 150

12.11 (c) E(send stock) = 4000

12.13 (a) E(combined rate) = 15.9%

12.15   E(partnership) = $113 750

12.17 (a) AW(optimistic, public accept) = $925 000

12.17 (b) E(new product) = $107 975

12.21 (c) Average PW ≃ −€160 600

12.23   Average PW ≃ −$5600

12.25   Average PW ≃ −£20 300

12.27   Average PW ≃ −$15 700

CHAPTER 13

13.1   4, 5, 6, 7, 8, and 10

13.3 (b) L4, 60.2

13.5 (a)
$$\begin{bmatrix} 1 & 1 & 5 & \frac{1}{3} \\ 1 & 1 & \frac{1}{3} & 2 \\ \frac{1}{5} & 3 & 1 & 2 \\ 3 & \frac{1}{2} & \frac{1}{2} & 1 \end{bmatrix}$$

13.7 (a) 1, 3, 7, 8

13.9   D, 57.5

13.11
$$\begin{bmatrix} 1 & 1 & \frac{1}{3} & 5 & \frac{1}{5} \\ 1 & 1 & \frac{1}{3} & 5 & \frac{1}{5} \\ 3 & 3 & 1 & 5 & \frac{1}{3} \\ \frac{1}{5} & \frac{1}{5} & \frac{1}{5} & 1 & \frac{1}{5} \\ 5 & 5 & 3 & 5 & 1 \end{bmatrix}$$

13.17   [0.13 0.13 0.56 0.13 0.05]^T

13.23   C, 0.277

13.25 (b) [0.05 0.17 0.17 0.61]^T

13.27 (b) [0.141 0.263 0.455 0.141]^T

13.29
$$\begin{bmatrix} 0.05 & 0.61 & 0.14 & 0.44 \\ 0.17 & 0.26 & 0.26 & 0.07 \\ 0.17 & 0.08 & 0.46 & 0.05 \\ 0.61 & 0.05 & 0.14 & 0.44 \end{bmatrix}$$

13.31   [0.24 0.20 0.17 0.40]^T

13.A1 CR = 0.1343

13.A3 CR = 0.0393

Glossary

acid-test ratio: The ratio of quick assets to current liabilities. Quick assets are cash, accounts receivable, and marketable securities — those current assets considered to be highly *liquid*. The acid-test ratio is also known as the **quick ratio**.

actual dollars: Monetary units at the time of payment.

actual interest rate: The stated, or observed, interest rate based on actual dollars. If the real interest rate is *i'* and the inflation rate is *f*, the actual interest rate *i* is found by: $i = i' + f + i'f$

actual internal rate of return, IRR$_A$: The internal rate of return on a project based on actual dollar cash flows associated with the project; also the real internal rate of return which has been adjusted upwards to include the effect of inflation.

actual MARR: The minimum acceptable rate of return for *actual dollar* cash flows. It is the real MARR adjusted upwards for inflation.

amortization period: The duration over which a loan is repaid. It is used to compute periodic loan payment amounts.

analytic hierarchy process (AHP): A multi-attribute utility theory (MAUT) approach used for large, complex decisions, which provides a method for establishing the criteria weights.

annual worth method: Comparing alternatives by converting all cash flows to a uniform series, i.e., an annuity.

annuity: A series of uniform-sized receipts or disbursements that start at the end of the first period and continue over a number, *N*, of regularly spaced time intervals.

annuity due: An annuity whose first of *N* receipts or disbursements is immediate, at time 0, rather than at the end of the first period.

arithmetic gradient series: A series of receipts or disbursements that start at the end of the first period and then increase by a constant amount from period to period.

arithmetic gradient to annuity conversion factor: Denoted by (*A/G,i,N*), gives the value of an annuity, *A*, that is equivalent to an arithmetic gradient series where the constant increase in receipts or disbursements is *G* per period, the interest rate is *i*, and the number of periods is *N*.

asset-management ratios: Financial ratios that assess how efficiently a firm is using its assets. Asset management ratios are also known as **efficiency ratios**. Inventory turnover is an example.

assets: The economic resources owned by an enterprise.

backward induction: See **rollback procedure.**

balance sheet: A financial statement that gives a snapshot of an enterprise's financial position at a particular point in time, normally the last day of an accounting period.

base period: A particular date associated with *real dollars* that is used as a reference point for price changes; also the period from which the expenditure shares are calculated in a Laspeyres price index.

base year: The year on which real dollars are based.

benefit–cost analysis (BCA): A method of project evaluation widely used in the public sector that provides a general framework for assessing the gains and losses associated with alternative projects when a broad societal view is necessary.

benefit–cost ratio (BCR): The ratio of the present worth of net user's benefits (social benefits less social costs) to the present worth of the net sponsor's costs for a project. That is,

$$\text{BCR} = \frac{\text{PW(user's benefits)}}{\text{PW(sponsor's costs)}}$$

bond: An investment that provides an annuity and a future value in return for a cost today. It has a *par* or *face* value, which is the amount for which it can be redeemed after a certain period of time. It also has a *coupon rate*, meaning that the bearer is paid an annuity, usually semiannually, calculated as a percentage of the face value.

book value: The depreciated value of an asset for accounting purposes, as calculated with a depreciation model.

break-even analysis: The process of varying a parameter of a problem and determining what parameter value causes the performance measure to reach some threshold or "break-even" value.

capacity: The ability to produce, often measured in units of production per time period.

capital cost: The depreciation expense incurred by the difference between what is paid for the assets required for a particular capacity and what the assets could be resold for some time after purchase.

capital cost allowance (CCA): In the Canadian tax system, the maximum depreciation expense allowed for tax purposes on all assets belonging to an asset class.

capital cost allowance (CCA) asset class: In the Canadian tax system, a categorization of assets for which a specified CCA rate is used to compute CCA. Numerous CCA asset classes exist in the CCA system.

capital cost allowance (CCA) rate: In the Canadian tax system, the maximum depreciation rate allowed for assets in a designated asset class within the CCA system.

capital expense: The expenditure associated with the purchase of a long-term depreciable asset.

capital recovery factor: Denoted by $(A/P,i,N)$, gives the value, A, of the periodic payments or receipts that is equivalent to a present amount, P, when the interest rate is i and the number of periods is N.

capital recovery formula: A formula that can be used to calculate the savings necessary to justify a capital purchase based on first cost and salvage value.

capitalized value: The present worth of an infinitely long series of uniform cash flows.

cash flow diagram: A chart that summarizes the timing and magnitude of cash flows as they occur over time. The X axis represents time, measured in periods, and the Y axis represents the size and direction of the cash flows. Individual cash flows are indicated by arrows pointing up (positive cash flows, or receipts) or down (negative cash flows, or disbursements).

challenger: A potential replacement for an existing asset. See **defender**.

comparison methods: Methods of evaluating and comparing projects, such as present worth, annual worth, payback, and IRR.

compound amount factor: Denoted by $(F/P,i,N)$, gives the future amount, F, that is equivalent to a present amount, P, when the interest rate is i and the number of periods is N.

compound interest: The standard method of computing interest where interest accumulated in one interest period is added to the principal amount used to calculate interest in the next period.

compound interest factors: Formulas that define the mathematical equivalence of specific common cash flow patterns.

compounding period: The interest period used with the compound interest method of computing interest.

consistency ratio: A measure of the consistency of the reported comparisons in a PCM. The consistency ratio ranges from 0 (perfect consistency) to 1 (no consistency). A consistency ratio of 0.1 or less is considered acceptable.

constant dollars: See **real dollars**.

consumer price index (CPI): The CPI relates the average price of a standard set of goods and services in some base period to the average price of the same set of goods and services in another period.

continuous compounding: Compounding of interest which occurs continuously over time, i.e., as the length of the compounding period tends toward zero.

continuous models: Models that assume all cash flows and all compounding of cash flows occur continuously over time.

corporation: A business owned by shareholders.

cost of capital: The minimum rate of return required to induce investors to invest in a business.

cost principle of accounting: Statement that assets are to be valued on the basis of their cost as opposed to market or other values.

cumulative distribution function (CDF): A function that gives the probability that the random variable X takes on a value less than or equal to x, for all x. The CDF is often denoted by $F(x)$ and is defined by $F(x) = Pr(X \leq x)$ for all values of x. Also known as **cumulative risk profile**.

cumulative risk profile: See **cumulative distribution function**.

current assets: Cash and other assets that could be converted to cash within a relatively short period of time, usually a year or less.

current dollars: See **actual dollars**.

current liabilities: Liabilities that are due within some short period of time, usually a year or less.

current ratio: The ratio of all current assets to all current liabilities. It is also known as the **working capital ratio**.

debt-management ratios: See **leverage ratios**.

decisional equivalence: A consequence of indifference on the part of a decision maker among available choices.

decision matrix: A multi-attribute utility theory (MAUT) method in which the rows of a matrix represent criteria, and the columns alternatives. There is an extra column for the weights of the criteria. The cells of the matrix (other than the criteria weights) contain an evaluation of the alternatives.

decision tree: A graphical representation of the logical structure of a decision problem in terms of a sequence of decisions and chance events.

declining-balance method of depreciation: A method of modelling depreciation where the loss in value of an asset in a period is assumed to be a constant proportion of the asset's current value.

defender: An existing asset being assessed for possible replacement. See **challenger**.

deflation: The decrease, over time, in average prices. It can also be described as the increase in the purchasing power of money over time.

depreciation: The loss in value of a capital asset.

discrete models: Models that assume all cash flows and all compounding of cash flows occur at the ends of conventionally defined periods like months or years.

discrete probability distribution function: A probability distribution function in which the random variable is discrete.

dominated: An alternative that is not efficient. See **inefficient**.

economic life: The service life of an asset that minimizes its total cost of use.

earnings before interest and taxes (EBIT): A measure of a company's operating profit (revenues less operating expenses), which results from making sales and controlling operating expenses.

effective interest rate: The actual but not usually stated interest rate, found by converting a given interest rate (with an arbitrary compounding period, normally less than a year) to an equivalent interest rate, with a one-year compounding period.

efficiency ratios: See **asset-management ratios**.

efficient: An alternative is efficient if no other alternative is valued as at least equal to it for all criteria and preferred to it for at least one.

efficient market: A market is efficient when decisions are made so that it is impossible to find a way for at least one person to be better off and no person would be worse off.

engineering economics: Science that deals with techniques of quantitative analysis useful for selecting a preferable alternative from several technically viable ones.

equity ratio: A financial ratio of total owners' equity to total assets. The smaller this ratio is, the more dependent the firm is on debt for its operations and the higher are the risks the company faces.

equivalence: A condition that exists when the value of a cost at one time is equivalent to the value of the related benefit received at a different time.

equivalent annual cost (EAC): An annuity that is mathematically equivalent to a more complex set of cash flows.

expected value: A summary statistic of a random variable that gives its mean or average value. Also known as the **mean**.

expenses: Either real costs associated with performing a corporation's business or a portion of the capital expense for an asset.

external rate of return (ERR): The rate of return on a project where any cash flows that are not invested in the project are assumed to earn interest at a predetermined rate (such as the MARR).

extraordinary item: A gain or loss that does not typically result from a company's normal business activities and is therefore not a recurring item.

financial accounting: The process recording and organizing the financial data of a business. The data cover both flows over time, like revenues and expenses, and levels, like an enterprise's resources and the claims on those resources, at a given date.

financial market: Market for the exchange of capital and credit in the economy.

financial ratio analysis: Comparison of a firm's financial ratios with ratios computed for the same firm from previous financial statements and with industry standard ratios.

financial ratios: Ratios between key amounts taken from the financial statements of a firm. They give an analyst a framework for answering questions about the firm's liquidity, asset management, leverage, and profitability.

fixed costs: Costs that remain the same, regardless of actual units of production.

flat tax rate: A taxation rate that requires that a fixed percentage of

income (for individuals) or profit (for corporations) be paid as tax.

future worth: See the definition of **interest rate**.

future worth method: Comparing alternatives by taking all cash flows to future worth.

geometric gradient series: A set of disbursements or receipts that change by a constant *proportion* from one period to the next in a sequence of periods.

geometric gradient to present worth conversion factor: Denoted by $(P/A,g,i,N)$, gives the present worth, P, that is equivalent to a geometric gradient series where the base receipt or disbursement is A, and where the rate of growth is g, the interest rate is i, and the number of periods is N.

goal: The decision to be made in AHP.

growth adjusted interest rate, $i°$: $i° = (1 + i)/(1 + g)$ so that $1/(1 + i°) = (1 + i)/(1 + g)$ where i is the interest rate and g is the growth rate. The growth adjusted interest rate is used in computing the geometric gradient to present worth conversion factor.

income statement: A financial statement that summarizes an enterprise's revenues and expenses over a specified accounting period.

independent projects: Two projects are independent if the expected costs and the expected benefits of each of the projects do not depend on whether or not the other one is chosen.

inefficient: An alternative that is not efficient.

inflation: The increase, over time, in average prices of goods and services. It can also be described as the decrease in the purchasing power of money over time.

inflation rate: The rate of increase in average prices of goods and services over a specified time period, usually a year; also, the rate of decrease in purchasing power of money over a specified time period, usually a year.

installation costs: Costs of acquiring capacity (excluding the purchase cost), which may include disruption of production, training of workers, and perhaps a reorganization of other production.

interest: The compensation for giving up the use of money.

interest period: The base unit of time over which an interest rate is quoted. The interest period is referred to as the compounding period when compound interest is used.

interest rate: If the right to P at the beginning of a time period exchanges for the right to F at the end of the period, where $F = P(1 + i)$, i is the interest rate per time period. In this definition, P is called the **present worth** of F, and F is called the **future worth** of P.

internal rate of return (IRR): That interest rate, i^*, such that, when all cash flows associated with a project are discounted at i^*, the present worth of the cash inflows equals the present worth of the cash outflows.

inventory-turnover ratio: A financial ratio that captures the number of times that a firm's inventories are replaced (or turned over) per year. It provides a measure of whether the firm has more or less inventory than normal.

Laspeyres price index: A commonly used price index that measures weighted average changes in prices of a set of goods and services over time as compared with the prices in a base period. The weights are the expenditure shares in the base period. The weights are then converted to percentages by multiplying by 100.

leverage ratios: Financial ratios that provide information about how

liquid a firm is, or how well it is able to meet its current obligations. Also known as **debt-management ratios**.

liabilities: Claims, other than those of the owners, on a business's assets.

liquidity ratio: A financial ratio that evaluates the ability of a business to meet its current liability obligations. The current ratio and quick ratio are two examples of liquidity ratios.

long-term assets: Assets that are not expected to be converted to cash in the short term, usually taken to be one year.

long-term liabilities: Liabilities that are not expected to draw on the business's current assets.

management accounting: The process of analyzing and recording the costs and benefits of the various activities of an enterprise. The goal of management accounting is to provide managers with information to help in decision making.

market: A group of buyers and sellers linked by trade in a particular product or service.

market equivalence: The ability to exchange one cash flow for another at zero cost.

market failure: Condition in which output or consumption decisions are made in which aggregate benefits to all persons who benefit from the decision are less than aggregate costs imposed on persons who bear costs that result from the decision.

market value: Usually taken as the actual value an asset can be sold for in an open market.

mathematical equivalence: An equivalence of cash flows due to the mathematical relationship between time and money.

mean: See **expected value**.

mean-variance dominance: Alternative X dominates alternative Y in a mean-variance sense if: $EV(X) \geq EV(Y)$ and $Var(X) \; Y)$ and $Var(X) \leq Var(Y)$.

minimum acceptable rate of return (MARR): An interest rate that must be earned for any project to be accepted.

modified benefit–cost ratio: The ratio of the present worth (or annual worth) of benefits minus the present worth (or annual worth) of operating costs to the present worth (or annual worth) of capital costs, that is, BCRM = [PW(user's benefits) − PW(sponsor's operating costs)] ÷ PW(sponsor's capital costs).

Monte Carlo simulation: A procedure that constructs the probability distribution of an outcome performance measure of a project by repeatedly sampling from the input random variable probability distributions.

multi-attribute utility theory (MAUT): An MCDM approach in which criteria weights and preference measures are combined mathematically to determine a best alternative.

multi-criterion decision making (MCDM): Methods that explicitly take into account multiple criteria and guide a decision maker to superior or optimal choices.

mutually exclusive projects: Projects are mutually exclusive if, in the process of choosing one, all the other alternatives are excluded.

net cash flow: The difference between cash inflows and outflows for the period. The net cash flow, A_t, is given by $A_t = R_t − D_t$, where R_t is cash inflow in period t, and D_t is cash disbursed in period t.

net-profit ratio: See **return-on-assets (ROA) ratio**.

nominal dollars: See **actual dollars**.

nominal interest rate: The conventional method of stating the annual interest rate. It is calculated by multiplying the interest rate per compounding period by the number of compounding periods per year.

one year principle: This principle states that if the EAC (capital costs) for the defender are small compared to the EAC (operating costs), and the yearly operating costs are monotonically increasing, the economic life of a defender is one year and its total EAC is the cost of using the defender for one more year.

operating and maintenance costs: Ongoing costs to operate and maintain an asset over its useful life. Include costs such as electricity, gasoline, parts, repair, and insurance.

outcome dominance: Alternative X is said to have outcome dominance over alternative Y when a) the worst outcome for alternative X is at least as good as the best outcome for alternative Y, or b) X is at least as preferred to Y for each outcome, and is better for at least one outcome.

owners' equity: The interest of the owner or owners of a firm in its assets.

pairwise comparison: An evaluation of the importance or preference of a pair of criteria (or alternatives), based on an AHP value scale.

pairwise comparison matrix (PCM): A device for storing pairwise comparison evaluations.

partnership: A business owned by two or more owners (partners).

par value: The price per share set by a firm at the time the shares are originally issued.

payback period: The period of time it takes for an investment to be recouped when the interest rate is assumed to be zero.

payback period method: A method used for comparing alternatives by comparing the periods of time required for the investments to pay for themselves.

performance measures: Calculated values that allow conclusions to be drawn from data.

policy instrument: A remedy for market failure in which the government imposes rules intended to modify behaviour of business organizations.

present worth: See **interest rate**.

present worth factor: Denoted by $(P/F,i,N)$, gives the present amount, P, that is equivalent to a future amount, F, when the interest rate is i and the number of periods is N.

present worth method: Comparing alternatives by taking all cash flows to present worth.

price index: A number, usually a percentage, that relates prices of a given set of goods and services in some period, t_1, to the prices of the same set of goods and service in another period, t_0.

principal eigenvector: The eigenvector associated with the largest eigenvalue, λ_{max}, of a PCM.

priority weights: Weights calculated for alternatives by normalizing the elements of a PCM and averaging the row entries.

probability: The limit of a long-run proportion or relative frequency; see Close-Up 12.1 for other views of probability.

probability distribution function: A set of numerical measures (probabilities) associated with outcomes of a random variable. Also known as a risk profile.

profitability ratios: Financial ratios that give evidence of how productively assets have been employed in producing a profit. Return on total assets (or net-profit ratio) is an example of a profitability ratio.

progressive tax rate: A taxation rate that increases as the income level increases.

project: A term used throughout this text to mean *investment opportunity*.

project balance: If a project has a sequence of net cash flows $A_0, A_1, A_2, \ldots, A_T$, and the interest rate is

i', there are $T + 1$ project balances, $B_0, B_1, \ldots, B_T$, one at the end of each period t, $t = 0,1, \ldots, T$. A project balance, B_t, is the cumulative future value of all cash flows, up to the end of period t, compounded at the rate, i'.

proportional tax rate: See **flat tax rate**.

public–private partnership (P3): A cooperative arrangement between the private and public sectors for the provision of infrastructure or services.

quick ratio: See **acid-test ratio**.

random variable: A random variable is a parameter or variable that can take on a number of possible outcomes. The probability of each outcome is given by its probability distribution.

real dollars: Monetary units of constant purchasing power.

real interest rate: The interest rate, i', is the interest rate that would yield the same number of real dollars in the absence of inflation as the actual interest rate yields in the presence of inflation at the rate f. It is given by $i' = (1 + i)/(1 + f) - 1$

real internal rate of return: The internal rate of return on a project based on real dollar cash flows associated with the project.

real MARR: The minimum acceptable rate of return when cash flows are expressed in real, or constant, dollars.

recovery period: The designated service life for depreciation calculation purposes in U.S. tax law.

related but not mutually exclusive projects: For pairs of projects in this category, the expected costs and benefits of one project depend on whether the other one is chosen.

repeated lives: Used for comparing alternatives with different service lives, based on the assumption that alternatives can be repeated in the future, with the same costs and benefits, as often as necessary. The life of each alternative is repeated until a common total time period is reached for all alternatives.

replacement: An asset may be replaced if there is cheaper way to get the service the asset provides or if the service provided by the asset is no longer adequate.

retained earnings: The cumulative sum of earnings from normal operations, in addition to gains (or losses) from transactions such as the sale of plant assets or investments that have been reinvested in the business (i.e., not paid out as dividends).

retire: To remove an asset from use without replacement.

return-on-assets (ROA) ratio: A financial ratio that captures how productively assets have been employed in producing a profit. It is also known as the **net-profit ratio**.

return-on-equity (ROE) ratio: A financial ratio that captures how much profit a company has earned in comparison to the amount of capital that the owners have tied up in the company.

risk profile: See **probability distribution function**.

rollback procedure: A procedure in decision tree analysis that computes an expected value at each chance node and selects a preferred alternative at each decision node; also known as **backward induction**.

salvage value: Either the actual value of an asset at the end of its useful life (when it is sold), or an estimate of the salvage value calculated using a depreciation model.

scenario analysis: The process of examining the consequences of several possible sets of variable values associated with a project.

scrap value: Either the actual value of an asset at the end of its physical

life (when it is broken up for the material value of its parts), or an estimate of the scrap value calculated using a depreciation model.

sensitivity analysis: Methods that assess the sensitivity of an economic measure to uncertainties in estimates in the various parameters of a problem.

sensitivity graph: A graph of the changes in a performance measure, holding all other variables fixed.

series present worth factor: Denoted by $(P/A,i,N)$, gives the present amount, P, that is equivalent to an annuity, A, when the interest rate is i and the number of periods is N.

simple interest: A method of computing interest where interest earned during an interest period is not added to the principal amount used to calculate interest in the next period. Simple interest is rarely used, except as a method of calculating approximate interest.

simple investment: A project that consists of one or more cash outflows at the beginning, followed only by one or more cash inflows.

sinking fund: Interest-bearing account into which regular deposits are made in order to accumulate some amount.

sinking fund factor: Denoted by $(A/F,i,N)$, gives the size, A, of a repeated receipt or disbursement that is equivalent to a future amount, F, when the interest rate is i and the number of periods is N.

sole proprietorship: A business owned by one person.

specialist company: A firm that concentrates on manufacturing a limited range of very specialized products.

standard deviation: The square root of the variance. A measure of dispersion for a random variable.

statement of changes in financial position: A financial statement that shows the amounts of cash generated by a company's operations and by other sources, and the amounts of cash used for investments and other non-operating disbursements.

stochastic dominance: If two decision alternatives a and b have outcome cumulative distribution functions $F(x)$ and $G(x)$, respectively, then alternative a is said to have first order stochastic dominance over alternative b if $F(x) \geq G(x)$ for all x.

straight-line depreciation: A method of modelling depreciation which assumes that the rate of loss in value of an asset is constant over its useful life.

study period: A period of time over which alternative projects are compared.

sunk costs: Costs that were incurred in the past and are no longer relevant in replacement decisions.

tax benefit factor (TBF): A value that summarizes the effect of the future benefit of tax savings due to depreciation.

tax credits: Real or nominal costs that are not taxed or are taxed at a reduced rate.

term: The duration over which a loan agreement is valid.

trend analysis: A form of financial analysis that traces the financial ratios of a firm over several accounting periods.

undepreciated capital cost (UCC): The remaining book value of assets subject to depreciation for taxation purposes. For any given year, the UCC balance can be calculated as follows: $\text{UCC}_{\text{opening}} + \text{additions} - \text{disposals} - \text{CCA} = \text{UCC}_{\text{ending}}$

uniform series compound amount factor: Denoted by

($F/A,i,N$), gives the future value, F, that is equivalent to a series of equal-sized receipts or disbursements, A, when the interest rate is i and the number of periods is N.

variable costs: Costs that change depending on the number of units produced.

variance: Variance measures the dispersion of a random variable about its mean.

weighted average cost of capital: A weighted average of the costs of borrowing and of selling shares. The weights are the fractions of total capital that come from the different sources.

working capital: The difference between total current assets and total current liabilities.

working capital ratio: See **current ratio.**

written down value (WDV): The remaining value in an asset pool used for British taxation calculations.

Index